Progress in Atomic Spectroscopy

Part D

PHYSICS OF ATOMS AND MOLECULES

Series Editors

P. G. Burke, *The Queen's University of Belfast, Northern Ireland*
H. Kleinpoppen, *Atomic Physics Laboratory, University of Stirling, Scotland*

Editorial Advisory Board

A Continuation Order Plan is available for this series. A continuation order will bring delivery of each new volume immediately upon publication. Volumes are billed only upon actual shipment. For further information please contact the publisher.

Progress in Atomic Spectroscopy

Part D

Edited by

H. J. Beyer

and

Hans Kleinpoppen

University of Stirling
Stirling, United Kingdom

Plenum Press · New York and London

PHYSICS

Library of Congress Cataloging in Publication Data

(Revised for Part D)

Progress in atomic spectroscopy.

(Physics of atoms and molecules)
Pts. C–   edited by H. J. Beyer and Hans Kleinpoppen.
Includes bibliographical references and indexes.
1. Atomic spectroscopy. I. Hanle, Wilhelm, 1901–   . II. Kleinpoppen, H. (Hans).
III. Beyer, H. J. 1940–   . IV. Series.
QC454.A8P76   543'.0858   78-18230
ISBN 0-306-42528-9

© 1987 Plenum Press, New York
A Division of Plenum Publishing Corporation
233 Spring Street, New York, N.Y. 10013

Printed in the United States of America

To Wilhelm Hanle

Professor Wilhelm Hanle (1927)

Contents of Part D

Chapter 1

Laser–Microwave Spectroscopy
R. Neumann, F. Träger, and G. zu Putlitz

Chapter 8
Analysis and Spectroscopy of Collisionally Induced Autoionization Processes
R. Morgenstern

Chapter 9
Near Resonant Vacancy Exchange between Inner Shells of Colliding Heavy Particles
N. Stolterfoht

Chapter 10

Polarization Correlation in the Two-Photon Decay of Atoms
A. J. Duncan

Contents of Part A

Contents of Part B

Contents of Part C

Introduction

H. J. BEYER AND H. KLEINPOPPEN

We are pleased to present Part D of *Progress in Atomic Spectroscopy* to the scientific community active in this field of research. When we invited authors to contribute articles to Part C to be dedicated to Wilhelm Hanle, we received a sufficiently enthusiastic response that we could embark on two further volumes and thus approach the initial goal (set when Parts A and B were in the planning stage) of an almost comprehensive survey of the current state of atomic spectroscopy.

As mentioned in the introduction to Parts A and B, new experimental methods have enriched and advanced the field of atomic spectroscopy to such a degree that it serves not only as a source of atomic structure data but also as a test ground for fundamental atomic theories based upon the framework of quantum mechanics and quantum electrodynamics. However, modern laser and photon correlation techniques have also been applied successfully to probe beyond the "traditional" quantum mechanical and quantum electrodynamical theories into nuclear structure theories, electroweak theories, and the growing field of local realistic theories versus quantum theories.

It is obvious from the contents of this volume and by no means surprising that applications of laser radiation again played a decisive role in the development of new and high-precision spectroscopic techniques. The method of laser–microwave spectroscopy can be considered as growing out of the atomic beam resonance method (Rabi-type experiments), the optical pumping method (as suggested by Brossel and Kastler), and the double resonance method (Brossel and Bitter). In present laser–microwave resonance experiments only moderate atomic beam collimation is required, resulting in a much larger atomic flux than in Rabi-type systems. The higher sensitivity of laser–microwave experiments compared to Rabi experiments is also due to the much better detectability of fluorescence photons compared to atoms of a deflected atomic beam. Zero-magnetic-field experiments are possible with the laser–microwave method, so that complications from the Zeeman effect can be avoided. The various laser–microwave methods have been applied to a large number of ions, atoms, molecules, and even solids. The sensitivity of these laser–microwave methods has now reached the ultimate limit, where a signal can be detected from a single atom (compared to typically 10^9 atoms in classical double resonance experiments or 10^{19} in nuclear or electron spin resonance experiments in solids!).

Even without the use of microwave transitions, high detection efficiency and good resolution can be achieved simultaneously in laser spectroscopy by collinear laser fast-beam arrangements. In these experiments the laser light is directed along a fast atom or ion beam so that the Doppler broadening of the resonance photon absorption profile is largely eliminated by the narrowing of the momentum distribution in the direction of the beam. This method is particularly suited to absorption–fluorescence studies on short-lived artificial isotopes as produced in high-energy physics centers (e.g., CERN in Geneva). In recent years such experiments have provided a wealth of hyperfine structure and isotope shift data and have led to important new information on nuclear spins, nuclear moments, nuclear shapes, and also parity violation effects in nuclei.

The study of atomic Rydberg states has been advanced to a very large extent through laser technology. In Rydberg atoms the outer electron spends most of its time far away from the ionic core. Accordingly, the electron of a Rydberg state can serve as a very sensitive probe of the properties of the ionic core. Fine structure, hyperfine structure, isotope shifts, singlet–triplet mixings, and other properties have been determined for a large number of Rydberg atoms. Combined laser and radiofrequency techniques have been applied with impressive success in spectroscopic studies of Rydberg atoms.

Atomic spectroscopy has substantially contributed to the advance of quantum mechanics and quantum electrodynamics. This important role of atomic physics in tests of fundamental theories has again been demonstrated by the detection of the "weak interaction" in atomic structure. This interaction manifests itself as a parity nonconservation effect in atomic absorption and emission lines and has been measured successfully in several laboratories. These parity violation effects in heavy atoms are almost orthogonal to those studied in high-energy physics and also provide direct tests of the Weinberg–Salam–Glashow theory of the weak interaction. Accordingly, high-energy and atomic physics experiments supplement each other in tests of the fundamental interaction theory.

The renewed interest in and success of the spectroscopy of highly ionized atoms during the last two decades was due to the detection and analysis of the extreme ultraviolet solar spectrum recorded by space vehicles and also due to the need for diagnostic tools for hot plasmas in fusion research. The development of powerful laboratory light sources such as the low-inductance spark and the laser-produced plasma was important for the analysis of extended spectral series of highly charged ions. The considerable recent progress of the experimental analysis of ion spectra has been matched by the development of theoretical methods used to interpret the experimental data.

The favorable characteristics of intense synchrotron radiation (e.g., from electron-storage rings with wigglers and undulators) permit spectroscopy in the extreme ultraviolet and x-ray regions. Studies of x-ray absorption, x-ray fluorescence, and Auger electron emission by free and bound atoms lead to inner shell spectroscopy not possible before. Scattering of hard synchrotron radiation has also made it possible to extend Rayleigh, Compton, and resonant Raman scattering processes into this spectral region.

Research in atomic collision physics has reached a state in which the analyzing methods are often related to nuclear physics (e.g., angular correlation and particle spin experiments) as well as to atomic spectroscopy. The study of atomic autoionizing states, for example, which in most cases are populated by collisions, combines methods from both nuclear physics and atomic spectroscopy. The decay of the quasistationary autoionizing states is in most cases dominated by ionization into an electron and an atomic ion. Accordingly, the study of the electrons from decaying autoionizing states leads to information about the configuration of the state and about postcollision interactions.

Electron spectroscopy of autoionizing states is analogous to photon spectroscopy of excited atomic states (although the kind of information from electron spectroscopy appears to be richer owing to the shorter de Broglie wavelength of the ejected electron and to the angular momentum of the electron not being limited to $l = 1$, as in the case of photon dipole radiation). Also, in heavy-particle–atom excitation of autoionizing states, the formation of short-lived quasimolecules with a lifetime of 10^{-12}–10^{-15} sec can be traced by electron spectroscopy of the decaying quasimolecules, whereas photon emission occurs long after the collision when the particles are well separated again.

The efficient production of inner shell vacancies and their exchange between colliding heavy atomic particles has been the object of a considerable number of x-ray spectroscopy studies. For inner shells of atoms that are close in energy, charge exchange is nearly resonant, so that vacancies are transferred with high probability. The theoretical analysis of near-resonant charge exchange between inner shells may be restricted to a few basic states.

Collision systems with nearly matching inner shells can thus be used as ideal test cases for few-state atomic models. Significant experimental and theoretical work on near-resonant vacancy exchange has been carried out, in particular for processes involving K and L shells in relatively light systems.

While two-photon absorption spectroscopy has been widely applied for precision measurements of atomic structure, the polarization correlation of the simultaneous two-photon emission from the metastable $2s$ state of atomic hydrogen has only been measured very recently. The emission of the coincident two photons can be described by a single state vector which determines the circular and linear two-photon polarization. Compared to the two-photon cascade experiments the polarization correlation of the simultaneous two-photon decay of metastable hydrogen is conceptually closer to the original proposals by Bell and Bohm for tests of the foundation of quantum mechanics. More than 50 years have elapsed since the famous Einstein-Bohr debate on microphysical reality and quantum formalism. The present and future outcome of the hydrogen two-photon correlation experiment is considered to be a most crucial test with regard to the rivalry between quantum mechanics and local realistic theories.

We conclude this introduction by referring again to Professor Wilhelm Hanle. The authors and editors dedicate this volume to Wilhelm Hanle in the year 1986 for his 85th birthday. The photograph on page vi shows Wilhelm Hanle as a young physicist in 1927. We consider it appropriate to list below some of the important and pioneering publications of Wilhelm Hanle from this time:

Hanle Effect (Level Crossing)

Über den Zeemaneffekt bei der Resonanzfluoreszenz (On the Zeeman effect in resonance fluorescence), *Naturwissenschaften* **11**, 690–691 (1923).

Über magnetische Beeinflussung der Polarization der Resonanzfluoreszenz (The influence of magnetic fields on the polarization of resonance fluorescence radiation), *Z. Phys.* **30**, 93–105 (1924).

Die magnetische Beeinflussung der Resonanzfluoreszenz (The influence of magnetic fields on the resonance fluorescence), *Ergeb. Exakten Naturwiss.* **4**, 214–232 (1925).

Die elektrische Beeinflussung der Polarisation der Resonanzfluoreszenz von Quecksilber (The influence of electric fields on the polarization of resonance fluorescence radiation in mercury) *Z. Phys.* **35**, 346–364 (1926).

Polarization of Fluorescence Light

Die Polarisation der Resonanzfluoreszenz von Natriumdampf bei Anregung mit zirkular polarisiertem Licht (The polarization of resonance fluorescence radiation of sodium vapor excited by circularly polarized light), *Z. Phys.* **41**, 164–183 (1927).

Polarisationserscheinungen bei der stufenweisen Anregung der Fluoreszenz von Quecksilber-dampf (Polarization effects of the fluorescence radiation in mercury following stepwise excitation of the 7^3S state), *Z. Phys.* **54**, 811–818 (1929) (with E. F. Richter).

Optical Excitation Functions

Die Anregungsfunktion der Quecksilberresonanzlinie 2537 (The excitation function of the mercury resonance line λ 2537 Å), *Z. Phys.* **54**, 848–851 (1929).

Die Anregungsfunktion von Spektrallinien (The excitation function of spectral lines), *Naturwissenschaften* **15**, 832–833 (1927).

Die Lichtausbeute bei Elektronenstossanregung (Measurement of the light intensity following electron impact excitation), *Phys. Z.* **30**, 901–905 (1929) (with W. Schaffernicht).

Messung der Lichtausbeute im Quecksilberspektrum bei Elektronenstossanregung (Measurement of the intensity of mercury spectral lines using electron impact excitation), *Ann. Phys.* (*Leipzig*) (5) **6**, 905–931 (1930) (by W. Schaffernicht).

Die optischen Anregungsfunktionen der Quecksilberlinien (Optical excitation functions of spectral lines of mercury), *Z. Phys.* **62**, 106–142 (1930) (by W. Schaffernicht).

Messung von Anregungsfunktionen im Heliumspektrum (Measurement of excitation functions of spectral lines of helium), *Z. Phys.* **56**, 94–113 (1929).

Anregungsfunktionen im Neonspektrum (Excitation functions of spectral lines of neon), *Z. Phys.* **65**, 512–516 (1930).

Die Leuchtausbeute in Abhängigkeit von der Voltgeschwindigkeit der Elektronen und die relativen Intensitäten von Cadmium- und Zinklinien bei Anregung durch Elektronenstoss (The intensity of spectral lines as a function of the electron energy and the relative intensities of cadmium and zinc lines excited by electron impact), *Z. Phys.* **67**, 440–477 (1931) (by K. Larché).

Neue Messungen der Lichtausbeute bei Elektronen- und Ionenstoss (New measurements of spectral line intensities using electron and ion impact excitation), *Phys. Z.* **33**, 245–247 (1932).

Elektronen-, Ionen- und Atomstossleuchten (Light emission in collisions with electrons, ions and atoms), *Phys. Z.* **33**, 884–887 (1932) (with K. Larché).

Über Polarisation bei Neon-Elektronenstossleuchten und Neon-Kanalstrahlleuchten (Polarization of spectral lines of neon excited by electron impact and in canal rays), *Z. Phys.* **54**, 819–825 (1929) (with B. Quarder).

Bemerkung über die Intensität von Spektrallinien (Note on the intensity of spectral lines), *Z. Phys.* **54**, 852–855 (1929).

Raman Effect

Über eine Anomalie bei der Polarisation der Ramanstrahlung (On the anomalous polarization of Raman lines), *Naturwissenschaften* **19**, 375 (1931).

Über zirkulare Polarisation beim Ramaneffekt (The circular polarization of Raman lines), *Phys. Z.* **32**, 556–558 (1931).

Untersuchungen über die zirkulare Polarisation der Ramanlinien (Study of the circular polarization of Raman lines), *Ann. Phys.* (*Leipzig*) (5) **11**, 885–904 (1931).

Laser–Microwave Spectroscopy

R. Neumann, F. Träger, and G. zu Putlitz

1. Introduction

When Brossel and Kastler[1] proposed the intermarriage of optical spectroscopy and radiofrequency resonance nearly four decades ago, they triggered off the development of numerous novel spectroscopic methods—a development that continues even today. The first generation of "optical pumping" and "optical double resonance" experiments not only yielded a wealth of very precise data on atoms, molecules, condensed matter, and nuclei alike, but they also initiated the detailed study of the interaction of atoms with radiation, particularly under the aspects of coherence, and of the modification of atomic states through radiation fields.

It was a logical consequence thereof that, after the invention of the laser in the 1960s, the problems mentioned above would lead to even more exciting experimental verifications. The laser, an unprecedented spectroscopic light source as compared to ordinary spectral lamps regarding its intensity, luminosity, spatial divergence, wavelength stability, and linewidth, emits its radiation even coherently in the optical wavelength region. However, it took another decade still—well into the 1970s—before lasers were widely used in spectroscopic experiments. Most of the interesting scientific questions were tied to the study of a particular atomic or molecular system, or to particular states in condensed matter. Here, the wavelengths of the radiation required for the experimental study were determined by the system itself. Hence, optical lasers had to be found and constructed exhibiting the same frequency tunability and stability as the microwave sources already available. Consequently, it was not until the advent of the

R. Neumann, F. Träger, and G. zu Putlitz • Physikalisches Institut der Universität Heidelberg, Philosophenweg 12, D-6900 Heidelberg, Federal Republic of Germany.

tunable dye laser and the tunable color center laser that the field wa suddenly opened up.

The advantages of the simultaneous excitation of optical lines by laser radiation and microwave transitions to observe directly the energy difference in question are still the same; optical quanta are detected with a large quantum efficiency; microwave transitions exhibit a very small Doppler broadening. Without questioning the merits of all the Doppler-free laser spectroscopic methods, one should keep in mind also the enormous progress in microwave technology.

This chapter assesses the state of the art in laser microwave spectroscopy, i.e., investigations where both laser radiation and microwave radiation are involved. The many concepts and methods applied to numerous different atomic, molecular, and solid state systems are outlined, not always with emphasis on completeness of all the later references to the same method. This chapter hopefully stimulates further development of laser microwave spectroscopic methods and applications, even crossing the border of different disciplines.

1.1. Classical Experiments with Radiofrequency Transitions

The spectra of atoms, ions, and molecules as well as larger aggregates thereof in the gaseous, liquid, and solid phase can be studied with high precision by the methods of microwave spectroscopy.* First experiments of that kind date back to the 1930s, when Rabi and co-workers introduced the method of atomic beam radiofrequency resonance.[2,3] This technique relies on the spatial separation of atoms or molecules in states with different magnetic quantum numbers by means of inhomogeneous magnetic fields. Particles evaporated from an oven travel across the inhomogeneous field A where they are deflected as shown schematically in Figure 1. Field C is homogeneous, so that the path of flight is not affected. The B field acts like the A field, the directions of $\mathbf{H}$ and $d\mathbf{H}/dz$ being the same. Therefore, the particles are "defocused," i.e., they do not reach the detector, which is installed on the axis of the apparatus. If, however, a radiofrequency field within C induces magnetic dipole transitions (we consider the case $m = 1/2 \rightarrow m = -1/2$) the force in the B field will be opposite to the one in A, so that the particles are refocused onto the detector ("flop-in" method). Thus, a signal can be generated if the rf field in C has the appropriate frequency to induce a microwave transition. Rabi's method has been applied to a large number of atoms and molecules and yielded a wealth of data, e.g., on hyperfine structure splittings of ground and metastable excited states.

* We refer to microwave radiation also as radiofrequency radiation.

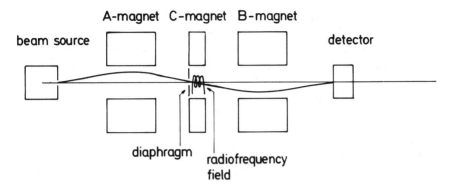

Figure 1. Apparatus for atomic beam magnetic resonance.

The combination of two electromagnetic fields, namely, of visible light and radiation in the radiofrequency range, for the purpose of spectroscopy dates back to 1949: the technique of "optical pumping" of ground states or metastable excited states was invented by Brossel and Kastler,[1] and the method of "double resonance" of short-lived excited states was introduced by Bitter,[4] and by Lamb and Skinner.[5] A typical experimental setup for classical double resonance[6] is illustrated in Figure 2. It consists of a light source such as a hollow-cathode or discharge lamp for the excitation of the

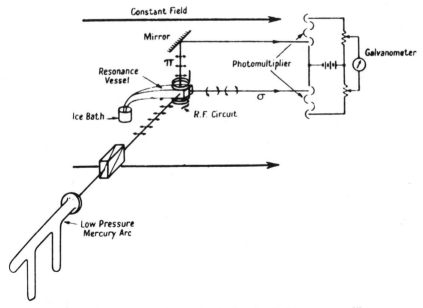

Figure 2. Apparatus of Brossel and Bitter for double resonance.[6]

species under study, a polarizer, and a resonance cell surrounded by a coil emitting the radiofrequency field. In the case of double resonance (see Figure 3a) the absorption of polarized light is followed by induced microwave transitions in an excited state, which usually results in a change of the polarization or alignment of the atomic state and therefore also of the polarization and angular distribution, in cases of large splittings even of the wavelength of the reemitted light. This change in the resonance radiation is monitored either directly by taking the difference of the currents of two photomultipliers installed at right angles with respect to each other, or equivalently, by a single photomultiplier with a polarizer mounted in front of it. In the case of optical pumping (see Figure 3b), the resonant absorption of polarized light is followed by the reemission of photons, which can result in a net population difference of the sublevels of the lower-lying state depending on the different matrix elements and selection rules for the excitation and decay of the excited state into these energy levels, respectively. Radiofrequency transitions then tend to equalize the population difference. If the rf frequency is varied around the exact rf resonance position, a change of the transmission of the vapor with respect to the pumping light beam can be observed, or alternatively, the changing intensity of the emitted resonance radiation can be recorded.

Two basic features of these methods have to be pointed out here: firstly, the resonant absorption of radiofrequency or microwave quanta with energies of $E/h \simeq 10^5$–10^{10} Hz is monitored via a change in the intensity of visible light with quantum energies in the range of $E/h \simeq 5 \times 10^{14}$ Hz. Thus, the detection sensitivity is increased by orders of magnitude. Moreover, the detecting quanta for microwave transitions are now far removed from thermal noise. For this reason, the number of atoms or spins indispensable for a signal have gone from 10^{18} in nuclear magnetic resonance to 10^9 in optical pumping experiments to literally one in recent laser rf experiments. Secondly, a major advantage of all rf spectroscopic experiments in comparison to optical spectroscopy is the reduction of the Doppler width $\Delta \nu_D$ of the rf induced transition because of its proportionality to the frequency.

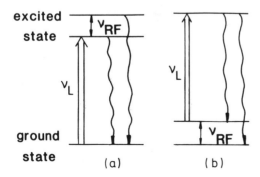

Figure 3. Optical and radiofrequency transitions for double resonance (a) and optical pumping (b) in three-level systems.

Usually, the Doppler width becomes negligibly small as compared to other mechanisms of line broadening, in particular to the natural linewidth. Therefore, double resonance and optical pumping combined with rf transitions allow for a much higher resolution than classical optical experiments. In addition, the frequency of the rf field can be controlled and measured with a much higher precision than one can determine the wavelength difference in the optical range with monochromators or etalons, even if the experimental resolution would be comparable. Nevertheless there is remarkable progress in replacing direct microwave radiation methods by laser heterodyne techniques, e.g., by difference frequency measurements in the optical range (see Section 7).

Finally, optical pumping and double resonance have opened the way to experiments where the species under study can be confined in closed resonance cells, whereas Rabi's technique relied on the spatial separation of atoms or molecules in a beam arrangement. Also, the new methods allowed the investigation of ions that cannot be studied with a Rabi-type apparatus.

During the past three decades an enormous amount of spectroscopic data of atoms, ions, and—to a limited extent—also of molecules has been obtained by means of the methods mentioned above. Finer details of the spectra not accessible before could be studied. In the years following, many different experimental schemes, all based on the optical–rf methods mentioned in the beginning, have been used. Besides the very numerous original publications, a considerable number of monographs have described the techniques, compiled the experimental results, and discussed the underlying physics. For more details the reader is referred to the books: *Molecular Beams* by N. Ramsey,[7] and *Nuclear Moments* by H. Kopfermann,[8] and to the articles "Radiofrequency Spectroscopy of Excited Atoms" by G. W. Series,[9] "Optical Pumping" by C. Cohen-Tannoudji and A. Kastler,[10] "Optical Pumping" by W. Happer,[11] and G. zu Putlitz on "Double Resonance."[12]

In addition to the Rabi-type experiments, a second important method of modern microwave spectroscopy was introduced in the 1930s. Gases confined in closed resonance cells were investigated by measuring the absorption of a propagating microwave field as a function of its frequency (see Figure 4). Since no pumping mechanism or other state selection was

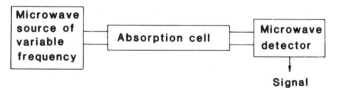

Figure 4. Arrangement for microwave spectroscopy.

applied, the generation of a signal by inducing transitions actually relied on the different occupation numbers of different energy levels at thermal equilibrium. Because the Boltzmann factor for levels separated by microwave frequencies differs only slightly, and because the detection sensitivity for microwave radiation is rather low as compared to optical photons, strong signals were not easily obtained, even if long absorption cells containing the gas under high pressure were used.

First experiments of that type were carried out by Betz[13] in 1932, who observed fine structure transitions in hydrogen gas, and by Cleeton and Williams[14] in 1934, who studied the absorption spectrum of ammonia in the wavelength range from $\lambda = 1$ cm to $\lambda = 4$ cm by measuring the transmitted microwave radiation with a bolometer. It was not until the late 1940s that new microwave oscillators and techniques became available, and that spectroscopy in this range was developed and applied further. Most experiments were carried out on molecules, and primarily yielded data on the frequencies associated with molecular rotation. A combination of microwave transitions with optical excitation has not been attempted very often. This is because the classical spectral light sources available at the time did not provide a sufficiently narrow spectral output, and a sufficiently high intensity to start from a weakly populated vibrational–rotational level and to populate only a single well-defined upper level. In total, this method has been promoted to real significance only since lasers have helped to overcome intensity and linewidth problems (see below).

Books on microwave spectroscopy have been published, e.g., by Gordy, Smith, and Trambarulo,[15] by Townes and Schawlow,[16] and by Chantry.[17]

1.2. Lasers versus Classical Light Sources in rf Spectroscopy Experiments

The invention of the laser introduced a totally new kind of light source, and within a short period of time opened completely new branches of spectroscopy. The advantages of the laser, namely, high monochromaticity, coherence, high output power, and small spatial divergence, made possible all kinds of novel experiments. For example, the spectral resolution in the optical range was increased enormously by Doppler-free techniques that can only be realized with lasers. Because of the narrow spectral output, atomic or molecular energy levels could be populated selectively without relying on dipole selection rules only. Also, the sensitivity could be increased remarkably, so that the excitation of very weak one- or multiquantum transitions became feasible and tiny quantities of particles could be studied. Therefore, more and more kinds of molecules and atoms including their higher-lying levels up to the series limit became accessible. A particular advantage of specific types of lasers, such as dye lasers or color center lasers, is that the output frequency can be tuned continuously and therefore matched with many molecular and atomic transition frequencies. Thus, one does not have to rely anymore on accidental coincidences of the lines of fixed-frequency lasers and transitions of the species to be investigated.

About a decade ago first experiments were performed that employed laser-induced transitions together with microwave-induced transitions. The techniques applied were basically those of optical double resonance and optical pumping combined with the advantages of laser excitation. However, only two or three years later new schemes were invented that also employed nonlinear phenomena. Since then, considerable progress has been made. The field has grown rapidly and branched off in many different applications. For example, laser–microwave spectroscopy has been used to study fundamental systems such as atomic hydrogen, but also complicated molecules such as chlorophyll, or very small samples of atoms in Rydberg states. Ions in fast beams but also ions confined in Penning traps were investigated. The underlying physical goals range from hyperfine structure measurements for the determination of nuclear moments, over studies of collisional energy transfer in molecules, work aiming for the development of new frequency standards, to investigations of spin-lattice relaxation rates in crystals.

There is also a variety of experimental methods, including, for example, "simple" optical pumping with subsequent rf transitions, but also photon-echo microwave nuclear double resonance, or the generation of multiphoton Lamb dips with a laser and a microwave field.

In order to handle efficiently the large number of experiments carried out by laser-microwave spectroscopy and to illustrate the techniques and their differences as compared to work utilizing, e.g., pure laser or pure radiofrequency excitation, the organization of this chapter is by experimental methods rather than by the physical goals pursued in the different experiments. Typical examples will be described in some detail, and reference to other or related work is made. However, the list of references may not be complete.

2. Classification of Laser–Microwave Spectroscopy

2.1. General Considerations

Spectroscopic experiments in which the species under study interacts with a laser light field and a radiofrequency field can usually be divided into three steps: (i) the preparation of an ensemble of particles such that the energy level population becomes noticeably different from thermal equilibrium, (ii) the stimulation of rf transitions, and (iii) the detection of the rf transitions. The steps are mostly carried out subsequently, which means that the interaction of the particles with the different fields can be described by rate equations without taking coherence effects into account. This either holds for beam experiments where the steps occur along the beam axis in well-defined time intervals, or for cell experiments if the laser light field and the rf field are weak. These types of experiments will be classified in the scheme that is treated in detail below.

A discussion of schemes that do not allow a clear distinction between the steps, that is, where coherent superpositions of energy levels are essential, can be found in Sections 2.2.2 and 5.

2.2. Classification

2.2.1. Three-Step Processes

A classification scheme for laser–microwave spectroscopy based on well-defined three-step processes is displayed in Figure 5. It illustrates the great variety of possible experimental methods. The steps can be carried out in many different ways depending, for example, on the physical properties of the species under study. To be more specific, details of the three steps are as follows:

Step 1. Preparation of an Ensemble of Particles. In the first step an ensemble of particles has to be "prepared" in a certain state. This means that the population of the energy levels in question must be altered in such a way that a distinct difference of the occupation numbers as compared to thermal equilibrium is created.*There are several ways to accomplish this:

(1) The laser can be used to excite particles by one-photon or multi-photon absorption. A sublevel a of the upper state under study can be populated selectively by making use of the selection rules for electric dipole radiation and/or—if the spectral bandwidth of the laser is sufficiently narrow in relation to the experimental linewidth—by driving only the transition, which connects the lower state and the excited state sublevel a. This rather simple process is usually followed by an rf transition within the excited state.

(2) A more complicated process is necessary if the population distribution within a ground or metastable excited state has to be changed. Here, the particles undergo one or more successive absorption–emission processes, i.e., an optical pumping cycle. Similarly to the situation described above, it creates differences of the occupation numbers most efficiently if not only the selection rules for electric dipole radiation are utilized, but also a narrow-band laser is applied that can depopulate a certain lower state sublevel selectively. Naturally, the degree of polarization or alignment that will ultimately be obtained depends on the spectral properties of the species under study, in particular on the angular momentum quantum numbers of the states involved.

---

* In certain cases the small difference already present for a thermal Boltzmann distribution can be sufficient.

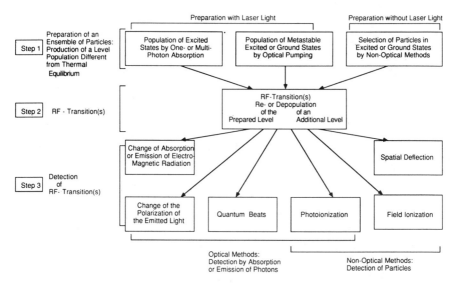

Figure 5. Classification scheme for laser–microwave spectroscopy.

(3) Finally, an ensemble of particles can be prepared in a certain state with a population different from equilibrium by nonoptical methods, e.g., by magnetic deflection in the first part of a classical Rabi apparatus. This method together with the technique of optical pumping is preferable if rf transitions in ground or metastable excited states are to be induced. The nonoptical methods also include excitation by electron bombardment or selective alignment of highly ionized states by stripping in foils. This opens the way to populate states that are not accessible by optical excitation either because suitable light sources are not available or because electric dipole selection rules prohibit population from the ground state. Microwave-optical experiments of this type have been reviewed by Wing and MacAdam (see Part A of this series, p. 491).

Step 2. Microwave Transitions. In the first step a certain level a had been over- or depopulated compared to other levels of the species under investigation. Now, in step 2 microwave transitions between a and another, mostly neighboring, level b are stimulated. This results in a de- or repopulation of a, i.e., in a transfer of population to the level that had a lower occupation after step 1. In most experiments the microwave field induces magnetic dipole transitions. There are measurements, however, where electric dipole transitions are stimulated. In this case the rf field acts as a coherent "light source" in the spectral range of up to tens of gigahertz.

Step 3. Detection of Microwave Transitions. For the detection of the microwave transitions many different experimental schemes based on a large variety of physical processes have been applied. Similar to the first step, one can distinguish between optical and nonoptical techniques. Both methods probe selectively the population of a or b that had been altered by the microwave field. In the case of optical detection and a beam arrangement, this change can be monitored with an additional light beam that probes the changes of absorption (resulting also in changes of the emitted fluorescence light intensity), going along with the altered occupation numbers of the energy levels in question. This technique applies particularly well to studies of ground and metastable excited states. On the other hand, microwave transitions in a short-lived excited state will result directly in a change of the intensity and/or the polarization of the emitted light, which can be detected with a photomultiplier. Electric dipole transitions between Rydberg states have also been detected by observing quantum beats in the fluorescence emitted from the two levels being investigated. In a wider sense, the optical methods also comprise the detection of the microwave radiation itself, and thereby the change of its intensity, as the frequency is swept through the resonance. However, for the reasons given above, such a scheme is not very effective.

Nonoptical methods can be used to probe ground and metastable excited states as well as short-lived levels. One of these techniques, for example, utilizes the backward part, i.e., the B magnet of a classical Rabi apparatus, to deflect molecules after laser excitation and interaction with the rf field, and leaves the other particles unaffected, or vice versa. Another method, which has been applied rather frequently since it provides a high sensitivity, is the detection of ions or photoelectrons after state selective ionization of atoms or molecules. This can be accomplished by photoionization or by field ionization, where the latter applies particularly well to very highly excited states, i.e., to Rydberg atoms.

2.2.2. Nonlinear Processes

So far, we have considered experiments where the laser and microwave transitions occur consecutively in time. This can be considered as a valid assumption if the fields are weak, and/or if they act on the particles in different regions of space.

In a second category of laser–microwave spectroscopy the fields are strong and applied simultaneously. Thus, coherent superpositions of two or more energy levels become possible. Saturation phenomena can be observed and a clear distinction between the steps is no longer possible. Simple rate equation models may not be sufficient for a theoretical description. This category of experiments also includes laser–rf spectroscopy with multiphoton processes where the individual electromagnetic fields need not be in resonance with atomic or molecular energy levels but rather the sum or difference of the optical and microwave photon energies coincides with the splittings to be investigated. Since high light intensities may be required, experiments utilizing nonlinear phenomena are often carried out with the sample located inside the cavity of a laser. More details of such experiments can be found in Section 5.

2.3. Resume

In designing a laser–rf spectroscopy experiment one has to decide which of the different methods that are possible in step 1 and step 3 are to be applied to achieve best performance of the setup. This choice, of course, strongly depends on the spectral properties of the particles to be studied and on the kind of information that the experiment should provide. Other important points are the chemical characteristics of the species, which, for example, can favor a beam experiment or measurements in a closed resonance vessel, the spectral resolution required, and the sensitivity necessary to achieve a signal-to-noise ratio sufficiently high. In principle, all possible combinations of the experimental techniques introduced in step 1 to change the initial population and to detect the rf transitions in step 3, can be chosen. If spatial deflection in *both* steps is used (a case that is clearly not the subject of this chapter), one comes back to the well-known Rabi apparatus; in other words, the classification scheme for laser–rf spectroscopy as displayed in Figure 5 contains this classical method as a special case. Its counterpart is to use laser radiation for the preparation of the ensemble of particles as well as for the detection of the microwave transitions in an atomic or molecular beam experiment. Figure 6 shows schematically an arrangement[18] that meets these conditions and which has been used in a large number of experiments. As in Rabi's experiments, it is adequate for the study of ground or metastable excited states. The atoms or molecules travel across a first laser beam. Here, the state in question is optically pumped via excitation to a higher-lying short-lived state. The pumping can readily be observed in a second interaction region where another light beam branched off from the first one, i.e., one with exactly the same frequency, intersects the molecular beam. Here, no or only a very small amount of light can be absorbed owing to the depletion of the corresponding sublevels in the first zone. Consequently, the intensity of the emitted resonance fluorescence light is very small. If, however, rf transitions are induced in between the two interaction regions, the level depleted initially is repopulated. This gives rise to an increase of absorption and of the emitted resonance fluorescence intensity in region II. Therefore, microwave transitions can be detected by changing the frequency of the rf field and simultaneously recording the light emitted from this second interaction region.

Besides this technique, numerous combinations of optical and nonoptical methods in step 1 and step 3 have been used. Examples will be given in the following text in order to illustrate the large variety of ways in which laser–microwave spectroscopy can be performed.

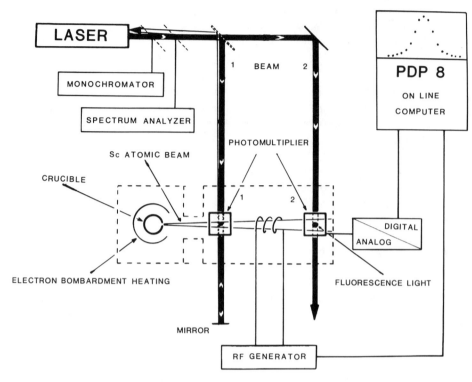

Figure 6. Schematic diagram of an atomic-beam laser–microwave experimental setup; after Ref. 18.

3. Measurements Based on Optical Pumping

3.1. Experiments with Resonance Cells

As already mentioned in Section 2, the classical microwave experiments on molecules were hampered by the fact that the absorption of microwave radiation and therefore the generation of a signal entirely relied on the different occupation numbers of the involved energy levels at thermal equilibrium. To improve the situation, optical pumping could hardly be carried out successfully, because the light sources available before the advent of lasers exhibited broad emission profiles that are not suitable for selective excitation in spectra with many closely spaced energy levels. In view of this situation, which is rather disadvantageous as compared to the clear-cut experiments on atoms, it is not surprising that laser–rf spectroscopy was first applied to a molecule: In 1967, Ronn and Lide[19] reported on infrared–microwave double resonance with a CO_2 laser, the $P(20)$ line of which nearly coincides with a vibrational transition of CH_3Br under study. They used a conventional waveguide absorption cell with a length of 5 ft and equipped with Irtran windows so that the CO_2 laser beam at about 10.6 μm could pass (see Figure 7). As in classical microwave experiments the absorption of the microwave field by the molecules was measured, and certain rotational transitions in the ground state were observed. Consecutively, Ronn and Lide investigated the influence of the laser light on the amplitude of these microwave transitions and thus obtained the first signals by infrared laser microwave double resonance. Also, the transmitted laser beam was directed into an infrared spectrometer to determine which of the four lines emitted by the laser was absorbed by the CH_3Br molecules. It was concluded from the experiment that the disturbance of the equilibrium population by the laser is transmitted over an appreciable range of J quantum numbers but not to levels of different K. This is consistent with other studies indicating that collisional transfer of rotational energy tends to follow optical selection rules.

The observations of Ronn and Lide were confirmed in a similar experiment by Lemaire and co-workers,[20] also carried out on CH_3Br. Fourrier et al.[21] studied the influence of CO_2 laser light on ground state microwave transitions in ammonia. Here, however, some doubt remained as to whether the infrared pumping did not cause a purely thermal effect, i.e., heating. Ammonia was also investigated by Shimizu and Oka,[22,23] where an N_2O laser served as the light source. The small mismatch of the laser frequency and the center frequency of the molecular transition could be decreased by applying a dc Stark field. Similar to the experiments reported earlier, Shimizu and Oka also detected the transmitted microwave radiation and studied the influence of the nearly resonant infrared laser light on the amplitudes of microwave absorption lines. Shimizu and Oka concluded that "hole burning" occurred, in the Doppler profile of the infrared line, i.e., only molecules with velocity component v in the direction of propagation of the laser light

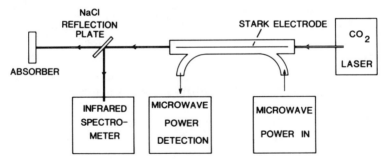

Figure 7. Apparatus for infrared-microwave spectroscopy with detection of the microwave absorption; after Ref. 19.

such that $\nu_0(1 + v/c) \simeq \nu_L$ are pumped efficiently; ν_0 stands for the center frequency of the infrared absorption, and ν_L is the frequency of the laser line. The authors point out that—even though this condition for hole burning is not strictly met because of velocity changes of the pumped molecules and instabilities of the laser frequency—the experimental results can be explained reasonably well by an approach based on the theory for saturated absorption by Karplus and Schwinger.[24] However, saturation by the laser light was not used as a special means for generating or detecting a signal. The latter kind of experiment, which employs saturation and other nonlinear phenomena, will be treated in Section 5.

In a similar experiment Frenkel *et al.*[25] investigated CH_3Cl^{35} and CH_3Cl^{37} and measured the influence of $P(26)$ CO_2 laser radiation on the microwave absorption. As was also reported in the earlier experiments, e.g., by Ronn and Lide, and by Shimizu and Oka, Frenkel and co-workers observed that in such a double resonance experiment not only the participating levels but also other rotational levels, vibrational levels, and even species of different isotopic composition are affected via collisional transfer. Frenkel *et al.* also determined relaxation times. This was accomplished by modulating the laser intensity with a chopper whose speed could be varied between 80 and 4000 Hz and by measuring the rise time of the double resonance signal from which vibrational relaxation rates can be deduced.

The first attempt to register the laser light intensity, i.e., its change after being transmitted through the sample, as a result of microwave absorption was made in 1971 by Takami and Shimoda.[26] Their experiment on formaldehyde with a Zeeman-tuned He–Xe laser at 3.5 μm constitutes an important step toward higher sensitivity since the classical method with detection of microwaves was given up. The interaction cell filled with H_2CO at a pressure of around 10 mTorr was placed inside the laser cavity. The laser frequency was tuned to the line center of the microwave absorption line at 2850.633 cm^{-1} by applying a magnetic field parallel to the laser tube. The higher-frequency component of the laser oscillation was separated with a quarter wave plate and a polarizer, and finally detected with an In–Sb detector as a function of the microwave frequency. This combination of laser light detection and the sample placed inside the resonator makes possible high-sensitivity experiments, even though the output power of the He–Xe laser is only of the order of a few hundred microwatts. For comparison it should be mentioned that detection schemes where the microwave absorption was measured typically required laser powers of several watts.

Takami and Shimoda extended their work with a similar apparatus to HDCO.[27] In this particular experiment saturated absorption was used to detect signals in the ground as well as in the excited state with the same experimental arrangement. This work therefore demonstrates the similarities between double resonance and optical pumping and shows how easily both can converge into techniques that employ nonlinear phenomena (see Section 5).

Optical detection of microwave transitions was also accomplished by Field and co-workers.[28] Important differences of their experiment as compared to the work of Takami and Shimoda[26] were (i) that the sample was not placed inside the laser cavity and (ii) that, instead of the influence of the microwave transitions on the transmitted power, the change of the emitted resonance fluorescence intensity as the microwave frequency was swept was registered. Measurements have been carried out with high sensitivity on BaO with Ar^+ laser excitation in the optical range at $\lambda = 496.5$ nm (see Figure 8). Later, similar experiments were also performed in *excited* states of BaO (see Section 4.1).

Experiments utilizing optical pumping with either microwave or optical detection have also been reported, e.g., on ammonia,[29-33] propynal (HCCCHO),[34] SrF,[35] CaF,[36] and NO_2.[37]

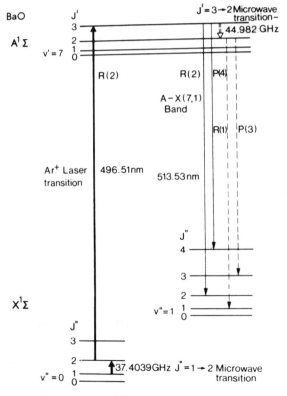

Figure 8. Energy level diagram of BaO.[28] The optical line with $\lambda = 496.51$ nm is indicated as well as the microwave transition in the ground state.

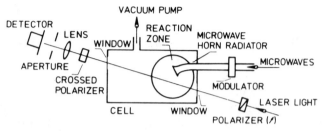

Figure 9. Experimental setup for microwave–optical polarization spectroscpy (MOPS).[38] The
BaO molecules under investigation were produced by a gas phase reaction of metallic
Ba with N_2O.

Recently, a completely different approach to register microwave transitions in ground states with high sensitivity has been introduced by Ernst
and Törring.[38] They applied a new technique called "microwave–optical
polarization spectroscopy" (MOPS) to BaO as an absorbing gas and later
also to CaCl[39] and to SrF.[40] The method has similarities to optical
polarization spectroscopy and polarization labeling but uses a single laser
light beam sent through the sample, which is located between two (nearly)
crossed polarizers (see Figure 9). Thus, any optical anisotropy in the sample
changes the light intensity transmitted through the second polarizer. In the
case of microwave optical polarization spectroscopy such an anisotropy is
created by polarized microwave radiation that pumps a rotational transition
in the ground state (see Figure 10). A signal is observed if this transition

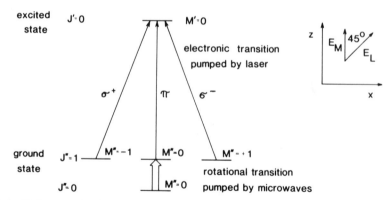

Figure 10. An example of a pumping scheme for microwave–optical polarization spectroscopy.
A rotational transition in the ground state is detected via an anisotropy in the optical
absorption. With the z axis chosen parallel to E_m, the microwaves only pump
$\Delta M = 0$ transitions. When the linear polarization of the laser light (E_L) is rotated
by 45° and its frequency tuned to the appropriate transition, the z and x components
(pumping π and σ transitions, respectively) will experience different absorptions
and refraction indices. This leads to microwave induced changes of the isotropic
optical absorption. After Ref. 39.

shares a common level with the transition induced by the laser probe beam. The authors point out that microwave–optical polarization spectroscopy is not simply a special case of polarization spectroscopy or labeling with the second light beam replaced by microwave radiation. Important differences are that the rf transitions are pressure broadened rather than Doppler broadened so that microwave pumping is not velocity selective like optical pumping. Secondly, the ground state microwave transitions are induced between levels that are almost equally populated at thermal equilibrium. Consequently, an observable alignment (which is proportional to the population difference) can only be created with the microwaves if this equilibrium population difference has been substantially increased by the laser. Thus, the light beam not only serves for "probing" the anisotropy but has to be strong enough to produce a noticeable effect on the level populations.

Major advantages of microwave–optical polarization spectroscopy are narrower linewidths as compared to "conventional" laser–rf double resonance and smaller intensities required for the laser light field and the microwaves, so that strongly saturating conditions can be avoided. Therefore, the sensitivity as well as the resolution can be greatly enhanced.

3.2. Particle-Beam Techniques

In 1952 Kastler and co-workers[41] replaced the A and B magnets of the classical Rabi apparatus by two light beams of a spectral lamp. The next step was done by Schieder and Walther,[42] who used two light beams originating from the same laser instead of light from a spectral lamp. They pointed out that in contrast to the Rabi-type magnetic deflection method, the configuration with two light beams also permits the investigation of ions. The scheme of two laser-beam–particle-beam crossing regions as already outlined in Section 2.3 (see Figure 6) was completed by additional application of a radiofrequency field between the two zones. This configuration was introduced by Rosner and co-workers and by Ertmer and Hofer.

Rosner and co-workers[43] replaced the A and B magnets of the conventional molecular beam magnetic resonance technique with two particle-beam–laser-beam crossing regions, in order to measure the zero-field hfs of a single vibrational–rotational level of Na_2 by additional radiofrequency transitions in a zone C located between A and B. Sodium dimers with an abundance of about 5% were produced in a supersonic beam evaporated from an oven at 620°C. The flux of Na_2 molecules through the B region of the apparatus was 3×10^{11}/sec. In region A the linearly polarized light of a single-mode Ar^+ laser at $\lambda = 476.5$ nm (power 0.1 W) excited molecules in the $v'' = 0$, $J'' = 28$ level of the $^1\Sigma_g^+$ ground state to the $B\,^1\Pi_u\,v' = 6$, $J' = 27$ level and depleted the population of the lower level by optical pumping. This depletion was monitored in zone B via a decrease of the fluorescence light intensity by a factor of 10. The rf field with a frequency of around 100 kHz was produced by a 36.2-cm-long solenoid and chopped at a frequency of 107 Hz to improve the signal-to-noise ratio. The resolution was limited to 3.7 kHz by the transit time of the molecules through the rf field. Blocking the laser beam in zone A destroyed the rf induced signal, verifying that optical pumping was the source of polarization in the molecules.

Independently from Rosner *et al.*, Ertmer and Hofer[18] applied the same technique to metastable states of ^{45}Sc atoms (see Figure 6). The states under investigation were populated by electron bombardment, which simultaneously served to evaporate scandium atoms from a crucible. The light beam of a continuous-wave (cw) single-mode dye laser in the yellow–red spectral region crossed the beam of excited atoms at right angles in a first zone, and excited the atoms from the $3d^2 4s\,^4F$ to the $3d^2 4p\,^4G^0$ state, thereby optically pumping a hyperfine sublevel. A fractional light beam, split off from the main laser beam, crossed the atoms in a second region and monitored the optical pumping effect which transferred 80% of the atoms to another ^4F hyperfine substate. The presence of metastable atoms (about 1% of all atoms in the beam) was strictly correlated with electron impact and not of thermal origin. This was verified by the fact that the fluorescence vanished as soon as the electron bombardment was interrupted, even though the crucible was still at high temperature. Ten different hyperfine transitions were stimulated with a radiofrequency loop inserted between the two laser zones and detected via the increase of fluorescence intensity in zone 2. The frequencies ranged from about 470 to 2300 MHz. A signal half-width of 55 kHz was obtained at 800 MHz. The spectroscopic results yielded constants of magnetic dipole and electric quadrupole hyperfine interaction.

Childs *et al.* performed laser–rf measurements of the hyperfine structure of ^{51}V.[44] A vanadium atomic beam containing a sufficient fraction of atoms in the desired metastable state was produced by heating a tantalum oven and at the same time exciting the effusing atoms by electron bombardment. The laser pump beam with typically 6 W/cm^2 and the probe beam with 6×10^{-2} W/cm^2 intersected the atomic beam orthogonally at a distance of about 20 cm. After a preliminary analysis of the hyperfine structure measured with a pure laser scanning method using a confocal Fabry–Perot interferometer for frequency calibration, the laser frequency was set to a particular transition, and rf was applied simultaneously between the laser regions. Since the hyperfine splittings under investigation were already known very precisely from the laser scan results, the radiofrequency tuning range could be limited to ± 2 MHz. The measurements yielded rf signal curves with a half-width of typically 20 kHz. Experimental errors as low as 5×10^{-7} for the splittings are reported.

Further laser–rf studies using the described scheme have been performed for a considerable number of species, namely, for Sm isotopes,[45] for ^{50}V,[46] ^{235}U,[47,48] ^{47}Ti, ^{49}Ti,[49] CaF,[50,51] CaCl,[52,53] CaBr,[54] ^{138}La, ^{139}La,[55] ^{141}Pr,[56,57] ^{159}Tb,[58] ^{95}Mo, ^{97}Mo,[59] ^{55}Mn,[60] ^{57}Fe,[61] ^{53}Cr,[62] ^{43}Ca,[63] ^{161}Dy, ^{163}Dy and ^{167}Er,[64,65] ^{165}Ho,[66] and ^{87}Sr.[67] Among the numerous aims of these investigations were hyperfine structure interaction constants, nuclear moments, effects of configuration interaction and core polarization in the electron shell of atoms, molecular data like effective spin rotation constants, etc. The combination of experimental data and advanced theoretical procedures, like relativistic self-consistent field calculations, provided many fruitful results regarding the reliability of electronic wave functions of atoms.

As was pointed out in several papers, the advantages of these laser–rf techniques compared to the classical Rabi method are the moderate atomic beam collimation ratio of typically 1:100 providing a much larger flux of atoms than in Rabi-type machines. The high sensitivity originates from at least one detected fluorescence photon per atom in comparison to a detection probability of only one atom in about 10^4 with electron impact ionization in deflection experiments. Also, such laser–rf measurements no longer rely on changes of the beam trajectory, so that transitions between substates with rather similar effective magnetic moments can be observed. The absence of certain requirements concerning quantum number selection rules is regarded to be a further advantage of laser–microwave spectroscopy. In addition, such measurements are very insensitive to small laser frequency drifts around the optical resonance center. Measurements at zero magnetic field are possible so that the influence of the Zeeman effect is eliminated. Also, ac Stark shifts (light shifts) due to the laser field are avoided, since the regions of laser light and microwave irradiation do not overlap.

Rosner and collaborators applied the laser–microwave scheme discussed above not only to atoms but for the first time also to ions, particularly to hyperfine structure measurements[68] of Xe^+. In this work the ion beam passed a coaxial waveguide where the rf transitions were induced. Since the ions moved in opposite direction but parallel to the rf wave propagation, the resonance frequencies had to be corrected for the Doppler shift.

Microwave techniques were combined with fast ion beam laser spectroscopy[69,70] (see Figure 11) in order to investigate the hyperfine structure of ^{235}U$^+$. A 50-keV, isotopically pure ^{235}U$^+$ ion beam with a current of typically 0.5 μm was generated. The important difference compared to the experiments discussed so far is that the ion beam overlaps *collinearly* with the light of a cw dye ring laser. The ion beam passes two postacceleration zones A and B. Here, appropriate voltages Doppler-tune one particular (F-F') hyperfine transition of the ions between a certain metastable configuration and a higher-lying level multiplet into resonance with the

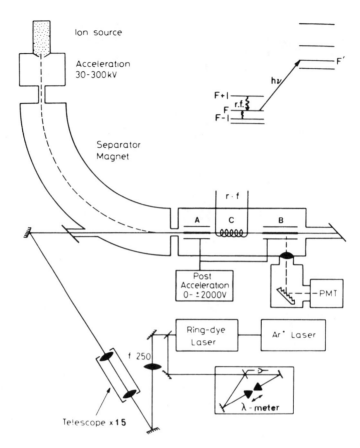

Figure 11. Arrangement for collinear laser–rf ion beam spectroscopy; after Ref. 70.

laser light. Thus, two well-defined interaction regions are created even though the ion and laser beams travel collinearly. Depletion of the initial metastable *F* sublevel by optical pumping in the A region is monitored via a decrease of fluorescence light intensity in the B region. An rf field between A and B induces $(F \pm 1 \leftrightarrow F)$ transitions and refills the *F* substate, thus causing a fluorescence increase in zone B. The rf field is created in a coaxial transmission line and consists of one TEM mode traveling along the beam axis. The measured rf transition frequencies are therefore Doppler-shifted by about 50–500 kHz.

In our laboratory a series of laser–microwave studies has been performed on metastable and short-lived excited states of heliumlike Li$^+$. Singly ionized lithium, like all members of the two-electron He isoelectronic sequence, belongs to the fundamental systems in atomic physics. Many of its spectroscopic and quantum-mechanical characteristics have been calculated very precisely with advanced methods of perturbation theory. These calculations together with high-precision measurements can serve as a stringent test of our understanding of two-electron systems. The investigations discussed in the following have been preceded by a considerable number of experiments performed with various methods. Most of the respective publications are quoted in Ref. 71. A first series of laser–microwave measurements[72,73] concerned the hfs splittings of the $1s2s\ ^3S_1$ multiplet of $^6Li^+$ and of $^7Li^+$. The respective part of the energy level diagram is given in Figure 12. The metastable $2\ ^3S_1$ (lifetime $\tau = 50$ sec[74]) and the short-lived $1s2p\ ^3P_{0,2,1}$ term (lifetime $\tau = 43$ nsec[75]) are connected via a resonant transition with $\lambda = 548.5$ nm. Because of the nuclear spin of $I = 1$ of $^6Li^+$ and $I = 3/2$ of $^7Li^+$, the $2\ ^3S_1$ state splits into three hfs sublevels with quantum numbers $F = 0, 1, 2$ and $1/2, 3/2,$ and $5/2$, respectively. The experimental arrangement is shown in Figure 13. A low-velocity Li^+ ion beam (velocity $v = 7.4 \times 10^6$ to 1×10^7 cm/sec corresponding to a kinetic energy of 200–400 eV) with a sufficient amount of ions in the $^3S_1$ state is crossed orthogonally by a continuous wave single-mode dye laser beam with $\lambda = 548.5$ nm. The laser light depletes one of the $2\ ^3S_1$ hfs sublevels by optical pumping via a suitable $2\ ^3P$ hfs state. The ions then pass a waveguide where $\Delta F = 1$ microwave transitions between the depleted substate and one of the adjacent levels are induced. The repopulation of the depleted level is monitored by the second laser beam, branched off from the first one, and detected through the increase of the fluorescence light intensity coming from the second crossing region.

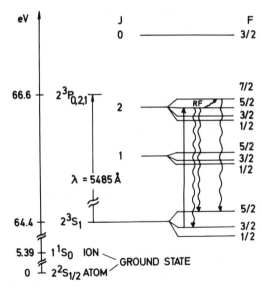

Figure 12. Energy level scheme of the $1s2s\ ^3S_1$ and $1s2p\ ^3P$ terms of $^7Li^+$. The diagram includes an example of a laser–microwave pumping cycle.

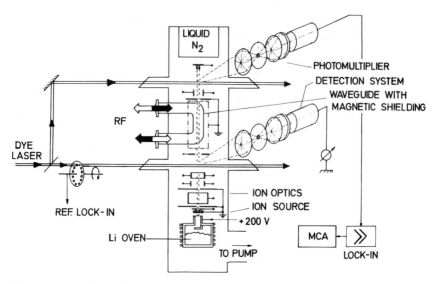

Figure 13. Experimental arrangement of the laser–microwave spectrometer and ion beam chamber for investigations of lithium ions.[72]

The ion beam is produced in the following way: Li atoms are evaporated from an oven at a temperature of typically 400°C, and are ionized and excited to the metastable $2\,^3S_1$ state by electron impact, when leaving the oven aperture. The electrons are emitted from a little ring-shaped tungsten wire cathode which is placed horizontally several millimeters above the oven exit. The cathode is held at ground potential, the oven at +200 V. The electrons are accelerated directly onto the oven aperture thus counterpropagating to the ions which are accelerated in the same electric field. The ions pass the cathode loop and are formed into a well-collimated beam by an electrostatic lens system.

The microwave field, generated with a synthesizer and amplified with a traveling wave tube, is reflected onto itself at one end of the waveguide thus providing two counterpropagating waves. One field travels with the ion motion, the other one in the opposite direction. Therefore, two fluorescence light signals which arise in the second laser ion-beam crossing region are caused by the Doppler-shifted microwave resonance frequencies ν^+ and ν^-. The center frequency ν^0 is the arithmetic average $(\nu^+ + \nu^-)/2$. The lower laser beam (the pump beam) is intensity modulated; the fluorescence signal from the upper region is processed with lock-in technique and stored in a multichannel analyzer. Figure 14 displays the $(F = 3/2 \leftrightarrow F = 5/2)$ transition of $^7Li^+$ with two Doppler-shifted signal curves. The results are shown in Table 1. The evaluation of the $2\,^3S_1$ measurements unveiled a nuclear structure effect in the hyperfine interaction constant of each isotope

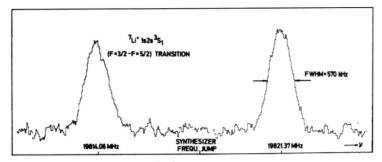

Figure 14. $^3S_1$, $F = 3/2$ to $F = 5/2$ microwave transition in $^7Li^+$ with two Doppler-shifted signal curves.[72]

and a tiny ($F = 1$) substate depression due to $2\,^3S_1$–$2\,^1S_0$ hyperfine mixing. The analysis of the experimental results and a theoretical treatment are found in Refs. 76 and 77.

An apparatus similar to the one described above but adapted to the investigation of rotational transitions in molecules has been reported recently[78] and applied to precise Stark effect measurements in the $X\,^2\Sigma$ ground state of CaCl.[79]

The description of the Li^+ experiments leads over to a particularly interesting subgroup of schemes based on optical pumping: microwave transitions in a short-lived excited state are detected via their influence on the transfer of population to sublevels in the lower-lying state from which laser light is absorbed initially. An example of such a laser-rf resonance cycle concerning transitions between the $2\,^3P_2$ ($F = 5/2 \leftrightarrow F = 7/2$) and the $2\,^3S_1$ states of $^7Li^+$ is shown in Figure 12. Radiofrequency transitions in the $2\,^3P$ state change the branching ratio of the decay back to the different $2\,^3S_1$ hfs sublevels and thus affect the population. Similar to the work described above, this "microwave-induced optical pumping" can be monitored in a second laser-beam–ion-beam crossing region via a change of the fluorescence intensity.

In the case of measurements in the $2\,^3P$ multiplet of $^7Li^+$,[73,76,80] however, modifications of the apparatus had to be made. Because of the short lifetime of the $2\,^3P$ energy levels, laser excitation from the $2\,^3S$ state

Table 1. Hfs Splittings of the 2^3S_1 State for $^6Li^+$ (a) and $^7Li^+$ (b), Measured by Laser–Microwave Spectroscopy with an Ion Beam[a]

(a)		(b)	
$F = 0 \leftrightarrow F = 1$	3001.780 (50)	$F = 1/2 \leftrightarrow F = 3/2$	11 890.018 (40)
$F = 1 \leftrightarrow F = 2$	6003.600 (50)	$F = 3/2 \leftrightarrow F = 5/2$	19 817.673 (40)

[a] All values in MHz, from Refs. 72, 73.

and the stimulation of rf transitions must take place in the same region in a microwave cavity. A signal obtained with this technique is displayed in Figure 15.

Laser–rf spectroscopy was also performed on the radioactive sodium isotopes ^{21}Na, $^{25-29}$Na produced by bombarding sodium in an oven with the 20-GeV proton beam of the proton synchrotron at CERN.[81] A cw single-mode dye laser beam crossed the atomic beam orthogonally and selectively induced ($3s\,^2S_{1/2}-3p\,^2P_{3/2}$) transitions (*D*2 line) between particular hyperfine substates. Since a weak magnetic field parallel to the laser beam was present, optical pumping changed the population of the different F, m_F sublevels of the ground state. An rf field was applied in the laser-beam atomic beam crossing region and caused magnetic dipole ($F = 3 \leftrightarrow F = 2$) transitions in the $2\,^2P_{3/2}$ state, thereby influencing the decay channels back to the atomic ground state and hence the ground state population distribution. This is similar to the Li$^+$ experiments described above. However, for the sodium isotopes a nonoptical detection scheme (see Figure 16) was applied: The atoms passed a strong inhomogeneous magnetic field of a hexapole magnet which focused atoms with $m_J = +1/2$ onto an ionizing hot rhenium surface and defocused atoms with $m_J = -1/2$. The ions were subsequently mass-separated and counted. The experimental curves exhibit a depth as large as 20%. The authors performed calculations of the width and depth of the rf resonance curves and also considered effects arising from a slight laser detuning. The measured $2\,^2P_{3/2}$ hfs splittings provided the electric quadrupole hyperfine interaction constants B from which the nuclear quadrupole moments of the six unstable isotopes ^{21}Na, $^{25-29}$Na could be extracted.

Ter Meulen *et al.* applied ultraviolet–microwave spectroscopy to OH radicals.[82] As in the $2\,^3P$ measurement of ^{7}Li$^+$,[80] the chopped laser light beam crossed the particle beam at right angles inside a tunable microwave

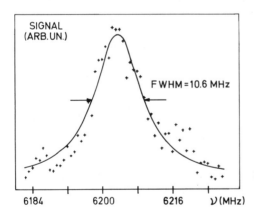

Figure 15. Signal curve of the 2^3P_2 ($F = 1/2$ to $F = 3/2$) microwave transition in ^{7}Li$^+$.[80]

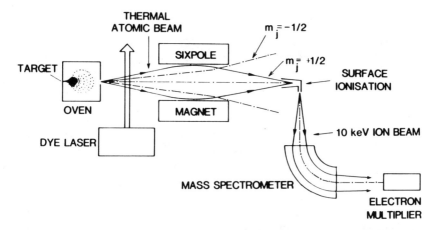

Figure 16. Scheme for laser–rf spectroscopy of sodium isotopes with nonoptical detection. Atoms with $m_J = -1/2$ are defocussed by a hexapole magnet; those with $m_J = +1/2$ reach a hot surface where they are ionized. The ions are mass separated and counted. After Ref. 81.

cavity. The uv radiation at $\lambda = 307$ nm, obtained through internally frequency-doubling the light of a ring dye laser, induced $X\,^2\Pi_{3/2}$ ground state to $A\,^2\Sigma_{1/2}^+$ excited state transitions and caused optical pumping of the ground state. Similar to the detection scheme used in Ref. 80, the optical pumping effect and thus also its change, introduced by magnetic dipole transitions within the $A\,^2\Sigma_{1/2}^+$ (lifetime $\tau \simeq 1\ \mu$sec), was monitored via the fluorescence light intensity in a second laser-beam–particle-beam crossing region. Five transitions were measured at frequencies between about 23 and 31 GHz with error bars of ±50 to ±80 kHz.

It is interesting to note that in certain molecular beam magnetic resonance experiments as described above, the state selection and the detection by strong optical fields can produce rf-magnetic resonances whose line shape differs drastically from the Rabi shape, even though the optical and rf fields do not act simultaneously on the molecules. Adam et al.[83] have investigated these effects and suggest that in the case of optically unresolved ground state sublevels—i.e., with splittings smaller than the inverse of the transit time across the laser field—ground state coherences are created by absorption and stimulated emission. Thus, interference effects become possible and the resonance curves may even acquire negative peaks. Adam and co-workers point out that these features cannot be explained if one only considers populations and the probabilities of their transfer among states by the optical and rf fields in the different interaction regions. They give a model calculation including coherence effects, which accounts for the essential features of the observed rf spectra.

3.3. Spectroscopy of Trapped Ions

Ions can be confined in "traps" created by potential wells of electric and magnetic fields. Depending on the quality of the vacuum, the trapping time can be very long—hours, days, even months. Such an arrangement has several advantages compared to experiments with ion beams or ions in resonance cells:

- The ions cannot escape from the resonance region. Therefore, transit time effects that cause line broadening can be neglected.
- Collisions with other particles are almost avoided. Thus, collisional broadening is of minor importance and extremely narrow linewidths of a few hertz and less can be achieved.
- Experiments with pulsed lasers can be carried out. This opens the way to measurements in spectral ranges, e.g., in the ultraviolet, where continuous-wave lasers are not available. The principle of such experiments with pulsed lasers is to create a population different from equilibrium with a first laser pulse, subsequently stimulate rf transitions and probe their effect on the population with the following pulse. This can hardly be realized with ions in resonance vessels or beams because they either escape from the resonance region in the time interval of typically 0.01–0.1 sec between the laser pulses or because the excitation and polarization are destroyed by collisions. On the other hand, a single laser pulse is usually not long enough to be used for pumping of the system as well as for probing of population changes induced by the microwaves.

However, the relatively small number of ions, 10^5–10^6, that can be stored in a trap restricts the signal-to-noise ratio. Besides that, an existing average ion velocity limits the accuracy of a microwave measurement by the second-order Doppler shift. The latter effect can be reduced by the application of laser side band cooling through resonant light pressure. Cooling the ions with laser light also attacks the problem of insufficient signal-to-noise ratio, since narrowing of the inhomogeneous Doppler profile results in a transition probability roughly equal for all ions.

Many experiments discussed below were motivated by the possibility of future frequency standards based on microwave measurements with trapped ions.

Starting with the work of Dehmelt *et al.*[84] high-precision experiments on trapped ions made use of various schemes of state selection and signal detection before lasers were introduced in ion trap spectroscopy. The first successful attempt of laser rf spectroscopy of trapped ions was made by Wineland and co-workers[85-87] on singly ionized Mg, which is isoelectronic with the Na atom. The authors measured the hyperfine splitting of the

$3s\,^2S_{1/2}$ ground state of $^{25}Mg^+$ with a nuclear spin of $I = 5/2$. They used the resonance transition between the first $3p\,^2P$ state and the ground state with a wavelength of $\lambda = 279.6$ nm in order to change the population conditions in the ground state by laser optical pumping. The light source was a frequency doubled single-mode cw dye laser with an intensity of $<40\,\mu W$. The beam was focused to a diameter of $50\,\mu m$. The ions were stored in a Penning trap (for the basic experimental apparatus see Refs. 88, 89) with a uniform magnetic field and a quadrupolar electrostatic potential. Space charge effects limited the maximum ion density in the volume, which overlapped with the light beam. Typically 10–100 ions were confined to a small volume near the trap center by laser-cooling to about 1 K. Shifts caused by the laser field were avoided by shutting the light off while the microwave transitions were induced. Thus the magnetic field instability was by far the dominating source of uncertainty in determining the resonance frequencies. Fluctuations of 2×10^{-7} in 1 sec and of 10^{-6} in several seconds were measured with an NMR probe. Other possible sources of shifts and broadenings of the rf and microwave resonances involved were also discussed. Some of them were recognized to be negligibly small:

- Frequency shifts due to the second-order effect which were estimated to be -1.1×10^{-13} at an ion temperature of $T = 20$ K;
- Frequency shifts due to collisions with background neutrals (collision rate less than 5/min);
- Spin exchange reactions between the ions, since the Coulomb force keeps the ions apart;
- Hyperfine structure Stark shifts produced by the Coulomb fields of neighboring ions;
- Hyperfine pressure shifts (estimated to be less than 10^{-15}).

The final result for the magnetic dipole hyperfine interaction constant as obtained from this experiment is

$$A(^{25}Mg^+) = -596.254\,376\,(54)\text{ MHz}$$

and is compared with a theoretical value of $-A = -553$ MHz by Lindgren (private communication in Ref. 86), including core polarization and relativistic corrections. Another experimental result was the nuclear-to-electronic g factor ratio of $g_I/g_J(^{25}Mg) = 9.299\,484\,(75) \times 10^{-5}$.

A similar ion trap experiment was performed with $^9Be^+$ ions.[90,91] The ions were cooled via the $2s\,^2S_{1/2}(M_I = -3/2, M_J = -1/2) \rightarrow 2p\,^2P_{3/2}(-3/2, -3/2)$ transition at $\lambda = 313$ nm with a frequency doubled dye laser, and were additionally optically pumped in the ground state. Measurements of the axial (ν_z), magnetron (ν_m), and electric field shifted cyclotron (ν_c')frequencies of the stored ions provide the free-space cyclotron frequency ν_c from the equation

$$qB_0/2\pi m = \nu_c = [(\nu_c')^2 + \nu_z^2 + \nu_m^2]^{1/2}.$$

Here B_0 is the applied magnetic field, and q and m are the ion charge and mass, respectively. Measuring ν_c for different species allows mass comparisons. Excitation of the ion motional frequencies by an oscillating electric field increases the size of the ion orbits in the trap. The fluorescence light emitted from the ions goes down in intensity when the electric field frequency gets into resonance, since the laser light beam is focused onto the ion cloud. This effect is used for detection of the motional resonance. The small cloud size requires only slight motional excitation intensity and minimizes effects of B field inhomogeneity, and of trap anharmonicity. The authors estimate that ion cyclotron resonances may ultimately be measured with an accuracy near one part in 10^{13}. They cite the interesting idea of R. D. Deslattes that a comparison of the mass difference between nuclear isomers with the respective γ-ray wavelength would provide a conversion factor from wavelengths to atomic mass units.

The ground state hyperfine structure splitting of $^9\text{Be}^+$ was measured by optical pumping and rf transitions between suitable (M_I, M_J) substates in magnetic fields of roughly 0.7–0.8 T. The respective transitions were induced with two coherent rf pulses of 0.5 sec duration separated by 19 sec. This Ramsey interference method provides signal linewidths dominated by the pulse separation time. The obtained magnetic hyperfine interaction constant of

$$A = -625\,008\,837.048\,(4)\ \text{Hz}$$

confirms the value of

$$A = -625.009\,(3)\ \text{MHz}$$

measured earlier in an optical pumping rf precision experiment using a hollow cathode Be^+ light source, and ions produced by a gas discharge and kept in a He buffer gas cell.[92] The comparison of the two results illustrates the enormous improvement in accuracy realized with the advent of ion trap techniques and laser microwave spectroscopy.

The ground state hyperfine structure splitting of $^{137}\text{Ba}^+$ [93,94] and of $^{135}\text{Ba}^+$ [95] was measured with microwave optical double resonance. The following description is restricted to $^{135}\text{Ba}^+$ since the experiments are similar for both isotopes. Pulsed laser light tuned to one of the hyperfine components of the $(6s\,^2S_{1/2}\text{–}6p\,^2P_{1/2})$ resonance line at $\lambda = 493.4$ nm produces a population difference between the F levels of the ground state by optical pumping.

A microwave field is irradiated between two laser pulses. It induces transitions and restores the statistical population equilibrium. The microwave signal is monitored via the change of the reemitted fluorescence light intensity after laser excitation. An N_2 laser pumped dye laser of about 10 kW peak power, 6 nsec pulse width, 1 GHz spectral band width, and a repetition rate of up to 50 Hz is used. The Ba^+ ions can also decay from the $6\,^2P_{1/2}$ state to the extremely metastable $5\,^2D_{3/2}$ states ($\tau = 17.5$ sec) and are thus eliminated from the optical pumping cycle. To avoid this effect, helium gas with a pressure of 10^{-5} mbar was added to the trap. This reduces the effective D state lifetime to about 100 msec. A microwave signal half-width of 0.87 Hz was reached. The final result for the hfs splitting, extrapolated to zero magnetic field strength, is

$$\nu = 7\,183\,340\,234.90\,(0.57)\ \text{Hz}$$

The 1σ error includes uncertainties due to magnetic field extrapolations (0.40) and to the second-order Doppler shift (0.40), and could be diminished by better shielding of the magnetic field and by optical sideband cooling. The result of a classical optical pumping experiment[96] was

$$\nu = 7183.3412\,(5)\ \text{MHz}$$

the error being three times the standard deviation. The respective values for $^{137}Ba^+$ are $\nu = 8\,037\,741\,667.69\,(0.36)\ \text{Hz}$[94] and $\nu = 8037.7422\,(8)$ MHz.[96]

In another trap experiment,[97,98] Werth and co-workers (see also Chapter 5 of Part C) measured the ground state hfs splitting of $^{171}Yb^+$, again using a N_2-laser pumped dye laser (see Figure 17). A broadcast 77.5-kHz

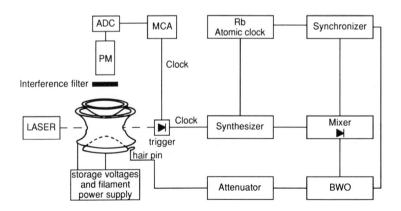

Figure 17. Configuration of a laser–microwave ion trap experiment; after Refs. 94 and 98.

frequency, locked to the primary Cs standard of the Physikalisch-Technische Bundesanstalt (PTB), provided the absolute frequency calibration and long-term stability. The result is

$$\nu = 12\ 642\ 812\ 124.2\ (1.4)\ \text{Hz}$$

The narrowest linewidth obtained was 33 mHz (see Figure 18), corresponding to a line Q factor of 3.8×10^{11}. The final accuracy was mainly determined by the stability of the reference frequency.

Laser–microwave spectroscopy has also been extended to trapped negative ions, which differ from neutral atoms, molecules, and positive ions in various aspects. The attached electron undergoes no long-range Coulomb interaction with the neutral, and usually there is only one bound state. Negative ions are regarded as particularly suitable objects for tests of atomic theories, e.g., concerning electron–electron interaction, binding energies, photodetachment behavior, and fine structure splittings. The work, briefly reviewed here, comprises measurements of bound state Zeeman splitting frequencies in O^- and S^- [99] as well as the determination of the $2\,^2P_{3/2}$ ground state g factor of $^{32}S^-$ [100] and of hyperfine parameters of the $^2P_{3/2}$ state in $^{33}S^-$. [101] About 10^4 negative ions are stored in a Penning trap with a magnetic field of roughly 1 Tesla. The ^2P ground state of O^- for example has a $J = 1/2 \leftrightarrow J = 3/2$ fine structure splitting of 5.3 THz. At a B field of 0.95 T the $^2P_{3/2}$ state exhibits a Zeeman pattern with a 18-GHz splitting frequency between neighboring m_J substates. Irradiation of π-polarized laser light with a power of a few milliwatts selectively photodetaches

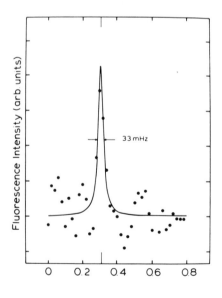

Figure 18. Microwave signal obtained with an ion trap experiment on $^{171}\text{Yb}^+$. [98]

electrons, e.g., from the $m_J = 3/2$ substate. A microwave field of 18 GHz with less than 140 μW simultaneously induces $m_J = 1/2 \leftrightarrow m_J = 3/2$ transitions and thus refills the $m_J = 3/2$ Zeeman level. After 1–3 sec of irradiation, the number of ions remaining is determined by driving their axial motion and measuring the current induced on the ring electrode of the ion trap. This procedure is repeated as a function of microwave frequency and provides the Zeeman resonances. The technique works also for the $m_J = -1/2 \leftrightarrow m_J = -3/2$ transition. The same kind of measurement was also performed for S$^-$. The ratio of the sum of the two Zeeman frequencies to the electron cyclotron frequency in the respective magnetic field gives the g_J factor. The result for ^{32}S$^-$ is

$$g_J = 1.333\,984\,(28)$$

and differs by $+651\,(28) \times 10^{-6}$ from the Lande value of 4/3. The correction of 773×10^{-6} due to the anomalous magnetic moment of the electron leaves a difference of $-122\,(28) \times 10^{-6}$, probably arising mainly from relativistic and diamagnetic effects. None of these effects, however, has been calculated for any negative ion.[100]

3.4. Experiments in Solids

Combined laser–microwave spectroscopy based on optical pumping was also performed in the solid state. Spectral line broadening caused, e.g., by strain and phonon interaction, can be overcome by extreme cooling and specific site selective procedures. Very narrow lines are attainable particularly in the spectra of rare earth ions doped to crystals in low concentration. Rare earth ions, therefore, play an important role in solid state spectroscopy, as will be illustrated in the course of this section.

One of the early subjects of laser rf studies in solids were color centers and concerned the relaxed excited state (RES) of the F center. F centers are created by trapping of an electron in a negative ion vacancy. The electron can be optically excited from its ground state ($1s$) into a higher-lying state from which it decays by the emission of phonons into a relaxed excited state. This relaxed excited state has a lifetime of the order of microseconds and decays by emission of light (see Figure 19). The spin polarization at thermal equilibrium of the electronic ground state of the F center can be changed by irradiation of circularly polarized light. The size of this optical pumping effect is determined by the spin mixing properties of the whole excitation/deexcitation cycle. Additional electron spin resonance (ESR) or electron–nuclear double resonance can influence the pumping cycle and therefore the spin polarization of the ground state under optical pumping. The F center shows magnetic circular dichroism (MCD), which means that

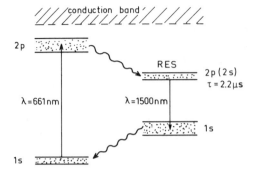

Figure 19. Level scheme of an *F* center in KI. RES stands for relaxed excited state; the wavy lines symbolize phonon transitions. After Ref. 109.

the absorption probability of a light beam parallel to an applied static magnetic field differs for left-handed and right-handed circularly polarized light.

Mollenauer and co-workers[102] used the MCD effect in the absorption band of an *F* center in order to detect, for the first time, changes in the ground state spin polarization induced by ENDOR in the relaxed excited state. A He–Ne laser served for optical pumping as well as for monitoring the magnetic circular dichroism, while several preceding experiments still had used conventional lamps.[103,104] The laser beam, which passed a quarter wave plate with a stress modulation frequency of 50 kHz, was irradiated onto a KI crystal that contained the *F* centers under study. The transmitted laser light was monitored with a photomultiplier and processed with lock-in technique.

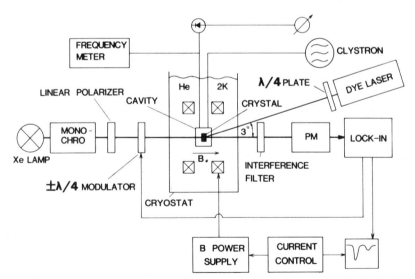

Figure 20. Experimental scheme for laser-rf spectroscopy of *F* centers; after Ref. 110.

In the time following, detailed studies of the optical pumping cycle were performed.[105,106] They showed that saturated optical pumping should improve the microwave induced signals. Thus, monitoring of electron spin resonances in the relaxed excited state via saturated optical pumping with a continuous wave dye laser was introduced.[107] The experimental setup is shown in Figure 20. The size of the ESR signal depends critically on the wavelength of the pumping light, which has to be optimized by tuning of the dye laser. It was concluded from the ESR linewidths[108] that the wave function of the relaxed excited state is very diffuse in comparison to that of the ground state. Numerical values for the mean radii were given for different crystals.

Besides the ESR ground state signal, and that of the relaxed excited state (RES1), a third resonance (RES2) was detected (see Figure 21), after the ESR frequency had been changed from 9 to 30 GHz.[109,110] The new signal could also be related to the relaxed excited state, since it appeared only when the F center was optically pumped. It also decreases strongly together with RES1, if the F center concentration is increased from 10^{16} cm^{-3} to more than 10^{17} cm^{-3}. This indicates that RES2 is connected with the isolated F center. The g factor of RES2 is closer to the free electron g factor than that of RES1. Finally, the RES2 lifetime seems to be larger than that of RES1, as concluded from the ESR saturation behavior.

Dieckmann et al.[111] observed rf transitions between ligand hyperfine structure levels in CaF_2:Tm^{2+} with a technique that replaces the role of the microwave transition in conventional ENDOR by cross-relaxation processes. The Tm^{2+} ions in the liquid-He-cooled crystal are optically pumped via the strong $4f$–$5d$ absorption band with circularly polarized light of a Kr^+ laser, the population transfer being monitored through the circularly polarized fraction of the fluorescence light. Sudden signal changes due to cross relaxation processes between ground state and/or excited state levels occur at certain values of an additional static magnetic field. The cross-

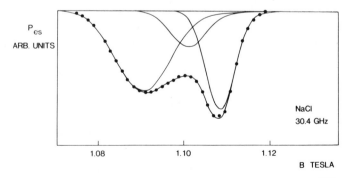

Figure 21. ESR signal for ground and relaxed excited states in KCl.[109]

relaxation signal increases when rf transitions are induced in the system of surrounding ligand nuclear spins. The signal analysis disclosed details of the spin dynamics, and provided insight on the time scale of the optical pumping of the Tm^{2+} ions and the relaxation of particular configurations of neighboring nuclear spins. An interesting feature of this method lies in the fact that cross relaxation between ground and excited states can be used to monitor rf transitions within the ligand hyperfine structure of optically excited states not accessible to conventional ENDOR.

In a second experiment,[112] ENDOR measurements were performed in the optically populated excited $^2F_{5/2}$, $E_{5/2}$ state of Tm^{2+} in CaF_2, using the same apparatus. The ENDOR transitions were monitored via the circular polarization of the fluorescence. The authors obtained the ligand hyperfine structure constants $A_s = \pm 4.83\,(3)$ MHz and $A_p = \pm 3.59\,(3)$ MHz of the first shell of fluorine neighbors, thus providing the first ENDOR results of an optically excited state of an impurity center.

Erickson illuminated a Pr^{3+} (0.01 at. %) doped LaF_3 crystal at 4.6 K with a linearly polarized cw single-mode dye laser beam parallel to the crystal c axis, in order to induce $^3H_4\,(0\,cm^{-1})\text{-}^1D_2\,(16\,874\,cm^{-1})$ σ transitions of trivalent praseodymium.[113] An ion returning from its excited state to a substate different from the initial quadrupole hyperfine level of the ground electronic state no longer absorbs the laser light. Thus, the laser burns a long-lived hole in the inhomogeneous spectral line profile, and thereby produces a significant site selective polarization of the nuclear spins. A radiofrequency magnetic field, overlapping in the crystal with the laser beam, excited transitions between adjacent Pr^{3+} ground state hyperfine terms. The rf resonance was monitored through an increase of the fluorescence from the $^1D_2\,(16\,872\,cm^{-1})\text{-}^3H_4\,(197\,cm^{-1})$ π-transition. In zero magnetic field the two observed resonance curves at 8.47 and 16.70 MHz had widths of about 200 and 180 kHz. A signal-to-noise ratio of 100 was measured for 1.5×10^{10} interacting Pr^{3+} ions. With the described method similar measurements on Pr^{3+} in $YAlO_3$ were performed.[114]

Sharma and Erickson applied the laser–rf technique developed in Ref. 113 to an excited state of Pr^{3+} embedded in a $LiYF_4$ crystal.[115] The authors excited the $^3H_4\Gamma_2\,(0\,cm^{-1})\text{-}^1D_2\Gamma_2\,(16\,740\,cm^{-1})$ magnetic dipole (zero-phonon) transition of Pr^{3+} by σ-polarized single-mode laser light at a crystal temperature of 4.5 K. Radiofrequency transitions from the laser-populated hyperfine sublevel to neighboring excited substates blocked the decay back to the initial substate, from which the laser started, thus reducing laser light absorption and also the monitored fluorescence. The resulting transition frequencies are 5123.5 (0.9) and 10 244.6 (0.8) kHz, the former being obtained from a signal curve with about 140 kHz FWHM. The authors extracted a hyperfine constant.

Shelby *et al.* extended the optical detection scheme for solid-state NMR, used by Erickson,[113] to the investigation of rf coherent transients in nuclear hyperfine substates of dilute Pr^{3+} in LaF_3 at 2 K.[116] They observed rf nutation, free-induction decay, and echos on the 16.70-MHz hyperfine transition of the Pr^{3+} ground state, and found the homogeneous linewidth to be 19 kHz. The same authors used a laser-rf method to measure spin-lattice relaxation rates of Pr^{3+} (0.01 to 2.0 at. %) doped into LaF_3.[117] Rate variations of about two orders of magnitude between 2 and 4.5 K were obtained. Various interaction processes are discussed to explain the data, like resonant two-phonon relaxation and praseodymium–lanthanum cross relaxation.

In contrast to the previous experiments with dilute rare-earth ions, quadrupole splittings in the ground ($^7F_0$) and excited ($^5D_0$) states of Eu^{3+} in the stoichiometric rare-earth compound EuP_5O_{14} were also determined by laser hole burning and optically detected NMR.[118]

Combining a laser switching technique demonstrated earlier,[119] with the irradiation of an appropriate rf field, Rand *et al.* observed spin decoupling and line narrowing in the optical $^3H_4$-$^1D_2$ transition of Pr^{3+} in LaF_3 at 2 K.[120] They were able to quench the fluctuating ^{19}F-^{19}F dipolar interaction, and thus reduce the optical linewidth from ~10 to ~2 kHz, as predicted by a spin diffusion theory.

Optical hole burning of Pr^{3+} in CaF_2 was studied with a cw dye laser of ~1 MHz bandwidth.[121] A hole was prepared with the lowest $^3H_4$-$^1D_2$ transition of $\lambda = 594.1$ nm via irradiation for ~1 sec with 20 mW intensity. Scanning the laser after this preparation with a 10^3 times weaker intensity through the transition, and monitoring the fluorescence revealed antiholes, i.e., profiles of increased absorption only about ±10 MHz outside the hole center. The authors proposed that the hole-burning mechanism involved optically induced spin-flips of neighboring ^{18}F nuclei rather than the Pr nuclei, since the Pr^{3+} hyperfine splitting frequencies are 2.7 GHz in the ground state, and only a few megahertz in the excited state, and thus cannot account for the phenomenon. This explanation was confirmed by application of an rf field which refilled the holes. A Zeeman splitting pattern corresponding to the ^{19}F nucleus appeared in the optically detected NMR spectral lines. A detailed description of these investigations is given in Ref. 122.

The same authors measured the nuclear magnetic dipole moment of ^{141}Pr in an excited electronic state with rf–laser optical double resonance in $CaF_2 : Pr^{3+}$.[123] The precision was increased by two orders of magnitude in comparison to the best previous (atomic beam) measurement. The improvement was based on an appropriate choice of electronic state, crystal site symmetry, and magnetic field direction which minimized the electronic contribution to the nuclear moment to 3×10^{-4}. One of the six resolved hyperfine substates of the excited $^1D_2$ state was selectively populated by a single-frequency cw dye laser at $\lambda = 594.097$ nm. Hole burning caused by optical pumping occurred in the ground state substructure, and was influenced by additional rf transitions between adjacent Zeeman components of the laser-excited hyperfine level. Since the rf was amplitude modulated, the resonance signal could be measured via phase-sensitive detection of the fluorescence. The corresponding energy-level diagram with the laser–rf optical pumping cycle and an optically detected nuclear resonance signal are displayed in Figures 22a and 22b, respectively.

Shelby *et al.* measured large pseudoquadrupole splittings of 33.4 and 41.6 MHz of the Pr^{3+} ground state in YAG, which were obtained by optical pumping and fluorescence detected nuclear quadrupole resonance.[124]

Laser–rf spectroscopy was also performed in molecular crystals. Dinse and collaborators pumped nuclear spins of ^{14}N in phenazine by exciting a spin-forbidden phononless singlet–triplet transition with a single-mode cw dye laser.[125] The crystal, which was immersed in liquid helium, could be irradiated with an rf field in the range of 1–40 MHz and exposed to a static magnetic field of up to 60 G. A static field of ~50 G was applied in order

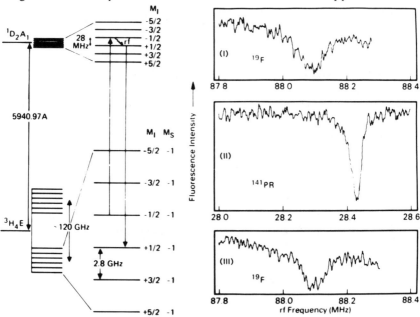

Figure 22. (Left) Energy level diagram for the 594.097-nm transition in $CaF_2 : Pr^{3+}$. The rf transition in the excited state measures the nuclear moment of ^{141}Pr and is detected by its contribution to optical hole burning since it results in population transfer in the ground state. (Right) Optically detected nuclear resonance signals. (I) and (III), ^{19}F resonances for calibration of the external magnetic field before and after measurement of the ^{141}Pr resonances. (II) ^{141}Pr resonance.[123]

to distinguish ground state nuclear quadrupole resonance transitions from those occurring in the triplet state. All three possible zero field NQR transitions were measured and exhibited a linewidth as small as 140 (20) Hz.

Similar measurements were carried out in ^{17}O-benzophenone.[126,127] The authors mention that in general only a few percent of the total intensity is emitted through the vibrationless 0, 0-transition because of coupling of intramolecular vibrational modes to the electronic transition. Therefore, the nuclear spin pumping rate is very small.

Hartmann and co-workers extended the spin echo microwave nuclear double resonance technique[128] to photon echos by replacing the microwave pulses by light pulses, and first applied the scheme to ruby.[129-131] Three resonant optical laser pulses, passing a sample, create a stimulated photon echo. If a radiofrequency field, applied to the sample between the second and third pulses, induces a nuclear spin-flip in the initial or final level of the echo transition, the echo will decrease. This occurs since the echo size is proportional to the population of one of the rf-coupled substates. The change of the echo amplitude is measured as a function of the rf. The rf interaction time rather than the laser linewidth limits the signal width. Gating the beam of an Ar^+ laser pumped ruby laser with Pockels-cell light modulators in order to produce the desired pulse sequence, the authors measured Cr–Al hyperfine and electric-quadrupole interaction parameters associated with Cr^{3+} in the ^2E $(\bar{E})$ optically excited state in ruby. They also observed resonances in the $^4A_2$ ground state.

Chiang *et al.* measured the nuclear hyperfine splittings in the excited $^1D_2$ state of $LaF_3:Pr^{3+}$ with the above-mentioned photon echo technique, utilizing two independently triggered nitrogen-laser-pumped dye lasers.[132] They found agreement with the results of Erickson based on laser saturation measurements.[133]

In the following we summarize several publications where heterodyne techniques were combined with various photon echo schemes in one or the other way, and were utilized for high-resolution spectroscopy in crystals. These subjects could have also been inserted in Section 7.

DeVoe and colleagues measured optical free-induction dephasing times[119] as long as 16 μsec in $LaF_3:Pr^{3+}$. This corresponds to an optical homogeneous linewidth of 10 kHz for the $^3H_4$-$^1D_2$ transition of Pr^{3+} ions in LaF_3 at 2 K. The cw laser beam, before irradiating the crystal, passed an acousto-optic modulator driven continuously at 110 MHz. Free induction decay occurred after a sudden rf frequency shift with 100 nsec rise time and 40 μsec duration from 110 to 105 MHz. A heterodyne beat frequency of 5 MHz was observed. A 2.0-kHz half-width was measured for the $\lambda = 610.5$ nm $^1D_2$-$^3H_4$ transition of Pr^{3+} ions in $YAlO_3$ in an external magnetic field, corresponding to an optical resolving power of 1.2×10^{11}.[134] By amplitude gating the cw dye laser beam with an acousto-optic modulator, the sample was coherently excited by a $\pi/2$-π pulse sequence. Driving the acousto-optic modulator with 80 MHz shifted the optical frequency of the $\pi/2$ and π pulses by this amount with respect to the laser frequency. Also the echo frequency contained a shift of 80 MHz. Applying to the modulator a 76-MHz rf pulse immediately before the echo appeared, gave rise to a 4-MHz heterodyne beat echo signal.

The hyperfine splittings of the lowest excited $^1D_2$ level of YAlO$_3$:Pr^{3+} were studied[135] by measuring two-pulse photon echoes with the scheme described in Ref. 134 but, in addition, simultaneously irradiating rf. The effect is to modulate the echo decay at the Rabi frequency. The separation between the $\pi/2$ and π pulses, produced with an amplitude-gated single-mode cw dye laser, was 20 μsec and an rf pulse of 50 μsec duration was applied, starting shortly before the $\pi/2$ pulse. The splittings obtained are 0.923 and 1.565 MHz. Delayed heterodyne photon echo measurements, as described in Refs. 134 and 135 were performed[136] for crystals of YAlO$_3$ doped with Pr^{3+} or Eu^{3+}. They concerned the ^{27}Al nuclear quadrupole resonance frequency shifts due to weak perturbations of the YAlO$_3$ crystal lattice introduced by substituting a Pr^{3+} or Eu^{3+} ion for an yttrium ion. MacFarlane and Shelby[137] also reported hole-burning and coherent transient effects in the $^3H_4$-$^1D_2$ absorption line of CaF$_2$:Pr^{3+} using photon echo and optical free induction decay techniques, as applied in Refs. 119 and 138. They performed time-dependent hole-width measurements which showed evidence for spectral diffusion.

4. Measurements Based on Double Resonance

The term "double resonance" as used in this section covers experiments where rf or microwave transitions within laser-optically excited short-lived (in contrast to metastable) states are detected via changes of the intensity, polarization, or wavelength of the reemitted fluorescence rather than via their influence on the population of the level to which the excited species decays. The short lifetime of the excited state of typically $\tau = 10^{-8}$ sec excludes experimental schemes in which a certain excited state population has to survive for a longer period of time; for example, probing the signal with another laser beam in a second interaction region is impossible as well as nonoptical registration methods.

Part of the literature reviewed is closely related to that on optical pumping as contained in Section 3.

4.1. Low-Lying (Short-Lived) Excited States

Double resonance (DR) spectroscopy in the sense defined above was introduced by Brossel and Bitter.[6] The work on atoms carried out before the advent or use of lasers was reviewed, e.g., by Series[9] and zu Putlitz.[12] Double resonance measurements with lasers in low-lying excited states concentrated on molecules. As already mentioned, the reason for this lies in the much simpler spectra of atoms, which had permitted such experiments already with classical light sources. In contrast, lasers with their narrow spectral bandwidth and high intensity are particularly useful for the excitation of well-defined molecular levels or sublevels in complicated spectra.

For representative molecular work with classical light sources, see, e.g., Refs. 139–141. Lasers were brought into practice for DR experiments on molecules by Field and co-workers[142] in an investigation on the excited $A\,^1\Sigma$ state of BaO. The 496.5-nm line of an Ar^+ laser induced the transition from the $X\,^1\Sigma(v'' = 0, J'' = 1)$ ground state to the $A\,^1\Sigma(v' = 7, J' = 2)$ excited state (see Figure 8). Fluorescence in the A–$X(7, 1)$ band was monitored as a function of $J' = 2 \rightarrow 3$ and $J' = 2 \rightarrow 1$ microwave transitions of the $A\,^1\Sigma(v' = 7)$ state (see Figure 23). In a further, more detailed investigation of BaO, Field et al.[143] for the first time used a dye laser in a DR experiment. Rotational constants could be calculated with the help of the measured transition frequencies.

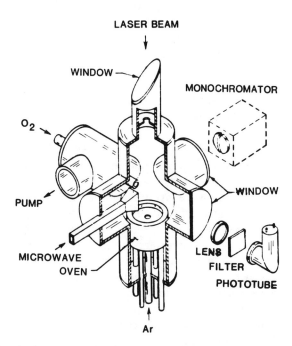

Figure 23. Photoluminescence region for the study of BaO in a microwave–optical double resonance experiment. BaO is produced when barium vapor entrained in argon reacts with oxygen. The microwave radiation is introduced with a horn radiator. Two photomultipliers view the photoluminescence region through a combination of filters and through a double monochromator. After Ref. 143.

Solarz and Levy[144] irradiated an NO_2 sample, contained in an X band waveguide, simultaneously with 600 MW of 488-nm argon ion laser light, and with microwave power from a 6-W travelling wave tube. They reported the first observation of a DR spectrum from an excited electronic state of NO_2. More detailed DR measurements of the $^2B_2$ electronic excited state of NO_2 with a tunable dye laser were performed by Tanaka and co-workers,[145] who determined spin and hyperfine structure splittings in this state. A slotted section of $K(18-26.5\ GHz)$, $R(26.5-40\ GHz)$, or $V(50-75\ GHz)$ band waveguides was used as double resonance cell. The observed microwave transitions were in these frequency regions.

The following subdivision refers to a series of more recent papers on laser–radiofrequency spectroscopy on NO_2. Weber and collaborators[146] measured the $^2B_2$ state g values, observing the change of fluorescence intensity, as the rf field was brought into resonance with the separation of magnetic sublevels. Subsequent work,[147-149] following the first low-precision measurements, yielded a deviation of the g factor from theoretically expected values (Hund's case b-coupling of the molecule) with a characteristic dependence on the rotational quantum numbers of the excited state. More recently, a correlation of this deviation with the lifetime of the excited state was reported.[150] In addition to the normal double resonance spectrum, the experiments on NO_2 also revealed a very broad resonance structure.[151] Detailed investigations[147,152] supported the conclusion, first reported in Ref. 153, that this structure is a direct consequence of a "light-induced stabilization effect."[154] This novel effect, also causing an inversion of the double resonance signal with increasing intensity of the exciting laser light, is explained by a model according to which the molecule evolves in an intramolecular process from the state excited initially to the state which decays radiatively. However, the process is impeded if the laser light creates a strong correlation between the ground state and the initially excited state of the molecule.

Extensive laser–microwave investigations have also been performed in the spectrum of NH_2. Hills and Curl, Jr.[155] observed strong electric dipole microwave transitions between a previously unobserved rovibronic level and the $J = 1/2$ and $J = 3/2$ spin-rotational levels of l_{10}, $\tilde{A}\,^2A_1\,\pi(0, 10, 0)$, respectively. The NH_2 molecules passed through a resonant half-wave microwave cavity with a 60-mW single-mode cw dye laser beam along the axis. The microwave field was amplitude modulated, and the laser-excited fluorescence signal was detected with a phase sensitive amplifier. Microwave transitions in that part of the NH_2 spectrum were used by Hills[156] to assign numerous optical transitions.

Further work on NH_2, concerning ground state microwave transitions[157-160] and microwave modulated saturation spectroscopy[161,162] can only be quoted by reference here.

Knöckel and Tiemann[163] performed laser–microwave DR measurements in the $v' = 2$ vibrational state of the excited electronic state $B\,^3\Pi(0^+)$ of the interhalide ICl. The gas, passing a waveguide section, which was used as vacuum chamber, was excited by cw dye laser light. The frequency of the amplitude modulated microwave field was swept, and DR signals were obtained by phase sensitive detection via the change of the spectral fluorescence distribution. The microwave transitions were induced in the lowest possible rotational quantum number, in order to obtain a well-resolved hyperfine structure splitting and concerned $\Delta F_1 = 0$ transitions of iodine. The observed linewidths were about 1.5 MHz, mainly due to unresolved chlorine hyperfine splittings. The hyperfine constants were computed by fitting the observed values to theoretical ones, based on a Hamiltonian that included perturbations to second order. A significant difference between the iodine coupling constants of the two isotopic species $I^{35}Cl$ and $I^{37}Cl$ was found.

Further work[164] on the rotational structure of the three lowest vibrational states $v' = 0, 1, 2$ and the hyperfine structure of $v' = 1, 2$ in the $B\,^3\Pi(0^+)$ state of ICl provided rotational and hyperfine parameters. They exhibit a large vibrational dependence. Hansen and Howard[165] observed several microwave transitions between hyperfine levels in $I^{35}Cl$ $B\,^3\Pi(0^+)$, $v = 0$, using DR spectroscopy of a supersonic beam. A focused, single-mode cw dye laser beam crossed the gas jet 1–3 cm downstream from the nozzle, while microwave radiation left an open-ended waveguide 5 cm away from the nozzle and counterpropagated the jet. Values were determined for the vibrational constant and the quadrupole coupling constant. A strong variation in the electric field gradient with the vibrational level was confirmed, as predicted in Ref. 163. Measurements with high resolution in the $A\,^3\Pi(1)$ state have been reported by Hansen et al.[166]

Remarkably, laser–microwave DR spectroscopy was also applied to large biomolecules. The investigations concentrated on the electronic structure and excited state dynamics of fluorescent organic systems via optical detection of magnetic resonance in the triplet state. Plenty of work was also performed utilizing classical light sources (see, e.g., Ref. 167).

Clarke and Hofeldt[168,169] determined the depopulation rates for the individual triplet state spin sublevels of chlorophyll a and chlorophyll b by microwave-modulated fluorescence intensity measurements. The species was dissolved in n-octane at a temperature of 2 K. The solvent n-octane is a low-temperature host matrix which allows high-resolution spectroscopy in the chlorophyll triplet state. Triplet absorption detection of magnetic resonance as well as fluorescence-microwave double resonance techniques were applied. The experimental arrangement was described in Ref. 167. In the case of fluorescence detection, chlorophyll b was irradiated with the 457.9-nm single-mode line of an Ar^+ laser. Microwave transitions were observed at 840, 1000, and 1080 MHz in zero magnetic field, and provided sharp lines of 5–20 MHz width. The middle spin sublevel was found to be the most active in triplet state intersystem crossing.

Laser–microwave spectroscopy was also reported of the triplet state of photosynthetic bacteria,[170,171] namely, of *Rhodospirillum rubrum, Rhodopseudomonas spheroides,* and *Chromaticum vinosum,* in chemically reduced cellular preparations at 2 K. The authors found similarities of the triplet state frequencies, spectral features, and intersystem crossing rates that suggest that the reaction centers in photosynthetic bacteria possess a common structure.

Clarke and co-workers extended the triplet state double resonance spectroscopy also to chlorophyll aggregate systems *in vitro,*[172] using the same spectroscopic techniques as in the references cited. The triplet-state properties found in the aggregates are interpreted within the framework of a triplet exciton model. In addition, first results of laser–microwave investigations of *in vivo* chlorophyll were reported.[173]

4.2. Rydberg States

4.2.1. Detection via Fluorescence

Within the last ten years the spectroscopic features of highly excited atoms, called Rydberg atoms, increasingly attracted the interest of many scientists. The general physics of Rydberg atoms which exhibits numerous peculiarities has been reviewed in several articles (see, e.g., Ref. 174–178), and in a recent book,[179] and therefore will not be discussed here in detail. Also, Chapters 11 and 16 of *Progress in Atomic Spectroscopy*, Parts A and B, deal with Rydberg atoms.

Laser–microwave spectroscopy in Rydberg states grew out of studies in lower-lying levels. Svanberg and collaborators[180] excited Rb and Cs atoms in two steps in order to reach S, D, and F levels. The first step to the lowest excited P states was realized by resonance absorption of light from a strong rf spectral lamp, the second step with a tunable cw dye laser. They performed optical double resonance as well as level-crossing measurements. The results were magnetic and electric hyperfine structure constants. Svanberg and Belin used the same experimental arrangement to measure more ^2D states in Rb and Cs,[181] and high-lying $^2P_{3/2}$ levels in these atoms.[182] Hyperfine structure splittings of $^2S_{3/2}$ states in K and Cs were obtained by Tsekeris and co-workers[183] with a similar experimental setup but a different excitation and detection geometry. Tsekeris and Gupta[184] extended these investigations to additional ^2S states in Rb and Cs. Further work concerned potassium,[185] rubidium,[186-189] cesium,[186,189] and sodium.[190,191]

The work of Lundberg and co-workers[191] on sodium can be regarded as representative for the numerous papers mentioned before, and will be described in more detail. The experimental setup is shown in Figure 24. An Ar^+ laser pumped cw dye laser excited Na atoms, enclosed in a quartz resonance cell with the yellow D line from the ground state to the $3\,^2P_{3/2}$ state as an intermediate level. A second dye laser, pumped by another Ar^+ laser, operating on uv lines or on the 488-nm line, respectively, further excited the atoms from the $3\,^2P_{3/2}$ to the 6, 7, or $8\,^2S_{1/2}$ state. The dye laser light beams passed the Na cell parallel to an applied static magnetic sweep field. The two light beams had circular polarization of opposite helicity and by two-step excitation to the $^2S_{1/2}$ state under investigation generated an orientation in this level. Atoms decaying to the $4\,^2P$ states transferred part of this orientation. Thus, the polarization of the $4\,^2P$–$3\,^2S$ transition lines at $\lambda = 330$ nm was influenced. The 330-nm fluorescence was detected with a lock-in amplifier after passing a $\lambda/4$ plate and a linear polarizer rotating with the lock-in reference frequency, thereby monitoring the intensity difference of σ^+ and σ^- light. Transitions between ^2S magnetic substates induced by an rf field perpendicular to the magnetic field were measured via a decrease of the 330-nm light polarization. Resonances were observed at low and high magnetic fields. The magnetic field was swept while keeping the rf frequency fixed. The signals were accumulated in a multichannel analyzer. Such a signal curve together with part of the energy level diagram

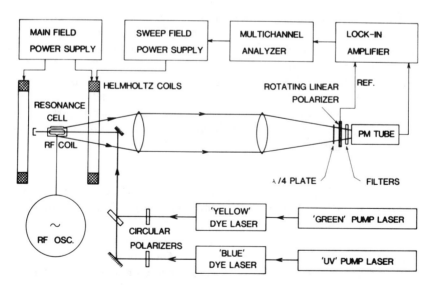

Figure 24. Schematic diagram of the experimental arrangement used in laser–rf spectroscopy of excited Na atoms; after Ref. 191.

is given in Figure 25. The measurements provided magnetic dipole interaction constants of the $^2S_{1/2}$ hyperfine structure, which are also discussed in the light of theoretical predictions.

Another group of microwave measurements, based on fluorescence light detection, concerned energy level splittings in alkali atoms between Rydberg levels with the same principal quantum number n but different orbital angular momentum quantum numbers l. These Δl splittings, called quantum defects, arise from interactions of the valence electron with the ionic core. Gallagher and co-workers[192] populated d levels with $n = 11$–17 in sodium, exciting the atoms in a cell from the $3s$ to $3p$ and then from $3p$ to nd by two synchronized pulsed dye lasers, both pumped with a N_2 laser. Microwave irradiation induced d to f electric dipole transitions at frequencies between 17.6 and 64.4 GHz. The microwave signals were detected via the nf-$3d$ and $3d$-$3p$ fluorescence in the far red beyond 800 nm. Besides quantum defect data, the authors obtained the f fine structure splittings which exhibit the normal level sequence in contrast to the inverted d doublets.

The measurements were extended to two-photon and three-photon absorption related to d-g and d-h transitions, respectively, in the $n = 13$–17 states of sodium.[193] In the case of the three-photon transitions, a microwave frequency was swept while an rf frequency was fixed, and two rf photons and one microwave photon were absorbed. The detection scheme was the same as in Ref. 192. The measurements provided accurate quantum defect values and revealed hydrogenic fine structures for $l < 2$.

Similar measurements of d-f and d-g intervals in lithium Rydberg states by Cooke *et al.*[194] yielded fine structure intervals in accordance with the hydrogenic theory. Ruff and co-workers[195] measured $n = 15$–17 f-g intervals in Cs and determined values for the effective Cs^+ dipole and quadrupole polarizabilities. To reach the Rydberg levels the first laser excited atoms from the $6s_{1/2}$ ground state to the $7p_{3/2}$ state, from which they cascaded to the $5d_{5/2}$ level. The second laser then induced the $5d_{5/2}$-$16f_{7/2}$ transition. This excitation technique was first used by Lundberg and Svanberg for lifetime measurements.[196]

4.2.2. Detection via Ionization

Detection of microwave transitions via fluorescence becomes more and more difficult with increasing principal quantum number n of the level under study, since the oscillator strength of an optical transition from a level with n to a lower level decreases as n^{-3}. The field ionization technique therefore favorably replaces fluorescence detection for large principal quantum numbers. Laser–microwave spectroscopy combined with field ionization was first realized by Gallagher *et al.*[197] and by Fabre *et al.*[198] Two-step photoexcitation and selective field ionization of $|m_J|$ fine structure (fs) levels of high-lying p (n = 16-19) and d (n = 15-17) states of sodium were used[197] to measure fine structure splittings and tensor polarizabilities by rf spectroscopy. The rf field was applied in the time interval between the laser pulses and the pulse of the ionizing electric field. Sweeping the rf frequency through the fine structure resonance increased the ionization current, since the ionizing field was set to ionize the final fs state but not the one pumped by the laser. Safinya et al.[199] measured n = 16-18, nf-nh and n = 17-18, nf-ni intervals in cesium via two-photon and three-photon microwave transitions, respectively. Delaying the field ionization pulse made it possible to discriminate between the f states and the longer-lived higher l states. Effective dipole and quadrupole polarizabilities of the Cs^+ core were extracted from the microwave data. Investigating core polarization in Ba, Gallagher and co-workers[200] used the delayed field ionization approach for radiofrequency resonance measurements of the Ba $6sng$-$6snh$-$6sni$-$6snk$ intervals for $n \simeq 20$. Two microwave photons and one rf photon were involved in the three-photon g-k transitions. The above-mentioned[188] preliminary results were followed by a detailed report[201] on millimeter spectroscopy in sodium Rydberg states. It concerns single-photon and two-photon microwave transitions with Δn = 0, 1, 2, Δl = 0, ±1 between levels with n = 23-41 in the 300-GHz range. The investigations yielded quantum-defect, fine structure, and polarizability data.

These measurements on sodium Rydberg states were continued and improved considerably by introducing a semiconfocal millimeter microwave cavity (see Figure 26) and stabilizing the carcinotron frequency via standard phase-lock techniques.[202,203] The standing wave in the cavity made it

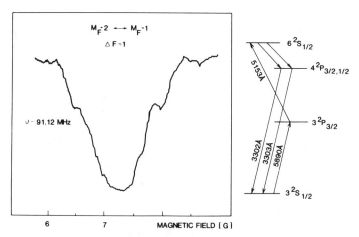

Figure 25. Sodium energy level scheme with transitions used for excitation and detection, and a resonance curve for the $6^2S_{1/2}$ state; after Ref. 191.

possible to discriminate the first-order Doppler shift of two-photon microwave resonances. Transit-time limited linewidths in the 10-kHz range were
observed for frequencies around 100 GHz. A two-photon microwave signal
is shown in Figure 27. The transition frequencies obtained are quoted with
an error of 4×10^{-8}. The authors pointed out that alkali Rydberg atoms
could provide a secondary standard for frequency calibration from the
microwave to the visible region, as long as the precision required does not
exceed 10^9–10^{10}. They also mentioned the possible extension of the experiments described to hydrogen or deuterium Rydberg states, envisaging a
precise value of the Rydberg constant, directly determined in frequency
units. Studies on maser oscillation and microwave superradiance of Rydberg
atoms with the experimental configuration and detection scheme displayed
in Refs. 202 and 203 will be described in the next section.

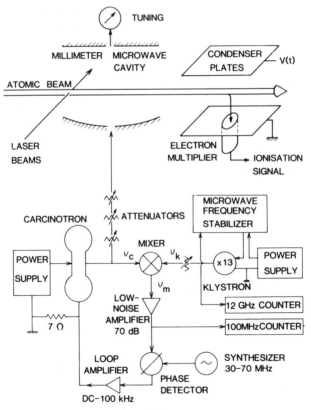

Figure 26. Schematic diagram of an experimental setup for the study of Rydberg states
displaying the atomic beam apparatus and the carcinotron phase-locking device;
after Ref. 202.

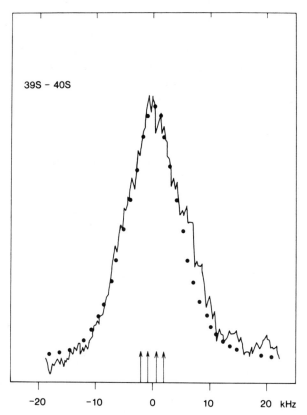

Figure 27. Two-photon microwave resonance 39S–40S with the line center frequency at 59 260 221 kHz. Circles represent a theoretical line shape computed without adjustable parameters except height. The four arrows indicate the unresolved hyperfine components. After Ref. 202.

Using the same type of setup and spectroscopic techniques Goy *et al.*[204] measured quantum defects, fine and hyperfine structure splittings in the S, P, D, and F Rydberg levels of cesium with $n = 23$–45, providing a precise mapping of the cesium Rydberg levels. This allows the prediction of any splitting frequency for $n \geq 20$ and $l \leq 3$, with an accuracy of about one megahertz.

4.2.3. Rydberg Atoms as Masers and Radiation Detectors

Rydberg atoms interact very strongly with microwave radiation since they possess particularly large electric dipole matrix elements. They easily undergo transitions between neighboring levels. This feature opens up the possibility of building microwave sources with stimulated coherent emission which work with a number of atoms, that is orders of magnitude lower than in the case of conventional masers. The idea was first expressed and realized by Haroche and co-workers.[205,206] The collinear beams of two nitrogen laser pumped dye lasers crossed a beam of alkali atoms (Cs and Na) at right angles, and excited atoms to a Rydberg S state ($n \simeq 25$). The intersection region of the atomic beam and the laser beams was located in a semiconfocal millimeter wave Fabry-Perot resonator (finesse 200), thus providing a maser configuration. Maser activity occurred when the cavity was tuned exactly on, e.g., the $27S-26P_{1/2}$ transition of Na. The 1-μsec-long microwave radiation pulses carried very small energy amounts of $1-10^2$ eV, equivalent to peak powers of $10^{-13}-10^{-11}$ W. The microwave emission was detected indirectly via field ionization monitoring the evolution of the Rydberg state population, after the atoms left the cavity.

Further experimental studies[206] concern microwave superradiance. It is interesting to mention that blackbody radiation rather than spontaneous emission initiates superradiance and maser activity in the systems described above. Using a beam of Rydberg atoms Ducas *et al.*[207] detected far-infrared radiation from a laser source at 496 and 118 μm. A dc electric field was applied to tune the splitting between two neighboring Rydberg states to the photon energy. Absorption was monitored by ionizing selectively the higher level with an electric field applied periodically.

Interaction of blackbody radiation with Rydberg atoms was first studied by Gallagher and Cooke,[208] and by Beiting *et al.*[209] The dependence of this interaction on radiation exposure time and blackbody temperature was investigated by Koch and co-workers.[210]

To increase further the sensitivity of microwave or far-infrared radiation detectors based on Rydberg atoms, two improvements were introduced by Figger *et al.*[211] They excited Na atoms inside a box to the $22\,^2D$ Rydberg state with continuous wave radiation of two dye lasers, which provides a more favorable duty cycle as compared to pulsed excitation, and cooled the chamber walls down to 14 K to minimize thermal background radiation. After leaving the box the atoms were field ionized and the ions accelerated and monitored with a channeltron. The electric field strength was adjusted not to ionize the $22\,^2D$ state, but any higher-lying level. Exposure of the atoms through a window in the box to microwave or far-infrared radiation from a heated wire caused transitions to higher Rydberg states and thus an increase of the channeltron signal. With this 14 K Rydberg detector a noise equivalent power of 10^{-17} W Hz$^{1/2}$ was obtained. This compares favorably with other types of detectors in the considered wavelength region.

Moi *et al.*[212] reported the first direct detection of Rydberg maser pulses. Na atoms were excited to the $33S$ level by two synchronous dye laser pulses inside a high-quality semiconfocal Fabry-Perot microwave cavity. Maser transitions $33S$–$32P$ occurred when the cavity resonance frequency coincided with the transition frequency. The emitted microwave radiation left the cavity through a hole in the spherical mirror and was down-converted from roughly 108 GHz to 130 MHz, using a very sensitive radioastronomy heterodyne receiver. Signals with a power of about 10^{-12} W and a duration of 0.3 μsec were detected. The number of radiating Rydberg atoms was estimated to be 10^5.

5. Measurements Based on Nonlinear Phenomena

5.1. General Considerations

Laser–microwave spectroscopy based on nonlinear phenomena developed from the type of experiments on molecules already discussed in Section 3.2 which make use of optical pumping or double resonance. Occasionally, the laser and the rf power were high enough to create the nonlinear phenomena mentioned above, i.e., to saturate the transitions involved and/or to induce multiphoton transitions. The intermediate level in, e.g., two-photon transitions did not have to be a *real* state but could be virtual as well. Therefore, a drawback often encountered in earlier infared laser–microwave experiments could be avoided: if the laser transition frequency did not exactly coincide with the molecular absorption line the Stark or Zeeman effect had to be used for tuning.[22] This results in an undesired line splitting. With laser–microwave multiphoton processes, however, the laser can be operated at its inherent transition frequency. Exact resonance with molecular lines is then achieved by using a nonlinear effect, i.e., a radiofrequency quantum is "added" to or "subtracted" from the laser frequency (see Figure 28).

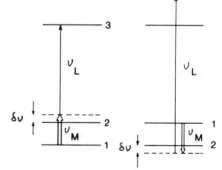

Figure 28. Energy level schemes for infrared-microwave two-photon transitions; after Ref. 213. ν_L and ν_M stand for the frequency of the laser and microwave induced transitions, respectively.

5.2. Experiments with Resonance Vessels

Oka and Shimizu first reported the observation of infrared–microwave two-photon transitions in NH_3.[213] They used an N_2O or CO_2 laser in the 10-μm region that could be tuned to a number of lines with a grating. The laser light passed through a 1-m absorption cell containing $^{15}NH_3$ at a pressure of between 10 mTorr and 1 Torr. The light intensity was detected with a Ge:Au detector. The high-power microwave radiation (20 W at 23 GHz) was generated with a floating drift tube klystron and coupled into the absorption cell. The absorption of the laser light was measured as a function of the microwave frequency giving two-photon absorption signals with large signal-to-noise ratio. With the microwave power on, about 60% of the laser light was absorbed.

Although this experiment was only a demonstration of infrared–microwave two-photon transitions because the laser frequency could not be controlled accurately enough for precision measurements, it actually constituted the first step in a new series of experiments with nonlinear phenomena.

Oka and Shimizu already pointed out that Lamb dip techniques[214] should be very useful for gaining a higher precision. This was accomplished by Freund and Oka in 1972.[215] The electric fields were strong enough to saturate two-photon transitions in ammonia isotopes with a difference frequency between the real and the intermediate level of $\Delta \nu = 500$ MHz. Similar to the experiment described above, the measurements were conducted with an N_2O or a CO_2 laser both operating in the 10-μm region. There are near-coincidences between the $P(15)$ line of the N_2O laser at 926.0 cm^{-1} and the $\nu_2[^qQ^-(4,4)]$ transition of $^{15}NH_3$, and between the $R(6)$ line of the CO_2 laser at 966.3 cm^{-1} and the $\nu_2[^qQ_+(5,4)]$ transition of $^{14}NH_3$. The absorption cell containing the ammonia was 30 cm long, and was constructed with a magic T to introduce the microwave power and with two waveguide extensions sealed with sodium chloride windows at the Brewster angle (see Figure 29). The cell was placed in the laser cavity. Two-photon Lamb dips

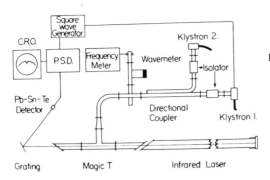

Figure 29. Setup for the observation of Lamb dips in infrared–microwave two-photon transitions. The cell containing the ammonia molecules under investigation is constructed of a magic T and placed inside the cavity of a CO_2 or N_2O laser. After Ref. 216.

were detected by monitoring the output power of the laser with a Pb–Sn–Te detector while tuning the frequency of the microwave field. Since the sum $\nu_l + \nu_m$ of the laser and rf frequency is well defined by the energies of the lower and upper (real) levels, the tuning of the laser frequency appears as a shift of the microwave frequency for the Lamb dip. Since the laser frequencies are known, the frequencies of the lower state splitting can be readily calculated after a measurement of the position of the dip. By applying a second microwave field with frequency ν'_m in addition to the first one, a coherency splitting of the two-photon dip was also observed. The same system was also described in a full-length paper.[216]

In 1972, Shimizu reported a radiofrequency infrared double resonance study between real levels in the saturated regime.[217] He induced rf transitions between Stark levels in the $P(3, 2)$ line of the ν_2 band in PH_3. Shimizu detected the absorption change of the laser light transmitted through a Stark cell containing PH_3 at a pressure of around 10 mTorr as a function of the dc Stark field. The signal was processed with a lock-in amplifier. The method can be applied to ground and excited states as well. The principle is, that, e.g., hyperfine resonances can be observed as a steady state absorption change in the total population between the excited and the ground state, if a resonant rf field is applied. In a three-level system (see Figure 30), for example, two levels in the excited state may form the hyperfine structure. The system is excited by very strong rf and laser fields. If only the laser is on and is resonant with one line of the hyperfine doublet, only one excited level is populated equally to the ground state. Thus, half of the molecules absorb one laser photon before leaving the three-level system, e.g., by collision or by spontaneous radiative decay. If, in addition, a strong rf field resonant with the hfs levels is applied, all three levels become equally populated. Therefore, two thirds of the particles absorb laser photons before relaxing. This increases the absorption by one third, thus giving a very large signal. One can argue similarly if the splitting is in the ground state. Shimizu points out that the situation does not change even if Doppler broadening

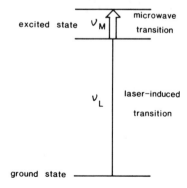

Figure 30. Microwave optical double resonance in a three-level system. Under strong saturation of both the optical and microwave fields all levels become equally populated.

is taken into account. This is in contrast to the case of linear absorption. Advantages of the technique are simplicity, excellent signal-to-noise ratio, and applicability to weak transitions. If the laser can distinguish between different hfs levels, i.e., if its spectral bandwidth is smaller than the hfs, the method can even be applied to systems in which all optical transitions have the same matrix element. This is not the case in classical optical pumping experiments even though they are rather similar to the scheme described above. Shimizu's measurements permitted the calculation of the dipole moment for the excited state which coincides within the experimental accuracy with the value for the ground state. An experiment with a similar excitation scheme was reported on CH_4 with a 3.39 μm He–Ne laser by Curl and Oka.[218] They also observed the influence of microwave transitions in an excited state on the amplitude and shape of the Lamb dip created inside the cavity of the infrared He–Ne laser.

Also in 1972, Shimoda and Takami[219] observed infrared–microwave signals between real levels of the ground state in H_2CO (see Figure 31) under strong saturation. They carried out double resonance experiments at various gas pressures in the absorption cell, and observed a change of sign of the signal as a function of the pressure. This can be explained by considering that laser radiation burns a hole in the molecular velocity distribution of the $5_{1,5}$ state around a certain value of the velocity. If the bottom of this hole is above the population of the $5_{1,4}$ level, the microwave pumping will decrease the population of the $5_{1,5}$ level, which results in a decrease of the absorption of the infrared radiation. If, however, the bottom of the hole is below the population of the $5_{1,4}$ level, the microwave radiation will increase the population of the $5_{1,5}$ level at the velocity corresponding to the laser frequency. This causes an increase of the infrared absorption. Since the depth of the hole is a monotonically decreasing function of the pressure, this leads to a sign change of the double resonance signal (see Figure 32) registered via the observation of the transmitted infrared power. Shimoda and Takami also presented a theoretical approach that explains the observed pressure dependence of the double resonance signal fairly

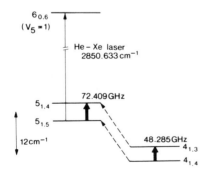

Figure 31. Energy level scheme of H_2CO; after Ref. 226.

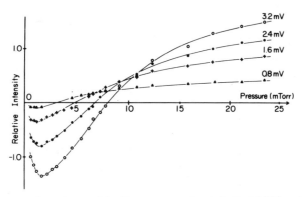

Figure 32. Pressure dependence of double resonance signals in H_2CO [226]; at low pressures a change of sign takes place. The numbers in mV give a measure of the laser power which was determined with an InSb detector.

well. In a second paper[220] published shortly afterwards they also reported the observation of Lamb dips in the same system. In addition, the method permits observation of the velocity dependence of collision-induced transitions.[221] In this experiment two microwave fields were applied.

The experimental development of infrared–microwave double resonance was paralleled by the elaboration of the theoretical framework. For example, Shimizu published a theory of two-photon Lamb dips,[222] and Takami theoretically described the optical–microwave double resonance with emphasis to fundamental aspects and the double resonance with microwave detection,[223] and with optical detection.[224] Detailed information on the theory of infrared–microwave double resonance can also be found in Refs. 216, 225–227.

Shimizu's theory served to explain characteristics of the line shape and fine structures of Lamb dips as observed by Freund et al.[228] The experiments provided the first experimental verification of "velocity-tuned three-photon processes" which are caused by combinations of ir radiation Doppler-shifted up (ν_+) and Doppler-shifted down (ν_-) with frequencies $\nu_\pm = \nu_l(1 + v/c)$. The resonance conditions with a molecular transition at ν_0 are

$$\nu_0 = \nu_l(1 \pm 3v/c)$$

If the saturation condition is met, the velocity-tuned three-photon processes burn holes in the velocity profile of molecules at one third the velocities $\pm v_0$ at which normal single-photon processes can create holes. Freund et al. studied these effects in CH_3F. The setup was similar to an earlier experiment carried out by the same group (see above, Ref. 215), and made use of a CO_2 laser with the absorption cell installed inside the cavity. Lamb dips were detected as sharp variations of the output power of the laser.

Other experiments employing laser–microwave induced nonlinear phenomena have been reported on HCCF,[229] on CH_3I,[230] where "pure" nuclear quadrupole resonances were observed, on GeH_4,[231] where in addition to the laser and microwave radiation a dc electric field was applied in order to observe the Stark pattern, and on CH_3I,[232] where a dressed-molecule formalism has been used to describe the Doppler-free resonances of the molecule "dressed" with the radiofrequency field.

Recent studies on molecules by infrared–microwave spectroscopy further involve, e.g., experiments where the species is contained in an intracavity absorption cell.[233] However, contrary to the experiments described above, the laser does not produce an anomalous population difference since it is operated off-resonance from an infrared transition. Similar to classical microwave experiments (see above), the generation of a signal rather relies on pure microwave pumping which produces a small variation of the population in certain rotational levels as compared to thermal equilibrium. This changes the bulk susceptibility of the gas through dispersion, and therefore also the effective cavity length of the laser. Consequently, the laser output power is also affected through which the signal is finally detected. Thus, microwave resonances can be observed through laser lines that are separated from molecular infrared transitions by several cm^{-1}. Arimondo and Oka point out the advantages of this method:

i. Similar to multiphoton processes (see above) one does not have to rely on rare accidental coincidences between laser lines and molecular transitions. This makes the technique more generally applicable.

ii. The accurate frequency of an infrared transition is not required. This avoids the two-dimensional search for a resonance.

iii. The theoretical analysis of the signal and its shape are relatively simple since the coupling of the infrared and microwave radiations is weak compared to a three-level system with both fields exactly in resonance.

A disadvantage of the method is that the sensitivity is much lower than in ordinary double resonance, because the laser does not produce a population anomaly. The experiment only utilizes the population difference present already for thermal equilibrium. Therefore, the technique is only applicable in the microwave and millimeter region where $h\nu/kT$ is sizable. It cannot be used effectively in connection with rf transitions.

Other experiments, carried out recently and based on nonlinear effects, involve laser–microwave double resonance measurements with intense Stark fields[234,235] on H_2CO, spectroscopy on $^{14}NH_3$ and $^{15}NH_3$ by intracavity infrared–microwave double resonance as well as by other methods,[236] work on $^{32}S^{18}O_2$ and $^{34}S^{18}O_2$,[237] investigations on $FC^{15}N$ including the observation of two-photon Lamb dips,[238] and infrared–microwave two-photon spectroscopy on $^{14}NH_3$,[239] $^{15}NH_3$,[240] and $^{12}CH_3F$ and $^{13}CH_3F$.[241]

So far, we have considered only experiments with continuous-wave lasers under steady state conditions. With time-resolved experiments, on the other hand, energy transfer rates and transition probabilities can be obtained. Such measurements were carried out by mechanically chopping the laser beam directed into an external absorption cell together with the microwave radiation.[25] Later, Levy et al. reported time-resolved infrared–microwave experiments with an N_2O laser Q-switched with a rotating mirror to produce pulses less than 1 μsec in duration.[242] They observed a transient nutation of the inversion levels of the molecule following the infrared laser pulse. Based on the Bloch equations, the observed phenomena could be explained quantitatively. From the decay envelope of the oscillations a value for the transverse relaxation time T_2 was determined. Similar effects were produced by rapidly switching a Stark field which brings the molecules into resonance with the cw microwave radiation.

Infrared–microwave transient phenomena have also been treated in Refs. 243–245.

Summaries and reviews on infrared laser microwave spectroscopy of molecules can be found in Refs. 225–227, 246–252.

While numerous experiments on molecules, performed under saturation conditions or with multiphoton transitions by simultaneously applying laser and microwave fields, can be found in the literature, only a few corresponding studies on atoms have been reported. For example, high-resolution measurements on atomic hydrogen generated in a discharge tube have been carried out by E. W. Weber and Goldsmith by "double quantum saturation spectroscopy."[253] Similar to Oka's and Shimizu's work, a laser photon and an rf photon "add up" in a two-photon process. However, the important difference here is that the frequencies are chosen such that the two fields are exactly in resonance with a real three-level system. Therefore, saturation is possible already at low light intensities and low rf power, so that two-photon saturation dips are generated. The correlated absorption of a laser and a radiofrequency photon results in negligibly small contributions to the width of the signal from the intermediate level thus providing narrower signals than ordinary single-quantum transitions in the same system. The authors point out that the technique combines advantages of saturation and two-photon spectroscopy. It was applied to the 2^2S-3^2S, 3^2D transitions in atomic hydrogen using a cw dye laser and an open, two-wire transmission line assembly for emission of the rf-field. The results of the experiment comprise a direct determination of the $3P_{3/2}$-$3D_{3/2}$ Lamb shift.

5.3. Particle-Beam Experiments

As has been pointed out already in Section 2.2, spectroscopy of atoms in the visible permits additional ways for the detection of signals as compared to experiments on molecules in the infrared. This results mainly from the fact that atoms usually emit fluorescence after excitation. Therefore, absorption measurements are not mandatory. If fluorescence quanta are detected, fewer particles are required for a sufficient signal-to-noise ratio, and atomic beam experiments with low particle densities but with all their advantages such as absence of collisional broadening become possible.

An example where nonlinear phenomena in connection with laser–rf spectroscopy have been used in atomic beams, is the recent work on calcium isotopes[254,255] carried out in our laboratory. The goal of these experiments was to determine nuclear electric quadrupole moments from precise hyperfine structure data of the atomic spectrum. This is of some importance in the case of calcium, since ^{40}Ca as well as ^{48}Ca are so-called double-magic nuclei, i.e., with closed proton and neutron shells. Radioactive ^{41}Ca ($\tau = 1.03 \times 10^5$ yr) and the stable isotope ^{43}Ca have been investigated by laser–rf spectroscopy. The measurements allow to study the influence of a single neutron and three neutrons, respectively, on the double-magic ^{40}Ca core.

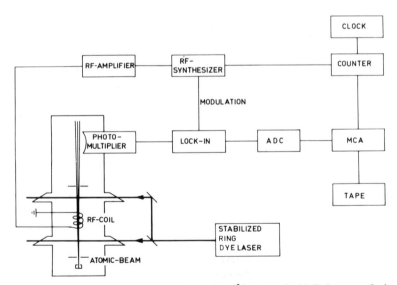

Figure 33. Setup for laser–rf spectroscopy in the $4s4p\ ^3P_1$ state of odd Ca isotopes. Owing to the long lifetime of the excited state the emitted fluorescence light can be observed in a region spatially separated from the laser interaction zone. Detection of the rf transitions is accomplished by saturating the Ca intercombination transition with $\lambda = 657.3$ nm.

^{41}Ca was produced by neutron capture from high-purity ^{40}Ca in the chemical form of carbonate. Since the cross section for this process is very small ($\sigma = 0.4$b), a ^{41}Ca concentration of only some 10^{-3} could be obtained even though the sample was irradiated with neutrons in a high flux reactor for one year. Subsequently, the carbonate was reduced with zirconium at 1400°C. The metallic calcium was collected on a tungsten disk which later on served as the source for a thermal atomic beam. Since the reduction is only feasible with small amounts of material, this beam was finally produced with no more than 10–15 μg of ^{41}Ca, and typically lasted for about 30 min. This rather limited measuring time and the low concentration of ^{41}Ca required a highly sensitive experimental technique. This also holds for the case of ^{43}Ca, which has an abundance of only 0.14% in the natural isotopic composition, which was used to generate the atomic beam. Furthermore, the hyperfine structure splitting had to be determined with high precision. This follows because rather small quadrupole moments have to be expected for isotopes in the vicinity of the double-magic nuclei, so that the quadrupole interaction constant B is of the order of no more than about 1% of the total hyperfine structure splitting. For these reasons a technique of combined laser radiofrequency saturation spectroscopy was applied. It has some similarities to experiments performed on molecules with infrared lasers,[217] but guarantees a better resolution together with the high sensitivity required here.

The experiments were carried out as follows; the thermal atomic beam, containing ^{41}Ca or ^{43}Ca is intersected by two light beams ($\lambda = 657.3$ nm) of a frequency-stabilized dye ring laser as shown schematically in Figure 33. In the first interaction region the optical transition between the $4s^2 \, ^1S_0$ ground state and one particular hfs level of the excited $4s4p \, ^3P_1$ state is saturated. Because of the $^3P_1$ radiative decay, a certain amount of resonance fluorescence light can be detected. If radiofrequency transitions are now induced between this hfs level already populated, and another hfs level unoccupied so far, additional laser light is absorbed in the second interaction region. This results in an increase of the emitted resonance fluorescence intensity of up to 25%, provided both the radiofrequency and the optical electromagnetic fields are sufficiently strong, i.e., saturate the two transitions.

In the present case the excited $4s4p \, ^3P_1$ state is very long-lived ($\tau = 0.39$ msec) so that the different optical and rf fields can interact with the atoms in different regions of space. Also, the emitted resonance radiation can be monitored readily in a region spatially separated from the excitation zone (see Figure 33). Thus, virtually all background light is suppressed.[256] It should be mentioned, however, that the method presented here can also be applied to short-lived excited states. In this case the atoms must interact simultaneously with the laser light field and the radiofrequency field.

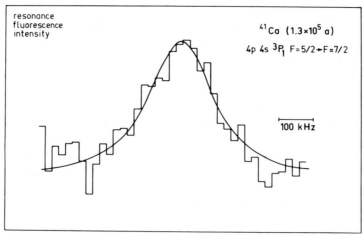

Figure 34. Rf-resonance signal of induced magnetic dipole transitions between hfs levels with
$F = 5/2$ and $F = 7/2$ of the $4s4p\,^3P_1$ state of radioactive ^{41}Ca measured with the
setup shown in Figure 33. The solid line represents the best fit to the experimental
data.[255]

The linewidth of the signals (for an example see Figure 34) is determined only by the interaction time of the atoms with the rf field, slightly broadened by the onset of saturation. Broadening of the resonances due to Zeeman splitting of the hfs levels could be avoided by a twofold Mu-metal shield. The following average center frequencies for the two hfs splittings of, e.g., ^{41}Ca have been obtained:

$$\Delta\nu(5/2 \leftrightarrow 7/2) = 847.219\,(19)\ \text{MHz}$$

$$\Delta\nu(7/2 \leftrightarrow 9/2) = 1079.519\,(4)\ \text{MHz}$$

where the error corresponds to three standard deviations. The precision is even higher for ^{43}Ca where more time for signal averaging was available due to the longer-lasting atomic beam of natural calcium.

From the experiment the nuclear electric quadrupole moments have been derived:

$$Q(^{41}\text{Ca}) = -80\,(8)\ \text{mb}$$

$$Q(^{43}\text{Ca}) = -49\,(5)\ \text{mb}$$

These values make it possible to test theoretical calculations in connection with different nuclear models.[255]

6. Other Schemes

This section describes several experiments in which the state preparation and signal detection procedure deviates from the schemes discussed in the previous sections.

M. Gustavson and co-workers[257] reported an atomic beam magnetic resonance (ABMR) measurement, where a cw tunable dye laser was used for signal detection. Barium atoms were excited to the metastable $6s5d\ ^1D$ and ^3D states in a plasma discharge and then traveled through an ABMR hexapole magnet. Only atoms which underwent a radiofrequency transition in the C field region, could pass the laser beam. They were excited to the $5d6p$ configuration with laser light of the wavelength $\lambda = 580$–610 nm and were monitored via the resonance fluorescence light. Individual states could thus be studied without interference from other states. To the authors' estimation, the method, especially advantageous for weakly populated states, seems to be superior to signal observation via surface ionization followed by mass separation.

In a second experiment,[258] using the same technique, the hyperfine splitting frequencies of the metastable $4s4p\ ^3P_2$ state of ^{43}Ca were measured. They provided the spectroscopic B factor which contains the electric quadrupole moment of the nucleus (see also Section 5.3).

With a similar experimental configuration Burghardt and collaborators[259] and Bürger et al.[260] measured hyperfine structure splittings of low-lying states in atomic ^{165}Ho, ^{185}Re, and ^{187}Re, respectively. After crucible-free evaporation from a target by electron impact, whereby the metastable states under investigation, were populated, the atoms passed flop-in type ABMR magnetic field regions. Those atoms that underwent an rf transition in the C field region, were focused selectively on the detector slit. Behind the slit, the beam of a cw single-mode dye laser excited the atoms to a short-lived state, thus allowing rf resonance detection via the reemitted fluorescence. The measurements yielded electronic g_J factors, magnetic and electric hyperfine interaction constants, and information on configuration mixing effects.

Rydberg states were also ionized by a microwave field where the ionization process itself was a subject of investigation. Koch and collaborators studied strong-field ionization of highly excited hydrogen atoms.[261] A fast beam of hydrogen atoms in the metastable $2s$ state was produced by electron transfer collisions of a proton beam passing through an H_2 gas target. An argon ion laser beam overlapped the $H(2s)$ beam collinearly and excited atoms to high-lying states with $40 < n < 55$. The atomic beam was Doppler tuned to the laser excitation of a desired value of n via acceleration voltage variation of the proton beam. The atoms were then ionized when passing a high-Q X-band microwave cavity or X-band waveguide. The resultant protons were energy analyzed by an electrostatic lens. The laser light beam was intensity modulated in order to allow for phase sensitive detection of the protons with an electron multiplier. Microwave ionization rates as a function of microwave power are presented which exhibit saturation after a steep increase over two orders of magnitude. The work was extended to experiments where the hydrogen Rydberg atoms first passed a waveguide and then a microwave cavity.[262] Microwave multiphoton transitions from the initial laser excited Rydberg state to even higher-lying bound states were induced in the waveguide. The microwave cavity field subsequently ionized these states and thereby provided the detection of the multiphoton processes. In comparison to the measurements with only one microwave field region, the cavity was voltage labeled rather than the waveguide. The authors saw evidence for an ionization mechanism involving chains of multiphoton transitions enhanced by the upper bound atomic state. Specifically, resonances of the ionization rate were observed as a function of the frequency. They were caused by strong transitions between neighboring energy states with absorption of 3–7 photons. Tunneling as the dominant intense-field ionization mechanism was ruled out.

On the basis of the same device and technique, Koch[263] precisely measured the intense-field Stark effect in highly excited states of atomic hydrogen. Atoms in the $n = 10$ state, produced by $H^+ + Xe$ electron transfer collisions, were excited to the $n = 25$ and $n = 30$ states with a CO_2 laser at $\lambda = 10\ \mu m$. It was demonstrated that the Rayleigh–Schrödinger perturbation theory does not describe the experimental results properly. After a convergent behavior up to a certain order, the theory diverges for higher orders.

The microwave ionization studies were extended to highly excited helium atoms,[264] with the electron configuration $1s(n \simeq 30)s\ ^3S_1$. It was found that the ionization rate does not increase monotonically as a function of the oscillatory electric field. The ionization curves of He exhibit one or more plateau regions and a bump near threshold. These structures are both absent in the respective curves of hydrogen. From this fact it can be concluded that these specific features are caused by the interaction between the highly excited He valence electron and the $1s$ electron. The authors considered a quasistatic model that relates the structures to anticrossing effects of the He state under study with adjacent Stark states.

The technique to shift molecular transitions into coincidence with fixed frequencies of a laser is called laser magnetic resonance (LMR). This type of spectroscopy, which is used for studying unstable molecules, was combined with the double resonance technique by Lowe and McKellar.[265] A radiofrequency double resonance cell was placed in the 6-cm gap of the electromagnet of an intracavity laser magnetic resonance apparatus.[266] The Zeeman magnetic field was oriented parallel to the laser light polarization ($\Delta M = 0$ infrared transitions) and perpendicular to the electric field vector of the radiofrequency field ($\Delta M = \pm 1$ rf transitions). The NO molecule was chosen to test the technique since it is paramagnetic and provides strong LMR spectra with a CO laser. In order to obtain double resonance signals the rf was swept and frequency modulated with $\simeq 30\,\text{kHz}$, and the laser output was processed with the lock-in technique. The measurements yielded detailed spectra concerning the $^2\Pi_{3/2}$ and $^2\Pi_{1/2}$ substates of NO and provided information on parameters in the Zeeman Hamiltonian. The combined LMR radiofrequency method using a CO_2 laser was also applied to the transient free radical HO_2, which was produced by mixing the products of an rf discharge in O_2 with allyl alcohol. When minimizing pressure and rf broadening the proton hyperfine structure splitting of $\simeq 4.5\,\text{MHz}$ of a single M_J component could be resolved. The authors mention that with this measurement an unstable free radical was detected for the first time by infrared–rf or infrared–microwave double resonance.

7. Laser–Microwave Heterodyne Techniques for Spectroscopic Purposes

The combined application of laser light and microwave fields for spectroscopic purposes—not necessarily acting in the same space region or at the same time—normally concerns the irradiation of the two electromagnetic fields independently of each other onto a species. The microwave field thus interacts directly with the species under investigation, and induces transitions, thereby changing level populations.

However, many cases can be considered where the experimental situation would not allow measuring the energy gap between levels or spectral lines via the absorption or induced emission of a microwave photon. Obvious examples are the isotope shift of a line, or cases where selection rules prevent transitions between the levels.

Techniques have been developed that transfer the precision of microwave equipment to lasers in the sense that the absolute frequency of laser light is changed with a feedback mechanism by exactly the same amount as the frequency of a microwave oscillator. The change of frequency of the low-frequency oscillator is thus exactly imposed on a radiation source of extremely high frequency, namely, e.g., visible laser light with roughly $\nu = 5 \times 10^{14}\,\text{Hz}$. An ultimate outcome of methods of that kind is the direct measurement of the frequency of visible light by means of long chains of different lasers and microwave oscillators.

Microwave fields are used in many different schemes in order to modify laser light in one or the other way, like side band production, laser pulse shaping, etc. Only a limited selection of the overwhelming number of papers concerning laser–microwave heterodyning and related topics can be presented here.

7.1. Frequency-Offset Locking

The expression "frequency-offset locking" was coined by Barger and Hall in 1969.[267] Two lasers oscillate with frequencies whose difference $\Delta\nu$ may lie in the microwave region. One laser is stabilized to keep its frequency fixed, e.g., by locking it to a spectral line. One compares the difference frequency $\Delta\nu$ with the frequency ν_{rf} of a radiofrequency oscillator. The frequency of the second laser can be scanned synchronously with a change of ν_{rf} via a ramp voltage proportional to $|\Delta\nu - \nu_{rf}|$ which is obtained with a linear frequency-to-voltage converter.

Numerous variations of this scheme have been used. Examples are; frequency stabilization of a cw dye laser,[268,269] and length control of a Fabry-Perot interferometer.[270] The latter procedure was performed in Ref. 271 by locking the interferometer to a "local oscillator" He–Ne laser which was in turn frequency offset locked to a Lamb dip stabilized He–Ne laser. By tuning the local oscillator He–Ne laser, the cavity length of the interferometer and thus its transmission peak pattern was changed, thereby scanning the frequency of a cw single-mode dye laser also locked to the reference cavity. The frequency change was calibrated by measuring the difference frequency of the two He–Ne lasers. This method, however, is frequency limited by the width of the He–Ne laser gain profile of roughly 1 GHz.

A microcomputer-controlled two-laser spectrometer based on the frequency-offset locking scheme was reported in Refs. 272 and 273: The difference frequency $\Delta\nu$ of two cw single-mode dye lasers is produced by a fast photodiode and after amplification is measured by a high-frequency counter. One laser is locked to a spectral line of molecular iodine via polarization spectroscopy. A series of difference frequency values is stored in a microcomputer. In order to realize a certain stored value, the computer compares $\Delta\nu$, given by the counter, with this value. It alters the frequency of the tunable laser by a digital-to-analog converter until $\Delta\nu$ and the respective values coincide. The computer then keeps the frequency of the tunable laser constant for a desired period of time. In this manner the laser spectrometer can be scanned in steps over an optical line profile, and the fluorescence light intensity can be measured as a function of $\Delta\nu$. Since the frequency steps are exactly reproducible, signal averaging with a multichannel analyzer is feasible. Using the Lamb dip technique, the two-laser spectrometer was tested at the $1s2s\,^3S_1$ ($F = 3/2 \leftrightarrow F = 5/2$) hyperfine structure splitting frequency of $^7Li^+$. The test result of 19 817.640 (800 = 3σ) MHz is in very good agreement with the value of 19 817.673 (40 = 3σ) MHz, obtained with a combined laser–microwave measurement.[72] The difference frequency range of the described spectrometer was limited to ± 18 GHz by the amplifier available. Improvements are under way concerning enlargement of the beat-frequency range, the data acquisition technique, and the on-line signal fit procedures for immediate elimination of systematic Lamb dip shifts, caused by slight misalignment of the counterpropagating laser beams.[274]

Beat frequency generation between visible lasers in the 80-GHz range using GaAs Schottky barrier diodes was published in Ref. 275.

The development of metal–insulator–metal (MIM) point contact diodes,[276,277] and of point contact and planar Schottky barrier mixers[278,279] allows the creation of difference frequencies up to the terahertz region between two visible laser lines and their measurement by additional mixing of the laser light with a microwave frequency. The difference between the laser beat frequency and the nth harmonic of, e.g., a klystron frequency may then be low enough to be handled more conventionally, i.e., mixed and measured with standard microwave equipment. Further information on MIM diodes is given in Refs. 280 and 281.

Application of a large-frequency-difference optical spectrometer consisting of two ring dye lasers was reported by Daniel et al.[279] and in more detail by Bergquist and Daniel.[282] Figure 35 illustrates their experimental configuration. One stabilized ring laser was locked by saturated absorption to a line of $^{129}I_2$ at about equal frequency distances from the $1s_5-2p_8$ neon line and the line $P(33)$ 6–3 in $^{127}I_2$, the latter being used as the reference transition of the $\lambda = 632.8$ nm He–Ne laser line. Fractional beams of both lasers with about 10–20 mW each were focused onto a planar Schottky barrier mixing diode together with microwave radiation from a stabilized 39-GHz klystron. The laser difference frequency and the sixth harmonic of the klystron frequency were both created and mixed by the diode. The resulting difference frequency was superimposed with a suitable synthesizer frequency in a second mixer which produced a 30-MHz beat frequency. By stepping the synthesizer frequency, the second dye laser was frequency scanned via feedback techniques such that the 30-MHz beat frequency remained unchanged. Linear spectroscopy of the ^{20}Ne, $1s_5-2p_8$ transition, and saturated absorption spectroscopy of the $P(33)$, 6–3 line in $^{127}I_2$ provided the frequency offset of the ^{20}Ne line from the $^{127}I_2$ reference line and finally its absolute frequency with an uncertainty of 1×10^{-9}, the absolute frequency of the I_2 line being known.

The combination of lasers and microwave sources also plays a very important role in metrology. The frequency of 473 THz of the iodine-stabilized HeNe laser at $\lambda = 633$ nm was measured directly against the cesium standard of time[283] with a chain of lasers and klystrons starting up from a Cs atomic clock. The authors give a total uncertainty of the

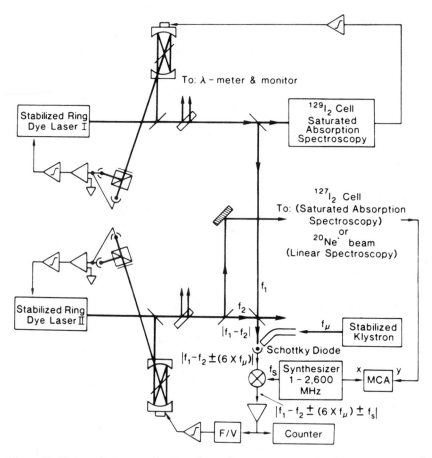

Figure 35. High-resolution two-dye laser heterodyne spectrometer for the measurement of large frequency differences.[282]

measurement of 1.6 parts in 10^{10}. With the new definition of the speed of light[284,285] being $c = 299\ 792\ 458$ m/sec, the frequency can be immediately converted into the wavelength of the He–Ne laser. This means the realization of the meter in wavelength units, based on a frequency measurement. Another direct frequency measurement of the same kind and with similar precision as the one of the He–Ne laser, concerned transitions at 520 THz (576 nm) in iodine, and at 260 THz (1.15 μm) in Ne.[286]

Investigations are under way[287] to combine a cesium beam atomic clock with laser optical pumping, replacing the A and B magnets of conventional clocks by two laser-beam–Cs-beam crossing regions. There would be no spatial selection of atoms with certain velocities as occurs with the A magnet. The preliminary results support the potential of this technique to eventually achieve an accuracy of the order of 10^{-14}. In particular, to approach this value, spurious light shifts caused by laser stray light and fluorescence from the atomic beam will have to be avoided.

7.2. Laser Light Modulation and Side Band Tuning

Sidebands with frequencies $\nu_0 \pm \nu_m$ arise from amplitude modulation of a carrier of frequency ν_0 with ν_m. This technique was used by Smaller[288] already in 1951 for nuclear magnetic resonance measurements, and later by Acrivos.[289] Similarly, by amplitude modulation of a laser light beam of frequency ν_0 with a radiofrequency ν_m—e.g., via an electrooptic crystal between two crossed polarizers—two sideband frequencies $(\nu_0 + \nu_m)$ and $(\nu_0 - \nu_m)$ appear in the beam behind the crossed polarizer. Using this effect, Burghardt and co-workers[290] tuned a dye laser in the following way: one of the two side frequencies, e.g., $(\nu_0 + \nu_m)$, passed a Fabry-Perot interferometer whose length was kept constant by locking it to an iodine-stabilized He–Ne laser (see Figure 36). When the radiofrequency was changed by $\Delta \nu$, a stabilization circuit that referred to constant interferometer transmission of the selected sideband frequency $(\nu_0 + \nu_m)$ pulled the dye laser frequency from ν_0 to $\nu_0 - \Delta \nu$ such that the side frequency remained constant. The frequency range of this elegant method is limited to twice the highest possible modulation frequency of the electro-optical device. Frequencies of up to 500 MHz are feasible. The precision that can be obtained with this scheme is discussed in Ref. 290, and is believed to approach that of typical rf methods. Bjorklund[291] probed weak spectral features in molecular iodine by a single frequency-modulated sideband of

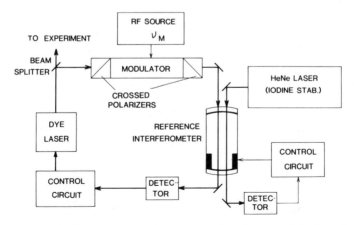

Figure 36. Schematic diagram of the tuning system, using rf-generated dye laser frequency sidebands; after Ref. 290.

a cw dye laser. He measured both the associated absorption and dispersion by mixing the signal from the sample with the oscillator frequency with which the laser was modulated.

7.3. Various Other Schemes

Many other schemes have been described in the literature, and only a fraction of them can be mentioned here. Harris and co-workers[292] stabilized a laser to the center of its gain profile by frequency-modulated sideband techniques. Frequency modulation of pulsed and cw lasers was used to produce pulse-shaping effects.[293-295] Brewer and Genack[138,296] reported that optical free induction decay in I_2 creates a beat signal with the light of a frequency-shifted laser. Various other authors reported frequency shifting of infrared laser light with a microwave field using the linear electro-optic effect in CdTe and GaAs.[297-300] Lamb dip and inverse Lamb dip spectroscopy of molecules was performed with such devices.[301]

Hall and co-workers[302] investigated line profiles of hyperfine structure components in I_2 with a refined optical phase modulation/radiofrequency sideband technique using a saturating and a probe laser beam.

Ezekiel and co-workers[303] stabilized a 1772-MHz microwave oscillator with reference to the Na ground state hyperfine structure splitting frequency via a resonance Raman transition in a beam of sodium atoms. The experiment involved a dye laser beam with frequency ω_1, from which a laser field of frequency ω_2 was generated by an acousto-optic frequency shifter driven with the oscillator.

8. Concluding Remarks

Spectroscopy utilizing tunable laser and microwave sources has been applied widely in exploring atoms, molecules, and condensed matter. Besides the classical areas of optical double resonance and optical pumping the extension of these or related methods to difference frequency measurements in the optical range seems to be of increasing importance. This includes heterodyne techniques. Laser microwave schemes can also play an essential role for the generation of modern frequency standards. Last but not least, there will be many technical applications like infrared detectors, wavemeters, magnetometers, etc.

Acknowledgment

This paper was sponsored by the Deutsche Forschungsgemeinschaft.

References

1. J. Brossel and A. Kastler, *C.R. Acad. Sci.* **229**, 1213 (1949).
2. I. Rabi, J. Zacharias, S. Millman, and P. Kusch, *Phys. Rev.* **53**, 318 (1938).
3. I. Rabi, S. Millman, P. Kusch, and J. Zacharias, *Phys. Rev.* **55**, 526 (1939).
4. F. Bitter, *Phys. Rev.* **76**, 833 (1949).
5. W. E. Lamb, Jr. and M. Skinner, *Phys. Rev.* **78**, 539 (1950).
6. J. Brossel and F. Bitter, *Phys. Rev.* **86**, 308 (1952).
7. N. F. Ramsey, *Molecular Beams*, Clarendon, London (1956).
8. H. Kopfermann, *Nuclear Moments*, Academic, New York (1958).
9. G. W. Series, *Rep. Prog. Phys.* **22**, 280 (1959).
10. C. Cohen-Tannoudji and A. Kastler, *Progress in Optics* (edited by E. Wolf), Vol. V, p. 1, North-Holland, Amsterdam (1966).
11. W. Happer, *Rev. Mod. Phys.* **44**, 169 (1972).
12. G. zu Putlitz, in *Ergebnisse der Exakten Naturwissenschaften* (edited by E. A. Niekisch), Vol. 37, p. 105, Springer, Berlin (1965).
13. O. Betz, *Ann. Phys. (Leipzig)* **15**, 321 (1932).
14. C. E. Cleeton and N. H. Williams, *Phys. Rev.* **15**, 234 (1934).
15. W. Gordy, W. V. Smith, and R. F. Trambarulo, *Microwave Spectroscopy*, Wiley, New York, Chapman and Hall, London (1953).
16. C. H. Townes and A. L. Schawlow, *Microwave Spectroscopy*, McGraw-Hill, New York (1955).
17. G. W. Chantry, Ed., *Modern Aspects of Microwave Spectroscopy*, Academic, New York (1979).
18. W. Ertmer and B. Hofer, *Z. Phys.* A **276**, 9 (1976).
19. A. M. Ronn and D. R. Lide, Jr., *J. Chem. Phys.* **47**, 3669 (1967).
20. J. Lemaire, J. Houriez, J. Bellet, and J. Thibault, *C.R. Acad. Sci. Ser.* B **268**, 922 (1969).
21. M. Fourrier, M. Redon, A. van Lerbergh, and C. Borde, *C.R. Acad. Sci. Ser.* B **270**, 537 (1970).
22. T. Shimizu and T. Oka, *J. Chem. Phys.* **53**, 2536 (1970).
23. T. Shimizu and T. Oka, *Phys. Rev.* A **2**, 1177 (1970).
24. R. Karplus and J. Schwinger, *Phys. Rev.* **73**, 1020 (1948).
25. L. Frenkel, H. Marantz, and T. Sullivan, *Phys. Rev.* A **3**, 1640 (1971).
26. M. Takami and K. Shimoda, *Jpn. J. Appl. Phys.* **10**, 658 (1971).
27. M. Takami and K. Shimoda, *Jpn. J. Appl. Phys.* **12**, 603 (1973).
28. R. W. Field, R. S. Bradford, D. O. Harris, and H. P. Broida, *J. Chem. Phys.* **56**, 4712 (1972).
29. W. A. Kreiner, M. Römheld, and H. D. Rudolph, *Z. Naturforsch.* **28a**, 1707 (1973).
30. W. A. Kreiner and H. D. Rudolph, *Z. Naturforsch.* **28a**, 1885 (1973).
31. H. Jones and A. Eyer, *Z. Naturforsch.* **28a**, 1703 (1973).
32. W. A. Kreiner and H. Jones, *J. Mol. Spectrosc.* **49**, 326 (1974).
33. S. Kano, T. Amano, and T. Shimizu, *J. Chem. Phys.* **64**, 4711 (1976).
34. M. Takami and K. Shimoda, *J. Mol. Spectrosc.* **59**, 35 (1976).
35. P. J. Domaille, T. C. Steimle, and D. O. Harris, *J. Mol. Spectrosc.* **68**, 146 (1977).
36. J. Nakagawa, D. J. Domaille, T. C. Steimle, and D. O. Harris, *J. Mol. Spectrosc.* **70**, 374 (1978).
37. B. Khoobehi and J. A. Roberts, *J. Mol. Spectrosc.* **97**, 227 (1983).
38. W. E. Ernst and T. Törring, *Phys. Rev.* A **25**, 1236 (1982).
39. W. E. Ernst and T. Törring, *Phys. Rev.* A **27**, 875 (1983).
40. W. E. Ernst, *Appl. Phys.* B **30**, 105 (1983).
41. J. Brossel, A. Kastler, and J. Winter, *J. Phys. Rad.* **13**, 668 (1952).
42. R. Schieder and H. Walther, *Z. Phys.* **270**, 55 (1974).

43. S. D. Rosner, R. H. Holt, and T. D. Gaily, *Phys. Rev. Lett.* **35**, 785 (1975).
44. W. J. Childs, O. Poulsen, L. S. Goodman, and H. Crosswhite, *Phys. Rev. A* **19**, 168 (1979).
45. W. J. Childs, O. Poulsen, and L. S. Goodman, *Phys. Rev. A* **19**, 160 (1979).
46. W. Ertmer, U. Johann, and G. Meisel, *Phys. Lett.* **85B**, 319 (1979).
47. W. J. Childs, O. Poulsen, and L. S. Goodman, *Opt. Lett.* **4**, 35 (1979).
48. W. J. Childs, O. Poulsen, and L. S. Goodman, *Opt. Lett.* **4**, 63 (1979).
49. J. Dembczynski, W. Ertmer, U. Johann, S. Penselin, and S. Stinner, Book of Abstracts, Sixth International Conference on Atomic Physics, Riga (1978), p. 19.
50. W. J. Childs and L. S. Goodman, *Phys. Rev. A* **21**, 1216 (1980).
51. W. J. Childs, G. L. Goodman, and L. S. Goodman, *J. Mol. Spectrosc.* **86**, 365 (1981).
52. W. J. Childs, D. R. Cok, L. S. Goodman, and O. Poulsen, *Phys. Rev. Lett.* **47**, 1389 (1981).
53. W. J. Childs, D. R. Cok, and L. S. Goodman, in *Laser Spectroscopy V* (edited by A. R. W. McKellar, T. Oka, B. P. Stoicheff), p. 150, Springer, Berlin (1981).
54. W. J. Childs, D. R. Cok, G. L. Goodman, and L. S. Goodman, *J. Chem. Phys.* **75**, 501 (1981).
55. L. S. Goodman and W. J. Childs, Symposium on Atomic Spectroscopy, Tucson, September 1979, Book of Abstracts, p. 166.
56. W. J. Childs, L. S. Goodman, S. A. Lee, and H. Crosswhite, Symposium on Atomic Spectroscopy, Tucson, September 1979, Book of Abstracts, p. 168.
57. W. J. Childs and L. S. Goodman, *Phys. Rev. A* **24**, 1342 (1981).
58. W. J. Childs and L. S. Goodman, *J. Opt. Soc. Am.* **69**, 815 (1979).
59. M. Dubke, W. Jitschin, and G. Meisel, *Phys. Lett.* **65A**, 109 (1978).
60. J. Dembczynski, W. Ertmer, U. Johann, S. Penselin, and P. Stinner, *Z. Phys. A* **291**, 207 (1979).
61. J. Dembczynski, W. Ertmer, U. Johann, and P. Stinner, *Z. Phys. A* **294**, 313 (1980).
62. W. Ertmer, U. Johann, and R. Mosmann, *Z. Phys. A* **309**, 1 (1982).
63. R. Aydin, W. Ertmer, and U. Johann, *Z. Phys. A* **306**, 1 (1982).
64. W. J. Childs, L. S. Goodman, and V. Pfeufer, *Z. Phys. A* **311**, 251 (1983).
65. W. J. Childs, L. S. Goodman, and V. Pfeufer, *Phys. Rev. A* **28**, 3402 (1983).
66. W. J. Childs, D. R. Cok, and L. S. Goodman, *J. Opt. Soc. Am.* **73**, 151 (1983).
67. P. Grundevik, M. Gustavsson, I. Lindgren, G. Olsson, T. Olsson, and A. Rosen, *Z. Phys. A* **311**, 143 (1983).
68. S. D. Rosner, T. D. Gaily, and R. A. Holt, *Phys. Rev. Lett.* **40**, 851 (1978).
69. N. Bjerre, M. Kaivola, U. Nielsen, O. Poulsen, P. Thorsen, and N. I. Winstrup, in *Laser Spectroscopy VI* (edited by H. P. Weber and W. Lüthy), p. 128, Springer, Berlin (1983).
70. U. Nielsen, O. Poulsen, P. Thorsen, and H. Crosswhite, *Phys. Rev. Lett.* **51**, 1749 (1983).
71. R. Bayer, J. Kowalski, R. Neumann, S. Noehte, H. Suhr, K. Winkler, and G. zu Putlitz, *Z. Phys. A* **292**, 329 (1979).
72. U. Kötz, J. Kowalski, R. Neumann, S. Noehte, H. Suhr, K. Winkler, and G. zu Putlitz, *Z. Phys. A* **300**, 25 (1981).
73. R. Neumann, J. Kowalski, F. Mayer, S. Noehte, R. Schwarzwald, H. Suhr, K. Winkler, and G. zu Putlitz, in *Laser Spectroscopy V* (edited by A. R. W. McKellar, T. Oka, and B. P. Stoicheff), p. 130, Springer, Berlin (1981).
74. R. D. Knight and M. H. Prior, *Phys. Rev. A* **21**, 179 (1980).
75. H. Schmoranzer, D. Schulze-Hagenest, and S. A. Kandela, Symposium on Atomic Spectroscopy, Tucson, Sep. 1979, Book of Abstracts, p. 195.
76. J. Kowalski, R. Neumann, S. Noehte, K. Scheffzek, H. Suhr, and G. zu Putlitz, *Hyperfine Interact.* **15/16**, 159 (1983).
77. J. Kowalski, R. Neumann, S. Noehte, H. Suhr, G. zu Putlitz, and R. Herman, *Z. Phys. A* **313**, 147 (1983).
78. W. E. Ernst and S. Kindt, *Appl. Phys. B* **31**, 79 (1983).
79. W. E. Ernst, S. Kindt, and T. Törring, *Phys. Rev. Lett.* **51**, 979 (1983).

80. M. Englert, J. Kowalski, F. Mayer, R. Neumann, S. Noehte, R. Schwarzwald, H. Suhr, K. Winkler, and G. zu Putlitz, *Appl. Phys. B* **28**, 81 (1982).
81. F. Touchard, J. M. Serre, S. Büttgenbach, P. Guimbal, R. Klapisch, M. de Saint Simon, C. Thibault, H. T. Duong, P. Juncar, S. Liberman, J. Pinard, and J. L. Vialle, *Phys. Rev. C* **25**, 2756 (1982).
82. J. J. Ter Meulen, W. Ubachs, and A. Dymanus, in *Laser Spectroscopy VI* (edited by H. P. Weber and W. Lüthy), p. 345, Springer, Berlin (1983).
83. A. G. Adam, S. D. Rosner, T. D. Gaily, and R. A. Holt, *Phys. Rev. A* **26**, 315 (1982).
84. H. A. Schuessler, E. N. Fortson, and H. G. Dehmelt, *Phys. Rev.* **187**, 5 (1969).
85. D. J. Wineland, J. C. Bergquist, W. M. Itano, and R. E. Drullinger, *Opt. Lett.* **5**, 245 (1980).
86. W. M. Itano and D. J. Wineland, *Phys. Rev. A* **24**, 1364 (1981).
87. W. M. Itano and D. J. Wineland, in *Laser Spectroscopy V* (edited by A. R. W. McKellar, T. Oka, and B. P. Stoicheff), p. 360, Springer, Berlin, (1981).
88. D. J. Wineland, R. E. Drullinger, and F. L. Walls, *Phys. Rev. Lett.* **40**, 1639 (1978).
89. R. E. Drullinger, D. J. Wineland, and J. C. Bergquist, *Appl. Phys.* **22**, 365 (1980).
90. D. J. Wineland, J. J. Bollinger, and W. M. Itano, *Phys. Rev. Lett.* **50**, 628 (1983).
91. J. J. Bollinger, D. J. Wineland, W. M. Itano, and J. S. Wells, in *Laser Spectroscopy VI* (edited by H. P. Weber and W. Lüthy), p. 168, Springer, Berlin (1983).
92. J. Vetter, H. Ackermann, G. zu Putlitz, and E. W. Weber, *Z. Phys. A* **276**, 161 (1976).
93. R. Blatt and G. Werth, *Z. Phys. A* **299**, 93 (1981).
94. R. Blatt and G. Werth, *Phys. Rev. A* **25**, 1476 (1982).
95. W. Becker and G. Werth, *Z. Phys. A* **311**, 41 (1983).
96. F. von Sichart, H. J. Stöckmann, H. Ackermann, and G. zu Putlitz, *Z. Phys.* **236**, 97 (1970).
97. R. Blatt, H. Schnatz, and G. Werth, *Phys. Rev. Lett.* **48**, 1602 (1982).
98. R. Blatt, H. Schnatz, and G. Werth, *Z. Phys. A* **312**, 143 (1983).
99. D. J. Larson and R. M. Jopson, in *Laser Spectroscopy V* (edited by A. R. W. McKellar, T. Oka, and B. P. Stoicheff), p. 369, Springer, Berlin (1981).
100. R. M. Jopson and D. J. Larson, *Phys. Rev. Lett.* **47**, 789 (1981).
101. R. M. Jopson, R. Trainham, and D. J. Larson, *Bull. Am. Phys. Soc.* **26**, 1306 (1981).
102. L. F. Mollenauer, S. Pan, and A. Winnacker, *Phys. Rev. Lett.* **26**, 1643 (1971).
103. H. Panepucci and L. F. Mollenauer, *Phys. Rev.* **178**, 589 (1969).
104. L. F. Mollenauer, S. Pan, and S. Yngvesson, *Phys. Rev. Lett.* **23**, 683 (1969).
105. A. Winnacker, K. E. Mauser, and B. Niesert, *Z. Phys. B* **26**, 97 (1977).
106. K. E. Mauser, B. Niesert, and A. Winnacker, *Z. Phys. B* **26**, 107 (1977).
107. K. Hahn, K. E. Mauser, H. J. Reyher, and A. Winnacker, *Phys. Lett.* **63A**, 151 (1977).
108. H. J. Reyher and A. Winnacker, *Z. Phys. B* **45**, 183 (1982).
109. K. Hahn, H. J. Reyher, T. Vetter, and A. Winnacker, *Phys. Lett.* **72A**, 363 (1979).
110. A. Winnacker, K. Hahn, H. J. Reyher, and T. Vetter, *J. Phys. (Paris)* **41**, C6 (1980).
111. A. Dieckmann, G. Strauch, T. Vetter, and A. Winnacker, *Radiat. Eff.* **73**, 1 (1983).
112. G. Strauch, T. Vetter, and A. Winnacker, *Phys. Lett.* **94A**, 160 (1983).
113. L. E. Erickson, *Opt. Commun.* **21**, 147 (1977).
114. L. E. Erickson, *Phys. Rev. B.* **19**, 4412 (1979).
115. K. K. Sharma and L. E. Erickson, *Phys. Rev. Lett.* **45**, 294 (1980).
116. R. M. Shelby, C. S. Yannoni, and R. M. MacFarlane, *Phys. Rev. Lett.* **41**, 1739 (1978).
117. R. M. Shelby, R. M. MacFarlane, and C. S. Yannoni, *Phys. Rev. B* **21**, 5004 (1980).
118. R. M. MacFarlane, R. M. Shelby, A. Z. Genack, and D. A. Weitz, *Opt. Lett.* **5**, 462 (1980).
119. R. G. DeVoe, A. Szabo, S. C. Rand, and R. G. Brewer, *Phys. Rev. Lett.* **42**, 1560 (1979).
120. S. C. Rand, A. Wokaun, R. G. DeVoe, and R. G. Brewer, *Phys. Rev. Lett.* **43**, 1868 (1979).
121. R. M. MacFarlane, R. M. Shelby, and D. P. Burum, *Opt. Lett.* **6**, 593 (1981).
122. D. P. Burum, R. M. Shelby, and R. M. MacFarlane, *Phys. Rev. B* **25**, 3009 (1982).
123. R. M. MacFarlane, D. P. Burum, and R. M. Shelby, *Phys. Rev. Lett.* **49**, 636 (1982).

124. R. M. Shelby, A. C. Tropper, R. T. Harley, and R. M. MacFarlane, *Opt. Lett.* **8**, 304 (1983).
125. A. M. Achlama, U. Harke, H. Zimmermann, and K. P. Dinse, *Chem. Phys. Lett.* **85**, 339 (1982).
126. U. Harke, G. Wäckerle, and K. P. Dinse, in *Photochemistry and Photobiology*, Proceedings of the International Conference, Alexandria, Egypt (edited by A. H. Zewail), Harwood Academic Publishing, Chur (1983).
127. K. P. Dinse, U. Harke, and G. Wäckerle, in *Laser Spectroscopy VI* (edited by H. P. Weber and W. Lüthy), p. 347, Springer, Berlin (1983).
128. W. B. Mims, *Proc. R. Soc. London* **283**, 452 (1965).
129. P. Hu, R. Leigh, and S. R. Hartmann, *Phys. Lett.* **40A**, 164 (1972).
130. P. F. Liao, R. Leigh, P. Hu, and S. R. Hartmann, *Phys. Lett.* **41A**, 285 (1972).
131. P. F. Liao, P. Hu, R. Leigh, and S. R. Hartmann, *Phys. Rev. A* **9**, 332 (1974).
132. K. Chiang, E. A. Whittacker, and S. R. Hartmann, *Phys. Rev. B* **23**, 6142 (1981).
133. L. E. Erickson, *Phys. Rev B* **16**, 4731 (1977).
134. R. M. MacFarlane, R. M. Shelby, and R. L. Shoemaker, *Phys. Rev. Lett.* **43**, 1726 (1979).
135. R. M. Shelby, R. M. MacFarlane, and R. L. Shoemaker, *Phys. Rev. B* **25**, 6578 (1982).
136. D. P. Burum, R. M. MacFarlane, and R. M. Shelby, *Phys. Lett.* **90A**, 483 (1982).
137. R. M. MacFarlane and R. M. Shelby, in *Laser Spectroscopy VI* (edited by H. P. Weber and W. Lüthy), p. 113, Springer, Berlin (1983).
138. R. G. Brewer and A. Z. Genack, *Phys. Rev. Lett.* **36**, 959 (1976).
139. K. German and R. Zare, *Phys. Rev. Lett.* **23**, 1207 (1969).
140. S. J. Silvers, T. H. Bergeman, and W. Klemperer, *J. Chem. Phys.* **52**, 4385 (1970).
141. R. W. Field and T. H. Bergeman, *J. Chem. Phys.* **54**, 2936 (1971).
142. R. W. Field, R. S. Bradford, H. P. Broida, and D. O. Harris, *J. Chem. Phys.* **57**, 2209 (1972).
143. R. W. Field, A. D. English, T. Tanaka, D. O. Harris, and D. A. Jennings, *J. Chem. Phys.* **59**, 2191 (1973).
144. R. Solarz and D. H. Levy, *J. Chem. Phys.* **58**, 4026 (1973).
145. T. Tanaka, R. W. Field, and D. O. Harris, *J. Chem. Phys.* **61**, 3401 (1974).
146. H. G. Weber, P. J. Brucat, W. Demtröder, and R. N. Zare, *J. Mol. Spectrosc.* **75**, 58 (1979).
147. F. Bylicki and H. G. Weber, *Phys. Lett.* **82A**, 456 (1981).
148. F. Bylicki and H. G. Weber, *Chem. Phys. Lett.* **79**, 517 (1981).
149. F. Bylicki and H. G. Weber, *J. Chem. Phys.* **78**, 2899 (1983).
150. F. Bylicki, H. G. Weber, H. Zscheeg, and M. Arnold, *J. Chem. Phys.* **80**, 1791 (1984).
151. H. G. Weber, P. J. Brucat, and R. N. Zare, *Chem. Phys. Lett.* **63**, 217 (1979).
152. F. Bylicki, M. Kolwas, and H. G. Weber, *Phys. Rev. A* **25**, 1550 (1982).
153. F. Bylicki, H. G. Weber, and H. Zscheeg, *Z. Phys. A* **314**, 123 (1983).
154. F. Bylicki and H. G. Weber, *Phys. Lett.* **100A**, 182 (1984).
155. G. W. Hills and R. F. Curl, Jr., *J. Chem. Phys.* **66**, 1507 (1977).
156. G. W. Hills, *J. Mol. Spectrosc.* **93**, 395 (1982).
157. G. W. Hills, J. M. Cook, R. F. Curl, Jr., and F. K. Tittel, *J. Chem. Phys.* **65**, 823 (1976).
158. J. M. Cook, G. W. Hills, and R. F. Curl, Jr., *Astrophys. J.* **207**, L139 (1976).
159. G. W. Hills and J. M. Cook, *Astrophys. J.* **209**, L157 (1976).
160. J. M. Cook, G. W. Hills, and R. F. Curl, Jr., *J. Chem. Phys.* **67**, 1450 (1977).
161. J. V. V. Kasper, R. S. Lowe, and R. F. Curl, Jr., *J. Chem. Phys.* **70**, 3350 (1979).
162. R. S. Lowe, J. V. V. Kasper, G. W. Hills, W. Dillenschneider, and R. F. Curl, Jr., *J. Chem. Phys.* **70**, 3356 (1979).
163. H. Knöckel and E. Tiemann, *Chem. Phys. Lett.* **64**, 593 (1979).
164. H. Knöckel and E. Tiemann, *Chem. Phys. Lett.* **70**, 345 (1982).
165. S. G. Hansen and B. J. Howard, *Chem. Phys. Lett.* **85**, 249 (1982).
166. S. G. Hansen, J. D. Thompson, C. M. Western, and B. J. Howard, *Mol. Phys.* **49**, 1217 (1983).

167. R. H. Clarke and J. M. Hayes, *J. Chem. Phys.* **59**, 3113 (1973).
168. R. H. Clarke and R. H. Hofeldt, *J. Am. Chem. Soc.* **96**, 3005 (1974).
169. R. H. Clarke and R. H. Hofeldt, *J. Chem. Phys.* **61**, 4582 (1974).
170. R. H. Clarke, R. E. Connors, J. R. Noris, and M. C. Thurnauer, *J. Am. Chem. Soc.* **97**, 7178 (1975).
171. R. H. Clarke and R. E. Connors, *Chem. Phys. Lett.* **42**, 69 (1976).
172. R. H. Clarke, D. R. Hobart, and W. R. Leenstra, *J. Am. Chem. Soc.* **101**, 2416 (1979).
173. R. H. Clarke, S. P. Jagannathan, and W. R. Leenstra, in *Lasers in Photomedicine and Photobiology* (edited by R. Pratesi and C. A. Sacchi), p. 171, Springer, Berlin (1980).
174. R. F. Stebbings, *Science* **193**, 537 (1976).
175. S. A. Edelstein and T. F. Gallagher, *Adv. At. Mol. Phys.* **14**, 365 (1978).
176. F. B. Dunnings and R. F. Stebbings, *Commun. At. Mol. Phys.* **10**, 9 (1980).
177. S. Haroche, in *Atomic Physics 7* (edited by D. Kleppner and F. M. Pipkin), p. 141, Plenum Press, New York (1981).
178. D. Kleppner, M. G. Littman, and M. L. Zimmerman, *Sci. Am.* **244**, 108 (1981).
179. R. F. Stebbings and F. B. Dunning, eds., *Rydberg States of Atoms and Molecules*, Cambridge University Press, Cambridge (1983).
180. S. Svanberg, P. Tsekeris, and W. Happer, *Phys. Rev. Lett.* **30**, 817 (1973).
181. S. Svanberg and G. Belin, *J. Phys. B* **7**, L82 (1974).
182. G. Belin and S. Svanberg, *Phys. Lett.* **47A**, 5 (1974).
183. P. Tsekeris, R. Gupta, W. Happer, G. Belin, and S. Svanberg, *Phys. Lett.* **48A**, 101 (1974).
184. P. Tsekeris and R. Gupta, *Phys. Rev. A* **11**, 455 (1975).
185. G. Belin, L. Holmgren, I. Lindgren, and S. Svanberg, *Phys. Scr.* **12**, 287 (1975).
186. S. Svanberg and P. Tsekeris, *Phys. Rev. A* **11**, 1125 (1975).
187. G. Belin, L. Holmgren, and S. Svanberg, *Phys. Scr.* **13**, 351 (1976).
188. J. Farley and R. Gupta, *Phys. Rev. A* **15**, 1952 (1977).
189. J. Farley, P. Tsekeris, and R. Gupta, *Phys. Rev. A* **15**, 1530.
190. P. Tsekeris, K. H. Liao, and R. Gupta, *Phys. Rev. A* **13**, 2309 (1976).
191. H. Lundberg, A.-M. Martensson, and S. Svanberg, *J. Phys. B* **10**, 1971 (1977).
192. T. F. Gallagher, R. M. Hill, and S. A. Edelstein, *Phys. Rev. A* **13**, 1448 (1976).
193. T. F. Gallagher, R. M. Hill, and S. A. Edelstein, *Phys. Rev. A* **14**, 744 (1976).
194. W. E. Cooke, T. F. Gallagher, R. M. Hill, and S. A. Edelstein, *Phys. Rev. A* **16**, 1141 (1977).
195. G. A. Ruff, K. A. Safinya, and T. F. Gallagher, *Phys. Rev.* **22**, 183 (1980).
196. H. Lundberg and S. Svanberg, *Z. Phys. A* **290**, 127 (1979).
197. T. F. Gallagher, L. M. Humphrey, R. M. Hill, W. E. Cooke, and S. A. Edelstein, *Phys. Rev. A* **15**, 1937 (1977).
198. C. Fabre, P. Goy, and S. Haroche, *J. Phys. B* **10**, L183 (1977).
199. K. A. Safinya, T. F. Gallagher, and W. Sandner, *Phys. Rev. A* **22**, 2672 (1980).
200. T. F. Gallagher, R. Kachru, and N. H. Tran, *Phys. Rev. A* **26**, 2611 (1982).
201. C. Fabre, S. Haroche, and P. Goy, *Phys. Rev. A* **18**, 229 (1978).
202. P. Goy, C. Fabre, M. Gross, and S. Haroche, *J. Phys. B* **13**, L83 (1980).
203. C. Fabre, S. Haroche, and P. Goy, *Phys. Rev. A* **22**, 778 (1980).
204. P. Goy, J. M. Raimond, G. Vitrant, and S. Haroche, *Phys. Rev. A* **26**, 2733 (1982).
205. S. Haroche, C. Fabre, P. Goy, M. Gross, and J. M. Raimond, in *Laser Spectroscopy IV* (edited by H. Walther and K. W. Rothe), p. 244, Springer, Berlin (1979).
206. M. Gross, P. Goy, C. Fabre, S. Haroche, and J. M. Raimond, *Phys. Rev. Lett.* **43**, 343 (1979).
207. J. W. Ducas, W. P. Spencer, A. G. Vaidyanathan, W. H. Hamilton, and D. Kleppner, *Appl. Phys. Lett.* **35**, 382 (1979).
208. T. F. Gallagher and W. E. Cooke, *Phys. Rev. Lett.* **42**, 835 (1979).
209. E. J. Beiting, G. F. Hildebrandt, F. G. Kellert, G. W. Foltz, K. A. Smith, F. B. Dunning, and R. F. Stebbings, *J. Chem. Phys.* **70**, 3551 (1979).

210. P. R. Koch, H. Hieronymus, A. F. J. van Raan, and W. Raith, *Phys. Lett.* **75A**, 273 (1980).
211. H. Figger, G. Leuchs, R. Straubinger, and H. Walther, *Opt. Commun.* **33**, 37 (1980).
212. L. Moi, C. Fabre, P. Goy, M. Gross, S. Haroche, P. Encrenaz, G. Beaudin, and B. Lazareff, *Opt. Commun.* **33**, 47 (1980).
213. T. Oka and T. Shimizu, *Appl. Phys. Lett.* **19**, 88 (1971).
214. C. Freed and A. Javan, *Appl. Phys. Lett.* **17**, 53 (1970).
215. S. M. Freund and T. Oka, *Appl. Phys. Lett.* **21**, 60 (1972).
216. S. M. Freund and T. Oka, *Phys. Rev. A* **13**, 2178 (1976).
217. F. Shimizu, *Chem. Phys. Lett.* **17**, 620 (1972).
218. R. F. Curl, Jr. and T. Oka, *J. Chem. Phys.* **58**, 4908 (1973).
219. K. Shimoda and M. Takami, *Opt. Commun.* **4**, 388 (1972).
220. M. Takami and K. Shimoda, *Jpn. J. Appl. Phys.* **11**, 1648 (1972).
221. M. Takami and K. Shimoda, *Jpn. J. Appl. Phys.* **12**, 934 (1973).
222. F. Shimizu, *Phys. Rev. A* **10**, 950 (1974).
223. M. Takami, *Jpn. J. Appl. Phys.* **15**, 1063 (1976).
224. M. Takami, *Jpn. J. Appl. Phys.* **15**, 1889 (1976).
225. K. Shimoda, in *Laser Spectroscopy* (edited by R. G. Brewer and A. Mooradian), p. 29, Plenum Press, New York (1974).
226. K. Shimoda, in *Laser Spectroscopy of Atoms and Molecules, Topics in Applied Physics* (edited by H. Walther), Vol. 2, p. 197, Springer, Berlin (1976).
227. T. Oka, in *Frontiers in Laser Spectroscopy*, Vol. 2, p. 529, North-Holland, Amsterdam (1977).
228. S. M. Freund, M. Römheld, and T. Oka, *Phys. Rev. Lett.* **35**, 1497 (1975).
229. T. Tanaka, C. Yamada, and E. Hirota, *J. Mol. Spectrosc.* **63**, 142 (1976).
230. E. Arimondo, P. Glorieux, and T. Oka, *Phys. Rev. A* **17**, 1375 (1978).
231. W. A. Kreiner, B. J. Orr, U. Andresen, and T. Oka, *Phys. Rev. A* **15**, 2298 (1977).
232. E. Arimondo and P. Glorieux, *Phys. Rev. A* **19**, 1067 (1979).
233. E. Arimondo and T. Oka, *Phys. Rev.* **26**, 1494 (1982).
234. T. Tanaka, A. Inayoshi, K. Kijima, K. Harada, and K. Tanaka, *Rev. Sci. Instrum.* **53**, 1552 (1982).
235. T. Tanaka, A. Inayoshi, K. Kijima, and T. Tanaka, *J. Mol. Spectrosc.* **95**, 182 (1982).
236. H. Sasada, Y. Hasegawa, T. Amano, and T. Shimizu, *J. Mol. Spectrosc.* **96**, 106 (1982).
237. J. Lindenmayer, H. Jones, and H. D. Rudolph, *J. Mol. Spectrosc.* **101**, 221 (1983).
238. J. Sheridan and H. Jones, *J. Mol. Spectrosc.* **98**, 498 (1983).
239. P. Shoja-Chaghervand, E. Bjarnov, and R. H. Schwendeman, *J. Mol. Spectrosc.* **97**, 287 (1983).
240. P. Shoja-Chaghervand and R. H. Schwendeman, *J. Mol. Spectrosc.* **97**, 306 (1983).
241. P. Shoja-Chaghervand and R. H. Schwendeman, *J. Mol. Spectrosc.* **98**, 27 (1983).
242. J. M. Levy, J. H.-S. Wang, S. G. Kukolich, and J. I. Steinfeld, *Phys. Rev. Lett.* **29**, 395 (1972).
243. J. M. Levy, J. H.-S. Wang, S. G. Kukolich, and J. I. Steinfeld, *Chem. Phys. Lett.* **21**, 598 (1973).
244. J. C. McGurk, T. G. Schmalz, and W. H. Flygare, in *Advances in Chemical Physics* (edited by I. Prigogine and S. A. Rice), Wiley, New York (1973).
245. J. C. McGurk, C. L. Norris, T. G. Schmalz, E. F. Pearson, and W. H. Flygare, in *Laser Spectroscopy* (edited by R. G. Brewer and A. Mooradian), p. 541, Plenum, New York (1974).
246. T. Oka, *Adv. At. Mol. Phys.* **9**, 127 (1973).
247. T. Oka, in *Laser Spectroscopy* (edited by R. G. Brewer and A. Mooradian), p. 413, Plenum, New York (1974).
248. T. Shimizu, in *Laser Spectroscopy* (edited by R. G. Brewer and A. Mooradian), p. 433, Plenum, New York (1974).

249. K. Shimoda, in *Laser Spectroscopy III* (edited by J. L. Hall and L. J. Carlsten), p. 279, Springer, Berlin (1977).

250. J. I. Steinfeld and P. L. Houston, in *Laser and Coherence Spectroscopy* (edited by J. I. Steinfeld), p. 1, Plenum Press, New York (1978).

251. H. Jones, in *Modern Aspects of Microwave Spectroscopy* (edited by G. W. Chantry), p. 123, Academic, New York (1979).

252. M. Takami, in *Laser Spectroscopy V* (edited by A. R. W. McKellar, T. Oka, and B. P. Stoicheff), p. 146, Springer, Berlin (1981).

253. E. W. Weber and J. E. M. Goldsmith, *Phys. Rev. Lett.* **41**, 940 (1978).

254. M. Arnold, E. Bergmann, P. Bopp, C. Dorsch, J. Kowalski, T. Stehlin, F. Träger, and G. zu Putlitz, *Hyperfine Interact.* **9**, 159 (1981).

255. M. Arnold, J. Kowalski, T. Stehlin, F. Träger, and G. zu Putlitz, *Z. Phys.* A **314**, 303 (1983).

256. E.-W. Otten, *Z. Phys.* **170**, 336 (1962).

257. M. Gustavsson, I. Lindgren, G. Olsson, A. Rosen, and S. Svanberg, *Phys. Lett.* **62A**, 250 (1977).

258. P. Grundevik, M. Gustavsson, I. Lindgren, G. Olsson, L. Robertsson, A. Rosen, and S. Svanberg, *Phys. Rev. Lett.* **42**, 1528 (1979).

259. B. Burghardt, S. Büttgenbach, N. Glaeser, R. Harzer, G. Meisel, B. Roski, and F. Träber, *Z. Phys.* A **307**, 193 (1982).

260. K. H. Bürger, B. Burghardt, S. Büttgenbach, R. Harzer, H. Hoeffgen, G. Meisel, and F. Träber, *Z. Phys.* A **307**, 201 (1982).

261. P. M. Koch, L. D. Gardner, J. E. Bayfield, in *Beam-Foil Spectroscopy*, Vol. 2 (edited by J. A. Sellin and D. J. Pegg), p. 829, Plenum Press, New York (1976).

262. J. E. Bayfield, L. D. Gardner, and P. M. Koch, *Phys. Rev. Lett.* **39**, 76 (1977).

263. P. M. Koch, *Phys. Rev. Lett.* **41**, 99 (1978).

264. D. R. Mariani, W. van de Water, P. M. Koch, and T. Bergeman, *Phys. Rev. Lett.* **50**, 1261 (1983).

265. R. S. Lowe and A. R. W. McKellar, in *Laser Spectroscopy V* (edited by A. R. W. McKellar, T. Oka, and B. P. Stoicheff), p. 341, Springer, Berlin (1981).

266. R. S. Lowe and A. R. W. McKellar, *J. Mol. Spectrosc.* **79**, 424 (1980).

267. R. L. Barger and J. L. Hall, *Phys. Rev. Lett.* **22**, 4 (1969).

268. R. L. Barger, M. S. Sorem, and J. L. Hall, *Appl. Phys. Lett.* **22**, 573 (1973).

269. H. Walther, *Phys. Scr.* **9**, 297 (1974).

270. R. L. Barger and J. L. Hall, *Appl. Phys. Lett.* **22**, 196 (1973).

271. H. Gerhardt and A. Timmermann, *Opt. Commun.* **21**, 343 (1977).

272. J. Kowalski, F. Mayer, R. Neumann, S. Noehte, R. Schwarzwald, H. Suhr, K. Winkler, and G. zu Putlitz, European Conference on Atomic Physics, Heidelberg (1981), Book of Abstracts, p. 1068.

273. M. Englert, J. Kowalski, F. Mayer, R. Neumann, S. Noehte, R. Schwarzwald, H. Suhr, and G. zu Putlitz, *Sov. J. Quantum Electron.* **12**, 664 (1982).

274. J. J. Snyder and J. L. Hall, in *Laser Spectroscopy* (edited by S. Haroche, J. C. Pebay-Peroula, T. W. Hänsch, and S. E. Harris), p. 6, Proceedings of the Second International Conference, Springer, Berlin.

275. B. Burghardt, H. Hoeffgen, G. Meisel, W. Reinert, and B. Vowinkel, *Appl. Phys. Lett.* **35**, 498 (1979).

276. H.-U. Daniel, M. Steiner, and H. Walther, *Appl. Phys. B* **26**, 19 (1981).

277. R. E. Drullinger, K. M. Evenson, D. A. Jennings, F. R. Peterson, J. C. Bergquist, L. Burkins, and H.-U. Daniel, *Appl. Phys. Lett.* **42**, 137 (1983).

278. H.-U. Daniel, B. Maurer, M. Steiner, H. Walther, J. C. Bergquist, in *Laser Spectroscopy VI* (edited by H. P. Weber and W. Lüthy), p. 432, Springer, Berlin (1983).

279. H.-U. Daniel, B. Maurer, and M. Steiner, *Appl. Phys. B* **30**, 189 (1983).

280. K. J. Siemsen and H. D. Riccius, in *Laser Spectroscopy VI* (edited by H. P. Weber and W. Lüthy), p. 435, Springer, Berlin (1983).

281. H. H. Klingenberg, in *Laser Spectroscopy VI* (edited by H. P. Weber and W. Lüthy), p. 437, Springer, Berlin (1983).

282. J. C. Bergquist and H.-U. Daniel, *Opt. Commun.* **48**, 327 (1984).

283. D. A. Jennings, C. R. Pollock, F. R. Petersen, R. E. Drullinger, K. M. Evenson, and J. S. Wells, *Opt. Lett.* **8**, 136 (1983).

284. T. Wilkie, *New Scientist* **100**, 258 (1983).

285. F. Bayer-Helms, *Phys. Bl.* **39**, 307 (1983).

286. C. R. Pollock, D. A. Jennings, F. R. Petersen, J. S. Wells, R. E. Drullinger, E. C. Beaty, and K. M. Evenson, *Opt. Lett.* **8**, 133 (1983).

287. M. Arditi, *Metrologia* **18**, 59 (1982).

288. B. Smaller, *Phys. Rev.* **83**, 812 (1951).

289. J. V. Acrivos, *J. Chem. Phys.* **36**, 1097 (1962).

290. B. Burghardt, W. Jitschin, and G. Meisel, *Appl. Phys.* **20**, 141 (1979).

291. G. C. Bjorklund, *Opt. Lett.* **5**, 15 (1980).

292. S. E. Harris, M. K. Oshman, B. J. McMurty, and E. O. Ammann, *Appl. Phys. Lett.* **7**, 185 (1965).

293. M. A. Duguay and J. W. Hansen, *Appl. Phys. Lett.* **14**, 14 (1969).

294. D. Grischkowsky, *Appl. Phys. Lett.* **25**, 566 (1974).

295. W. E. Bicknell, L. R. Tomasetta, and R. H. Kingston, *IEEE J. Quantum Electron* **QE-11**, 308 (1975).

296. R. G. Brewer, *Phys. Today* **30**(5), 50 (1977).

297. G. Magerl and E. Bonek, *J. Appl. Phys.* **11**, 4901 (1976).

298. J. E. Bjorholm, E. W. Turner, and D. B. Pearson, *Appl. Phys. Lett.* **26**, 564 (1975).

299. G. M. Carter, *Appl. Phys. Lett.* **32**, 810 (1978).

300. G. Magerl and E. Bonek, *Appl. Phys. Lett.* **34**, 452 (1979).

301. G. Magerl, J. M. Frye, W. E. Kreiner, and T. Oka, *Appl. Phys. Lett.* **42**, 656 (1983).

302. J. L. Hall, L. Hollberg, T. Baer, and H. J. Robinson, *Appl. Phys. Lett.* **39**, 680 (1981).

303. P. R. Hemmer, S. Ezekiel, and C. C. Leiby, Jr., *Opt. Lett.* **8**, 440 (1983).

2

Collinear Fast-Beam Laser Spectroscopy

R. Neugart

1. Introduction

The progress in atomic and molecular spectroscopy has gone hand in hand with improvements of the resolution. Before the tunable narrow-band lasers led to the invention of Doppler-free techniques, spectral lines from cooled hollow-cathode discharges[1] had typical widths larger than 300 MHz, and high resolution was achieved only in rf spectroscopy, e.g., within hyperfine structure multiplets, by the classical techniques like atomic beam magnetic resonance,[2] optical pumping,[3] or double resonance.[4] While the Doppler broadening

$$\delta\nu_D = \nu_0\left(\frac{8kT \ln 2}{mc^2}\right)^{1/2} \tag{1}$$

is negligible for resonance frequencies ν_0 in the rf regime, the narrowing of optical resonances by cooling to low temperatures, T, has obvious limits. On the other hand, Eq. (1) holds for an isotropic thermal velocity distribution, and a narrow Doppler width requires a reduction in the spread of velocity components only along the direction of observation. It is well known from ion optics that this can be achieved by electrostatic acceleration. If an ion beam propagates along the z direction, the constant phase–space volume requires that

$$\delta z \delta p_z = \text{const} \tag{2}$$

R. Neugart • Institut für Physik, Universität Mainz, D-6500 Mainz, Federal Republic of Germany.

and the broadening of the z distribution—due to acceleration—corresponds to a narrowing of the momentum distribution. Photons emitted along the beam have a considerably reduced Doppler width. This remained unobserved because of inefficient excitation mechanisms, which cause intensity problems considering the extremely small solid angle required for the observation, and because of insufficient stability of the acceleration voltage.

With laser beams, the effect can be observed in absorption. This is the basis for collinear fast-beam laser spectroscopy. Among the Doppler-free techniques (described in Part A, Chapter 15 by W. Demtröder) it is the only one using linear absorption without velocity selection as in collimated atomic beams.

This offers the great advantage of optical resonance with the whole ensemble of atoms and yields an extremely high sensitivity, if the fluorescence is detected with a large solid angle. This advantage, and the fact that on-line isotope separators deliver many radioactive nuclides in the form of ion beams, have played the decisive role for the success of this technique in studies of atomic isotope shifts and hyperfine structures. On the other hand, the high-resolution spectroscopy on ions has been provided with a powerful experimental tool.

The roots of the experimental development can be traced back to the fast-ion-beam laser excitation performed by Andrä et al.,[5] to measure lifetimes or quantum beats more precisely than by the unspecific beam-foil excitation (see Part B, Chapter 20). Here, the Doppler effect was exploited in tuning the laser frequency to an atomic resonance by the variation of the intersection angle between the ion and laser beams. The collinear or superimposed beam geometry was first used by Gaillard et al.[6] Their Ba^+ beam from a standard 350-keV implantation unit had a large kinetic energy spread, and high resolution was achieved by probing the hole-burning effect of optical pumping with narrow-band laser light in a second interaction zone at a variable electrical potential. This elegant utilization of Doppler-tuning yielded the excited-state hyperfine structure in the $6s\,^2S_{1/2}-6p\,^2P_{3/2}$ resonance line of Ba II[7] with an accuracy similar to previous quantum beat experiments,[8] but it avoided the need for high laser power and good spatial resolution in the differential detection along the beam.

Isotope shift and hyperfine structure studies on short-lived radioactive nuclides depend crucially on the sensitivity and the conditions of on-line production in nuclear reactions. Here, the search for a workable concept of experiments with fast isotope-separated beams was initiated by Otten. It was Kaufman who pointed out that the narrow velocity distribution along these beams corresponds to nearly Doppler-free conditions. The experimental scheme, worked out in early 1975,[9] included cw-laser excitation with high resolution and efficiency of a neutralized fast atomic beam and detection of the resonance fluorescence. A detailed discussion of the basic features[10] was followed by an experimental study on beams of the stable isotopes ^{23}Na and ^{133}Cs,[11] and led to the wealth of new results on radio-active isotopes that will be discussed in Section 5.

However, the first experimental achievement of a narrow Doppler width in fast-beam spectroscopy had already been reported independently by Wing et al.[12] They performed the first precision measurements in the infrared vibrational–rotational spectrum of the simplest molecule HD$^+$ by crossing the ion beam with the beam from a single-mode CO laser at a very small intersection angle.

These pilot experiments were joined by the studies on the metastable Xe II, $5p^4 5d\ ^4D_{7/2}$ state in Xe$^+$ beams,[13,14] from which the transition to $5p^4 6p\ ^4P_{5/2}$ is covered by the spectral range of the then standard laser dye Rhodamin 6G.

The present contribution complements several chapters of the earlier volumes: it describes an additional technique of Doppler-free spectroscopy (Part A, Chapter 15 by W. Demtröder) for application with fast beams of ions or neutral atoms (Part B, Chapter 20 by H. J. Andrä), which has solved many problems of a systematic study of isotope shifts and hyperfine structures on short-lived nuclides (Part B, Chapter 17 by H.-J. Kluge). The need for a reliable isotope shift analysis adds weight to the discussion of Part A, Chapter 7 by K. Heilig and A. Steudel. The present chapter will be devoted exclusively to the spectroscopy of atomic systems, i.e., neutral atoms and singly charged ions, although the technique has many applications in molecular ion spectroscopy and reaction studies.[15-21]

2. Basic Concept and Experimental Realization

The crucial parameter for the resolution in fast-beam laser experiments is the kinetic energy spread of the ions. For low-energy beams ($E \leqslant 100$ keV) it is determined by the ion source and typically of the order 1 eV, if the acceleration voltage is well stabilized. To meet this value, plasma sources have to be operated with special care with regard to the pressure and potential distribution within the source. Surface ionization sources may almost reach the thermal energy spread of kT, but this requires very homogeneous and clean surfaces.

In estimating the velocity spread of an extracted beam, we assume an energy spread δE, not considering details of the distribution which strongly depend on the particular ion source. The dimensionless parameter $\beta = v/c$ is used for the velocity. For its dependence on the acceleration voltage U and the atomic mass m, it is sufficient to take the nonrelativistic expression

$$\beta = (2eU/mc^2)^{1/2} \tag{3}$$

The kinetic energy is $E = eU = \frac{1}{2}mc^2\beta^2$, and the conservation of the energy spread

$$\delta E = mc^2\beta\,\delta\beta \tag{4}$$

involves a narrowing of the velocity distribution, proportional to the increase in the average velocity. The Doppler width is $\delta\nu_D = \nu_0\,\delta\beta$ for an optical frequency ν_0, and, using Eqs. (3) and (4), we obtain

$$\delta\nu_D = \nu_0\,\delta E/(2eUmc^2)^{1/2} \tag{5}$$

Here we have assumed a parallel beam for which the energy spread is transformed into a purely longitudinal spread of velocities. As an example, we obtain $\delta\nu_D = 5$ MHz for $\delta E = 1$ eV in a 50-keV beam of medium-mass ions ($A = 100$, $\beta = 10^{-3}$) and an optical frequency of 5×10^{14} Hz (6000 Å). This may be compared with the natural linewidth of strong atomic transitions, which is of the order 10 MHz. That is to say, the homogeneous width matches the inhomogeneous Doppler broadening, and all atoms in the beam interact simultaneously with the laser light once the frequency is tuned to resonance. This provides the basic sensitivity for experiments with extremely low beam intensities, and in particular with radioactive isotopes. If necessary, an additional homogeneous power broadening can be achieved quite easily.

In practice, one has to work with beams of a finite divergence depending on the emittance and the required beam diameter. For the present case, formula (73) of Part B, Chapter 20 by H. J. Andrä gives a vanishing contribution to the linewidth, because it includes only first-order terms in the angular dependence of the Doppler shift. For a symmetric full angular divergence $\delta\theta$ around the beam axis ($\theta = 0$) the additional broadening becomes $\nu_0\beta[1 - \cos(\delta\theta/2)]$ or, using the lowest-order term of the expansion,

$$\delta'\nu_D = \tfrac{1}{8}\nu_0\beta(\delta\theta)^2 \tag{6}$$

In our example, for a beam divergence of 10 mrad, this gives an additional contribution of 6 MHz to the linewidth. It is thus obvious that the Doppler width in collinear geometry is fairly insensitive to beam divergences. Correspondingly, the Doppler-shifted line positions are insensitive to small changes in the beam shape. This is a very important feature for Doppler-tuning experiments. For comparison, an intersection angle of 10° between the laser and the ion beam of the assumed divergence of 10 mrad would give a linewidth of almost 1 GHz, which means that high-resolution experiments of this type require more parallel and stable beams.

Of course, the superimposed-beam geometry involves a considerable Doppler shift of the transition frequency ν_0 in the rest frame of the atoms. In the laboratory frame this is given by

$$\nu_\pm = \nu_0 \frac{(1-\beta^2)^{1/2}}{1 \mp \beta} = \frac{1 \pm \beta}{(1-\beta^2)^{1/2}} \tag{7}$$

or, to first order,

$$\nu_\pm \approx \nu_0(1 \pm \beta) \tag{8}$$

for parallel $(+)$ or antiparallel $(-)$ interaction, respectively. It is interesting to note that the velocity narrowing factor is proportional to the shift $\nu_0\beta$, i.e., the product of Doppler width and Doppler shift is constant, namely, $\nu_0^2 \, \delta E / mc^2$. This shift requires a rather precise knowledge of the beam energy and the atomic mass for experiments in which frequencies in the atomic rest frame are to be determined. On the other hand, the Doppler shift offers great advantages: (i) Simple postacceleration or deceleration can be used to tune the frequency seen by the fast-moving ions. In our example of a 50-keV beam with $\beta = 10^{-3}$, the total Doppler shift is about 500 GHz, and a tuning range of 100 GHz is obtained with a $\pm$10-keV change in kinetic energy. (ii) As long as the atoms are ionized, the Doppler shift can be changed along the beam, and the interaction between the laser light and the ions can be switched on and off, or changed in frequency.

Still, many collinear fast-beam experiments have been performed on neutral atoms. For this purpose, the ions are neutralized by charge transfer, preferably in passing them through an alkali vapor cell. The charge-transfer cross sections for the reaction

$$B^+ + A \rightarrow B + A^+ + \Delta E$$

where B^+ represents the fast ion and A the alkali atom, are generally of the order 10^{-15}–10^{-14} cm^2 and particularly large, if the energy defect ΔE is small (near-resonant electron transfer).[22-25] These collisions occur with negligible energy transfer of the order $(\Delta E)^2/eU$ to the target atoms, and therefore the velocity distribution in the neutralized beam remains nearly unchanged. This also means that all arguments and estimates for the resolution and sensitivity remain valid.

Moreover, the missing energy ΔE is converted completely into additional kinetic energy of the fast beam. If several states of the atom B are populated in the electron capture, the beam is composed of atoms with several discrete velocities that can be observed in the Doppler-shifted optical resonances.[11] These velocity spectra give direct access to the final-state distribution of the charge transfer process, provided that the atoms decay rapidly into the ground state from which they are excited by the laser light. The components are well resolved only for beam energies up to a few keV, where the change in the Doppler shift for a separation of a few eV is still larger than the natural linewidth. At higher beam energies, the effect may contribute to an additional line broadening.

If metastable states are among those favored by the resonance condition $\Delta E \approx 0$, the neutral beam contains a large fraction of metastable atoms. This gives access to transitions that are difficult to observe otherwise. It is a particularly important feature that many elements with high excitation energies, or uv resonance lines from the ground state, have metastable states with binding energies close to the ground state of the alkali atoms between -4 and -6 eV. The selective population of these states can be controlled by the choice of the alkali vapor target (Na: -5.1 eV, Cs: -3.9 eV) and the optical excitation can be induced by cw laser light in the visible.[26,27] Such procedures have contributed a lot to the versatility of the technique and they have been used extensively in the isotope shift and hyperfine structure studies on long sequences of unstable isotopes[27] (see Section 5).

Although the details of the experimental setup have to be tailored for the special purpose, we have a simple basic design that is slightly different for spectroscopy on ions and neutral atoms. This is outlined in Figure 1. The ion beams may be provided by any source that has a low energy spread,

(a)

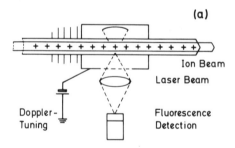

Ion Beam

Laser Beam

Doppler-Tuning

Fluorescence Detection

(b)

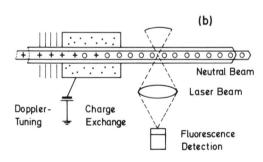

Neutral Beam

Laser Beam

Doppler-Tuning   Charge Exchange

Fluorescence Detection

Figure 1. Schematic setup for collinear laser spectroscopy: (a) The superimposed laser and ion beams pass through the interaction and fluorescence detection region at a variable potential. (b) Neutralized beams are post-accelerated or decelerated by a potential at the charge-exchange cell. The beams may travel either in the same or in opposite directions.

and a sufficiently stable acceleration voltage (typically between 1 and 100 kV). For ion spectroscopy (a), this beam, merged with the laser beam, is transmitted through the observation chamber which forms a Faraday cage at a variable potential. The setup for neutral atoms (b) includes a charge-exchange cell[28,29] in which the alkali vapor pressure of about 10^{-3} Torr is maintained by heating. The Doppler-tuning potential is applied to this cell and the observation chamber has to be put as close as possible to avoid optical pumping to nonabsorbing states.

Omitted in Figure 1 is all ion optics, and in particular the deflection into the laser beam axis, except for the acceleration/deceleration system in front of the observation chamber or charge-exchange cell. The latter has to minimize lens and steering effects of the tuning potential. It usually consists of a set of cylindrical electrodes, which produce a smoothly varying potential along the beam axis.

The collection of fluorescence photons can be performed by lenses, mirrors, light pipes, or a combination of them, and the design has to be optimized for the special conditions of the experiment. In particular, if the atomic spectrum excludes the rejection of laser light background by optical filters, one has to find a compromise between large solid angle and low acceptance for the scattered light.

A few examples of practical versions of this basic setup will be given in the review of experiments. Most of them use clean mass-separated beams, although the laser-induced fluorescence is selective for the element and the resonances of different isotopes are well separated by the Doppler effect.

3. Experiments Based on the Doppler Effect

The considerable Doppler shift in all collinear-beam experiments has opened up a few general applications beyond the spectroscopy of particular atomic or molecular systems. The scope of such applications ranges from simple beam velocity analyses to precision experiments related to metrology or problems of fundamental physics. These latter include the calibration of high voltages and measurements of the relativistic Doppler effect, for which the atomic transition frequency provides an intrinsic clock.

3.1. Beam Velocity Analysis

The measurement of the velocity distribution in (low-energy) ion beams is conventionally performed using electrostatic analyzers or the time-of-flight technique which also works with neutral beams. Similarly to the time of flight, the Doppler shift essentially measures the beam velocity v. However, the measured time interval decreases, whereas the Doppler shift increases proportionally with v. This gives decisive advantages of the Doppler shift technique for higher beam energies.

A more quantitative comparison is easily performed by use of Eq. (4). For time-of-flight measurements we assume that a fixed time interval δt can be resolved at the detector placed at a distance L from a chopper slit. Sufficiently fast choppers consist of electric deflection plates for the initial ion beam and a slit defining the entrance of the drift space, e.g., after a collision chamber where neutral products may arise from specific reactions. Then the energy resolution is given by

$$\delta E = \frac{(2E)^{3/2}}{m^{1/2}} \frac{\delta t}{L} \tag{9}$$

If the chopper (instead of the detector) determines the time resolution, it can be shown[30] that for an optimized electric field amplitude of the deflector

$$\delta E = 2\sqrt{2}E \frac{s}{(DL)^{1/2}} \tag{10}$$

where s is the width of the slit and D the distance of the deflector from the slit. A practical example for helium and $L = 2.5$ m is given in Ref. 30. For $\delta t = 7.5$ nsec, δE is increasing from 1 to 7 eV for beam energies E between 800 eV and 3 keV. This is well consistent with Eq. (9).

For the Doppler shift measurement we obtain

$$\delta E = (2Emc^2)^{1/2} \frac{\delta \nu_0}{\nu_0} \tag{11}$$

where $\delta \nu_0$ is the natural (or power-broadened) linewidth of the transition involved. For the quoted example this would give $\delta E \approx 0.1$ eV, if the atoms could be excited by an optical transition of typically $\delta \nu_0 / \nu_0 = 4 \times 10^{-8}$. This is possible only for the metastable $1s2s$ states, whereas the helium atoms in the ground state are inaccessible, because their resonance absorption lies in the far uv. We note here a general difference of both techniques: The time-of-flight spectra include all atoms in the beam, while the optical spectra only show the part of the atoms that are in the absorbing level. Therefore, apart from energy resolution arguments, both methods can be complementary to each other.

Coming back to our comparison we find that beyond mass $A \approx 40$ (with all other parameters unchanged) the time-of-flight resolution will become superior to the optical resolution. But the energy dependence shows that for beam energies above a few keV the Doppler shift measurement becomes definitely more favorable. Moreover, the choice of a weak transition with small natural linewidth can give another gain in resolution at the expense of fluorescence intensity. The basic resolution can be exploited in studies of the primary beam velocity distribution from the ion source. Fluctuations arising from instabilities of the acceleration voltage or the plasma potential may be eliminated by regulating the velocity with the Doppler-tuned resonance signal. This was demonstrated by Koch,[31] who excited a neutralized ^{4}He beam containing $1s10p$ Rydberg atoms to a higher Rydberg level by a CO_2-laser line and used the ionization in a microwave cavity to produce the reference signal.

In studies of secondary collision processes and their energy-loss spectra, it is clear that the resolution is limited by the energy spread of the original ion beam. An example[32] is given in Figure 2, which shows the energy-loss spectrum of the charge transfer from a 7-keV Cs^+ beam to Cs vapor, on top of the hyperfine structure of the transition $6s\,^2S_{1/2}$–$7p\,^2P_{3/2}$ at 4555 Å. The energy resolution of 0.7 eV is mainly determined by the ion source. The small satellite peaks at low vapor pressure are due to the nonresonant channel

$$Cs^+ + Cs(6s) \to Cs(6p) + Cs^+ - 1.4\,eV$$

With increasing vapor pressure these peaks become stronger, and additional equidistant peaks become observable. This is ascribed to secondary excitations, according to

$$Cs(6s) + Cs(6s) \to Cs(6p) + Cs(6s) - 1.4\,eV$$

From such a sequence of curves, one can extract the branching ratio of the primary electron capture into $6s$ and $6p$, as well as the cross section for collisional excitation in the fast atomic beam. Similar examples of state-resolved charge transfer are given in Ref. 11 for Na^+ on Na and K and in Ref. 27 for Yb^+ on Na. The time-of-flight technique, on the other hand, has yielded the metastable composition of a $2\,^3S$ He beam formed by charge exchange on alkali atoms.[33]

3.2. High-Voltage Measurement and Calibration

So far we have disregarded the excellent internal velocity calibration which is provided by the transition or the laser frequency. It can be helpful for precision spectroscopy experiments, but also used to calibrate voltages beyond a few keV with unprecedented accuracy. Voltages up to a few hundred keV are measured by compensated dividers, consisting of wire resistors, with relative uncertainties down to 3×10^{-5}. Commercial instruments reach the 1×10^{-4} level.

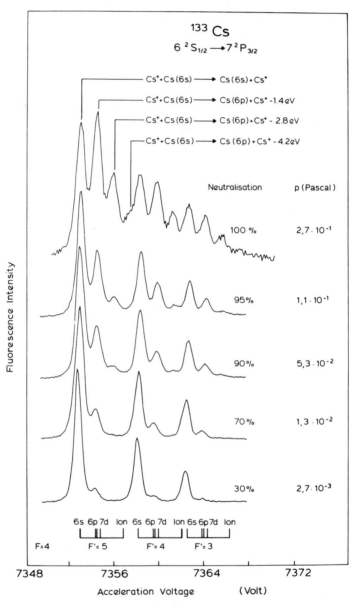

Figure 2. Energy-loss spectrum of neutralized Cs atoms, on top of the hyperfine structure of
^{133}Cs $(6s\,^{2}S_{1/2}\text{-}7p\,^{2}P_{3/2})$, for different Cs vapor pressures in the charge-exchange
cell: (1) resonant charge transfer, (2) nonresonant charge transfer and collisional
excitation $6s\text{-}6p$, (3) and (4) multiple excitation $6s\text{-}6p$.

In this context there have been two approaches of absolute beam energy measurement, (i) by accurate measurements of the Doppler-shifted resonance frequencies (or wavelengths) with parallel and antiparallel beams, and (ii) by using two-photon interactions of a three-level system in counterpropagating laser fields. As atomic masses are known very precisely, we do not differentiate between velocity and kinetic energy.

(i) The straightforward solution simply consists in a measurement of the Doppler shift. The limited use of fixed-frequency lasers with known wavelength has been circumvented by the choice of ir transitions between Rydberg states of which the lower is populated in a charge-transfer reaction.[31] Still, this gives rather poor accuracy, even if both the laser and the Rydberg transition frequencies are known precisely. Optical transition frequencies from tunable dye lasers can be measured to about 10^{-8} using fringe-counting wavemeters[34,35] calibrated against an I_2-stabilized He–Ne laser. This precision corresponds roughly to the linewidths of both the laser and the atomic transition. From a measurement of the Doppler-shifted resonance excitation frequencies for parallel and antiparallel laser and atomic beams and from Eq. (7) it follows that the beam velocity

$$\beta = \frac{\nu_+ - \nu_-}{\nu_+ + \nu_-} \tag{12}$$

is determined by the two wave numbers ν_+ and ν_-. Assuming a sufficiently monoenergetic beam one obtains an accuracy $\delta\beta = 2^{-1/2}\,\delta\nu/\nu_0$, independent of β. Since the relative accuracy increases with β, one should use light atoms. Here, the outstanding example for laser spectroscopy is the Balmer-α line of atomic hydrogen. In a pilot study[26] the metastable $2s$ state was populated by near-resonant charge transfer with cesium, and the Balmer-α excitation was detected as a flop-out signal in the Lyman-α radiation induced by an electric quenching field. For a 7.5-keV deuterium beam the width of the laser-induced resonance $2s$–$3p$ was 234 MHz, owing to the 2.3-V energy spread of the ion source and the hyperfine structure. This modest resolution already corresponds to an energy determination to 2×10^{-5}, if one assumes that the line centers can be determined within 10% of the width. As the natural linewidth is 29 MHz, a considerable gain in resolution is expected[36] from the installation of an electrostatic velocity filter which can supply beams of $\delta E = 50$ meV. With a resolved hyperfine structure, such a setup should reach an uncertainty of 1×10^{-6} in the beam energy. The precision may also be interesting for a redetermination of the Rydberg constant, of which the latest values are given with an error of 1×10^{-9}.[37,38]

(ii) An alternative possibility to determine beam energies[39] is based exclusively on atomic transition frequencies and special features of a particular atomic level scheme. The simultaneous resonance of the fast atoms with the direct and retroreflected laser light on two different transitions (resonant two-photon absorption, V or Λ configuration) determines the velocity to

$$\beta = \frac{\nu_2 - \nu_1}{\nu_2 + \nu_1} \tag{13}$$

where ν_1 and ν_2 are the transition frequencies in the rest frame of the atoms. Small detunings from the simultaneous resonance can be measured sufficiently accurately in laser or voltage scans and used to correct formula (13). However, most of the atomic transition frequencies have to be known more precisely than tabulated. They can be improved by new laser measurements. A more decisive drawback of this technique is the small voltage range covered by a specific atomic level scheme. Poulsen[39] gives a list of candidates for calibration points over a large energy range, from 50 keV up to several MeV. At such high energies the disadvantage of heavier atomic systems (ranging from Ne to U) is unimportant for the relative precision. An advantage is the possibility to set up secondary voltage standards based exclusively on atomic parameters.

The high precision of beam energy measurements does not necessarily correspond to a high precision of the voltages. This requires also a detailed knowledge of the ion source parameters, and in particular the ion formation at a well-defined potential. With some precaution, a conversion from the energy to the voltage scale should be possible within less than 1 eV, whereas for higher absolute accuracy even surface ionization requires a careful control of contact potentials.

3.3. Relativistic Doppler Shift

The measurement of absolute beam velocities, or the calibration of voltages, is already quite sensitive to the relativistic quadratic term in the Doppler shift formula (7). In fact, this "transverse" Doppler shift, caused by the time dilatation factor $\gamma = (1 - \beta^2)^{-1/2}$, was first observed in the spectral lines of fast-moving hydrogen atoms from a 30-keV beam of H_2^+ ions, viewed along and opposite the direction of propagation.[40,41] Comparable accuracy in the percent range was also achieved in Mössbauer experiments,[42] and more recently the time dilatation factor on the muon lifetime was determined to 1×10^{-3}.[43]

New laser experiments with fast atomic beams could simply measure the resonance frequencies in the laboratory frame

$$\nu_\pm = \nu_0 \gamma (1 \pm \beta) \tag{14}$$

for parallel and antiparallel beam–laser interaction, which gives

$$\nu_+ + \nu_- = 2\nu_0\gamma \tag{15a}$$

$$\nu_+ - \nu_- = 2\nu_0\gamma\beta \tag{15b}$$

For H_α spectroscopy ($\nu_0 = 4.57 \times 10^{14}$ Hz) on a 100-keV hydrogen beam ($\beta = 1.5 \times 10^{-2}$) one could imagine an uncertainty of the order 10^{-4} in the second-order shift of 50 GHz ($\gamma - 1 \approx \frac{1}{2}\beta^2 = 1.1 \times 10^{-4}$) from a state-of-the-art wavemeter.[34,35,44] However, this contribution appears on top of a huge first-order shift, which makes the measurement extremely sensitive to beam energy fluctuations. Nevertheless, Juncar et al.[45] have shown that an accuracy comparable to the best previous experiments can be achieved with moderate experimental effort.

The alternative is an elimination of the first-order shift, which also helps to avoid the clumsy measurement of absolute wavelengths for the determination of a relatively small difference. Such an alternative had already beeen sought in the first-generation laser experiment by Snyder and Hall.[46] A beam of fast neon atoms in the metastable $3s$ $[3/2]_2$ state was transversely excited to $3p'$ $[1/2]_1$ in a standing-wave laser field. The saturation peaks, observed in fluorescence, are free from first-order shifts, provided that the optical alignment is perfect. Even a small deviation 2α from perfect retroreflection introduces a significant residual first-order shift $\nu_0\alpha\beta$. Therefore, an improvement in the transverse shift far beyond the reported 6×10^{-3} appears very difficult.

The advantages of the collinear geometry and a purely relativistic Doppler effect can be combined in a resonant two-photon absorption experiment.[47] The two photons are absorbed from the direct and the retroreflected laser beam of frequency ν_L, and a few atomic systems[39,47] have intermediate states that can be tuned into resonance at a beam velocity selected by

$$\nu_1 = \nu_L\gamma(1 + \beta) \tag{16a}$$

$$\nu_2 = \nu_L\gamma(1 - \beta) \tag{16b}$$

where ν_1 and ν_2 are the frequencies connecting the intermediate level with the lower and upper ones. Figure 3, taken from Poulsen and Winstrup,[47] shows such an example in the energy-level diagram of NeI, and Figure 4 compares the narrow two-photon resonance in a laser-frequency scan with the broad detuning curve for the intermediate level in a voltage scan. The width of 4.7 MHz for the two-photon resonance is mainly composed of the laser bandwidth, second-order Doppler broadening, the natural width of

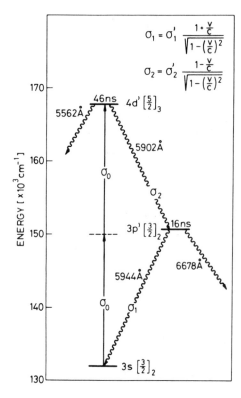

Figure 3. Excitation scheme for resonant two-photon spectroscopy on Ne I with a beam energy of 119.141 keV. (From Ref. 47.)

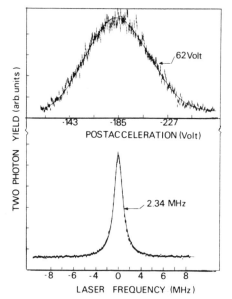

Figure 4. Two-photon resonance in Ne I, detected in the decay $4d'$-$3p$. The upper part shows the detuning of the intermediate level from the resonance condition as a function of the postacceleration voltage. (From Ref. 47.)

the upper level, and some power broadening. Poulsen and Winstrup also demonstrate that the ac Stark shift vanishes at resonance, which is in accordance with theoretical predictions.[48]

A recent upgraded version[49] of this experiment led to a new precision measurement of the relativistic Doppler shift in a 120-keV neon beam. The resonantly enhanced two-photon absorption in the beam was compared with the nonresonant two-photon absorption of neon atoms in a low-pressure discharge cell. With the very precisely known transition frequencies ν_1 and ν_2 [44] there was no need to measure the beam velocity: The two-photon resonances in the cell at $\nu_C = (\nu_1 + \nu_2)/2$ and in the beam at $\nu_B = (\nu_1\nu_2)^{1/2} = \gamma\nu_C$ were induced by two frequency-stabilized laser beams whose beat frequency of about 3 GHz was measured on a fast photodiode. This gives the experimental Doppler shift of 3235.94 (14) MHz, in excellent agreement with the prediction 3235.89 (5) MHz of special relativity. All possible error contributions are considerably smaller than the statistical error of 0.11 MHz for the beat frequency, obtained with a width of 3 MHz (FWHM) for both resonances and extrapolation to zero of the neon pressure in the cell. In particular, the errors for a slight angular misalignment θ give negligible frequency shifts of the order $\nu_0\beta\theta^2$. A significant improvement over the present accuracy of 4×10^{-5} would require higher-energy beams of excellent velocity stability and better absolute values of the transition frequencies.

4. Spectroscopic Studies

Many of the spectroscopic studies were performed to demonstrate the capability of the technique and of a number of variants which are specific for the combination of laser spectroscopy with fast beams of ions or atoms. An example has already been discussed in Section 3.3: Resonant two-photon excitation becomes possible by adjusting the Doppler shifts for interaction with the direct and retroreflected laser beam to the atomic transition energies. Other features include the preparation of otherwise inaccessible atomic states, the separation of hyperfine structures from different isotopes by the Doppler shift, or the observation of time-resolved transient phenomena along the beam. The extensive research on nuclear moments and radii from the hyperfine structure and isotope shift constitutes a self-contained program, which will be discussed separately in Section 5.

The experiments are not discussed systematically, as their scope ranges from the development of the instrumentation to the study of particular atomic systems. This indicates that during the first decade the technique has provided a playground for the experimentalist rather than establishing a standard procedure of atomic structure studies.

4.1. Experimental Details

Many experiments were performed with a setup which is similar to the one outlined in Figure 1. The beam energy ranges from 1 to 200 keV, sometimes given by the particular layout of an existing machine or isotope separator. Even though the essential parameters like Doppler shift and narrowing as well as the charge-transfer cross sections depend on the ion velocity, the choice of the beam energy is not very critical for most applications. Generally, the Doppler broadening from the ion source energy spread drops with increasing beam energy, while the stability and precision measurement of the acceleration voltage becomes more important. The charge-transfer neutralization at higher beam energies gives a broader distribution over the final states with decreasing total cross sections only for the near-resonant cases.[22-25] On the other hand, the broader energy spectrum of the neutral products is again compensated by the improved narrowing.

Surface ion sources are used for elements with ionization energies up to 6 eV. With clean surfaces they give beams of small energy spread and a well-defined starting potential for the ions. ($\delta E < 0.5$ eV were achieved for Rb^+ and Cs^+ from a tungsten surface.[50,51]) Various types of discharge ion sources have been used for elements with higher ionization energy. Their energy spread depends crucially on the operating conditions and can be reduced to a few eV for most standard-type sources, by minimizing the potential gradient over the effective source volume. In general, the optical linewidths lie between 10 and 100 MHz, and only for very light elements or narrow lines is the residual Doppler width an important drawback for the resolution.

A calibration problem for Doppler tuning arises from the poorly defined absolute beam energy, due to the potential of the plasma surface. Several techniques have been applied to circumvent this problem: (i) A calibrated laser-frequency scan is used to compensate a certain Doppler detuning of the resonance, and with known $\Delta\nu$ and ΔU the unknown beam energy eU—in first-order approximation

$$eU = \frac{e^2}{2mc^2}\nu\frac{\Delta U}{\Delta\nu} \tag{17}$$

—can be eliminated.[52] (ii) The absolute Doppler shift can be measured by comparison with the atlas of the I_2 or Te_2 absorption spectra[53] or by a precision wavemeter. (iii) The measurement at different beam energies, and with parallel and antiparallel laser and ion beams, can be used to eliminate an offset voltage from the relations (7) and (3), if the main acceleration voltage is measured precisely.

This can be checked independently by the remeasurement of a well-known and sufficiently large atomic energy difference. (iv) With calibrated laser-frequency tuning for parallel and antiparallel beams the constant beam energy essentially cancels out in the averaging of the measured frequency intervals.[54] For example, the isotope shift $\delta\nu^{AA'}$ between the two isotopes A and A' is given to first order by the expression

$$\delta\nu^{AA'} = \tfrac{1}{2}[(\nu_+^{A'} + \nu_-^{A'}) - (\nu_+^A + \nu_-^A)] \tag{18}$$

A different combination of the measured frequencies

$$\frac{\nu_+^A - \nu_-^A}{\nu_+^{A'} - \nu_-^{A'}} = \frac{m_{A'}^{1/2}}{\left(1 + \dfrac{\delta\nu^{AA'}}{\nu_0}\right) m_A^{1/2}} \tag{19}$$

gives the atomic mass of the isotope A' from the known mass of the isotope A. Although such a measurement can be fairly precise (of the order 10^{-6}), the practical application to unstable isotopes may play a minor role, as there are new developments of rf mass spectrometers especially designed for this application.[55,56]

For historical reasons, a few experiments were performed with beams intersecting at a small angle. This geometry avoids the optical pumping outside the observation region at the expense of resolution. The standard solution of this problem is the Doppler switching, i.e., the tuning into resonance within an insulated Faraday cage at nonzero potential. As this works only with ions, the fluorescence observation on neutralized beams should be close to the charge-exchange region. To some extent the excitation and optical pumping in front of the observation volume can be avoided by Zeeman-detuning of the participating energy levels,[53] or exploited to achieve strong signals by the use of two lasers.[57]

As the spectroscopy on ions often involves high-lying metastable states, the metastable population in the ion source is a crucial point. It is difficult to measure, and the estimates for various ion sources scatter between 10^{-6} and 10^{-2}.[58–60] A more quantitative systematic investigation of the best conditions for metastable beam production would certainly be useful. The situation is better under control for neutral atoms whose metastable states are populated by specific charge-transfer reactions. Here the population can easily reach a few tenths of the total beam intensity.

The commonly used fluorescence detection has to cope with a considerable background from scattered laser light. In fact, most experiments avoid detecting on the excitation wavelength and use the decay branch to a third state, selected by optical filters. If such a decay is weak or does not exist, the background has to be rejected by careful shaping of the laser beam profile and a suitable detection geometry that eliminates the scattered light with blackened diaphragms.[11] It has been shown in many experiments on weak beams of radioactive isotopes that a good photon-collection efficiency is compatible with a rather low background (see Section 5 and in particular Ref. 61). The Doppler shift between laser light and fluorescence can be discriminated by a narrow-band interference filter,[58] but this implies a very small angular acceptance of the detection system.

4.2. Laser-rf and Related Techniques

In classical atomic spectroscopy the hyperfine structure of ground or metastable states is measured with very high precision by rf techniques like atomic beam magnetic resonance (ABMR), which has no analog in ion spectroscopy. Optical pumping-rf experiments have been performed only on a few alkalilike ions stored in a buffer gas[62] or an ion trap (cf. Part C, Chapter 5 by G. Werth). On the other hand, the laser interaction with a fast ion beam can be used to detect rf resonances. The standard setup (Figure 5) corresponds to a Rabi-type ABMR experiment in which the A magnet is replaced by an optical pumping zone and the B magnet by a fluorescence detector. Both the A and B zones can be tuned into optical resonance by an electrical potential on the Faraday cages. A coaxial rf transmission line gives a sufficiently uniform rf field with a transit time of a few microseconds. Experiments of this type were performed on the hyperfine structure of a few low-lying odd-parity states in U II[63] and the metastable $5d\ ^2D_{3/2}$ state in Ba II,[64] while earlier approaches on $5d\ ^4D_{7/2}$

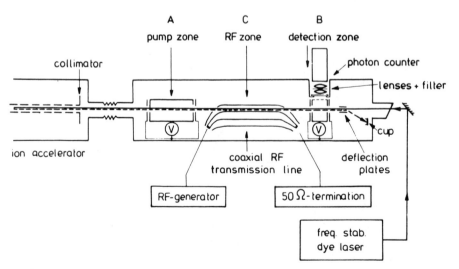

Figure 5. Experimental setup for fast-beam laser-rf double resonance spectroscopy. (From Ref. 64.)

in Xe II[59] and $2\,^3S_1$ in Li II[65] used intersecting beams at small or right angles. In all cases the setup is rather similar to the original proposals for an optical pumping rf technique by Kastler[3] and to the ABMR-LIRF technique on thermal atomic beams (cf. Part A, Chapter 10 by S. Penselin). The ultimate resolution limit with fast beams is given by the transit-time broadening of the order 100 kHz.

Of course, the possible combinations of laser–rf double resonance include all schemes that have been applied to thermal atomic beams. Thus for neutral atoms one can apply the laser and rf fields simultaneously as in conventional optical pumping experiments. This has been studied in the Na–D lines,[66] where Zeeman transitions in the ground-state hyperfine structure were recorded. Because of the well-defined interaction time of the atoms with the rf field the line shape was in very good agreement with the Majorana formula. The monoenergetic beams even allow the observation of Zeeman structures in a static-field experiment[66,67]: The alignment of the Na I ground state or the Ba II metastable states by polarized laser light reduces the fluorescence intensity, and the free Larmor precession in a static magnetic field gives rise to a modulation of this fluorescence as a function of the field (Figure 6). Both the rf transition frequency and the Larmor fringes essentially measure the g_F factor $g_F = g_J/2I + 1$ in the low-field limit, which may help to obtain unambiguous nuclear spins (cf. Section 5). Similarly, this procedure can be applied to diamagnetic atoms with $^1S_0$ ground states where the free nuclear precession is observed in fields of a few kilogauss.[68] Such experiments are important for radioactive elements such as radium, where the magnetic moments in a long sequence of isotopes (cf. Section 5) can be calibrated by a recently performed direct g_I measurement on ^{213}Ra and ^{225}Ra.

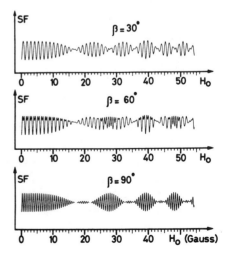

Figure 6. Fluorescence signal versus static magnetic field applied between the optical pumping and detection regions, from the $D1$ transitions $^2S_{1/2}$, $F = 2 - {}^2P_{1/2}$, $F' = 1$ in Na I. The relative contributions of the single and double Larmor frequency, ω_L and $2\omega_L$, depend on the angle β between the alignment axis and the observation. (From Ref. 66.)

4.3. Nonlinear Spectroscopy

The inherent resolution of collinear-beam spectroscopy is still limited by the residual Doppler broadening. In beams with a broad velocity distribution the labeling of one velocity class by optical pumping, probed in a second Doppler-tuning zone, was exploited already before narrow Doppler widths were achieved.[6,7,69,70] The complete elimination of the first-order Doppler effect in resonant two-photon absorption on Ne I has been discussed in Section 3.3, in connection with a precision measurement of the relativistic Doppler effect. A similar experiment was performed on In I, where the $29p$ Rydberg state was excited from $5p\,{}^2P_{3/2}$ via $6s\,{}^2S_{1/2}$ and detected by field ionization.[71] The linewidth caused by the laser jitter can be reduced to the transit-time limit of a few hundred kilohertz.

Classical saturated absorption experiments on fast beams require different laser wavelengths to saturate and probe the same transition. However, a three-level system in V or Λ configuration can be realized to use the same direct and retroreflected laser beam[72] interacting with the velocity class β under the condition

$$\beta = \frac{\nu_2 - \nu_1}{\nu_2 + \nu_1} \qquad (20)$$

where ν_1 and ν_2 are the two atomic transition frequencies. This situation is shown in Figure 7. Simultaneous resonance of the atoms on both transitions gives a reduction of population of both the excited levels of the V configuration. The Λ configuration includes two metastable states, and the laser field burns holes into their velocity distribution, which appear as peaks on the other side. At simultaneous resonance the strong coupling of all three states gives rise to increased emission. In both configurations the saturation signals are Doppler-free to first order, with a homogeneous linewidth given by $1/2(1/\tau + 2/T)$, where T is the transit time and τ is the excited-state lifetime. Experimental curves for the V case in Ne I $3s\,[3/2]_2 - (3p'\,[3/2]_2,$ $[3/2]_1)$ are shown in Figure 8, which demonstrates that the first-order Doppler effect is eliminated in the hole-burning dip, while the Doppler-broadened structure is shifted by postacceleration. The Ne I experiment[72] yielded precision wave numbers of three transitions between the $2p^5\,3s$ and $2p^5\,3p'$ configurations of Ne I. Similarly, the Λ case was realized in the Ca I transitions $(4s4p\,{}^3P_2,\ {}^3P_1)-4s5s\,{}^3S_1.$[72] Another detailed study of this system with two copropagating laser fields and the involved coherent two-photon processes[73] gave a quantitative description of the observed dispersive and absorptive line shapes.

$$\sigma_1 = \sigma_L \ \gamma(\beta)(1-\beta)$$
$$\sigma_2 = \sigma_L \ \gamma(\beta)(1+\beta)$$

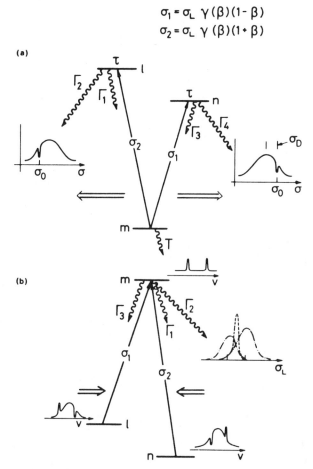

Figure 7. V and Λ configurations for resonant three-level saturation spectroscopy. (From Ref. 72.)

4.4. Transient Phenomena

The observation of transient effects is closely connected to the feature of Doppler switching by sudden changes of the ion velocity. Applications include cascade-free lifetime and quantum beat measurements by observation of the free decay after a short excitation region. Following the first experiments by Andrä,[5] the alternative to produce a short excitation pulse by crossing the ion and laser beams has been applied extensively (cf. Part B, Chapter 20 by H. J. Andrä).

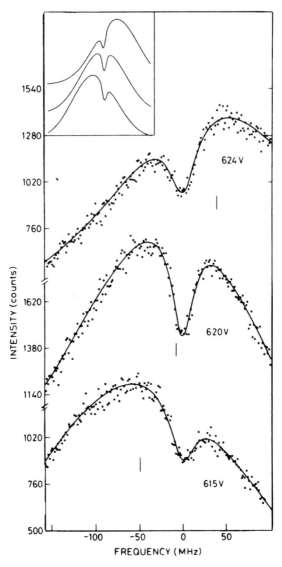

Figure 8. Saturation signal for the *V* configuration in Ne I, with the broad Doppler profile shifted by different postacceleration voltages around the exact resonance at 61.763 keV. (From Ref. 72.)

Silverans *et al.*[74] studied interference effects between ground (or metastable) and excited states in a setup that is analogous to Ramsey's separated microwave field method. In comparison with other optical experiments, the interference conditions are easy to establish with fast ion beams, and the observed fringes are consistent with a Fourier analysis of the evolution in time of the laser frequency probed by the ions.[75]

4.5. Spectroscopic Results

It is not surprising that most results in atomic spectroscopy were obtained on singly charged ions which are difficult to prepare for the usual Doppler-free techniques on thermal samples. On the other hand, the fast-beam technique has certain advantages also on neutral atoms, such as the availability of metastable beams, the sensitivity, and the Doppler-tuning.

4.5.1. Few-Electron Systems

Experiments of rather fundamental importance are those performed in the few-electron systems for which theory can provide accurate predictions of spectroscopic data. The first collinear-beam experiment on the simplest molecule HD^+,[12] and the possibility of a redetermination of the Rydberg constant on H atoms[26] point to this category of experiments. Also the two-electron atom He and the members of its isoelectronic sequence Li^+, Be^{++}, etc. have been treated theoretically by advanced perturbation methods, and the $2\,^3P$ fine structure of He is the source of one of the most accurate values of the fine structure constant α.[76] While the He I spectrum, and in particular the fine and hyperfine structures of the low-lying triplet states are known with high precision, the spectroscopy on the He-like ions was handicapped by the poor resolution achieved under plasma conditions or with beam-foil excitation. Quantum beat measurements, which have a limited scope in the scale of atomic frequencies, yielded first precise· hyperfine structures of the $1s2p\,^3P$ levels of Li II (cf. Part B, Chapter 20 by H. J. Andrä). In a Doppler-tuned ion beam laser experiment Fan *et al.*[58,77] resolved the complete hyperfine structure of the triplet $2\,^3S_1$–$2\,^3P_J$ at 5485 Å for both stable isotopes ^6Li and ^7Li, using a beam of metastable Li^+ ions from a sputter source. The linewidth was 450 MHz, and an obvious improvement in the beam energy spread would make this experiment competitive with a more recent one[78] using saturation spectroscopy on a low-energy (200 eV) Li^+ beam. In contrast to earlier measurements all three approaches confirm the many-body calculations[79] which were believed to be accurate within the present experimental errors of 10^{-4}.

An absolute wavelength measurement on the same transitions $2\,^3S_1$–$2\,^3P_J$ was performed by Holt *et al.*[60] to improve the experimental value of the differential Lamb shift (cf. Part A, Chapter 4 by A. M. Ermolaev). As in fast-beam experiments the absolute wave numbers depend critically on the knowledge of the Doppler shift,

$$\nu_0 = \gamma(1 \pm \beta \cos \theta)\nu_{a,b} \tag{21}$$

for a small intersection angle θ which was chosen for technical reasons, β from the beam energy was largely eliminated by the use of two counterpropagating laser beams with the frequencies ν_a, ν_b. Stabilizing these frequencies on suitably chosen I_2 absorption lines required a small Doppler detuning $|\beta_1 - \beta_2| \ll \beta$, but still the expression for the wave number

$$\nu_0 = \frac{\gamma_a\gamma_b(\beta_a + \beta_b)}{(\gamma_b\beta_b/\nu_a) + (\gamma_a\beta_a/\nu_b)} \tag{22}$$

is quite insensitive to an unknown offset in the absolute beam energy. The results, after elimination of the hyperfine structure, are in good agreement with the theoretical values for both the fine structure[80] and the Lamb shift.[81]

More recently, the extensive studies of the system $2\,^3S_1 - 2\,^3P_J$ in $^6Li^+$ and $^7Li^+$ by the Heidelberg group[65,78,82-84] have confirmed and improved all the earlier results from collinear laser spectroscopy—including the absolute wavelengths, the fine and hyperfine structures, and the isotope shift. These experiments use the alternative technique of crossed-beam saturation spectroscopy combined with microwave resonance, taking advantage of the long interaction times in a low-energy Li^+ beam. Still, it seems that the considerable improvements are rather due to the skilful design of the experiments than due to a principal advantage of the technique.

4.5.2. Transition Energies and Lifetimes

Only in exceptional cases can precise atomic transition energy measurements add relevant information to the tabulated energy levels.[85,86] Examples were given in connection with the few-electron system Li^+,[60] and the Ne I spectrum[72] which has a considerable importance as a secondary wavelength standard and for the precision measurement of the relativistic Doppler shift.[49] In complex spectra the selective excitation may facilitate the classification of lines, in particular if combined with lifetime measurements. This has been demonstrated by a fairly systematic study[87] of U II transitions from the low-lying odd configurations f^3s^2, f^3ds, and f^3d^2. The excited-state lifetimes were measured by cw-laser modulation. The access to long-lived excited states by charge transfer collisions gives interesting possibilities to explore the poorly known parts of a level scheme.

For highly charged ions it has been pointed out[88] that the charge-transfer electron capture leads to Rydberg levels whose separation energies are in the optical spectral range.

Lifetime measurements where the conventional beam-foil is replaced by laser excitation are discussed by H. J. Andrä (Part B, Chapter 20).

A special wavelength measurement was recently performed in the second, $7s-8p$ resonance doublet of Fr I,[89] of which the first optical lines, $7s-7p$, were discovered only a few years ago.[90,91] Here, the motivation for a collinear-beam experiment was the sensitivity to locate the narrow resonances within the estimated range of 100 Å, on a beam of 10^9 atoms/sec for the isotope ^{212}Fr, which has a half-life of 20 min. The francium ion beam was produced at ISOLDE from a thorium carbide target by spallation with 600-MeV protons, and neutralized by charge exchange with cesium. The optical spectroscopy on the heaviest alkali element provides a stringent test of the present theory of heavy atomic systems including strong relativistic contributions.[92,93]

4.5.3. Hyperfine Structure and Isotope Shift

Following the experimental tests on ions which have accessible transitions within the range of the first cw laser dye Rhodamin 6G, a series of experiments yielded results on the hyperfine structure and isotope shift of Xe$^+$ $(5p^4 5d\ ^4D_{7/2} - 5p^4 6p\ ^4P_{5/2})$[14,59,94,95] and Ba$^+$ $(5d\ ^2D_{3/2,5/2} - 6p\ ^2P_{3/2})$.[96,97,52,98,64] Of these, the alkalilike Ba II spectrum is of special interest for the theoretical or semiempirical description of the atomic structure. The considerable field shift in the $5d-6p$ transitions[96,52] points to a large screening of the core s shells by the $5d$ electron in agreement with theoretical calculations.[99] A precision measurement[98] even reveals a slight spin dependence of this effect within the doublet. Correspondingly, the $5d$ hyperfine structures in $^2D_{3/2}$ and $^2D_{5/2}$ are distorted by a nonvanishing contact term from core polarization. Precision measurements of these hyperfine structures by laser–rf spectroscopy,[64,100] including a thorough investigation of possible light shifts,[101] are sensitive to the second-order corrections due to a mutual perturbation of the fine structure levels. They indicate a nonvanishing hyperfine anomaly for the $^2D_{3/2}$ state. Many-body perturbation theory including correlation contributions reproduces rather well the experimental A-factors and can be used to determine independent values of the nuclear quadrupole moments from the B-factors. They agree well with the adopted semiempirical values including the Sternheimer correction.[61]

A more transparent situation is met in the resonance doublet $6s-6p$ of Ba II where the isotope shift is dominated by the direct contribution of $6s$ in the $^2S_{1/2}$ ground state. The experimental observation of different field shifts in the transitions to $^2P_{1/2}$ and $^2P_{3/2}$ [102] gives direct access to the relativistic $p_{1/2}$ electron density at the nucleus which is in accordance with a semiempirical estimate.

The motivation for hyperfine structure and isotope shift studies in the isoelectronic case of Ra II and in Ra I came originally from nuclear physics. After early measurements of the atomic energy levels on the long-lived ^{226}Ra,[103] the isotope shifts and the hyperfine structures of the odd-A isotopes remained inaccessible until recently.[104] The extraction of nuclear moments required the analysis of the electronic hyperfine fields which was performed in on-line collinear-beam studies of several transitions involving the $7s\,^2S_{1/2}$ and $7p\,^2P_{1/2}$ states of Ra II and all states of the $7s7p$ configuration of Ra I.[105] The semiempirical modified Breit–Wills theory[106] or equivalently the effective operator formalism[107] are shown to be valid also for the heaviest two-valence-electron atom with strong relativistic contributions to the wave function. The experiment initiated several *ab initio* atomic structure calculations,[108–111] which can be tested in additional experiments including the direct g_I-factor measurement (cf. Section 4.2) and the spectroscopy in the near-uv Ra II resonance line to $7p\,^2P_{3/2}$.

Various other hfs investigations have been carried out on ions that have a complex electronic structure. They are usually treated within the effective operator formalism[107] whose parameters are related to radial integrals $\langle r^{-3} \rangle$ (see, e.g., Part A, Chapter 10 by S. Penselin) for which theoretical or semiempirical values are introduced to determine in particular the nuclear quadrupole moments. Thus for Eu II the hfs of the multiplet $4f^7(^8S)5d\,^9D$, measured in transitions to $4f^76p$,[112] yield another independent value of the first discovered nuclear quadrupole moment.[113] Similarly, the analysis of the hfs of the $5f^37s^2$, $5f^36d7s$, and $5f^36d^2$ configurations from the laser–rf experiment on U II[63] gives the $7s$ contact interaction term and a new value for the magnetic moment of ^{235}U. All metastable $5d^2$ and $5d6s$ levels of La II were studied in a number of transitions[114] and the extracted hfs parameters including two-body far-configuration mixing were compared with *ab initio* evaluations.[115] The analysis also gives a corrected value of the quadrupole moment of ^{139}La. In one of the transitions $(5d^2\,^3F_4 - 5d4f\,^3F_4)$ a new type of $\Delta F = \pm 2$ transitions was observed and ascribed to a three-quantum dipole process.[116] In the rare-gas spectrum of Cs II the hfs of $6p\,[3/2]_2$, excited from the metastable $5d\,[3/2]_2$ level,[117] was compared with a semiempirical evaluation. A similar measurement and analysis had been performed using the in-flight hole-burning technique of Gaillard on the $5s\,[3/2]_2 - 5p[1/2]_1$ transition of Rb II.[70] Transitions from $3d^2$ to $3d4p$ in Sc II were examined within the scope of stellar abundance analysis.[118] The measured hfs shows that theoretical predictions involve considerable uncertainties for the interpretation of line shapes in the spectra from celestial objects.

The isotope shifts of light elements give absolute values of the specific mass effect caused by electron momentum correlations. Studies of transitions of the type $3p^{n-1}3d-3p^{n-1}4p$ in Ar II ($n = 5$), Cl II ($n = 4$), and S II ($n = 3$)[119,120] indicate a strong correlation of the $3d$ electron with the inner p shells that is much larger than in the neutral spectra. (The same transitions in Ar II were also used to study optical pumping and the Hanle effect, which yielded the lifetimes of the $4p$ levels.[121]) The experimental mass shifts are somewhat overestimated by calculations based on Hartree–Fock wave functions, although the trends are very well reproduced. Generally, it seems that only refined theoretical methods may give quantitative predictions of the mass shift, which, in the intermediate mass range, limits the accuracy of changes in the mean square nuclear charge radius extracted from the isotope shifts (cf. Section 5.1).

5. Spectroscopy on Unstable Isotopes

It may be surprising to find the most extensive application of collinear laser fast-beam spectroscopy in a field[122] that a priori has little connection with the special features of this technique. Neither the Doppler shift nor the accessibility of ionic spectra plays a decisive role for the on-line experiments on radioactive isotopes from nuclear reactions. However, most of the problems encountered in the preparation of a sample of free atoms (cf. Part B, Chapter 17 by H.-J. Kluge) are solved by a combination of the fast-beam technique with the well-established concept of on-line isotope separation. The isotope separators (with ISOLDE at CERN[123] as an outstanding example) provide the unstable species in the form of ion beams whose phase-space volume is well matched to the requirements of collinear spectroscopy.

Obvious advantages allow the investigation of nearly any element which is available from the isotope separator:

i. Convenient optical transitions—including those from metastable states—can be chosen from the spectra of the neutral atom or the singly charged ion.

ii. The sensitivity, owing to efficient excitation within the narrow Doppler profile, is essential for the low production rate of isotopes far from stability.

iii. The Doppler effect provides a large tuning range and easy control of the effective wavelength of the laser light.

iv. Fluorescence detection gives the greatest versatility, and specialized particle detection schemes may further improve the sensitivity.

v. The high resolution yields sufficiently accurate hyperfine structure parameters and isotope shifts.

5.1. Hyperfine Structure and Isotope Shifts

We shall briefly review the relationship between spectroscopic observables and nuclear properties.[124-126] These properties include spins, magnetic dipole and electric quadrupole moments, as well as the variation in the mean square charge radius within a sequence of isotopes. They manifest themselves in the hyperfine structures and isotope shifts. The hyperfine energies of the different F states within a hyperfine multiplet $|J - I| \leq F \leq J + I$, given by the well-known formula

$$W_F = \frac{1}{2} KA + \frac{3/4K(K+1) - I(I+1)J(J+1)}{2I(2I-1)J(2J-1)} B \qquad (23)$$

with $K = F(F+1) - I(I+1) - J(J+1)$, are determined by the nuclear spin I, the magnetic dipole interaction constant

$$A = \mu_I H_e(0)/IJ \qquad (24)$$

and the electric quadrupole interaction constant

$$B = eQ_s \varphi_{JJ}(0) \qquad (25)$$

The nuclear moments μ_I and Q_s can be extracted from A and B using empirical or theoretical values for the magnetic hyperfine field $H_e(0)$ and the electric field gradient $\varphi_{JJ}(0)$ at the nucleus. Apart from hyperfine anomaly corrections $H_e(0)$ is usually known from direct g_I-factor measurements ($g_I = \mu_I/I\mu_N$) on the stable isotopes. $\varphi_{JJ}(0)$ is taken from semi-empirical or theoretical analyses, unless precise values are available from the hfs of muonic atoms.

Similarly, the isotope shift $\delta\nu^{AA'}$ of an optical transition is related to the change in the nuclear mean square charge radius $\delta\langle r^2 \rangle^{AA'}$ between the isotopes A and A' by

$$\delta\nu^{AA'} = F\delta\langle r^2 \rangle^{AA'} + M \frac{A' - A}{A'A} \qquad (26)$$

where the electronic factor, F, of the field shift term is proportional to the change of the electron charge density at the nucleus in the optical transition. The second term represents the mass shift due to the change of nuclear recoil energy. In simple (s–p or s^2–sp) transitions of heavy elements, where F is large and $(A' - A)/A'A$ is small, one can usually neglect the specific mass effect of the electron momentum correlation and assume $M = \nu/1836(1 \pm 1)$ for the transition frequency ν.[126] Isotope shifts in other

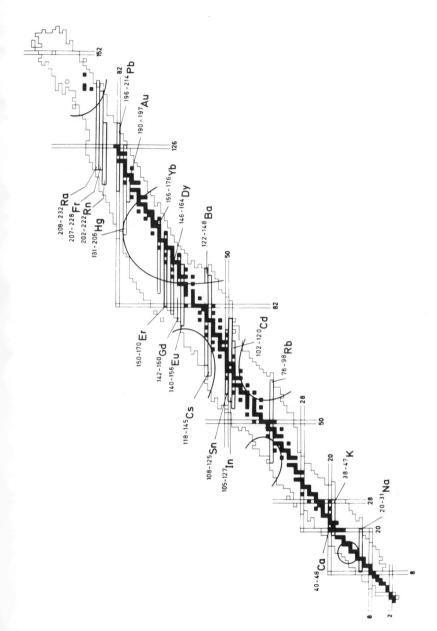

Figure 9. Chart of the nuclides with optically investigated chains of unstable isotopes marked by solid frames. Techniques other than collinear laser spectroscopy (see Table 1) were applied (partly) for Na, K, Ca, (Rb), Sr, Cd, (Sn), (Cs), (Ba), (Eu), (Hg), Pb, and (Fr).

Table 1. Measurements on Radioactive Nuclides by Collinear Fast-Beam Laser Spectroscopy

Element	Stable isotopes	Investigated radioactive isotopes	Production reactions	Production facility	Transition	Wavelength (Å)	Reference
Rb I	85, 87	89–93	$^{235}U(n_{th}$, fission)	Mainz reactor	$5s\,^2S_{1/2}-6p\,^2P_{3/2}$	4202	51
In I	113, 115	104–111	$^{92-100}Mo(^{16}O, pxn)$	GSI	$5p\,^2P_{3/2}-6s\,^2S_{1/2}$	4511	129, 130
					$5p\,^2P_{1/2}-$	4102	131
Sn I	112, 114–120, 122, 124	111–127	$^{238}U(p$, fission)	ISOLDE	$5p^2\,^1S_0-5p6s\,^1P_1$	4525	131
		108–111	$^{98-100}Mo(^{16}O, xn)$	GSI	$^3P_1$	5632	132
Cs I	133	137–142	$^{235}U(n_{th}$, fission)	Mainz reactor	$6s\,^2S_{1/2}-7p\,^2P_{3/2}$	4555	50, 133
Ba I	130, 132, 134–138	122, 123, 125, 127, 131, 137, 139–146	$^{139}La(p$, spallation)	ISOLDE	$6s^2\,^1S_0-6s6p\,^1P_1$	5535	134, 61
Ba II		140	$^{238}U(p$, fission) radioactive source	Leuven	$5d\,^2D_{3/2}-6p\,^2P_{3/2}$	5854	135
		139–146, 148	$^{238}U(p$, fission)	ISOLDE	$6s\,^2S_{1/2}-6p\,^2P_{3/2}$	4554	(see 61)
Sm II	144, 147–150, 152, 154	146, 151	Sm(n capture)	LNPI Gatchina/Mol	$4f^65d\,^8H_{17/2}-4f^66p;\,15/2$	6569	136, 137
Eu I	151, 153	140–150, 152, 142–150	$^{181}Ta(p$, spallation)	ISOLDE	$4f^76s^2\,^8S_{7/2}-4f^76s6p\,^8P_{7/2}$	4627	141, 142
					$^8P_{9/2}$	4594	
Eu II		147, 149, 152, 154–156	$^{181}Ta(p$, spallation), Sm, Eu(n capture)	LNPI Gatchina	$4f^75d\,^9D_4-4f^76p(7/2, 3/2)_4$	6049	137–140

Gd I	152, 154-158, 160	142, 144-151, 153	^{181}Ta(p, spallation)	ISOLDE	$4f^7 5d6s^2\,{}^9D_6 - 4f^7 5d6s6p;\ 7$	4226	144, 145
Dy I	156, 158, 160-164	146-155, 157, 159	^{181}Ta(p, spallation)	ISOLDE	$4f^{10}6s^2\,{}^5I_8 - 4f^{10}6s6p(8,1)_9$	4212	143-145
Er I	162, 164, 166-168, 170	150-160 (even), 154-160 (even), 152-161, 163, 165	^{181}Ta(p, spallation)	ISOLDE	$4f^{12}6s^2\,{}^3H_6 - 4f^{11}5d6s^2;\ 5$ $-4f^{12}6s6p(6,1)_7$ $-4f^{12}6s6p(6,1)_5$	4409 5827 4151	143-145
Yb I	168, 170-174, 176	158-169	^{181}Ta(p, spallation)	ISOLDE	$6s^2\,{}^1S_0 - 6s6p\,{}^3P_1$ $6s6p\,{}^3P_2 - 6s7s\,{}^3S_1$	5556 7699	27, 143-145
Hg I	196, 198-202, 204	156-166 (even), 182-198 (even)	Pb(p, spallation)	ISOLDE	$6s6p\,{}^3P_2 - 6s7s\,{}^3S_1$	5461	147
Tl I	203, 205	207	^{232}Th(p, spallation) $\to\,^{219}$Fr $\xrightarrow{a}\,^{207}$Tl ^{181}Ta(^{16}O, xn)	ISOLDE	$6p\,{}^2P_{3/2} - 7s\,{}^2S_{1/2}$	5350	148
Rn I	189, 191, 193			UNISOR	$7s[3/2]_2 - 7p[5/2]_3$	7450	149
Fr I	—	202-212, 218-222 212-213, 220-221	^{232}Th(p, spallation)	ISOLDE	$7s\,{}^2S_{1/2} - 8p\,{}^2P_{1/2}$ ${}^2P_{3/2}$	4325 4225	150, 151 89
Ra I	—	208-214, 220-230, 232 212, 214, 222-226	^{238}U(p, spallation)	ISOLDE	$7s^2\,{}^1S_0 - 7s7p\,{}^1P_1$ ${}^3P_1$ $7s7p\,{}^3P_2 - 7s7d\,{}^3D_3$	4826 7141 6446	104, 105 152, 153
Ra II	208, 210-214, 220-230	212, 214, 221-224, 226	^{238}U(p, spallation)	ISOLDE	$7s\,{}^2S_{1/2} - 7p\,{}^2P_{1/2}$	4683	

transitions can be related to these by the King plot procedure. Another advantage of the s–p transitions consists in the reliability of semiempirical or theoretical values of F which are mainly determined by the s-electron density. Shielding, relativistic and finite nuclear size effects, and the contribution of higher-order moments of the nuclear charge distribution are accounted for by corrections. These and the complications from configuration mixing are discussed in more detail by K. Heilig and A. Steudel in Part A, Chapter 7.

5.2. Review of Experiments

The virtues of collinear-beam spectroscopy for nuclear systematics are best demonstrated by the chart of nuclides (Figure 9) on which the investigated longer chains of unstable isotopes are marked by solid frames. The majority of the results were obtained from collinear-beam measurements, while the rest is shared by a number of more specialized techniques (see, e.g., Refs. 127–128). In a few cases the measurements cover nearly all known isotopes, and examples of nuclear physics results will be given in Section 5.4.

The individual experiments are compiled in Table 1, which is arranged in the order of elements and gives the isotopes and transitions that were investigated. A separate column shows the production reactions and the accelerator or reactor facility used for the on-line work. As the most extensive experimental program is concentrated at ISOLDE, the standard setup for these measurements will be outlined in the following.

5.3. Experimental Setup and Procedure

Figure 10 shows the essential parts of the ISOLDE apparatus.[61] Indicated are the target-ion source assembly and the mass-separator magnet from which the ion beam is guided to the particular experimental setup. For collinear laser spectroscopy the 60-kV acceleration voltage is stabilized[154] to fluctuations and long-term drifts of less than 10^{-5}. This is particularly important because of the high pulsed current load (10 mA) due to ionization of the target area by the 2-μA proton beam of 600 MeV.

Inside the apparatus the ion beam is deflected into the laser beam axis and neutralized in a charge-exchange cell at a variable potential of ± 10 kV. This gives a Doppler-tuning range for the fast atoms of typically 80 GHz. The fluorescence photons are collected along 20 cm by cylindrical optics and a light pipe. For spectroscopy on ions the potential is applied to the electrically insulated observation chamber instead of the charge-exchange cell.

The measurement comprises the comparison of the resonance positions of different isotopes. The result of such a measurement[143,145] is shown in

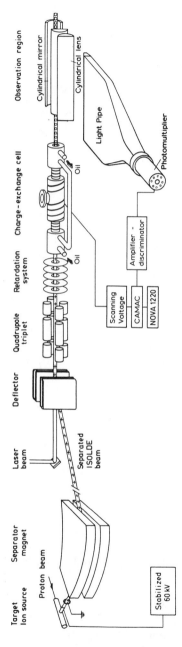

Figure 10. Essential components of the ISOLDE setup for collinear fast-beam laser spectroscopy. The postacceleration and scanning voltage on the charge-exchange cell ranges from +10 to −10 kV.

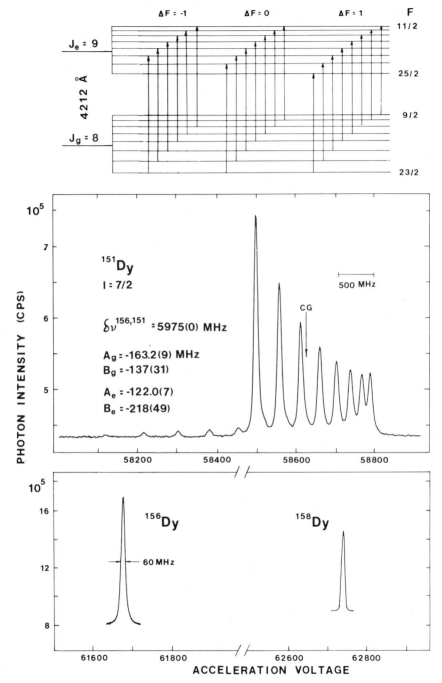

Figure 11. Hyperfine structure pattern of ^{151}Dy ($I = 7/2$) in the strong resonance transition at 4212 Å, with the resonances of the stable doubly even isotopes ^{156,158}Dy in a common voltage scale for measuring the isotope shift. The linewidth of 60 MHz includes some power broadening.

Figure 11 for the example of Dy I (4212 Å). The hfs of the odd-A isotope, [151]Dy, and the single components of the stable doubly even isotopes, [156]Dy and [158]Dy, are displayed versus the total acceleration voltage U, which is the sum of the constant isotope separator voltage and the Doppler-tuning potential. To first order the change ΔU in U is related to the change Δm in the atomic mass m and the isotope shift $\Delta \nu^{AA'}$ for a transition with frequency ν_0 by

$$\frac{\Delta U}{U} = \frac{\Delta m}{m} - \frac{\Delta \nu^{AA'}}{\nu_0} \left(\frac{2mc^2}{eU}\right)^{1/2} \qquad (27)$$

This indicates that the same relative precision (of, e.g., 10^{-4}) of both the small ΔU and the large U is adequate for the measurement of $\Delta \nu$, and only the stability of U has to correspond to the initial energy spread of the ions to avoid broadening. The cyclic measurement on a sequence of isotopes and fast scanning help to reduce further the errors introduced by fluctuations or drifts of the voltages, the laser frequency, or the beam intensities.

The sensitivity achieved in favorable transitions is illustrated in Figure 12 for the example of the $7s^2\,^1S_0$-$7s7p\,^1P_1$ resonance line of Ra I (4826 Å). The weakest beam of an even radium isotope, used in this experiment,[104,153] was about 10^4 atoms/sec. Odd isotopes require 10 to 100 times stronger beams, depending on their hyperfine structure. For [208]Ra, the recording of a resonance with a signal-to-noise ratio of 10 in 50 channels took 10^3 sec, which means that about 10^7 atoms, or 4 fg of radium, have passed through the apparatus during the measurement. The signal was 200 counts/sec, arising from 10 excitations per incident atom and a total photon-detection efficiency of 0.2%. This has to be compared to a background of 10^4 counts/sec of which the dominant part is due to stray light. In general, the sensitivity is reduced by smaller transition probabilities, hyperfine structure splittings, and optical pumping. Depending on these conditions, the practical limits lie between 10^4 and 10^7 atoms/sec.

5.4. Results in Nuclear Physics

To show the impact of laser spectroscopy on nuclear physics we discuss selected results from the aspect of information about the nuclear structure. Owing to collinear-beam spectroscopy it has now been possible to study a number of interesting regions all over the chart of nuclides.

5.4.1. Spins

The direct measurement of nuclear ground (and isomeric) state spins from the hfs is of basic importance for all indirect spin assignments of nuclear spectroscopy. The spins provide a stringent test of theoretically

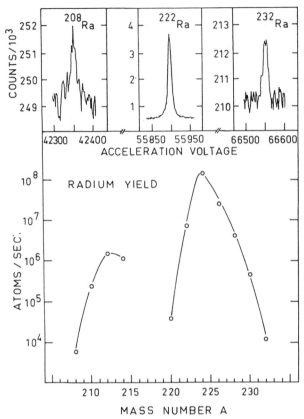

Figure 12. Yield curve of Ra isotopes from ISOLDE, and $7s^2\,^1S_0\text{–}7s7p\,^1P_1$ resonances from the strongest and weakest beams of isotopes. ^{232}Ra was observed here for the first time. ^{220}Ra has the shortest half-life of 20 msec and $^{215\text{–}219}$Ra are too short-lived to diffuse out of the UC$_2$ target.

predicted single-nucleon states which are obtained by spherical or deformed shell-model calculations. Very often the spins have given first evidence of unexpected phenomena. A recent example is the spin sequence of the heavy odd-A radium isotopes $^{221\text{–}229}$Ra,[104] which fits perfectly into the picture of parity-mixed neutron orbitals in a reflection asymmetric nuclear potential.[155] In other words, nuclei may have intrinsic octupole-deformed shapes, although static electric octupole moments are forbidden by parity conservation.

5.4.2. Magnetic Moments

The magnetic dipole moments are sensitive to details of the single-nucleon wave functions. Thus the mixture of opposite-parity orbitals in the octupole-deformed potential gives small magnetic moments in between the two Schmidt lines for $j = l_1 + 1/2$ and $j = l_2 - 1/2$ nucleons. Quantitative calculations[155,156] reproduce fairly well the experimentally observed moments[104,105] of $^{221-229}$Ra. In such calculations the single particle is coupled to a rotor core and the numerous modes of core polarization are usually subsumed under an effective gyromagnetic factor $g'_s \approx 0.6-0.7g_s$(free), where the latter refers to the free nucleon.

A quantitative theoretical understanding of the magnetic moments is established only for nuclei that have one particle or hole in a doubly magic core. For these nuclei an effective operator for the moment can be written as[157]

$$\mu_{\text{eff}} = (g_s + \delta g_s)s + (g_l + \delta g_l)l + g_p[s \times Y^2]^{(1)} \tag{28}$$

The spin and orbital gyromagnetic ratios g_s and g_l are the free nucleon values; δg_s and δg_l are caused by both core polarization and meson exchange. The last term, arising from the dipole–dipole interaction, is a rank-one tensor product of the spherical harmonic of order 2, Y^2, and the spin operator. The separate contributions of δg_s, δg_l, and g_p can be extracted from experimental data that include high-spin and low-spin states and can also be calculated by theory. For the rather extensive evaluation of the neighborhood of ^{208}Pb,[157] the only gap in the systematics has been closed recently by the hfs measurement on ^{207}Tl,[148] which is the only known heavy nucleus with a single $s_{1/2}$ particle or hole. In this case both the g_l and g_p terms vanish, and the isolated δg_s accounts for mainly core polarization effects which reduce the magnetic moment from the free-proton value of $2.79\mu_N$ to $1.88\mu_N$, within a few percent of the best theoretical predictions.

Since ^{146}Gd ($Z = 64$, $N = 82$) behaves very much like a doubly magic nucleus,[158] the neighboring ^{145}Eu ($Z = 63$) is of similar nature as the nuclei around ^{208}Pb, and to either side the long sequence of $d_{5/2}$ proton states in the odd-A isotopes $^{141-151}$Eu show a gradual quenching of the single-particle moments due to the admixture of collective states (Figure 13). The onset of strong deformation between the two stable isotopes ^{151}Eu and ^{153}Eu gives a drop in the magnetic moment, because the $d_{5/2}$ shell model state becomes a nearly pure [413 5/2] Nilsson state with a deformation of $\beta_2 \approx 0.3$. This goes along with the sudden increase of the nuclear quadrupole moment which was discovered 50 years ago in the famous work of Schüler and Schmidt.[113] Figure 13 combines the results of two collinear-beam experiments performed at ISOLDE[141] and at the Gatchina isotope separator.[139]

5.4.3. Quadrupole Moments

Most spectroscopic quadrupole moments are considerably larger than the single-particle values of the spherical shell model. This is true even for

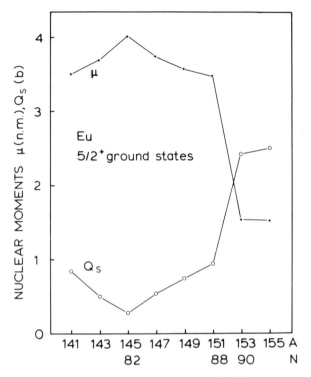

Figure 13. Nuclear magnetic dipole (μ) and electric quadrupole (Q_s) moments in the sequence of $5/2^+$ ground states of $^{141-155}$Eu.

nuclei near double shell closures. For example, $Q(^{145}\text{Eu}) = 0.29$ b exceeds the roughly estimated single-particle value $Q_{sp} = 0.13$ b by more than a factor of 2. Collective effects described by dynamic or static deformation become dominant as one goes away from magic proton or neutron numbers. However, the spectroscopic quadrupole moment Q_s is not directly related to the intrinsic quadrupole moment Q_0 which describes the nuclear shape in terms of a deformation parameter β_2,

$$Q_0 = \frac{3}{\sqrt{5\pi}} ZR^2\beta_2 \tag{29}$$

where R is the nuclear radius. Only in the limit of large deformation the strong coupling between the spin of the odd particle and the nuclear axis gives the simple projection

$$Q_s = \frac{I(2I-1)}{(I+1)(2I+3)} Q_0 \tag{30}$$

In the transitional regions one finds the situation that Q_s changes systematically and reverses the sign under sequential filling of a neutron shell. This is described by a change from the decoupled to the strongly coupled scheme, going along with a change of the projection Q_s/Q_0 from negative to positive values for the same nuclear spin. In such a sequence the intrinsic quadrupole moment Q_0 may be almost constant. This phenomenon was first observed in the $i_{13/2}$ isomers of $^{185-199}$Hg[159] and explained quantitatively by Ragnarsson.[160] Similar cases are pointed out by Ekström,[161] and another example is given in Figure 14 for the $f_{7/2}$ states of the rare-earth nuclides above the shell closure $N = 82$,[144] which have been the subject of systematic measurements by collinear laser spectroscopy. Thus the quadrupole moments can be understood as an interplay between the collective and single-particle structure of the nucleus.

5.4.4. Mean Square Charge Radii and Nuclear Shape

According to Eq. (26) the isotope shifts yield the changes of the mean square nuclear charge radii within the sequence of isotopes. These are largely insensitive to the single-particle structure and mainly reflect the collective properties of the nuclei. Even as a rough overall description of

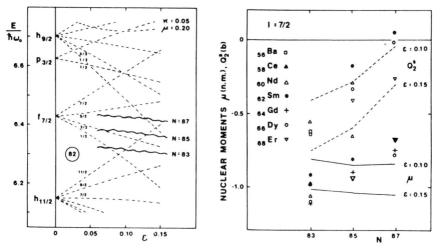

Figure 14. Nilsson diagram for odd neutrons close to $N = 82$. The Fermi levels for $N = 83$, 85, and 87 indicate a successive filling of the $f_{7/2}$ shell. On the right, the experimental magnetic dipole and electric quadrupole moments are compared with the results from particle-rotor calculations assuming deformations of $\varepsilon = 0.10$ and 0.15. The trend of the quadrupole moments reproduces the increase of coupling to the collective motion, while the discrepancy in the trend for the magnetic moments is understood as a change of core polarization.

the radii the liquid drop formula $R = R_0 A^{1/3}$, with $R_0 = 1.2\,\text{fm}$, and $\langle r^2 \rangle = \frac{3}{5}R^2$ is oversimplified, because protons and neutrons act differently on the charge radius. This is taken into account by the phenomenological droplet model,[162,163] which allows for nuclear compressibility and neutron-skin effects: For nuclei of constant shape the variation $\delta \langle r^2 \rangle$ with the neutron number is nearly half the value given by the $A^{1/3}$ law. Deviations from this global behavior arise from the shell structure and can be largely ascribed to deformation effects: The mean square radius of a spheroid which has the same volume as a sphere with $\langle r^2 \rangle_0$ is given by

$$\langle r^2 \rangle = \langle r^2 \rangle_0 \left(1 + \frac{5}{4\pi} \langle \beta_2^2 \rangle \right) \tag{31}$$

$\langle \beta_2^2 \rangle$ represents an average deformation parameter arising from either zero-point vibrations, or static deformation with $\langle \beta_2^2 \rangle = \beta_2^2$. This simple picture

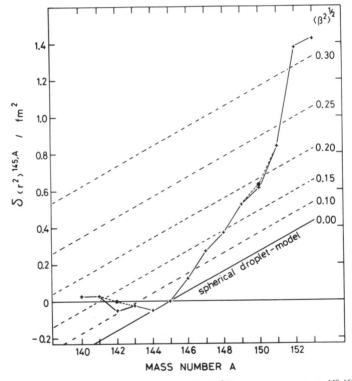

Figure 15. Change of the mean square charge radii $\delta \langle r^2 \rangle$ in the isotopic chain $^{140-153}$Eu as a function of the neutron number N, with ^{145}Eu ($N = 82$) taken as reference point. The change in deformation $\delta \langle \beta^2 \rangle^{1/2}$, with respect to $N = 82$, is indicated by the parallel lines, the slope of which is given by the droplet model.

describes fairly coherently most of the presently known $\delta\langle r^2 \rangle$ curves within long sequences of isotopes by

$$\delta\langle r^2 \rangle = \delta\langle r^2 \rangle_0 + \frac{5}{4\pi}\langle r^2 \rangle_0 \delta\langle \beta_2^2 \rangle \tag{32}$$

An example is given in Figure 15, again for the case of europium.[141] Assuming $\beta_2 = 0$, i.e., spherical shape for ^{145}Eu next to the double magic ^{146}Gd, one obtains $\beta_2 = 0.32$ for the strongly deformed ^{153}Eu by the decomposition of $\delta\langle r^2 \rangle^{145,153}$ into the droplet and deformation contributions. This is in perfect agreement with deformation values from reduced transition probabilities $B(E2)$ and from the spectroscopic quadrupole moment using Eqs. (29) and (30). The detailed calculations include small corrections and higher-order terms in Eqs. (29) and (32), which have been omitted for the sake of clarity.

Europium, which has been studied also in the first on-line experiments using multistep photoionization from a thermal atomic beam,[164,165] continues the large series of rare-earth elements investigated systematically at

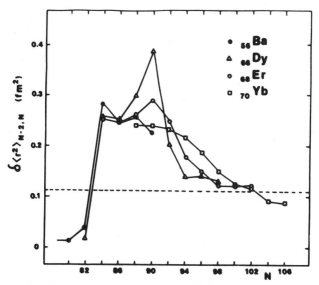

Figure 16. Differences in the nuclear mean square charge radii between even isotopes of neutron numbers $N - 2$ and N (Brix–Kopfermann diagram) for the rare-earth region. The curve for Dy is similar to those for Nd, Sm, and Gd which also show the peak from the sudden onset of deformation at $N = 90$. This effect disppears in the lighter and heavier elements. Data points above (below) the dashed droplet model line indicate increasing (decreasing) deformation from $N - 2$ to N, and the reversal of this effect is due to the $N = 82$ shell closure.

ISOLDE.[143-145] The main subject of these studies was the development of a strong deformation around $N = 90$ which occurs suddenly, i.e., between $N = 88$ and 90 in Nd ($Z = 60$), Sm ($Z = 62$), and Gd ($Z = 64$), which—like Eu—have stable isotopes in this region. On the other hand, the neutron-rich Ba ($Z = 56$) isotopes[61] and the neutron-deficient Yb ($Z = 70$) isotopes[27] behave smoothly. Including the measurements on Dy ($Z = 66$) and Er ($Z = 68$), Figure 16[143-145] shows a clear Z dependence in the onset of the strong deformation which is represented by the peak value of the differential $\delta\langle r^2\rangle^{N-2,N}$ (Brix–Kopfermann diagram). The "jump" into deformation, known from the early work of Brix and Kopfermann,[166,167] appears to be the exception which is related to the neighborhood of the semimagic proton number $Z = 64$, stabilizing the spherical shape for the neutron numbers $N \leqslant 88$. It is important to note that this first systematic study on the unstable isotopes of a sequence of different elements has become possible by the flexibility of the collinear beam technique.

The simplification in the two-parameter description of Eq. (32) becomes apparent with the prediction of a sizable octupole deformation for the neutron-rich radium isotopes. In fact, the experimental $\delta\langle r^2\rangle$ values between the nearly spherical ^{214}Ra ($N = 126$) and the heavier isotopes $^{220-232}$Ra[152,153] suggest a considerable higher-order deformation contribution. This can be taken care of by the addition of terms $(5/4\pi)\langle r^2\rangle_0$ $(\delta\langle\beta_3^2\rangle + \delta\langle\beta_4^2\rangle)$ to the expression of Eq. (32), but the isotope shift cannot distinguish between the different modes of deformation. It seems, however, that the inversion of the odd–even staggering, observed for ^{221,223,225}Ra[152,153] (Figure 17), and also for the neighboring ^{220,222}Fr[168]

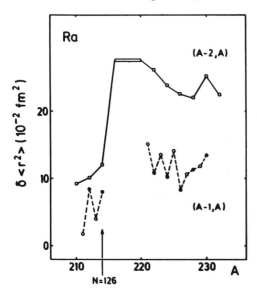

Figure 17. Differences of mean square charge radii and odd-even staggering in Ra. The upper curve corresponds to the plot of Figure 16. The lower curve represents the changes between neighboring isotopes with solid dots (open circles) for the steps from odd to even (even to odd) neutron number.

and $^{219,221}Rn^{(150,151)}$ can be understood qualitatively as an effect of octupole correlations which are particularly pronounced in the odd-neutron isotopes.[155] A conclusive interpretation of this effect implies the quantitative understanding of the generally observed normal odd–even staggering. One can expect that the key information for this will be provided by the wealth of new data in rather different regions of the chart of nuclides.

A by-product of the isotope shift measurements on long isotopic chains and in different transitions is the possibility of drawing King plots[126] of unprecedented accuracy. These provide an extremely sensitive test of the validity of Eq. (26), which implies that the isotope shifts in different transitions, multiplied by the inverse mass shift factor $AA'/(A' - A)$, or more precisely $m_A m_{A'}/(m_{A'} - m_A)$, depend linearly on each other. This procedure also makes it possible to relate the electronic factors F_i and the mass shift constants M_i of different transitions. A beautiful example of such a King plot is shown in Figure 18 for the 4683-Å line ($7s\,^2S_{1/2}-7p\,^2P_{1/2}$) in Ra II and the 4826-Å line ($7s^2\,^1S_0-7s7p\,^1P_1$) in Ra I.[105] The accuracy is

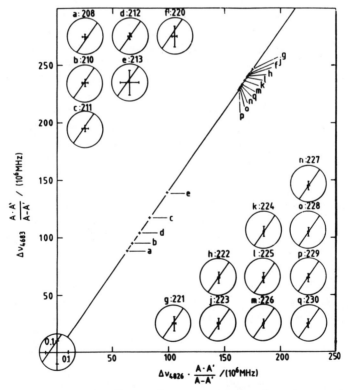

Figure 18. King plot of the 4683-Å line in Ra II against the 4826-Å line in Ra I. The inserts show the error bars enlarged by a factor of 100.

not only due to the inclusion of 17 isotopes between $A = 208$ and 230, but also to the considerably different isotope shifts on both sides of the magic neutron number $N = 126$ and the odd–even staggering.

5.5. Higher Sensitivity by Ionization

In all the experiments described the sensitivity limit of 10^4–10^7 atom/sec (cf. Section 5.3) arises from the low optical detection efficiency and the background due to stray light. Therefore, nonoptical detection schemes are considered most promising for experiments on the isotopes with very low production rates. The stepwise laser resonance ionization with subsequent ion counting is well developed for pulsed lasers with high power (but low duty cycle) acting on samples of thermal atoms.[169,170] With fast beams this technique gives improved isotopic selectivity, owing to the large Doppler shift and narrowing,[171,172] but considerable duty cycle losses in sensitivity. These may be minimized by the high repetition rate of copper-vapor pumped dye lasers. Single-mode cw lasers are limited in power, and attempts are being made to achieve efficient ionization by the excitation to low Rydberg states.[173,71] They have to cope with optical pumping losses and resolved hyperfine splittings in the intermediate states.

A straightforward alternative is found in the state-selective collisional ionization of metastable atoms. By laser excitation they can be pumped to the ground state, which leads to a reduction of the ionization rate. Similar techniques were used in the early collinear-beam experiments on HD^+ by Wing et al.[12] and on Ba^+ by Gaillard et al.,[69] who observed small changes in the ion beam current, due to slightly different neutralization or double-ionization cross sections for the states involved in the optical excitation and decay. On the other hand, there are examples of strongly state-selective processes, such as the near-resonant charge transfer neutralization. Thus, Silverans has shown recently[174] that the excitation of Sr^+ in the $5s$–$5p$ resonance lines, followed by the decay to the metastable $4d$ levels, strongly affects the beam composition of ions and neutral atoms after passage through sodium or cesium vapor.

The collisional ionization scheme[175,176] is especially suitable for the rare gases whose first excited states lie about 10 eV above the $np^6\,^1S_0$ ground states: The metastable $J = 2$ states of the configuration $np^5\,(n+1)s$—designated $(n+1)s\,[3/2]_2$—about -4 eV from the ionization threshold are efficiently populated in charge exchange with cesium.[24] They are connected to the states of the configuration $np^5(n+1)p$ by strong transitions in the red. In a test experiment on krypton[175,176] the neutralized beam, superimposed with the laser beam, passed a differentially pumped gas cell, after which the reionized part was deflected into a Faraday cup or an ion counter. The laser excitation from the $5s$ $[3/2]_2$ metastable to the $5p$ $[3/2]_2$ state at 7602 Å was followed by the decay cascade via $5s$ $[3/2]_1$ to the ground state (Figure 19). Since the cross section for electron stripping from the ground state (-14 eV) is much smaller than from the metastable state (-4 eV), this pumping leads to a reduction of the ion signal, which can thus be used to detect the optical resonance.

Among various stripping gases, SO_2 gave the highest reionization efficiencies of 10%–20%, mainly limited by secondary processes, such as charge-transfer neutralization. At resonance, the flop-out signal in the ion current was about 40%. In Figure 19 a direct comparison with the simultaneously recorded fluorescence signal shows a gain in detection efficiency of more than a factor of 10^3. In principle, this ion signal is free of background. However, for the planned experiments on weak radioactive isotopes it will be essential to have low contamination of the beams with molecular ions or strong neighboring masses from the isotope separator. In favorable cases, the radioactive decay may serve for detection, thus removing the possible background from stable beams.

5.6. Implantation of Polarized Atoms

The classical RADOP experiments are described by H.-J. Kluge in Part B, Chapter 17. Their sensitivity is based on the detection of the asymmetry in the radioactive β-decay, which provides an inherent spin detector for polarized nuclei. A combination of fast-beam spectroscopy with the RADOP

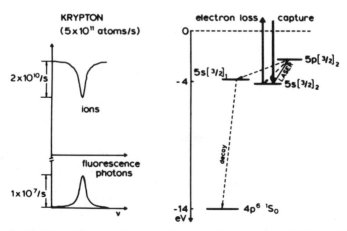

Figure 19. Simplified energy-level diagram of Kr I and the resonance $5s[3/2]_2$–$5p[3/2]_2$ in ^{84}Kr detected by fluorescence as well as state-selective collisional ionization of Kr atoms in the metastable state.

scheme can be realized by optical pumping with circularly polarized light and subsequent implantation of the atoms into a suitable crystal lattice where the nuclear polarization is detected in their decay.[177] For short-lived isotopes this can offer interesting advantages including the sensitive detection of optical resonance, direct g_I-factor measurements by NMR on the implanted nuclei, and solid-state relaxation and hyperfine field studies. This technique was developed by the use of neutron-rich Rb isotopes from fission[178] and applied recently to determine the spin and magnetic moment of ^{11}Li.[179] About 10^3 atoms/sec of this exotic nuclide are produced at ISOLDE from fragmentation of ^{181}Ta or ^{238}U. They were polarized by optical pumping in the resonance lines $2s\,^2S_{1/2}$-$2p\,^2P_{1/2,3/2}$ (6707 Å) and implanted into a gold foil placed in a static magnetic field of about 1 kG. The optical resonances were detected by asymmetries of a few percent and yielded the ground-state hyperfine splitting. The result indicates a magnetic moment close to the Schmidt value of the $p_{3/2}$ shell-model proton state. It seems to be incompatible with theoretical predictions of a strongly deformed nuclear system of spin $I = 1/2$ and favors the shell-model description of a $p_{3/2}$ proton plus a doubly magic ($Z = 2$, $N = 8$) core. More detailed information is expected from NMR experiments yielding the g_I factor and the spectroscopic quadrupole moment. Because of its short half-life of 8.7 msec and long relaxation time, ^{11}Li is an ideal candidate for this technique, and for solid-state studies the much more strongly produced isotopes ^{8}Li ($T_{1/2} = 842$ msec) and ^{9}Li ($T_{1/2} = 178$ msec) can be used as convenient probes.

6. Conclusion

It has been shown by means of examples that the collinear laser fast-beam technique has introduced many interesting aspects into the classical field of atomic spectroscopy. This discussion has not touched upon the promising applications to molecular ions including spectroscopy and reaction studies, as the physics involved is beyond the scope of this contribution. To date, it appears that a systematic application in atomic spectroscopy has been established in the work on radioactive nuclides, owing to the sensitivity and resolution, but even more the ideal adaptation of the spectroscopic method to the conditions of production.

The progress in the development of tunable and narrow-band uv lasers will give a new impact on ion spectroscopy, and future experiments may extend high-resolution spectroscopy to multiply charged ions. In this context the laser cooling and spectroscopy of ions in a storage ring[180,181] is a particularly attractive possibility, although it will need a lot of ingenuity to solve both the physical and the technical problems.

References

1. H. Schüler, *Z. Phys.* **59**, 149 (1930).
2. I. I. Rabi, S. Millman, P. Kusch, and J. R. Zacharias, *Phys. Rev.* **55**, 526 (1939).
3. A. Kastler, *J. Phys. Radium* **11**, 255 (1950). J. Brossel, A. Kastler, and J. Winter, *J. Phys. Radium* **13**, 668 (1952).
4. J. Brossel and A. Kastler, *C.R. Acad. Sci. (Paris)* **229**, 1213 (1949). J. Brossel and F. Bitter, *Phys. Rev.* **86**, 308 (1952).
5. H. J. Andrä, A. Gaupp, and W. Wittmann, *Phys. Rev. Lett.* **31**, 501 (1973).
6. M. Dufay, M. Carré, M. L. Gaillard, G. Meunier, H. Winter, and A. Zgainski, *Phys. Rev. Lett.* **37**, 1678 (1976).
7. H. Winter and M. L. Gaillard, *J. Phys. B* **10**, 2739 (1977).
8. H. J. Andrä, in *Beam-Foil Spectroscopy*, Vol. 2 (edited by I. A. Sellin and D. J. Pegg), Plenum Press, New York (1976), p. 835.
9. S. L. Kaufman, R. Neugart, and E. W. Otten, unpublished report, DFG Colloquium on Laser Spectroscopy, Bad Nauheim (May 1975).
10. S. L. Kaufman, *Opt. Commun.* **17**, 309 (1976).
11. K.-R. Anton, S. L. Kaufman, W. Klempt, G. Moruzzi, R. Neugart, E. W. Otten, and B. Schinzler, *Phys. Rev. Lett.* **40**, 642 (1978).
12. W. H. Wing, G. A. Ruff, W. E. Lamb, Jr., and J. J. Spezeski, *Phys. Rev. Lett.* **36**, 1488 (1976).
13. Th. Meier, H. Hühnermann, and H. Wagner, *Opt. Commun.* **20**, 397 (1977).
14. R. A. Holt, S. D. Rosner, and T. D. Gaily, *Phys. Rev. A* **15**, 2293 (1977).
15. W. H. Wing, in *Laser Spectroscopy III* (edited by J. L. Hall and J. L. Carlsten), Springer, Berlin (1977), p. 69.
16. J. T. Moseley, P. C. Cosby, J. Durup, and J. B. Ozenne, *J. Phys. (Paris)* **40**, C1-46 (1979).
17. A. Carrington and P. J. Sarre, *J. Phys. (Paris)* **40**, C1-54 (1979).
18. M. Horani, H. H. Bukow, M. Carré, M. Druetta, and M. L. Gaillard, *J. Phys. (Paris)* **40**, C1-57 (1979).
19. A. Carrington, *Ann. Isr. Phys. Soc. (Israel)* **4**, 231 (1980).
20. M. L. Gaillard, in *Atomic Physics 8* (edited by I. Lindgren, A. Rosén, and S. Svanberg), Plenum Press, New York (1982), p. 467.
21. A. Carrington and T. P. Softley, in *Molecular Ions: Spectroscopy, Structure and Chemistry* (edited by T. A. Miller and V. E. Bondybey), North-Holland, Amsterdam (1983), p. 49.
22. H. S. W. Massey and H. B. Gilbody, *Electronic and Ionic Impact Phenomena*, Vol. 4, Clarendon, Oxford (1974).
23. D. Rapp and W. E. Francis, *J. Chem. Phys.* **37**, 2631 (1962).
24. G. E. Ice and R. E. Olson, *Phys. Rev. A* **11**, 111 (1975).
25. F. W. Meyer, C. J. Anderson, and L. W. Anderson, *Phys. Rev. A* **15**, 455 (1977).
26. E. Arnold, T. Kühl, E. W. Otten, and L. von Reisky, *Phys. Lett.* **90A**, 399 (1982).
27. F. Buchinger, A. C. Mueller, B. Schinzler, K. Wendt, C. Ekström, W. Klempt, and R. Neugart, *Nucl. Instrum. Methods* **202**, 159 (1982).
28. M. Bacal, A. Truc, H. J. Doucet, H. Lamain, and M. Chrétien, *Nucl. Instrum. Methods* **114**, 407 (1974).
29. M. Bacal and W. Reichelt, *Rev. Sci. Instrum.* **45**, 769 (1974).
30. J. C. Brenot, J. Pommier, D. Dhuicq, and M. Barat, *J. Phys. B* **8**, 448 (1975).
31. P. M. Koch, *Opt. Commun.* **20**, 115 (1977).
32. B. Schinzler, Thesis, Mainz (1980).
33. C. Reynaud, J. Pommier, Vu Ngoc Tuan, and M. Barat, *Phys. Rev. Lett.* **43**, 579 (1979).
34. C.-J. Lorenzen, K. Niemax, and L. R. Pendrill, *Opt. Commun.* **39**, 370 (1981).
35. J. Kowalski, R. Neumann, S. Noehte, R. Schwarzwald, H. Suhr, and G. zu Putlitz, *Opt. Commun.* **53**, 141 (1985).

36. E. W. Otten, private communication.
37. J. E. M. Goldsmith, E. W. Weber, and T. W. Hänsch, *Phys. Rev. Lett.* **41**, 1525 (1978).
38. S. R. Amin, C. D. Caldwell, and W. Lichten, *Phys. Rev. Lett.* **47**, 1234 (1981).
39. O. Poulsen, *Nucl. Instrum. Methods* **202**, 503 (1982).
40. H. E. Ives and G. R. Stilwell, *J. Opt. Soc. Am.* **28**, 215 (1938); **31**, 369 (1941).
41. H. I. Mandelberg and L. Witten, *J. Opt. Soc. Am.* **52**, 529 (1962).
42. D. C. Champeney, G. R. Isaak, and A. M. Khan, *Proc. Phys. Soc.* **85**, 583 (1965).
43. J. Bailey, K. Borer, F. Combley, H. Drumm, F. Krienen, F. Lange, E. Picasso, W. von Rüden, F. J. M. Farley, J. H. Field, W. Flegel, and P. M. Hattersley, *Nature (London)* **268**, 301 (1977).
44. P. Juncar and J. Pinard, *Rev. Sci. Instrum.* **53**, 939 (1982).
45. P. Juncar, C. R. Bingham, J. A. Bounds, D. J. Pegg, H. K. Carter, R. L. Mlekodaj, and J. D. Cole, *Phys. Rev. Lett.* **54**, 11 (1985).
46. J. J. Snyder and J. L. Hall, in *Laser Spectroscopy* (edited by S. Haroche, J. C. Pebay-Peyroula, T. W. Hänsch, and S. E. Harris), Springer, Berlin (1975), p. 6.
47. O. Poulsen and N. I. Winstrup, *Phys. Rev. Lett.* **47**, 1522 (1981).
48. M. Kaivola, N. Bjerre, O. Poulsen, and J. Javanainen, *Opt. Commun.* **49**, 418 (1984).
49. M. Kaivola, O. Poulsen, E. Riis, and S. A. Lee, *Phys. Rev. Lett.* **54**, 255 (1985).
50. B. Schinzler, W. Klempt, S. L. Kaufman, H. Lochmann, G. Moruzzi, R. Neugart, E. W. Otten, J. Bonn, L. von Reisky, K. P. C. Spath, J. Steinacher, and D. Weskott, *Phys. Lett.* **79B**, 209 (1978).
51. W. Klempt, J. Bonn, and R. Neugart, *Phys. Lett.* **82B**, 47 (1979).
52. R. E. Silverans, G. Borghs, G. Dumont, and J. M. Van den Cruyce, *Z. Phys. A* **295**, 311 (1980).
53. G. Ulm, J. Eberz, G. Huber, H. Lochmann, R. Menges, R. Kirchner, O. Klepper, T. Kühl, P. O. Larsson, D. Marx, D. Murnick, and D. Schardt, *Z. Phys. A* **321**, 395 (1985).
54. G. Borghs, P. De Bisschop, J. M. Van den Cruyce, M. Van Hove, and R. E. Silverans, *Opt. Commun.* **38**, 101 (1981).
55. H. Schnatz, G. Bollen, P. Dabkiewicz, P. Egelhof, F. Kern, H. Kalinowsky, L. Schweikhard, H. Stolzenberg, and H.-J. Kluge, *Nucl. Instr. Meth. A* **251**, 17 (1986).
56. A. Coc, R. Fergeau, C. Thibault, G. Audi, M. Epherre, P. Guimbal, M. de Saint-Simon, F. Touchard, E. Haebel, H. Herr, R. Klapisch, G. Lebée, G. Petrucci, and G. Stefanini, Proc. 7th Int. Conf. on Atomic Masses and Fundamental Constants (AMCO-7) (edited by O. Klepper), Darmstadt (1984), p. 661.
57. N. Bendali, H. T. Duong, J. M. Saint Jalm, and J. L. Vaille, *J. Phys. (Paris)* **44**, 1019 (1983).
58. B. Fan, A. Lurio, and D. Grischkowsky, *Phys. Rev. Lett.* **41**, 1460 (1978).
59. S. D. Rosner, T. D. Gaily, and R. A. Holt, *Phys. Rev. Lett.* **40**, 851 (1978).
60. R. A. Holt, S. D. Rosner, T. D. Gaily, and A. G. Adam, *Phys. Rev. A* **22**, 1563 (1980).
61. A. C. Mueller, F. Buchinger, W. Klempt, E. W. Otten, R. Neugart, C. Ekström, and J. Heinemeier, *Nucl. Phys.* **A403**, 234 (1983).
62. F. von Sichart, H. J. Stöckmann, H. Ackermann, and G. zu Putlitz, *Z. Phys.* **236**, 97 (1970).
63. U. Nielsen, O. Poulsen, P. Thorsen, and H. Crosswhite, *Phys. Rev. Lett.* **51**, 1749 (1983).
64. M. Van Hove, G. Borghs, P. De Bisschop, and R. E. Silverans, *Z. Phys. A* **321**, 215 (1985).
65. U. Kötz, J. Kowalski, R. Neumann, S. Noehte, H. Suhr, K. Winkler, and G. zu Putlitz, *Z. Phys. A* **300**, 25 (1981).
66. N. Bendali, H. T. Duong, J. M. Saint-Jalm, and J. L. Vaille, *J. Phys. (Paris)* **45**, 421 (1984).
67. G. Borghs, P. De Bisschop, M. Van Hove, and R. E. Silverans, *Phys. Rev. Lett.* **52**, 2030 (1984).
68. M. Carré, J. Lermé, and J. L. Vialle, *J. Phys. B* **19**, 2853 (1986).
69. F. Beguin, M. L. Gaillard, H. Winter, and G. Meunier, *J. Phys. (Paris)* **38**, 1185 (1977).
70. F. Beguin-Renier, J. Désesquelles, and M. L. Gaillard, *Phys. Scr.* **18**, 21 (1978).

71. U. Dinger, J. Eberz, G. Huber, H. Lochmann, and G. Ulm, *Z. Phys.* **D1**, 137 (1986).
72. O. Poulsen, P. Nielsen, U. Nielsen, P. S. Ramanujam, and N. I. Winstrup, *Phys. Rev. A* **27**, 913 (1983).
73. M. Kaivola, P. Thorsen, and O. Poulsen, *Phys. Rev. A* **32**, 207 (1985).
74. R. E. Silverans, P. De Bisschop, M. Van Hove, J. M. Van den Cruyce, and G. Borghs, *Phys. Lett.* **82A**, 70 (1981).
75. G. Borghs, P. De Bisschop, J. M. Van den Cruyce, M. Van Hove, and R. E. Silverans, *Phys. Rev. Lett.* **46**, 1074 (1981).
76. M. L. Lewis and P. H. Serafino, *Phys. Rev. A* **18**, 867 (1978).
77. B. Fan, D. Grischkowsky, and A. Lurio, *Opt. Lett.* **4**, 233 (1979).
78. R. Bayer, J. Kowalski, R. Neumann, S. Noehte, H. Suhr, K. Winkler, and G. zu Putlitz, *Z. Phys. A* **292**, 329 (1979).
79. A. N. Jette, T. Lee, and T. P. Das, *Phys. Rev. A* **9**, 2337 (1974).
80. B. Schiff, Y. Accad, and C. L. Pekeris, *Phys. Rev. A* **8**, 2272 (1973).
81. A. M. Ermolaev, *Phys. Rev. A* **8**, 1651 (1973); *Phys. Rev. Lett.* **34**, 380 (1975).
82. J. Kowalski, R. Neumann, S. Noehte, R. Schwarzwald, H. Suhr, and G. zu Putlitz, in *Laser Spectroscopy VI* (edited by H. P. Weber and W. Lüthy), Springer, Berlin (1983), p. 40.
83. U. Kötz, J. Kowalski, R. Neumann, S. Noehte, H. Suhr, K. Winkler, and G. zu Putlitz, in *Precision Measurement and Fundamental Constants II* (edited by B. N. Taylor and W. D. Phillips), Natl. Bur. Stand. (U.S.), Spec. Publ. 617 (1984), p. 159.
84. J. Kowalski, R. Neumann, S. Noehte, H. Suhr, G. zu Putlitz, and R. Herman, *Acta Phys. Hung.* **56**, 199 (1984).
85. Ch. E. Moore, *Atomic Energy Levels*, Vols. 1–3, NSRDS-NBS 35, National Bureau of Standards, Washington (1971).
86. W. C. Martin, R. Zalubas, and L. Hagan, *Atomic Energy Levels—The Rare-Earth Elements*, NSRDS-NBS 60, National Bureau of Standards, Washington (1978).
87. O. Poulsen, T. Andersen, S. M. Bentzen, and U. Nielsen, *Phys. Rev. A* **24**, 2523 (1981).
88. G. Huber, private communication.
89. H. T. Duong, P. Juncar, S. Liberman, A. C. Mueller, R. Neugart, E. W. Otten, B. Peuse, J. Pinard, H. H. Stroke, C. Thibault, F. Touchard, J. L. Vialle, and K. Wendt, *Europhys. Lett.* **3**, 175 (1987).
90. S. Liberman, J. Pinard, H. T. Duong, P. Juncar, J. L. Vialle, P. Jacquinot, G. Huber, F. Touchard, S. Büttgenbach, A. Pesnelle, C. Thibault, and R. Klapisch, *C.R. Acad. Sci. (Paris)* **286B**, 253 (1978).
91. N. Bendali, H. T. Duong, P. Juncar, S. Liberman, J. Pinard, J.-M. Saint-Jalm, J. L. Vialle, S. Büttgenbach, C. Thibault, F. Touchard, A. Pesnelle, and A. C. Mueller, *C.R. Acad. Sci. (Paris)* **299** Ser. II, 1157 (1984).
92. H. Lundberg and A. Rosén, *Z. Phys. A* **286**, 329 (1978).
93. V. A. Dzuba, V. V. Flambaum, and O. P. Sushkov, *Phys. Lett.* **95A**, 230 (1983).
94. G. Borghs, P. De Bisschop, R. E. Silverans, M. Van Hove, and J. M. Van den Cruyce, *Z. Phys. A* **299**, 11 (1981).
95. C. R. Bingham, M. L. Gaillard, D. J. Pegg, H. K. Carter, R. L. Mlekodaj, J. D. Cole, and P. M. Griffin, *Nucl. Instrum. Methods* **202**, 147 (1982).
96. C. Höhle, H. Hühnermann, Th. Meier, and H. Wagner, *Z. Phys. A* **284**, 261 (1978).
97. E. Alvarez, A. Arnesen, A. Bengtson, R. Hallin, M. Niburg, C. Nordling, and T. Noreland, *Phys. Scr.* **18**, 54 (1978).
98. M. Van Hove, G. Borghs, P. De Bisschop, and R. E. Silverans, *J. Phys. B* **15**, 1805 (1982).
99. M. Wilson, *Phys. Lett.* **65A**, 213 (1978).
100. R. E. Silverans, G. Borghs, P. De Bisschop, and M. Van Hove, *Phys. Rev. A* **33**, 2117 (1986).

101. G. Borghs, P. De Bisschop, J. Odeurs, R. E. Silverans, and M. Van Hove, *Phys. Rev. A* **31**, 1434 (1985).

102. K. Wendt, S. A. Ahmad, F. Buchinger, A. C. Mueller, R. Neugart, and E. W. Otten, *Z. Phys. A* **318**, 125 (1984).

103. E. Rasmussen, *Z. Phys.* **86**, 24 (1933); **87**, 607 (1934).

104. S. A. Ahmad, W. Klempt, R. Neugart, E. W. Otten, K. Wendt, and C. Ekström, *Phys. Lett.* **133B**, 47 (1983).

105. K. Wendt, S. A. Ahmad, W. Klempt, R. Neugart, E. W. Otten, and H. H. Stroke, *Z. Phys. D.* **4**, 227 (1987).

106. A. Lurio, *Phys. Rev.* **142**, 46 (1966).

107. P. G. H. Sandars and J. Beck, *Proc. R. Soc. London Ser. A* **289**, 97 (1965).

108. J. L. Heully and A. M. Martensson-Pendrill, *Phys. Scr.* **31**, 169 (1985).

109. V. A. Dzuba, V. V. Flambaum, and O. P. Sushkov, *Phys. Scr.* **32**, 507 (1985).

110. T. P. Das, private communication and to be published.

111. G. Torbohm, B. Fricke, and A. Rosén, *Phys. Rev. A* **31**, 2038 (1985).

112. A. Arnesen, A. Bengtson, R. Hallin, C. Nordling, Ö. Staaf, and L. Ward, *Phys. Scr.* **24**, 747 (1981).

113. H. Schüler and Th. Schmidt, *Z. Phys.* **94**, 457 (1935).

114. C. Höhle, H. Hühnermann, and H. Wagner, *Z. Phys. A* **304**, 279 (1982).

115. J. Bauche, J.-F. Wyart, Z. Ben Ahmed, and K. Guidara, *Z. Phys. A* **304**, 285 (1982).

116. C. Höhle, H. Hühnermann, and M. Elbel, *Z. Phys. A* **295**, 1 (1980).

117. A. Bengtson, E. Alvarez, A. Arnesen, R. Hallin, C. Nordling, and Ö. Staaf, *Phys. Lett.* **76A**, 45 (1980).

118. A. Arnesen, R. Hallin, C. Nordling, Ö. Staaf, L. Ward, B. Jelénković, M. Kisielinski, L. Lundin, and S. Mannervik, *Astron. Astrophys.* **106**, 327 (1982).

119. A. Eichhorn, M. Elbel, W. Kamke, and J. C. Amaré, *Opt. Commun.* **31**, 306 (1979).

120. A. Eichhorn, M. Elbel, W. Kamke, R. Quad, and J. Bauche, *Z. Phys. A* **305**, 39 (1982).

121. A. Eichhorn, M. Elbel, W. Kamke, R. Quad, and H. J. Seifner, *Physica* **124C**, 282 (1984).

122. E. W. Otten, *Nucl. Phys. A* **354**, 471c (1981).

123. H. L. Ravn, *Phys. Rep.* **54**, 201 (1979).

124. H. Kopfermann, *Nuclear Moments*, Academic, New York (1958).

125. H. G. Kuhn, *Atomic Spectra*, Longman, London (1969).

126. K. Heilig and A. Steudel, *At. Data Nucl. Data Tables* **14**, 613 (1974).

127. H.-J. Kluge, *Hyperfine Interact.* **24–26**, 69 (1985).

128. C. Thibault, *Hyperfine Interact.* **24–26**, 95 (1985).

129. H. Lochmann, U. Dinger, J. Eberz, G. Huber, R. Menges, G. Ulm, R. Kirchner, O. Klepper, T. Kühl, D. Marx, and D. Schardt, *Z. Phys. A* **322**, 703 (1985).

130. J. Eberz, U. Dinger, T. Horiguchi, G. Huber, H. Lochmann, R. Menges, R. Kirchner, O. Klepper, T. Kühl, D. Marx, E. Roeckl, D. Schardt, and G. Ulm, *Z. Phys. A* **323**, 119 (1986).

131. J. Eberz, U. Dinger, G. Huber, H. Lochmann, R. Menges, R. Neugart, R. Kirchner, O. Klepper, T. Kühl, D. Marx, G. Ulm, and K. Wendt, *Nucl. Phys. A.* **464**, 9 (1987).

132. J. Eberz, U. Dinger, G. Huber, H. Lochmann, R. Menges, G. Ulm, R. Kirchner, O. Klepper, T. Kühl, and D. Marx, *Z. Phys. A.* **326**, 121 (1987).

133. J. Bonn, W. Klempt, R. Neugart, E. W. Otten, and B. Schinzler, *Z. Phys. A* **289**, 227 (1979).

134. R. Neugart, F. Buchinger, W. Klempt, A. C. Mueller, E. W. Otten, C. Ekström, and J. Heinemeier, *Hyperfine Interact.* **9**, 151 (1981).

135. R. E. Silverans, G. Borghs, and J. M. Van den Cruyce, *Hyperfine Interact.* **9**, 193 (1981).

136. K. Dörschel, H. Hühnermann, E. Knobl, Th. Meier, and H. Wagner, *Z. Phys. A* **302**, 359 (1981).

137. G. D. Alkhazov, E. Ye. Berlovich, V. P. Denisov, V. N. Panteleev, V. I. Tikhonov, K. Dörschel, W. Heddrich, H. Hühnermann, E. W. Peau, and H. Wagner, Proc. 7th Int. Conf. on Atomic Masses and Fundamental Constants (AMCO-7) (edited by O. Klepper), Darmstadt (1984), p. 327.

138. K. Dörschel, W. Heddrich, H. Hühnermann, E. W. Peau, H. Wagner, G. D. Alkhazov, E. Ye. Berlovich, V. P. Denisov, V. N. Panteleev, and A. G. Polyakov, *Z. Phys. A* **312**, 269 (1983).

139. K. Dörschel, W. Heddrich, H. Hühnermann, E. W. Peau, H. Wagner, G. D. Alkhazov, E. Ye. Berlovich, V. P. Denisov, V. N. Panteleev, and A. G. Polyakov, *Z. Phys. A* **317**, 233 (1984).

140. G. D. Alkhazov, E. Ye. Berlovich, V. P. Denisov, V. N. Panteleev, V. I. Tikhonov, K. Dörschel, W. Heddrich, H. Hühnermann, E. W. Peau, and H. Wagner, *Z. Phys. A* **316**, 123 (1984).

141. S. A. Ahmad, W. Klempt, C. Ekström, R. Neugart, and K. Wendt, *Z. Phys. A* **321**, 35 (1985).

142. S. A. Ahmad, C. Ekström, W. Klempt, R. Neugart, and K. Wendt, Proc. 7th Int. Conf. on Atomic Masses and Fundamental Constants (AMCO-7) (edited by O. Klepper), Darmstadt (1984), p. 341.

143. R. Neugart, in *Lasers in Nuclear Physics* (edited by C. E. Bemis, Jr. and H. K. Carter), Nuclear Science Research Conference Series, Vol. 3, Harwood, London (1982), p. 231.

144. R. Neugart, K. Wendt, S. A. Ahmad, W. Klempt, and C. Ekström, *Hyperfine Interact.* **15/16**, 181 (1983).

145. W. Klempt, R. Neugart, F. Buchinger, A. C. Mueller, K. Wendt, and C. Ekström, submitted to *Z. Phys. A.*

146. S. A. Ahmad, C. Ekström, W. Klempt, R. Neugart, and K. Wendt, to be published.

147. G. Ulm, S. K. Bhattacherjee, P. Dabkiewicz, G. Huber, H.-J. Kluge, T. Kühl, H. Lochmann, E. W. Otten, K. Wendt, S. A. Ahmad, W. Klempt, and R. Neugart, *Z. Phys. A.* **325**, 247 (1986).

148. R. Neugart, H. H. Stroke, S. A. Ahmad, H. T. Duong, H. L. Ravn, and K. Wendt, *Phys. Rev. Lett.* **55**, 1559 (1985).

149. J. A. Bounds, C. R. Bingham, P. Juncar, H. K. Carter, G. A. Leander, R. L. Mlekodaj, E. H. Spejewski, and W. M. Fairbank, Jr., *Phys. Rev. Lett.* **55**, 2269 (1985).

150. R. Neugart, E. W. Otten, H. T. Duong, G. Ulm, and K. Wendt, European Conference Abstracts (2'ECAMP, Amsterdam), Vol. 9B, 15 (1985).

151. G. Ulm, H. T. Duong, K. Wendt, W. Borchers, R. Neugart, E. W. Otten, to be published.

152. S. A. Ahmad, C. Ekström, W. Klempt, R. Neugart, E. W. Otten, and K. Wendt, Proc. 7th Int. Conf. on Atomic Masses and Fundamental Constants (AMCO-7) (edited by O. Klepper), Darmstadt (1984), p. 361.

153. S. A. Ahmad, W. Klempt, R. Neugart, E. W. Otten, G. Ulm, and K. Wendt, submitted to *Nucl. Phys. A.*

154. K.-H. Georgi, *Nucl. Instrum. Methods* **186**, 271 (1981).

155. G. A. Leander and R. K. Sheline, *Nucl. Phys.* **A413**, 375 (1984).

156. I. Ragnarsson, *Phys. Lett.* **130B**, 353 (1983).

157. A. Arima and H. Hyuga, in *Mesons in Nuclei* (edited by M. Rho and D. H. Wilkinson), North-Holland, Amsterdam (1979), p. 683.

158. M. Ogawa, R. Broda, K. Zell, P. J. Daly, and P. Kleinheinz, *Phys. Rev. Lett.* **41**, 289 (1978).

159. P. Dabkiewicz, F. Buchinger, H. Fischer, H.-J. Kluge, H. Kremmling, T. Kühl, A. C. Mueller, and H. A. Schüssler, *Phys. Lett.* **82B**, 199 (1979).

160. I. Ragnarsson, in *Future Directions in Studies of Nuclei Far from Stability*, Proc. Int. Symp. Nashville 1979 (edited by J. H. Hamilton, E. H. Spejewski, C. R. Bingham, and E. F. Zganjar), North-Holland, Amsterdam (1980), p. 367.

161. C. Ekström, in Proc. 4th Int. Conf. on Nuclei Far from Stability, Helsingör 1981 (edited by P. G. Hansen and O. B. Nielsen), CERN 81-09, Geneva (1981), p. 12.

162. W. D. Myers, *Phys. Lett.* **30B**, 451 (1969).

163. W. D. Myers and K.-H. Schmidt, *Nucl. Phys.* **A410**, 61 (1983).

164. A. N. Zherikhin, O. N. Kompanets, V. S. Letokhov, V. I. Mishin, V. N. Fedoseev, G. D. Alkhazov, A. E. Barzakh, E. E. Berlovich, V. P. Denisov, A. G. Dernyatin, and V. S. Ivanov, *Zh. Eksp. Teor. Fiz.* **86**, 1249 (1984) [English transl.: *Sov. Phys. JETP* **59**, 729 (1984)].

165. V. N. Fedoseev, V. S. Letokhov, V. I. Mishin, G. D. Alkhazov, A. E. Barzakh, V. P. Denisov, A. G. Dernyatin, and V. S. Ivanov, *Opt. Commun.* **52**, 24 (1984).

166. P. Brix and H. Kopfermann, *Z. Phys.* **126**, 344 (1949).

167. P. Brix and H. Kopfermann, *Rev. Mod. Phys.* **30**, 517 (1958).

168. A. Coc, C. Thibault, F. Touchard, H. T. Duong, P. Juncar, S. Liberman, J. Pinard, J. Lermé, J. L. Vialle, S. Büttgenbach, A. C. Mueller, and A. Pesnelle, *Phys. Lett.* **163B**, 66 (1985).

169. G. S. Hurst, M. G. Payne, S. D. Kramer, and J. P. Young, *Rev. Mod. Phys.* **51**, 767 (1979).

170. V. I. Balykin, G. I. Bekov, V. S. Letokhov, and V. I. Mishin, *Usp. Fiz. Nauk* **132**, 293 (1980) [English transl.: *Sov. Phys. Usp.* **23**, 651 (1980)].

171. Yu. A. Kudryavtsev and V. S. Letokhov, *Appl. Phys.* **B29**, 219 (1982).

172. Yu. A. Kudryavtsev, V. S. Letokhov, and V. V. Petrunin, *Pis'ma Zh. Eksp. Teor. Fiz.* **42**, 23 (1985) [English transl.: *JETP Lett.* **42**, 26 (1985)].

173. N. Bendali, H. T. Duong, P. Juncar, J. K. P. Lee, J. M. Saint-Jalm, and J. L. Vialle, *J. Phys.* (Paris) **47**, 1167 (1986).

174. R. E. Silverans, G. Borghs, P. De Bisschop, and M. Van Hove, *Hyperfine Interact.* **24–26**, 181 (1985).

175. R. Neugart, *Hyperfine Interact.* **24–26**, 159 (1985).

176. R. Neugart, W. Klempt, and K. Wendt, *Nucl. Instrum. Methods B*, **17**, 354 (1986).

177. E. W. Otten, *Hyperfine Interact.* **21**, 43 (1985).

178. E. Arnold, D. Bauer, J. Bonn, R. Gegenwart, T. Reichelt, and K. P. C. Spath, to be published.

179. E. Arnold, J. Bonn, R. Gegenwart, W. Neu, R. Neugart, E. W. Otten, G. Ulm, and K. Wendt, submitted to *Phys. Lett. B*.

180. J. Javanainen, M. Kaivola, O. Poulsen, and E. Riis, *J. Opt. Soc. Am.* **B2**, 1768 (1985).

181. T. Kühl, D. Marx, G. Huber, S. Schröder, and B. Fricke, in GSI 87-7 Report (1987).

3

Radiofrequency Spectroscopy of Rydberg Atoms

T. F. GALLAGHER

1. Introduction

One of the most compelling reasons for studying highly excited, or Rydberg, atoms is that in such atoms the Rydberg electron spends most of its time far from the ionic core. This one property has many interesting implications. For example, collisions of excited atoms with other neutral atoms and molecules may be treated as collisions with a separate ion and electron. From a spectroscopic point of view, the valence electron serves as a very gentle probe of the properties of the ionic core. Thus, one of the interesting features of Rydberg states of any atom but hydrogen is the comparison with hydrogenic behavior.

The most apparent difference between the energy levels of an arbitrary atom and those of hydrogen is the depression of the energies of the low l states below the hydrogenic levels. This is easily observed by purely optical techniques and is well known. Here we shall focus our attention upon the finer details of the higher l states of alkali and alkaline earth atoms which may be explored by radiofrequency spectroscopy. There are several motivations for such experiments. Valuable information can be obtained regarding the core polarization effects in these atoms. In fact the alkali and alkaline earth atoms are good examples of atoms that may or may not be treated by the static core polarization model.[1,2] In alkali atoms the fine structure intervals for states of high l are indistinguishable from those of hydrogen,

T. F. GALLAGHER • Molecular Physics Laboratory, SRI International, Menlo Park, California 94025. Present address: Department of Physics, University of Virginia, Charlottesville, Virginia 22901.

and as l is decreased the intervals begin to deviate from the hydrogenic values, apparently at the same l where penetration of the ion core by the outer electron becomes important.[3]

In alkaline earth atoms the high resolution of radiofrequency spectroscopy easily reveals irregularities due to perturbations in the spectra which might go unnoticed otherwise.

Aside from what these experiments reveal about the atoms, they provide in some cases an excellent means of studying the interaction of radiation with the atoms and may lead to some interesting applications such as far infrared detection.[4]

2. Rydberg Atoms

Very simple notions can be used to deduce the basic properties of Rydberg atoms. Let us consider a hydrogenic atom in which the electron moves in the $1/r$ Coulomb potential of the nucleus, and the energy W (in atomic units) of a level of principal quantum number n is given by $W = -1/2n^2$.[5] For an electron in a classical circular orbit in a Coulomb potential it is evident that r, the orbital radius of the Rydberg electron, is given by n^2 so that even for the relatively low $n = 10$ state $r = 100a_0$, where a_0 is the radius of the first Bohr orbit. Even if we consider noncircular orbits corresponding to states of low l, in a Coulomb potential the valence electron must spend most of its time at large orbital radius where it has low kinetic energy. Thus as a first approximation it is useful to consider Rydberg atoms as a valence electron in a large orbit about an ionic core. Many of the interesting properties of Rydberg atoms can be deduced from such a simple picture. For example, it is clear that an electron separated from the ionic core by a distance n^2 constitutes an electric dipole moment of order n^2, which suggests that Rydberg atoms should be highly polarizable, and that it should be quite easy to drive transitions between Rydberg states. Similarly, it is easy to see that there is very poor spatial overlap between the Rydberg state wave function and the ground state wave function so that the dipole moment connecting these states is necessarily small. Thus the radiative decay rates of Rydberg atoms are very slow in spite of the fact that an atom in a Rydberg state may have ~ 10 eV of energy.

Recalling a few properties of a hydrogenic wave function we can immediately make quantitative statements about the properties of Rydberg atoms. For example, for a hydrogen atom

$$\langle r \rangle \sim n^2 \tag{1}$$

and as $r \to 0$

$$\psi \propto n^{-3/2} \tag{2a}$$

$$\psi \propto l^{-5/2} \tag{2b}$$

From (2a), for example, it is clear that for the Rydberg states the radiative lifetimes scale as n^3 for high n since the frequency of the bluest transition does not change, but the transition matrix element decreases approximately as $n^{-3/2}$, reflecting the small r behavior of the wave function. Similar reasoning leads to the properties listed in Table 1, in which we give the n dependence of various properties as well as their magnitudes for the Na $10d$ state.

The level structure of all other atoms differ from that of hydrogen because the low l states are depressed in energy below the hydrogenic values. This can be represented by a quantum defect μ,[6] that is, $W = -1/2(n - \mu)^2$. Thus it is clear that the hydrogenic wave functions are not really applicable for, say, an alkali atom. However, we know that the Rydberg electron spends most of its time far from the ionic core where the potential is Coulombic. Thus in this region the wave function must be represented by a Coulomb wave function,[7] and properties that depend on the large r behavior of the wave function are calculable using Coulomb wave functions. Thus, as $r \to \infty$ we require that $\psi \to 0$, just as for hydrogen. However, for hydrogen we must also require that as $r \to 0$, ψ be bounded. It is the combination of these two boundary conditions that leads to the requirement that $W = -1/2n^2$ in hydrogen. In an alkali, for example, we can no longer apply the boundary condition that $\psi \to 0$ as $r \to 0$ and we simply require that the energy and boundary condition at $r \to \infty$ be used to determine empirically the phase shift δ of the wave function relative to a hydrogenic wave function as shown in Figure 1. It is straightforward to show that $\delta = \pi\mu$.[10,11] Note that this is the same phase shift experienced by a low-energy electron in scattering from the ionic core. The phase shift is independent of energy over some range of energy both above and below the ionization limit. This is not surprising in light of the fact that the kinetic energy of the electron when it scatters from the core at small orbital radius

Table 1. Properties of Rydberg Atoms

Property	n dependence	Na ($10d$)
Binding energy	n^{-2}	0.14 eV
Energy between adjacent n states	n^{-3}	0.023 eV
Orbital radius	n^2	$147a_0$
Geometric cross section	n^4	$68\,000a_0^2$
Dipole moment $\langle nd|er|nf \rangle$	n^2	$143ea_0$
Polarizability	n^7	0.21 MHz/(V/cm)^2
Radiative lifetime	n^3	1.0 μsec
Fine-structure interval	n^{-3}	-92 MHz

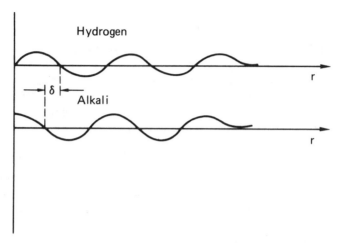

Figure 1. Radial wave functions for hydrogen and an alkali atom showing the phase shift δ.

is orders of magnitude larger than the binding energy of a Rydberg state. Thus, quantum defect theory treats in a unified fashion the bound and continuous portions of the spectrum. Efficient numerical procedures for calculation of Coulomb wave functions have been developed by Zimmerman et al.[8] and Bhatti et al.[9]

As suggested by Figure 1, an interesting aspect of the quantum defect theory is that the value of the quantum defect determines not only the energy of the state but also the spatial character of the wave function. Thus the properties in Table 1 may be applied to any atom if the effective quantum number $n^* = n - \mu$ is used instead of n.

3. Core Polarization and Penetration

To put rf spectroscopy in perspective, we show in Figure 2 a term diagram for Cs. As is evident in Figure 2, the $l < 3$ levels are depressed far below the hydrogenic values, whereas the $l \geq 3$ values are nearly hydrogenic, at least on the scale of Figure 2. The depression of the energy levels arises from two phenomena, the penetration and polarization of the ionic core by the outer electron.[6] In a penetrating orbit the outer electron spends a substantial fraction of the time inside the ionic core where it is exposed to more of the unshielded nuclear charge, thus depressing the energy of the state. Core penetration occurs only for the lowest l states, which have highly eccentric orbits. The higher l states, with more nearly circular orbits, exhibit very little core penetration, but still polarize the ionic core. To provide a

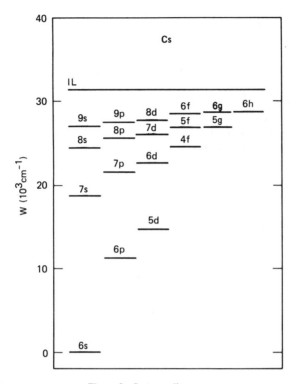

Figure 2. Cs term diagram.

feeling for the magnitude of this effect recall that an electron at a distance of 100 Å produces a field at the ion core of $\sim 10^5$ V/cm, a large field by laboratory standards, approximately comparable to those used in atomic deflection experiments.[12]

The phenomenon of core penetration is evidently a many-body effect, and its treatment is far beyond the scope of the present work. Thus we only note that since core penetration effects reflect the probability of finding the valence electron at the ionic core, the center of the atom, they scale as n^{-3} reflecting the $n^{-3/2}$ normalization of the wave function of the valence electron at the core.

On the other hand, core polarization is easily treated by a fairly simple model due to Mayer and Mayer which reduces the l dependence of the energies of high-l states to a few parameters.[1,2] The essence of the model is that the electric field of the electron at the ionic core distorts the polarizable core leading to an energy shift. Explicitly the energy level is given by

$$W = -1/2n^2 - W_{\text{pol}} \tag{3}$$

where the polarization energy W_{pol} is given by

$$W_{\text{pol}} = \frac{\alpha_d}{2}\left\langle\frac{1}{r^4}\right\rangle + \frac{\alpha_q}{2}\left\langle\frac{1}{r^6}\right\rangle \tag{4}$$

where α_d and α_q are the dipole and quadrupole polarizabilities of the ionic core and $\langle 1/r^4\rangle$ and $\langle 1/r^6\rangle$ are expectation values of $1/r^4$ and $1/r^6$ of the valence electron. Note that $\langle 1/r^4\rangle$ and $\langle 1/r^6\rangle$ give the squares of the electric field and gradient, respectively. The utility of Eq. (4) stems from the fact that $\langle 1/r^4\rangle$ and $\langle 1/r^6\rangle$ may be expressed analytically as functions of n and l using hydrogenic wave functions.

For a singly charged ion core[2]

$$\left\langle\frac{1}{r^4}\right\rangle = \frac{\frac{1}{2}[3n^2 - l(l+1)]}{n^5(l+3/2)(l+1)(l+1/2)l(l-1/2)} \tag{5}$$

and

$$\left\langle\frac{1}{r^6}\right\rangle = \frac{35n^4 - 5n^2[6l(l+1) - 5] + 3(l-1)l(l+1)(l+2)}{8n^7(l-3/2)(l-1)(l-1/2)l(l+1/2)(l+3/2)(l+2)(l+5/2)} \tag{6}$$

Thus the energies of all high-l, nonpenetrating, states may be expressed by α_d and α_q.

For the often encountered case of $n \gg l$[13]

$$W_{\text{pol}} \approx \frac{3\alpha_d}{4n^3 l^5} \tag{7}$$

From Eq. (7) it is clear that for high l the dipole polarizability dominates and $W_{\text{pol}} \propto l^{-5}$.

The limitation of the polarization model is that it treats the valence electron as though it is fixed, which is obviously not true. A measure of the validity is the ratio of the binding energy of the valence electron to the excitation energy of the ionic core. Consider an $n \approx 20$ state with a binding energy of $300\ \text{cm}^{-1}$. In the case of an alkali atom, with a closed shell ionic core, the excitation energies of the core are $\sim 200{,}000\ \text{cm}^{-1}$, whereas for an alkaline earth atom, with an open shell ionic core, they are as low as $5000\ \text{cm}^{-1}$. Thus the model may be expected to work better for alkali than alkaline earth atoms. Effects due to the motion of the electron are termed "nonadiabatic" effects. The magnitude of the effects can be determined, at least crudely, from the discrepancies between the calculated and observed polarizabilities. However, since the alkali ions have closed shells it is not a simple matter to calculate the ionic polarizability, so it is difficult to be confident regarding the source of any discrepancies.[14]

In alkaline earth atoms, on the other hand, the static core polarization model clearly does not reproduce the energies of the atomic levels, because of the magnitude of the nonadiabatic effects.[1] Several theories have been formulated for the nonadiabatic effects. For example, the validity of the approach of Eissa and Opik[15] has recently been verified by Vaidyanathan and Shorer[16] by comparing calculated and measured quantum defects of Ca.

The physical origin of the nonadiabatic correction is perhaps clearest in the earliest treatment of this problem by Van Vleck and Whitelaw,[17] who specifically addressed the problem of core polarization in atoms with two valence electrons. Here we outline their treatment.

Consider an atom with two spinless valence electrons. Its Hamiltonian is given by

$$H = \nabla_1^2 + \nabla_2^2 + f(r_1) + f(r_2) + 1/r_{12} \tag{8}$$

where ∇_1^2 and ∇_2^2 are the kinetic energy operators and r_1 and r_2 are the positions of the two electrons, $1/r_{12}$ is the Coulomb repulsion of the two electrons, and $f(r)$ is the potential due to the doubly charged core. As $r \to \infty$, $f(r) \to -2/r$. We can expand the interaction between the two electrons in multipoles as

$$\frac{1}{r_{12}} = \frac{1}{r_>} + \frac{r_<}{r_>^2} \cos \theta_{12} + \frac{r_<^2}{2r_>^3}(3\cos^2 \theta_{12} - 1) + \cdots \tag{9}$$

where θ_{12} is the angle between $\bar{r}_1$ and $\bar{r}_2$, and $r_>$ and $r_<$ are the larger and smaller of r_1 and r_2, respectively.

Let us focus our attention on an alkaline earth atom in which the outer electron is in a high-n, high-l state. In such atoms the outer electron is, to a reasonable approximation, always outside the inner electron. Thus if we only keep the first term of the $1/r_{12}$ expansion, and label the two electron positions as r_o and r_i (outer and inner), in Eq. (8) there is no longer any connection between the inner and outer electrons. The solutions then are products of one-electron wave functions that satisfy the equations

$$\nabla_i^2 + f(r_i) = W_i \tag{10a}$$

and

$$\nabla_o^2 + f(r_o) + 1/r_o = W_o \tag{10b}$$

The energies of the states are given by the sums of the two electron energies W_i and W_o and are thus relative to two free electrons and a doubly charged ion core. Note that Eq. (10a) corresponds to the energy levels of the singly charged ion and Eq. (10b) to the energy levels of the Rydberg electron. Recalling our previous assumption that $r_o > r_i$ we shall assume that $f(r_o)$ is given by its asymptotic expression $-2/r_o$ and thus Eq. (10b) reduces to the equation for neutral atomic hydrogen.

As a concrete example let us consider a specific atomic system, Ba. The first few energy levels of Ba^+, the states of the inner electron are shown by the bold lines of Figure 3, and the states of the outer electron are shown as a lighter series of lines converging to each state of Ba^+. Shaded areas above the ionic levels correspond to continua. It is to be understood that the lighter lines represent all values of l of the outer electron. Typical state vectors are $|6s\rangle|20g\rangle$ and $|5d\rangle|20g\rangle$, which would represent the $20g$ states converging to the $6s$ and $5d$ states of Ba^+. We note that the total angular momentum, L, is equal to 4 for the $|6s\rangle|20g\rangle$ state but may be anywhere from 2 to 6 for the $|5d\rangle|20g\rangle$ state. (Recall that we have assumed spinless electrons.) Since we are only considering interactions within the atom the total angular momentum L is conserved and only states of the same L interact. Although we shall not write L or its projection M on the axis of quantization explicitly we must bear in mind that they do not change.

In this approximation the states $6s20l$ are all degenerate; however, the degeneracy is removed by including the higher multipole terms in the $1/r_{12}$ expansion of Eq. (9). If we confine our attention for the moment to the $6snl$ states it is clear that the second term on the right-hand side of Eq. (9) leads, by second-order perturbation theory, to a shift $W_{d_{nl}}$ given by

$$W_{d_{nl}} = \sum_{n'',n'l'=l\pm1} \frac{|\langle n''p|\langle n'l'|\frac{r_i \cos\theta_{12}}{r_o^2}|6s\rangle|nl\rangle|^2}{W_{6snl} - W_{n''pn'l'}} \tag{11}$$

where $W_{n''pn'l'}$ and W_{6snl} are the energies of the $n''pn'l'$ and $6snl$ states. In addition the summation over n'' and n' is understood to include the continua, although this is not shown explicitly.

From Eq. (11) it is clear that the dependence on the inner electrons comes from the matrix elements $\langle n''p|r_i|6s\rangle$ and the energy separations as shown in Figure 3. Since the matrix element $\langle 6p|r_i|6s\rangle$ is more than two orders of magnitude larger than the matrix elements connecting $6s$ to the higher p states, we omit all the $n''p$ terms in Eq. (11) other than $6p$. Thus it is not necessary to sum over n''. The angular part of Eq. (11) (or of the analogous higher multipole terms) may be reduced using either the approach of Van Vleck and Whitelaw[17] or that of Edmonds.[18] The latter consists of evaluating the scalar product of spherical harmonics for the inner and

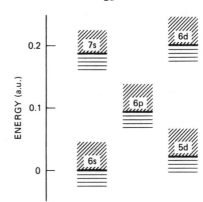

Figure 3. Low-lying levels of Ba$^+$ (——),
Rydberg series of Ba converging to each state
of Ba$^+$ (——) and continua above the Ba$^+$
state (///).[48]

outer electrons. This leads to

$$W_{d_{nl}} = \frac{|\langle 6p|r|6s\rangle|^2}{3} \sum_{n'} \frac{l|\langle n'l-1|1/r_o^2|nl\rangle|^2}{(2l+1)\,W_{6snl} - W_{6p'l-1)}}$$

$$+ \sum_{n'} \frac{(l+1)|\langle n'l+1|1/r_o^2|nl\rangle|^2}{(2l+1)(W_{6snl} - W_{6pn'l+1})} \tag{12}$$

If the spread in energies of the $n'l'$ ($l' = l \pm 1$) states involved in the sums
of Eq. (12) is small compared to the ion separation $W_{6p} - W_{6s}$ then we
may approximate $W_{6pn'l\pm1} - W_{6snl}$ by the ionic separation and remove it
from the sum. This yields finally

$$W_{d_{nl}} = -\frac{|\langle 6p|r_i|6s\rangle|^2}{3(W_{6p} - W_{6s})}$$

$$\times \frac{l}{2l+1}\sum_{n'} |\langle n'l-1|1/r_o^2|nl\rangle|^2 + \frac{l+1}{2l+1}\sum_{n'} |\langle n'l+1|1/r_o^2|nl\rangle|^2 \tag{13}$$

Note that $|\langle 6p|r_i|6s\rangle|^2/3(W_{6p} - W_{6s})$ is twice the dipole polarizability of the
Ba$^+$ $6s$ state. Van Vleck[19] has shown that

$$\langle nl|r^{2s}|nl\rangle = \sum_{n'}{}' |\langle n'l'|r^s|nl\rangle|^2 \tag{14}$$

for any value of l'. Thus Eq. (13) may be rewritten as

$$W_{d_{nl}} = \frac{-\alpha_d\langle nl|1/r_o^4|nl\rangle}{2} \tag{15}$$

which is the first term of Eq. (4). Here r_o corresponds to r in Eq. (4).

From the preceding development it is clear that the nonadiabatic effects arise from the fact that, in general, $W_{6pn'l'} - W_{6snl} \neq W_{6p} - W_{6s}$. Thus it is the energy *distribution* of the matrix elements $\langle n'l'|1/r_o^2|nl\rangle$ that leads to the nonadiabatic effects, and the magnitude of the nonadiabatic effects will be determined by the spread in energies of $\langle n'l'|1/r_o^2|nl\rangle$ relative to the energy of the ionic transition.

To provide a feeling for the energy spreads, the values of $|\langle n'l \pm 1|1/r^2|nl\rangle|^2$ calculated using hydrogenic wave functions are shown in Figure 4 for $n = 18$ and $l = 5$, the $6s18h$ state. The values shown in Figure 4 are normalized per unit energy (each bound state value is multiplied by n'^3 and given an energy width of n'^{-3} as is done by Fano and Cooper in their treatment of oscillator strengths[20]). Thus, as indicated by Eq. (14), the area under each curve equals $\langle 18h|1/r^4|18h\rangle$. As can be seen from Figure 4, the values of $|\langle n'l - 1|1/r^2|nl\rangle|^2$ lie below the Ba$^+$ $6p$ limit and the values for $|\langle n'l + 1|1/r^2|nl\rangle|^2$ lie above the Ba$^+$ $6p$ limit. Not surprisingly, the energy spread of the matrix elements is considerably greater for lower l than for higher l, and correspondingly, the nonadiabatic effects are larger.

It is useful to connect the results of the presumably accurate expression, Eq. (13), with the adiabatic polarization formula, Eqs. (4) and (15), by defining the factor k_d:

$$k_d\langle nl|1/r^4|nl\rangle = (W_{6p} - W_{6s})\frac{l}{2l+1}\sum_{n'}\frac{|\langle n'l - 1|1/r|nl\rangle|^2}{W_{6pn'l-1} - W_{6snl}}$$

$$+ \frac{l+1}{2l+1}\sum_{n'}\frac{|\langle n'l + 1|1/r^2|nl\rangle|^2}{W_{6pn'l+1} - W_{6snl}} \qquad (16)$$

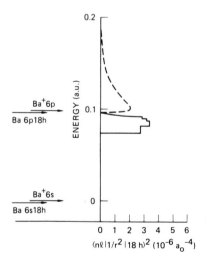

Figure 4. $\langle ng|1/r^2|18h\rangle^2$ (——) and $\langle ni|1/r^2|18h\rangle^2$ (- - -) plotted versus energy to show the energy distribution of the matrix elements.[48]

The next term in the expansion of $1/r_{12}$, the quadrupole term, equal to $(r_i^2/2r_o^3)(3\cos^2\theta_{12} - 1)$, may be treated in exactly the same manner. The only difference is that $6snl$ states are coupled to $n''dn'l'$ states where $l' = l - 2, l, l + 2$. We assume that only the low-lying $5d$ state of Ba$^+$ shown in Figure 3 leads to an appreciable energy shift, which is given by[17]

$$\Delta W_{q_{nl}} = \langle 5d|r_i^2|6s\rangle^2 \frac{3}{10(4l^2 - 1)(2l + 3)}$$

$$\times \left\{ (2l - 1)(l + 1)(l + 2) \sum_{n'} \frac{|\langle n'l + 2|1/r_o^3|nl\rangle|^2}{W_{6snl} - W_{5dn'l+2}} \right.$$

$$+ \frac{2(l^2 + l)(2l + 1)}{3} \sum_{n'} \frac{|\langle n'l|1/r_o^3|nl\rangle|^2}{W_{6snl} - W_{5dnl}}$$

$$\left. + (2l + 3)(l^2 - l) \sum_{n'} \frac{|\langle n'l - 2|1/r_o^3|nl\rangle|^2}{W_{6snl} - W_{5dn'l-2}} \right\} = |\langle 5d|r_i^2|6s\rangle|^2 \chi \qquad (17)$$

We estimate that ignoring the higher d states of Ba$^+$ leads to an error $\sim 10\%$. To a first approximation the numerical weightings of the $|\langle n'l'|1/r_o^3|nl\rangle|^2$ sums are given by the ratio $3:2:3$ for $l' = l + 2, l, l - 2$.

If the energy distributions of the $|\langle n'l'|1/r_o^3|nl\rangle|^2$ matrix elements cover a small range compared to the ionic $6s$–$5d$ separation the energy denominators may be taken out of the summations, and each summation, by Eq. (14), reduces to $\langle nl|1/r^6|nl\rangle$. Thus Eq. (17) reduces to

$$\Delta W_{q_{nl}} = \frac{-1}{5} \frac{\langle 6s|r_i^2|5d\rangle^2}{W_{5d} - W_{6s}} \langle nl|1/r_o^6|nl\rangle \qquad (18)$$

With

$$\alpha_q = \frac{2}{5} \frac{|\langle 6s|r_i^2|5d\rangle|^2}{W_{5d} - W_{6s}} \qquad (19)$$

Eq. (18) reduces to the quadrupole part of Eq. (4).

As for the dipole interaction the range of energies covered by $|\langle n'l'|1/r_o^3|nl\rangle|^2$ determines the magnitude of the nonadiabatic effects. For $l' = l - 2, l,$ and $l + 2$ the major contributions to $|\langle n'l'|1/r_o^3|6snl\rangle|^2$ come from the lowest-lying $5dn'l$ states, $n' = \infty$ (the vicinity of the Ba$^+$ $5d$ limit), and above the Ba$^+$ $5d$ ionization limit, respectively. Owing to the enormous spread in energy of the Ba $5dnl \pm 2$ states it is not surprising that, much more so than for the dipole terms, it is a poor approximation to use $W_{5dn'l'} - W_{6snl} = W_{5d} - W_{6s}$.

We again make connection with the adiabatic polarization formula by defining k_q, analogous to k_d

$$k_q\langle nl|1/r_o^6|nl\rangle = -5\chi(W_{5d} - W_{6s}) \tag{20}$$

where χ is defined implicitly in Eq. (17). Using k_d and k_q we can represent the polarization energy as

$$\Delta W_{pol} = -\frac{\alpha_d\langle 1/r^4\rangle}{2k_d} - \frac{\alpha_q\langle 1/r^6\rangle}{2k_q} \tag{21}$$

A similar expression is obtained by Vaidyanathan and Shorer[16] using the more sophisticated approach of Eissa and Opik.[15] Explicitly they obtain

$$\Delta W_{pol} = -\frac{\alpha_d}{2}(y_0^{(1)}\langle 1/r^4\rangle + y_2^{(1)}\langle 1/r^6\rangle)$$

$$-\frac{\alpha_d}{2}(y_0^{(2)}\langle 1/r^6\rangle + y_2^{(2)}\langle 1/r^8\rangle) \tag{22}$$

The y_2 terms do not appear in the Van Vleck and Whitelaw approach since it is only zeroth order in the wave function.

3.1. Fine Structure

In hydrogen the spin–orbit or fine structure interaction depends on the separation between the electron and the proton, specifically on $\langle 1/r^3\rangle$. In an alkali atom this interaction is present as well as interactions that occur at small orbital radius where the valence electron's orbit penetrates the core. Both of these effects scale as n^{-3} and are thus experimentally indistinguishable. However, the penetration effects occur only for low l states. Thus, it is not surprising that the penetrating states have fine structure intervals radically different, many times differing in sign, from those of hydrogen, while for the high-l states the fine structure intervals are apparently hydrogenic. The most interesting region to study is the boundary between the penetrating and nonpenetrating states where the fine structure intervals are frequently inverted. The calculation of such fine structure intervals has intrigued theorists, and several theories have been advanced for the behavior at the boundary between the penetrating and nonpenetrating states.[3,21,22]

3.2. The Stark Effect in Alkali Atoms

In an electric field the spherical symmetry of the atom is broken, but the azimuthal symmetry remains; thus if we ignore the spin of the valence electron $|m|$ remains a good quantum number although l does not. In atomic hydrogen each n manifold of states with azimuthal quantum number $|m|$ is split into $n - |m|$ states, each of which exhibits a nearly linear Stark shift out to a field where that state ionizes.

Of the $|m| = 0$ states the bluest member of the n manifold shifts up in energy by $\frac{3}{2}n(n-1)E$ and the reddest member shifts down by $\frac{3}{2}n(n-1)E$.[5] The other $n - 2$ members are evenly spread between these two. Ionization occurs at a field near the classical field required for ionization, $E \sim (16n^4)^{-1}$. Since the separation of the levels is given by $1/n^3$ it is apparent that adjacent n manifolds must cross at $E = (3n^5)^{-1}$, a field much lower than that required for ionization. In an alkali atom there are two major differences that arise from the presence of the finite-sized core. These effects are shown in Figure 5, experimental and calculated plots of the energy levels of the Na $|m| = 0$ and 1 states near $n = 15$.[23] First, the s, p, and d states have nonzero quantum defects and are well removed from the nearly degenerate manifold of high l states. Thus at low fields they exhibit a quadratic Stark shift. Second, states of the n and $n + 1$ manifolds repel each other because the levels are coupled by the core. The magnitude of the coupling is reflected in the magnitudes of the avoided crossings. In Figure 5 both of these effects are quite visible, and a resolution of ~ 1 cm^{-1}, which is easily obtained optically, is adequate to make useful measurements. However, in less well resolved cases a meaningful measurement can only be made by other techniques.

The Stark effect in alkali atoms is of more than academic interest since the high field behavior plays a crucial role in how atoms are ionized by a temporally varying field and the low field behavior is important for collision processes.

4. Experimental Techniques

A wide variety of techniques has been developed to measure radiofrequency intervals of Rydberg states. Some of these techniques are not truly radiofrequency techniques, but since they yield the same information we mention them here, although not in the same detail as the radiofrequency techniques.

4.1. Quantum Beats

The method of quantum beats has found application in the measurement of fine structure intervals, polarizabilities, and hyperfine intervals in several species. The basis of the approach is to prepare a coherent superposition of at least two states by pulsed excitation and observe the subsequent

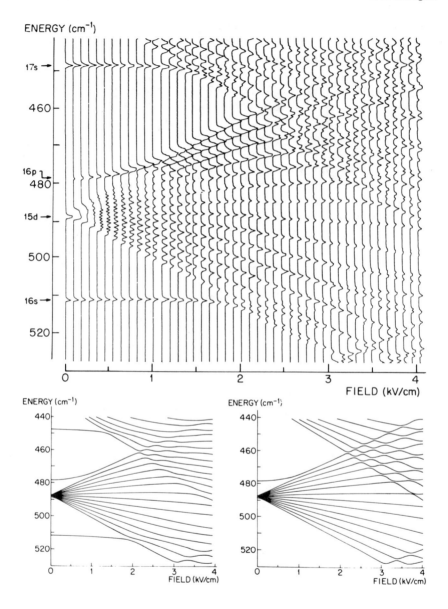

Figure 5. Stark structure of Na. (a) Experimental excitation curves for Rydberg states of Na
in the vicinity of $n = 15$. A tunable laser was scanned across the energy range
displayed (vertical axis). The zero of energy is the ionization limit. A signal generated
by ionizing the excited atoms appears as horizontal peaks. Scans were made at
increasing strengths (92 V/cm increments) and are displayed at the corresponding
field values. Both $|m| = 0$ and 1 states are present. Arrows identify zero-field position
of levels nl. (b) Calculated energy levels for the $|m| = 0$ states displayed above. (c)
Calculated energy levels for the $|m| = 1$ states displayed above.[23]

oscillation at the frequency of the interval between the two states in the evolution of the wave function using some sort of polarization detection. Good examples of such experiments are the measurement of the Na nd fine structure intervals using totally different techniques to observe the quantum beats. In these experiments a coherent superposition of the $nd_{3/2}$ and $nd_{5/2}$ states is created by a pulsed linearly polarized laser. After exciting Na in a vapor cell, Haroche et al.[24,25] used optical polarizers to select reradiated fluorescence polarized parallel and perpendicular to the polarization of the exciting light to observe the quantum beats for $9 \leqslant n \leqslant 16$. The beats occur in the time-resolved fluorescence as shown in Figure 6.

Two different approaches have been used to observe quantum beats of the Na nd states using electric field ionization of Rydberg atoms in an atomic beam. These experiments rely on the fact that states of the same n and l but different m are ionized by different electric fields.[26] In the experiments of Leuchs and Walther[27] an ionizing field was applied at a variable time after the laser excitation. The ion signal versus delay time yields a signal similar to those observed by Haroche et al.[24,25] and Jeys et al.[28], who have used the ingenious technique of applying a modest "freezing" field to stop the evolution of the fine structure states at a variable time after excitation leading to the same variation in the signal. Using this technique the Na nd fine structure intervals have been measured up to $n \sim 40$.

The major limitations of the approach are two. First the excitation must coherently excite both levels. For laser excitation this usually means

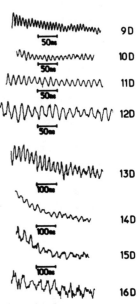

Figure 6. Time-resolved $(nd-3p)$ fluorescence signals from the Na nd states. The quantum beats occur at the nd fine structure intervals. Note the change in scale between $12d$ and $13d$.[24]

states of the same l which are separated by less than the inverse of the laser pulse duration. This is not usually a difficult requirement to fulfill. In addition the interval cannot be larger than the bandwidth of the detector. For fluorescence detection this is the photomultiplier bandwidth $\sim$300 MHz. For field ionization the rise time of the field ionization or freezing pulse sets a similar limit.

4.2. Optical Spectroscopy

Using the high resolution attainable with cw lasers it is possible to measure radiofrequency intervals, especially fine and hyperfine intervals using purely optical techniques. This approach has the obvious advantage that it only requires the discrimination between a Rydberg state and a low-lying state. As an example, we consider the determination by Fredriksson et al.[29] of the fine structure intervals of the Cs nf states by cw laser spectroscopy. Atoms were excited to the $7p$ state, and about 10% of these atoms cascade to the $5d$ state, from which they are excited to the nf states. The fact that the atoms are excited to the nf states is detected by observing the nf–$5d$ fluorescence. Since optical spectroscopy may be used with any detector that discriminates between Rydberg states and low-lying levels, it is very effective when used with very simple ionization detection methods such as a space charge limited diode.[30,31]

The major limitation of the purely optical approach is that it is limited to states of low l. It is most effective for fine structure intervals, which may be determined to high accuracy by the relative wavelength measurements of the transitions to the two fine structure levels.

4.3. Radiofrequency Resonance

The radiofrequency resonance experiments are analogous to molecular (or atomic) beam resonance experiments in which there is a state selector, rf transition region, and state analyzer.[32] In a typical Rydberg atom resonance experiment an exciting laser plays the role of the state selector, the radiofrequency transitions are induced in the usual fashion, and the final state analysis is done by a variety of techniques, which are described below. An experiment thus consists of populating one Rydberg state, exposing the atoms to a radiofrequency field, and looking for either a decrease in the population of the initial state or an increase in the final state (approaches termed "flop-out" or "flop-in" in conventional molecular beam terminology). This underscores the essential difficulty of such experiments, discriminating between two Rydberg states that have nearly the same energy. We note in passing that a practical difference between the familiar molecular beam resonance experiment and Rydberg atom experiments is that in the usual molecular beam experiment the three functions are separated in space and the molecule travels from state selector to rf region to state analyser. Rydberg atoms are relatively short lived, ~ 1 μsec, so such a scheme is in many cases not practical for thermal atoms. Thus the separation of the three functions is usually done in time, imposing some design constraints.

4.4. Optical Detection

Optical detection of the Rydberg atoms is a very attractive technique simply because the dominant fluorescence for two states that are nearly degenerate (≤ 1 cm^{-1} apart) can be red for one state and blue for the other. Consequently it is a straightforward matter to distinguish between two states, and optical detection has been used successfully with several alkali atoms, Li, Na, and Cs.[33-37] A typical experiment is the study of the Cs nf-ng transitions carried out by Ruff et al.[37] In Figure 7 we show the relevant Cs levels. The excitation of Cs atoms in a vapor cell to the nf states was accomplished with pulsed lasers using the cascade approach used by Fredriksson et al.[29] The transitions from the nf states to the ng states were detected by either observing the increase in the cascade $5f$-$5d$ fluorescence (flop in) or the decrease in the nf-$5d$ fluorescence (flop out) as the rf frequency was swept through the nf-ng resonance, as shown by Figure 8.

There are certain limitations to this technique that are worth noting. First, if the rf is on continuously, the size of the resonance signal S, the fraction of atoms that decay detectably from the final state, is related to the decay rates of the two states by

$$S = \alpha \frac{A_f}{A_i + A_f} \tag{23}$$

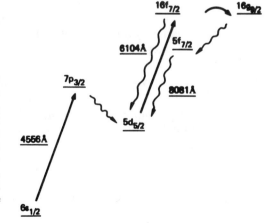

Figure 7. Energy levels for the microwave resonance experiment on the Cs $16f_{7/2}$-$16g_{9/2}$ transition. The straight upward arrows indicate the two laser pumping steps, the wavy downward arrows indicate the fluorescent decays, and the curved horizontal arrow indicates the microwave transition.

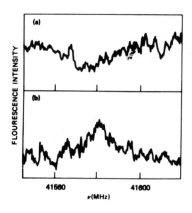

Figure 8. (a) Flop out Cs $16f_{7/2}$–$16g_{7/2}$ resonance on a sweep of decreasing frequency, (b) flop in Cs $16f_{5/2}$–$16g_{7/2}$ resonance on a sweep of increasing frequency. The slight offset is due to the time constant used in signal averaging.[37]

where A_i and A_f are equal to the radiative decay rates of the two states (presumably such experiments are done in a collision-free environment). The branching ratios of the transitions involved are incorporated into the numerical factor α which typically ranges from 0.03 and 0.5.[38] Thus observing a transition to a final state that is much longer lived than the initially populated state is difficult. In practice it is frequently more sensitive to look for an increase in fluorescence (such as the $5f$–$5d$ fluorescence of Figure 7) resulting from population of the final state. In this case the fraction of final state atoms that leads to the emission of such a photon is a major consideration. For the Cs case shown the fraction of ng atoms yielding a 8081-Å $5f$–$5d$ photon is 5%, whereas for the analogous Na nf states the yield of 8200-Å $3d$–$3p$ photons is 20% of the number of nf atoms.[38] The difference of a factor of 4 makes the Cs experiments considerably more difficult to carry out.

4.5. Selective Field Ionization

Selective field ionization (SFI) of Rydberg atoms has been used to observe radiofrequency transitions of both Na[39-43] and K.[44] The selectivity comes from the fact that the electric field required to ionize the valence electron increases approximately as the square of the binding energy of the atom. While this may not sound like a technique that would be sensitive enough to be useful in discriminating between nearly degenerate levels, in fact it is. For low $|m_l|$ states ($|m_l| \leq 2$) of Na it is straightforward to show that the fractional change in ionizing field for a $\Delta n = 0$, $\Delta l = 1$, $\Delta |m_l| = 0$ transition is given by[45]

$$\frac{\Delta E}{E} = \frac{4}{n^2} \qquad (24)$$

At $n = 20$, $\Delta E/E = 1\%$, a usable difference. Even greater sensitivity exists for changes in $|m_l|$. For example, $\Delta n = 0$, $\Delta l = 0$ and $|m_l| = 0 \rightarrow 1$ transitions lead to a fractional change of 3% in the ionizing fields required. Using this selectivity radiofrequency transitions at frequencies as low as 20 MHz $(6 \times 10^{-4} \text{ cm}^{-1})$ have been observed.[39]

A good illustration of the approach is given by the measurement of Na $ns\text{-}np$ and $nd\text{-}np$ intervals by Fabre *et al.*[41] As shown in Figure 9, Na atoms in an atomic beam are excited to ns states by a pulsed laser in the region between two plates and are subsequently exposed to microwave radiation for a period of ~ 1 μsec. Then a voltage ramp is applied to the lower plate producing a linearly increasing field which ionizes atoms in each state at its threshold field for ionization. Since these fields are reached at different times in the pulse, the signals from each state are temporally resolved. As shown in Figure 9, the $n - (n + 1)$ resonance, for example, is observed by monitoring the np signal at time t_{n+1} as the frequency of the microwave oscillator is swept. If radiative decay is negligible we note that the time integrated ionization signal (encompassing $t < t_{n+1}$ and $t > t_n$) is constant on or off resonance. Note that we could also observe the $n - (n + 1)$ resonance by observing the time integrated signal if we stopped the ramp before it reached V_n; such an approach has been used in measurements of Na p and d fine structure intervals and polarizabilities.[39]

While selective field ionization technique is an elegant and powerful technique for radiofrequency resonance experiments it has two inherent limitations. First, it has only been useful for $n > 15$, for at lower values of n, fields in excess of 10 kV/cm are required to ionize the atoms. Second, it is essential that the atomic states under study ionize at clearly different fields. In Na this condition is met at $n \sim 20$ by states of the same n but differing in l by one. In the heavier atoms Rb and Cs such selectivity is not possible,[46] and measuring Δl intervals using SFI is impossible. The reason for the difference becomes clear when we recall that the process of field ionization is inherently dynamic. As the field rises the atoms go from zero-field LS coupled fine structure states to high-field states in which the spin and orbit are essentially uncoupled. Exactly how this occurs depends on the atomic level separations relative to the slew rate of the ionizing field. In Na it happens that experimentally convenient slew rates yield diabatic crossings of the spin–orbit avoided level crossings in high field and very clean SFI behavior. In heavier alkalis with larger spin–orbit interactions, the avoided spin–orbit crossings are very large, as shown by Zimmerman *et al.*,[8] and this is no longer true. The resulting indiscriminant field ionization behavior precludes the use of SFI as a selective detector.

4.6. Delayed Field Ionization

In the preceding discussions of optical and SFI detection we have assumed that the radiofrequency field was always on. Here we describe a

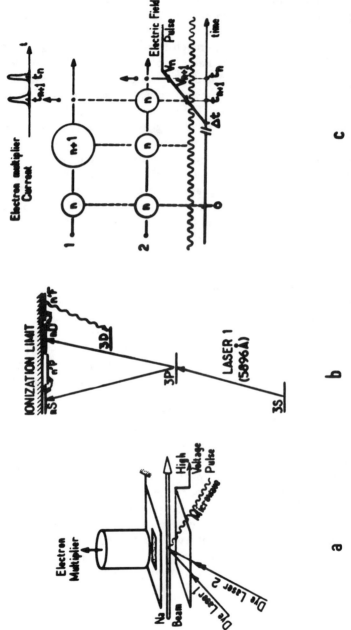

Figure 9. (a) Field ionization–rf experimental arrangement; (b) energy levels showing stepwise laser excitation and microwave transition; (c) the sequence of events in the experiment for two atoms excited by the laser excitation to level n. One of the atoms undergoes the microwave transition to $n+1$ and ionizes at time t_{n+1}. The atom remaining in n ionizes at t_n. [41]

technique in which the rf field is turned off and the detection of the resonance is based solely on the difference in the radiative lifetimes of the two states. This is particularly useful in the study of transitions to states of higher l which have longer lifetimes. A very important aspect of this technique is that no selectivity is required in the field ionization of the two states under study. It has thus been particularly useful for alkaline earth atoms, which do not have well-resolved field ionization behavior.[47-49]

The method was first employed by Safinya et $al.$[50] in their study of the Cs f-g, $f - h$, and f-i transitions using up to three rf photons. Cs atoms in a beam are excited to the nf states by two temporally separated pulsed lasers using the technique described earlier. The atoms are exposed to the rf field for 1 μsec following the laser excitation and the rf is then turned off. About 20 μsec later, three nf lifetimes, a field ionization pulse is applied, which ionizes all the Rydberg atoms present, and the total signal is recorded. Thus when the rf frequency is tuned to a resonance between the nf state and a higher l state, atoms undergo the nf-nl transition while the rf is on and are left in the more slowly decaying higher l state when the rf is turned off. This results in a larger ion signal when the field ionization pulse is applied $\sim$20 μsec later. Consequently as the rf frequency is swept through the resonance an increase in the ionization signal is observed. In principle the magnitude of the signal scales as $\exp(\Delta kT)$, where Δk is the difference in the decay rates and T is the delay time before the application of the ionizing pulse.

Although this method is quite general, its dependence on the difference in lifetimes introduces other problems. The fact that it is necessary that the delay between turning off the rf and the ionization pulse be several lifetimes of the initial state suggests that for very high-lying states with lifetimes of $\sim$10 μsec very long delay times will be needed, leading to possible transit time effects. The long delay times also mean that unless the apparatus is cooled to low temperatures the atoms are exposed to the background 300 K thermal radiation during this whole time, which drives transitions to states of higher l and tends to destroy the sensitivity, especially if the decay rates of the two states are nearly the same.[51,52] To get a feeling for the magnitude of this problem recall that for a 20s or 20d state the 300 K blackbody radiation transition rate is $\sim$20% of the 0 K radiative decay rate.[53]

While this technique has found its widest application in conjunction with field ionization detection, it is worth noting that it is equally useful in optical detection. In Eq. (23) we noted that with a cw rf source the optical signal corresponding to radiofrequency transition depends upon the relative radiative decay rates of the two levels, and the sensitivity decreases as the radiative decay rate of the final state. If the final state decays more slowly than the initial state, the signal is nearly proportional to the final state decay rate. In fact, if the pulsed laser excitation is used, and the rf is turned off about 1 μsec after the laser excitation, it should be possible to drive all the atoms in the initial state to the final state during the time the rf is on. Once the rf is turned off all the atoms are trapped in the final state; thus all the atoms must decay by radiating from the final state. Thus a 100% decrease in the fluorescence of the initial state and a 100% increase in the fluorescence of the final state is in principle possible, regardless of their relative decay rates. In fact significant improvements in the Cs $nf-ng$ resonances were observed using this approach.[54]

4.7. Selective Resonance Ionization

In selective resonance ionization we take advantage of the fact that there exist strong optical transitions from bound Rydberg states of alkaline earth atoms to autoionizing states and that the wavelengths and widths of these transitions depend upon the l of the valence electron of the bound Rydberg state as shown by Cooke et al.[55] Consider the Sr $5snl$ bound Rydberg states and their autoionizing $5pnl$ analogs. These are connected by the optical transition $5s-5p$ of the Sr$^+$ ion. This transition occurs at a frequency near the $5s-5p$ frequency of the Sr$^+$ ion, the exact frequency depending upon the values of the relative quantum defects of the $5snl$ and $5pnl$ states. The transition also has an intrinsic width determined by the autoionization rate of the $5pnl$ states, which decrease with increasing l, approximately as l^{-5}. Either of these effects can in principle be used to detect transitions between the bound $5snl$ states.

The former technique has been used by Cooke and Gallagher[56] to measure the intervals between Sr $5s(n + 2)d$ states and $5snf$ states. In Sr the quantum defects of the $5pnd$ states differ from those of the $5snd$ states by $1/2$, with the result that there are two strong optical transitions, corresponding to $5snd \rightarrow 5pn'd$, $n' = n \pm 1/2$, which straddle the location of the Sr$^+$ $5s-5p$ transition. For the $5snf$ states and $5pnf$ states the quantum defects are the same, so the transition occurs at the wavelength of the ion transition. Thus the wavelength falls between the wavelengths of the two $5snd-5pn'd$ transitions. The experiment was carried out by first exciting Sr atoms in a beam to the $5snd$ levels. The atoms are continuously exposed to a tunable radiofrequency field. About 50 nsec later a third laser pulse tuned to the Sr$^+$ $5s-5p$ transition crossed the atomic beam. Note that the third laser will not excite the $5snd$ states to the autoionizing $5pn'd$ states, but it will excite the $5snf$ states to the $5pnf$ states.

After the third laser pulse a small field is applied to extract ions resulting from excitation to the $5pnl$ autoionizing states. If the rf is off or not tuned to a Sr $5s(n + 2)d-5snf$ resonance we see very little ion signal. However, a sharp increase in the ionization signal is observed as the rf frequency is swept through the $5s(n + 2)d-5snf$ resonance.

It is quite apparent that this technique is very restricted in its applicability since we must have a laser to drive the transition to an autoionizing state. Thus while it is potentially interesting for alkaline earth atoms it is useless for alkali atoms. In addition, it is most useful when there is a l-dependent difference in the wavelength of the transitions to the autoionizing states, and experience to date indicates that this is not a common occurrence.

4.8. Applicability

From the preceding discussion of approaches to radiofrequency spectroscopy it is apparent that a variety of means exist to observe transition between any two Rydberg states. Since there is no fundamental limit to the frequency range over which the method is applicable it is the most general approach to measuring small intervals in Rydberg state energies and has found wide application in the measurement of Δl intervals, fine and hyperfine levels, and polarizabilities.

5. Overview of the Results Obtained

In this section we survey the results obtained for quantum defects and fine structure intervals in such a way as to highlight the unifying features of the observations.

5.2. Quantum Defects—Core Polarization

Early we noted that if we could measure several Δl intervals we could parametrize all the higher Δl intervals in terms of the polarizability of the ionic core. As an example of the application of the static polarization model let us consider the Cs Δl intervals. It is clear that Eq. (4) may be recast in terms of the intervals between adjacent l states. Let us label the difference in the polarization energies of the l and l' states, $W_{\text{pol}\,l} - W_{\text{pol}\,l'}$ as $\Delta_{ll'}$. It is convenient to adopt the convention of Edlen[2] and write $\Delta_{ll'}$ in cm^{-1} and define for l the terms $P_l = R\langle 1/r^4\rangle_l$, where R is the Rydberg constant and $Q_l = \langle 1/r^6\rangle_l/\langle 1/r^4\rangle_l$. If we define $\Delta P = P_l - P_{l'}$ and $\Delta PQ = P_l Q_l - P_{l'} Q_{l'}$ we may write $\Delta_{ll'}$ in the form used by Safinya et al.[51]:

$$\frac{\Delta_{ll'}}{\Delta P} = \alpha_d + \frac{\alpha_q \Delta QP}{\Delta P} \tag{25}$$

In this form Eq. (25) yields α_d and α_q, in a_0^3 and a_0^5, as the intercept and slope of a graph. In Figure 10 we show the Cs $nh-ni$ and $ng-nh$ intervals

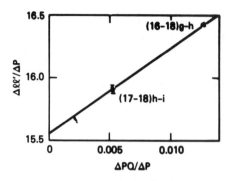

Figure 10. Plot of the scaled Cs Δl intervals $\Delta ll'/\Delta P$ for the Cs ng–nh and nh–ni intervals. The intercept and slope give the values of α_d and α_q directly in a_0^3 and a_0^5, respectively.[50]

plotted using Eq. (25). The manipulations leading to Eq. (25) essentially remove the variation due to the dipole polarizability. The g–h and h–i transitions lie on the line of Figure 10, whereas the f–g transitions lie distinctively above the line of Figure 10 reflecting of the core penetration of the Cs nf states which increases the nf–ng interval by about 20% above the value expected on the basis of core polarization alone. This seems to imply the desirability of using the highest l states possible. However, we note that as l is increased the states of adjacent l lie closer together and are thus more susceptible to systematic effects such as Stark shifts.

It is interesting to compare the results of the core polarization measurements from Rydberg atom Δl intervals with calculated values. As we mentioned earlier the polarizabilities from the measurements are expected to differ from the calculated values α_d and α_q because of the nonadiabatic effects. In Table 2 we present the measured [13,33,50] and calculated[57-60] values of the polarizabilities of the alkali ions.

Since the quantum defects of low angular momentum states are well known from optical spectroscopy and the high l quantum defects can be easily derived from the core polarization model, it is useful to present all the results together. In Figure 11 we show a plot of the alkali quantum defects vs. l on a logarithmic scale modified to include the s and p states.

Table 2. Observed and Calculated Alkali Ion Polarizabilities

	Observed		Calculated	
	α_d	α_q	α_d	α_q
Li$^+$	$0.1884\,(20)a_0^3$ [a]	$0.046\,(7)a_0^5$ [a]	$0.189a_0^3$ [b]	$0.1122a_0^5$ [c]
Na$^+$	$1.0015\,(15)a_0^3$ [d]	$0.48\,(15)a_0^5$ [d]	$0.9459a_0^3$ [e]	$1.53a_0^5$ [e]
Cs$^+$	$15.544\,(30)a_0^3$ [f]	$70.7\,(29)a_0^5$ [f]	$19.03a_0^3$ [g]	$118.26a_0^5$ [h]

[a] See Ref. 33.   [d] See Ref. 13.   [g] See Ref. 60.
[b] See Ref. 57.   [e] See Ref. 59.   [h] See Ref. 14.
[c] See Ref. 58.   [f] See Ref. 50.

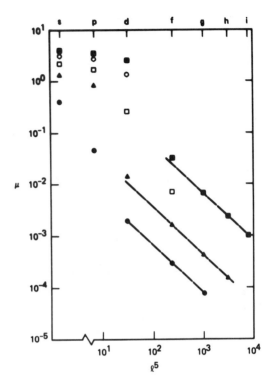

Figure 11. Plot of the quantum defects μ of the alkali atoms vs. l^5. Li ($\bullet$), Na ($\blacktriangle$), K ($\square$), Rb ($\bigcirc$), and Cs ($\blacksquare$). Note the l^{-5} dependence for high l where the dipole polarizability dominates.

Of particular interest is the fact that for the nonpenetrating high l states the quantum defect varies as l^{-5}, the leading term in the core polarization energy shift. In fact it is the departure from the l^{-5} behavior as l is decreased that indicates the onset of core penetration. This is particularly apparent for the Cs f states and Na d states, although no evidence for this is seen in the Li quantum defects.

5.2. The Nonadiabatic Effects in Alkaline Earth Atoms

Nonadiabatic polarization effects have been observed in both Ca[49] and Ba.[48] An analysis of the Ba Δl intervals using the adiabatic core polarization model as expressed in Eq. (25) yields a negative value of α_q, which is physically impossible and an indication of the magnitude of the nonadiabatic effects. The inclusion of the correction factors k_d and k_q, from Eqs. (16) and (20), to account for the nonadiabatic effects allows the extraction of α_d and α_q. Explicitly, using $P_l' = k_d P_l$ and $Q_l' = k_q Q_l$ leads to the results shown in Figure 12, which is analogous to Figure 10. Note that in Figure 10 the intervals between lower l states occur at larger values of $\Delta QP/\Delta P$ and that $\Delta QP/\Delta P > 0$, whereas for Ba, as shown by Figure 12,

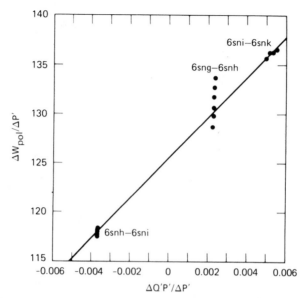

Figure 12. A plot of the scaled Ba Δl intervals using the factors k_d and k_q to correct for the nonadiabatic effects. The $6sng$-$6snh$ frequencies have been corrected to account for a local perturbation.[48]

neither of these is true, which is a direct consequence of the large non-adiabatic effects.

In Table 3 we tabulate the dipole and quadrupole polarizabilities of Ca^+ and Ba^+ which were calculated and derived from rf data. It is apparent that the more sophisticated approach of Eissa and Opik[15] gives better agreement than the method of Van Vleck and Whitelaw.[17]

5.3. Fine Structure Intervals of Alkali Atoms

In the earlier discussion the fact that the fine structure intervals converge to the hydrogenic value with increasing l was pointed out as an interesting

Table 3. Observed and Calculated Ca^+ and Ba^+ Polarizabilities

	Observed		Calculated	
	α_d	α_q	α_d	α_q
Ca^+	$87\,(3)a_0^3$ [a]	—	$89a_0^3$ [b]	$987a_0^5$ [b]
Ba^+	$125.5\,(10)a_0^3$ [c]	$2050\,(100)a_0^5$ [c]	$123a_0^3$ [c]	$2590a_0^5$ [c]

[a] See Ref. 49.
[b] See Ref. 16.
[c] See Ref. 48.

Table 4. $n = 15$ Fine Structure Intervals for H and Ratios for the Alkali Atoms

	H (MHz)	Li/H	Na/H	K/H	Rb/H	Cs/H
p	25.97 [a]	—	73.19(1) [c]	—	—	—
d	8.65 [a]	0.998(2) [b]	−3.20(5) [d]	−43.39(4) [g]	471(4) [h]	3460(30) [i]
f	4.33 [a]	0.99(2) [b]	0.95(2) [e]	—	—	−63.8(13) [j]
g	2.60 [a]	1.00(2) [b]	0.8(10) [f]	—	—	−0.3(20) [k]

[a] See Ref. 5.   [e] See Ref. 36.   [i] See Ref. 31.
[b] See Ref. 33.   [f] See Ref. 35.   [j] See Ref. 29.
[c] See Ref. 39.   [g] See Ref. 45.   [k] See Ref. 52.
[d] See Ref. 25.   [h] See Ref. 30.

aspect of alkali fine structure. This suggests that a useful way to present the observed intervals is in terms of the hydrogenic interval. Thus in Table 4 we present the ratio of the alkali fs intervals to the hydrogen intervals for $n \sim 15$. To provide a feeling for the magnitudes of the intervals we also give the hydrogenic intervals at $n = 15$. We note again that the fine structure intervals scale roughly as n^{-3}. Negative entries in Table 4 indicate an inverted fine structure doublet (i.e., the $j = l - 1/2$ level lies above the $j = l + 1/2$ level).

There are several general features in Table 4 that are worth mentioning. First, for the low l states, especially the p states, the fine structure intervals are orders of magnitude larger than for hydrogen for all but Li. This is hardly surprising in light of the substantial core penetration of the p states. At the other extreme, for high l there is not a measurable difference between the alkali and hydrogenic fs intervals. However, at the border between the penetrating and nonpenetrating states there are several notable features. First, in Na and Cs it appears that the d and f states, which are the highest l states to exhibit core penetration in the quantum defects, as shown by Figure 11, have inverted fine structure intervals. However, when l is increased by one to the lowest nonpenetrating state the fs intervals are at least close to hydrogenic. In the Na nf states the interval is only 5% less than in hydrogen, the Li nd states are hydrogenic to 1%, and in the Cs ng states the magnitude, if not the sign, is clearly hydrogenic. Thus it is clear that the fs interval is a better indication of core penetration than the quantum defect.

Thus it appears that the highest l state that is penetrating has inverted fine structure. This suggests that the K nd states and Rb nf states should be highest l penetrating states. Recall that K and Rb would fall between Na and Cs on Figure 11. Consequently we expect the fine structure intervals of the K nf states and Rb ng states to be hydrogenic. To our knowledge, though, experimental investigations of these states have not been carried out.

5.4. Applications

Radiofrequency spectroscopy of Rydberg atoms is useful in other investigations of Rydberg atoms, such as collision experiments, for it supplies useful values of energy level separations and coupling strengths. In addition, rf resonance techniques can be used to carry out such experiments, although such techniques have not been widely used.[61]

Perhaps the most intriguing application of Rydberg atoms is their use as an infrared detector.[4] Observing the absorption of a far infrared or microwave photon in the transition from one Rydberg state to another requires precisely the same techniques described here for rf spectroscopy, and of course a knowledge of the energy levels, Stark shifts. In fact experiments have already been done that demonstrate the sensitivity of the Rydberg atom detector.[62-64]

Acknowledgment

This work was supported by the National Science Foundation.

References

1. J. E. Mayer and M. G. Mayer, *Phys. Rev.* **43**, 605 (1933).
2. B. Edlen, *Handbuch der Physik*, Springer, Berlin (1964).
3. R. M. Sternheimer, J. E. Rodgers, T. Lee, and T. P. Das, *Phys. Rev. A* **14**, 1595 (1976).
4. D. Kleppner and T. W. Ducas, *Bull. Am. Phys. Soc.* **21**, 600 (1976).
5. H. A. Bethe and E. A. Salpeter, *Quantum Mechanics of One and Two Electron Atoms*, Academic, New York (1957).
6. E. U. Condon and G. H. Shortley, *The Theory of Atomic Spectra*, Cambridge Univ. Press, Oxford (1935).
7. D. R. Bates and A. Damgaard, *Phil. Trans. R. Soc. London* **249**, 101 (1949).
8. M. L. Zimmerman, M. G. Littman, M. M. Kash, and D. Kleppner, *Phys. Rev. A* **20**, 2251 (1980).
9. S. A. Bhatti, C. L. Cromer, and W. E. Cooke, *Phys. Rev. A* **24**, 161 (1981).
10. M. J. Seaton, *Proc. Phys. Soc. London* **88**, 801 (1966).
11. U. Fano, *Phys. Rev. A* **2**, 353 (1970).
12. T. M. Miller and B. Bederson, *Advances in Atomic and Molecular Physics*, Vol. 13, Academic, New York (1977).
13. R. R. Freeman and D. Kleppner, *Phys. Rev. A* **14**, 1614 (1976).
14. R. M. Sternheimer, *Phys. Rev. A* **1**, 321 (1970).
15. H. Elissa and U. Opik, *Proc. Phys. Soc. London* **92**, 556 (1967).
16. A. G. Vaidyanathan and P. Shorer, *Phys. Rev. A* **25**, 3108 (1982).
17. J. H. Van Vleck and N. G. Whitelaw, *Phys. Rev.* **44**, 551 (1933).
18. A. Edmonds, *Angular Momentum in Quantum Mechanics*, Princeton Univ. Press, Princeton, New Jersey (1960).
19. J. H. Van Vleck, *Proc. Natl. Acad. Sci. USA* **15**, 757 (1929).
20. U. Fano and J. W. Cooper, *Rev. Mod. Phys.* **40**, 441 (1968).

21. L. Holmgren, I. Lindgren, J. Morrison, and J.-M. Martensson, *Z. Phys. A* **276**, 179 (1976).
22. E. Luc-Koenig, *Phys. Rev. A* **13**, 2114 (1976).
23. M. G. Littman, M. L. Zimmerman, T. W. Ducas, R. R. Freeman, and D. Kleppner, *Phys. Rev. Lett.* **36**, 788 (1976).
24. S. Haroche, M. Gross, and M. P. Silverman, *Phys. Rev. Lett.* **33**, 1063 (1974).
25. C. Fabre, M. Gross, and S. Haroche, *Opt. Commun.* **13**, 393 (1975).
26. T. F. Gallagher, L. M. Humphrey, R. M. Hill, and S. A. Edelstein, *Phys. Rev. Lett.* **37**, 1465 (1976).
27. G. Leuchs and H. Walther, *Z. Phys. A* **293**, 93 (1979).
28. T. H. Jeys, K. A. Smith, F. B. Dunning, and R. F. Stebbings, *Phys. Rev. A* **23**, 3065 (1981).
29. K. Fredriksson, H. Lundberg, and S. Svanberg, *Phys. Rev. A* **21**, 241 (1980).
30. Y. Kato and B. P. Stoicheff, *J. Opt. Soc. Am.* **66**, 490 (1976).
31. S. M. Curry, C. B. Collins, M. Y. Mirza, D. Popescu, and I. Popescu, *Opt. Commun.* **16**, 251 (1976).
32. N. F. Ramsey, *Molecular Beams*, Oxford Univ. Press, London (1956).
33. W. E. Cooke, T. F. Gallagher, R. M. Hill, and S. A. Edelstein, *Phys. Rev. A* **16**, 1141 (1977).
34. T. F. Gallagher, R. M. Hill, and S. A. Edelstein, *Phys. Rev. A* **13**, 1448 (1975).
35. T. F. Gallagher, R. M. Hill, and S. A. Edelstein, *Phys. Rev. A* **14**, 744 (1976).
36. T. F. Gallagher, W. E. Cooke, S. A. Edelstein, and R. M. Hill, *Phys. Rev. A* **16**, 273 (1977).
37. G. A. Ruff, K. A. Safinya, and T. F. Gallagher, *Phys. Rev. A* **22**, 183 (1980).
38. A. Lindgard and S. E. Nielsen, *At. Data Nucl. Data Tables* **19**, 534 (1977).
39. T. F. Gallagher, L. M. Humphrey, R. M. Hill, W. E. Cooke, and S. A. Edelstein, *Phys. Rev. A* **15**, 1937 (1977).
40. C. Fabre, P. Goy, and S. Haroche, *J. Phys. B* **10**, L183 (1977).
41. C. Fabre, S. Haroche, and P. Goy, *Phys. Rev. A* **18**, 229 (1978).
42. P. Goy, C. Fabre, M. Gross, and S. Haroche, *J. Phys. B* **13**, L83 (1980).
43. W. E. Cooke, T. F. Gallagher, R. M. Hill, and S. A. Edelstein, *Phys. Rev. A* **16**, 2473 (1977).
44. T. F. Gallagher and W. E. Cooke, *Phys. Rev. A* **18**, 2510 (1978).
45. T. F. Gallagher, L. M. Humphrey, W. E. Cooke, R. M. Hill, and S. A. Edelstein, *Phys. Rev. A* **16**, 1098 (1977).
46. T. F. Gallagher, B. E. Perry, K. A. Safinya, and W. Sandner, *Phys. Rev. A* **24**, 3249 (1981).
47. T. F. Gallagher, W. Sandner, and K. A. Safinya, *Phys. Rev. A* **23**, 2969 (1981).
48. T. F. Gallagher, R. Kachru, and N. H. Tran, *Phys. Rev. A* **26**, 2611 (1982).
49. A. G. Vaidyanathan, W. P. Spencer, J. R. Rubbmark, H. Kuiper, C. Fabre, D. Kleppner, and T. W. Ducas, *Phys. Rev. A* **26**, 3346 (1982).
50. K. A. Safinya, T. F. Gallagher, and W. Sandner, *Phys. Rev. A* **22**, 2672 (1980).
51. E. J. Beiting, G. F. Hildebrandt, F. G. Kellert, G. W. Foltz, K. A. Smith, F. B. Dunning, and R. F. Stebbings, *J. Chem. Phys.* **70**, 3553 (1979).
52. T. F. Gallagher and W. E. Cooke, *Phys. Rev. Lett.* **42**, 835 (1979).
53. W. E. Cooke and T. F. Gallagher, *Phys. Rev. A* **21**, 588 (1980).
54. K. A. Safinya, private communication.
55. W. E. Cooke, T. F. Gallagher, S. A. Edelstein, and R. M. Hill, *Phys. Rev. Lett.* **40**, 178 (1978).
56. W. E. Cooke and T. F. Gallagher, *Opt. Lett.* **4**, 173 (1979).
57. H. D. Cohen, *J. Chem. Phys.* **43**, 3558 (1966).
58. J. Lahiri and A. Mukherji, *Phys. Rev.* **141**, 428 (1966).
59. J. Lahiri and A. Mukherji, *Phys. Rev.* **153**, 386 (1967).
60. J. Heinrichs, *J. Chem. Phys.* **52**, 6316 (1970).
61. T. F. Gallagher and W. E. Cooke, *Phys. Rev. A* **19**, 820 (1979).
62. T. F. Gallagher and W. E. Cooke, *Appl. Phys. Lett.* **34**, 369 (1979).
63. H. Figger, G. Leuchs, R. Straubinger, and H. Walther, *Opt. Commun.* **33**, 37 (1980).
64. T. W. Ducas, W. P. Spencer, A. G. Vaidyanathan, W. H. Hamilton, and D. Kleppner, *Appl. Phys. Lett.* **35**, 382 (1979).

4

Rydberg Series of Two-Electron Systems Studied by Hyperfine Interactions

H. Rinneberg

1. Introduction

For almost a decade, the study of atomic Rydberg states, i.e., states with one highly excited electron orbiting around an ion core, has been a fascinating and actively pursued field of atomic physics.[1-3] Considerable experimental and theoretical efforts have been spent to elucidate the electronic structure of such states. Term values, fine structure splittings, atomic lifetimes, or oscillator strengths, as well as the effect of external fields, such as Zeeman and Stark splittings, may be used to probe the change in electronic structure occurring along a Rydberg series. Generally, these quantities vary regularly with increasing effective principal quantum number n^*. Deviations from an expected trend may be indicative for perturbations of the electronic structure. This situation is frequently encountered in two-electron systems where configuration interactions may lead to dramatic changes of the electronic structure of states within a given Rydberg series.

In this chapter we concentrate on one particular aspect of Rydberg atom spectroscopy: the measurement, analysis, and interpretation of hyperfine structures of Rydberg states of two-electron systems. Although hyperfine interactions have traditionally been used to probe the electronic structure of atomic ground states and low-lying excited levels,[4,5] isotope

H. Rinneberg • Institut für Atom- und Festkörperphysik, Freie Universität Berlin, D-1000 Berlin 33, West Germany

shifts and hyperfine splittings of Rydberg states have been exploited only recently to this end. From both the experimental and theoretical point of view, two-electron systems like He, the alkaline-earth elements, group IIB metals (Zn, Cd, Hg), as well as Yb are especially attractive for such studies. In all these cases the hyperfine structure (hfs) of singly excited Rydberg states, i.e., states of configuration $msnl$, is dominated by the strong Fermi contact interaction between the inner ms valence electron and the nuclear magnetic moment. In contrast, the contribution of the nl Rydberg electron decreases as n^{*-3} and can be neglected in most cases. Hence the hfs of $msnl$ Rydberg states can be resolved even at high principal quantum numbers, employing techniques of Doppler-free laser spectroscopy,[6] whereas much higher precision is required to measure hyperfine structures of Rydberg states of alkalilike systems.[7] The coupling between both valence electrons is another factor that decisively influences the hfs of Rydberg states of two-electron systems.[8] Indeed, the singlet or triplet character is most crucial for the magnetic properties of Rydberg states, and hyperfine structure measurements turn out to be one of the most direct and sensitive ways to determine singlet–triplet mixing. With increasing excitation of the outer electron, singlet–triplet separations and fine structure splittings of $msnl$ Rydberg states decrease proportionally to n^{*-3} while the hyperfine coupling of the ms electron remains essentially constant. At a point where the Fermi contact interaction is of the same order as or even larger than fine structure splittings, strong mixing between $msnl$ fine structure components, corresponding to a recoupling of the various angular momenta S, L, J, I to the total angular momentum F, causes conspicuous shifts and intensity variations in the recorded hyperfine spectra. Throughout this article, we shall refer to "hyperfine-induced" mixing of fine structure components, whereas such phenomena have also been called "off-diagonal"[9] or "second-order"[10] hyperfine interaction in the literature. At sufficiently high principal quantum numbers, even hyperfine-induced n-mixing may occur. Because of hyperfine-induced singlet–triplet mixing, corresponding hyperfine components of $msnl\,^1L_l$ and $msnl\,^3L_l$ Rydberg states repel each other and move toward the neighboring $ms(n+1)l\,^3L_l$ and $ms(n-1)l\,^1L_l$ Rydberg states, respectively. Avoided crossings are observed due to off-diagonal elements of the hyperfine Hamiltonian, representing a mixing of Rydberg states with different principal quantum numbers.

Besides singly excited Rydberg states, both valence electrons may occupy higher orbits. Because of the low-lying $(m-1)d$ orbitals of Ca, Sr, and Ba the lowest members of $(m-1)dn'l'$ Rydberg series fall below the first ionization limits of these elements. Such doubly excited (valence) states may strongly perturb $msnl$ Rydberg series by configuration interaction, leading to displacements of the Rydberg levels as well as modifications of their wave functions. The hyperfine structure of perturbed Rydberg states is affected in two ways: (i) The admixture of $(m - 1)dn'l'$ character into the Rydberg states reduces the density of the ms valence electron at the nucleus, and consequently isotope shifts and the Fermi contact interaction are changed. (ii) Configuration interaction may profoundly modify the coupling between both valence electrons. Isotopes with and without nuclear spin are affected in the same way, contrary to hyperfine-induced mixing of fine structure components. However, hyperfine structure lends itself to monitor singlet–triplet mixing caused by configuration interaction.

Although hyperfine-induced mixing of neighboring Rydberg states accounts to a large extent for the striking variations in the recorded hyperfine spectra with increasing principal quantum number n, the use of isotope shifts and hyperfine structures for measuring configuration interactions is most important. A theoretical description of Rydberg series of two-electron systems in terms of a number of semiempirical parameters is provided by multichannel quantum defect theory (MQDT).[11,12] Besides oscillator strengths and electronic g factors,[13-15] experimental term values have traditionally been used as input data for such analyses. However, a large variety of different experimental data is necessary to determine all MQDT parameters unambiguously.[13,16] In particular, term values alone represent a too limited data set.[13,15,17] It has been shown recently[18] that the reliability of MQDT analyses can be improved considerably by including hyperfine structure data in addition to term values.

High-resolution laser spectroscopy of Rydberg states of two-electron systems has grown tremendously within the last five years. Indeed, nearly all the material discussed in this chapter stems from this period. In 1980, Liao et al.[19] determined hyperfine-induced singlet–triplet mixing in 2^3P and 3^3D states of ^3He, using intermodulated fluorescence spectroscopy. In the same year, Barbier and Champeau[20] reported a systematic study of hyperfine structures and isotope shifts of $6snd\ ^3D_1$ Rydberg states of ytterbium. Since then the heavier alkaline earth elements have been studied extensively. For this reason the emphasis of this chapter will be on a discussion of the results on Ca, Sr, and Ba, presented in Sections 3 and 4. In Section 5, we will briefly present the results reported for He, Mg, and Yb. The experimental techniques commonly used for measuring hfs of Rydberg states are introduced in Section 2.

Previous reviews devoted to the spectroscopy of one- and two-electron Rydberg atoms by Feneuille and Jacquinot[2] and by Fabre and Haroche[7] do not contain any of the material presented in this chapter. However, a wealth of information concerning the general physics of Rydberg states can be found in these articles. Furthermore, brief reports of the results obtained for Ca, Sr, and Ba Rydberg states by high-resolution laser spectroscopy have recently been published by several authors.[21-26] The interplay of laser spectroscopy of Ca, Sr, and Ba and multichannel quantum defect analysis has been the subject of a recent article by Aymar.[27]

2. Experimental Techniques

Although transitions to Rydberg states of sodium were observed as early as in 1906, systematic investigations of many atomic Rydberg series became feasible by the advent of tunable dye lasers. Up to now mostly pulsed dye lasers have been employed to populate Rydberg states either by coherent multiphoton excitation or by resonant, stepwise absorption of several photons of different frequency. Ionization of the excited electron and subsequent counting of free electrons or remaining ions in a charged particle detector represents a widely used, sensitive detection scheme for Rydberg atoms. Since the Rydberg electron is only weakly bound, pulsed electric fields of moderate field strength are sufficient for ionization. Alternatively, in a space charge limited diode the large cross sections of Rydberg states for collisional ionization are exploited for their detection. Detailed discussions of the various ways to populate and detect Rydberg states can be found in Refs. 2 and 7.

For high-resolution spectroscopy of Rydberg states generally narrow-band, tunable cw dye lasers are required, except for double resonance experiments[7] in which pulsed dye laser radiation is combined with microwave radiation. In this section we present the more important methods by discussing specific experiments aimed at measuring hyperfine structures and isotope shifts of Rydberg states of two-electron systems. For our purposes the different techniques are grouped into those employing an atomic beam and experiments using vapor cells. The suppression of the Doppler effect and the excitation and detection schemes are important aspects of the discussion.

2.1. Atomic Beam Experiments

2.1.1. One-Step Excitation and Fluorescence Detection

As is well known, the first-order Doppler effect can be eliminated by irradiating a well-collimated atomic beam perpendicular to a laser beam.[6] Since tunable, narrow-band cw dye laser radiation of sufficiently short wavelength is not yet available, in most cases the excitation of atomic Rydberg states by a one-photon transition from the ground state is not feasible. The population of Rydberg states, however, might proceed from excited, metastable or even short-lived states which are continuously pumped by an additional laser beam. The latter approach was chosen by

Grafström *et al.*,[28] who measured hyperfine structures and isotope shifts of $6snd\ ^1D_2$ Rydberg states of barium, using the experimental setup shown in Figure 1. A multimode dye laser oscillating at $\lambda = 553.5$ nm was employed to excite barium atoms from the $6s^2\ ^1S_0$ ground state to the $6s6p\ ^1P_1$ intermediate level. In order to avoid changes in the population of that state due to the fluctuating mode structure, the cavity length of the multimode dye laser was modulated, resulting in an effectively white output of the laser over a range of about 0.05 nm, wide enough to excite all barium isotopes simultaneously. The second step was accomplished by tuning a single-mode cw dye laser, operating in the blue spectral range, across the transition $6s6p\ ^1P_1 \rightarrow 6snd\ ^1D_2$. The excitation was detected by monitoring the fluorescence originating from the decay of the Rydberg states. A typical spectrum obtained in this way is shown in Figure 2, revealing the hyperfine structure of the $6s6p\ ^1P_1$ and $6s14d\ ^1D_2$ states for the isotopes ^{135}Ba $(I = 3/2)$ and ^{137}Ba $(I = 3/2)$. The isotope shift between ^{136}Ba and ^{138}Ba is barely resolved. In addition to the lifetime broadening of the $6s6p\ ^1P_1$ intermediate level of 20 MHz, residual Doppler broadening contributed to an overall linewidth of about 40 MHz. Since the lifetimes of Rydberg states increase proportionally to n^{*3}, the sensitivity of fluorescence detection decreases progressively, and only Rydberg states up to $n = 24$ were observed.

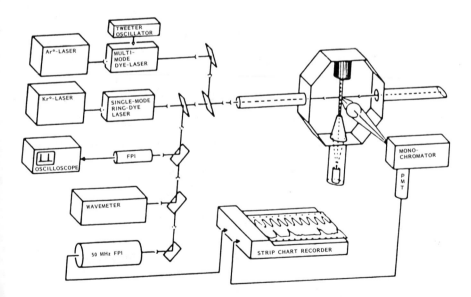

Figure 1. Experimental setup for hyperfine structure and isotope shift measurements of Ba Rydberg states. (Taken from Ref. 28.)

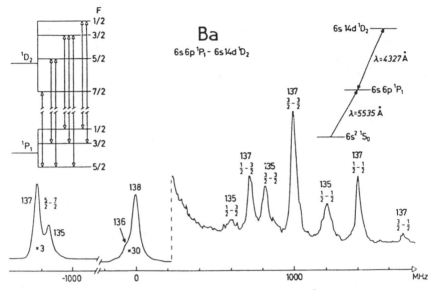

Figure 2. Spectrum of the $6s6p\ ^1P_1 \rightarrow 6s14d\ ^1D_2$ transition at 4327 Å. The excitation scheme and a schematic diagram showing the hyperfine transitions of ^{135}Ba and ^{137}Ba are included. (Taken from Ref. 28.)

2.1.2. One-Step Excitation and Field Ionization

Compared to fluorescence detection, field ionization is a considerably more sensitive technique to observe highly excited Rydberg states. In order to avoid broadening and shifts of spectral lines caused by the Stark effect, the excitation of Rydberg states has to be carried out in the absence of external electric fields. This can be achieved by either spatial or temporal separation of the laser radiation and the electric fields used for ionization. Temporal separation was employed by Barbier and Champeau[20] to detect $6snd\ ^3D_1$ Rydberg states of Yb. Starting from the $4f^{14}\ 6s6p\ ^3P_0$ metastable state, these Rydberg levels were excited by means of the frequency-doubled output of a pulsed single-mode dye laser, and a delayed electric field pulse was applied to ionize them. The ions produced were counted by an electron multiplier. The bandwidth (20 MHz) of the laser and residual Doppler broadening amounted to a total linewidth of about 60 MHz, sufficiently narrow to resolve isotope shifts and hyperfine structures of the heavy element ytterbium. The metastable state was populated by a discharge at the aperture of the atomic beam source. Figure 3 displays the excitation spectrum of the $4f^{14}\ 6s30d\ ^3D_1$ state. The shifts between the even isotopes (^{170,172,174,176}Yb) and the hyperfine structure of the odd ones [^{171}Yb ($I = 1/2$), ^{173}Yb ($I = 5/2$)] are clearly visible. Rydberg states up to $n = 53$ were observed in this way.

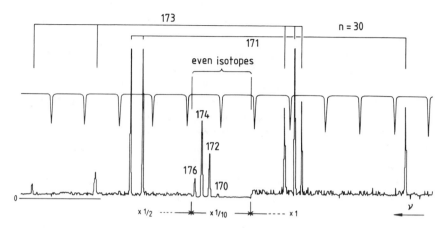

Figure 3. High-resolution spectrum of the transition $4f^{14}6s6p\,^3P_0 \rightarrow 4f^{14}6s30d\,^3D_1$ of Yb, obtained by delayed field ionization. (Taken from Ref. 20.)

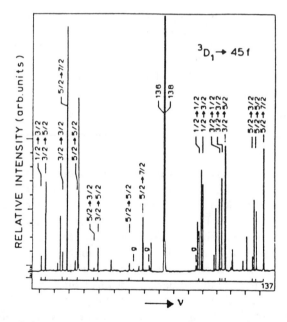

Figure 4. High-resolution spectrum of the transition $6s5d\,^3D_1 \rightarrow 6s45f\,^3F$ obtained by spatially separated field ionization. The hyperfine components of ^{137}Ba have been labeled according to the total angular momentum of the lower and upper states, respectively. The lines labeled "g" are spurious signals. The frequency scale corresponds to 613 MHz/division. (Taken from Ref. 30.)

Whereas delayed field ionization is used in conjunction with pulsed lasers, cw dye lasers require spatial separation of the laser fields and static electric fields. Using this technique, Eliel and Hogervorst[29] investigated $6snf\,^3F$ Rydberg states of barium. The frequency-doubled output of a single-mode cw dye laser excited Ba atoms from the $6s5d\,^3D_{1,2}$ metastable states to the $6snf\,^3F$ Rydberg levels. The excitation took place in a shielded, field-free region of the atomic beam. Because of their long lifetimes, Rydberg states could be ionized by a static electric field several millimeters downstream. Whereas the excitation is state-selective, the detection is not. Blackbody radiation can cause a redistribution[31] of the Rydberg atoms among neighboring states during the transit time of the atoms from the excitation to the ionization region. Figure 4 shows a recording of the $6s5d\,^3D_1 \rightarrow$ $6s45f\,^3F$ transition. Because of angular momentum selection rules the signals of ^{136}Ba and ^{138}Ba correspond to the $^3F_2$ fine structure level. Strong hyperfine-induced mixing of the ^3F fine structure components occurs for the odd isotopes.

2.1.3. Quasiresonant Two-Step Excitation and Field Ionization

Coherent two-photon spectroscopy[6,32] is a well-known technique that eliminates the Doppler effect, but suffers from the low cross sections σ_{fg} for the coherent absorption of two photons of equal frequency. For quasiresonant two-step excitation[33] the overall absorption cross section is considerably larger, but the Doppler effect is only partially removed. For example, a residual Doppler width (FWHM)

$$\Delta \nu_d^{\text{res}} = \left| \frac{\omega_2 - \omega_1}{\omega_2 + \omega_1} \right| \Delta \nu_d \qquad (1)$$

is encountered when two photons are absorbed coherently from two counter-propagating laser beams of frequencies $\omega_1/2\pi$ and $\omega_2/2\pi$. In Eq. (1) $\Delta \nu_d$ is the full Doppler width (FWHM) of the transition from the initial ($|g\rangle$) to the final state ($|f\rangle$). Therefore, similar to one-photon transitions, an atomic beam may have to be used to suppress the Doppler effect sufficiently.

Quasiresonant two-step excitation was employed by Eliel and Hogervorst to study hyperfine structures and isotope shifts of $6sns\,^{1,3}S$ [34] and $6snd\,^{1,3}D$ [35] Rydberg states of barium. Beams from two actively stabilized cw dye lasers intersected an atomic beam perpendicularly. The first laser was locked to a particular component of the barium resonance line $6s^2\,^1S_0 \rightarrow$ $6s6p\,^1P_1$. The second laser was scanned across the upper atomic transition $6s6p\,^1P_1 \rightarrow 6sns\,^{1,3}S$ or $6snd\,^{1,3}D$. Again the Rydberg states were detected by field ionization in a separate, shielded portion of the atomic beam and by subsequent counting of the electrons. Figures 5a and 5b show high-resolution spectra of the $6s20d$ $^1D_2$ Rydberg state obtained in this way. The hyperfine structures displayed in Figure 5 correspond to the $6s20d$ $^1D_2$ Rydberg state only; the intermediate $6s6p$ $^1P_1$ level does not contribute to the measured splittings. The upper and lower trace were recorded with the first laser locked to the $F_g = 3/2 \rightarrow F_i = 5/2$ (^{135}Ba) and $F_g = 3/2 \rightarrow F_i = 3/2$ (^{137}Ba) hyperfine transitions, respectively. Because of lifetime broadening of the $6s6p$ $^1P_1$ state, the signals of the barium resonance line corresponding to ^{134}Ba and ^{136}Ba strongly overlap with its ^{135}Ba $F_i = 5/2$ hyperfine component. Hence the excitation of these isotopes to the $6s20d$ $^1D_2$ state (cf. Figure 5a) proceeds resonantly as do the transitions to the hyperfine levels $F_f = 3/2$, $5/2$, and $7/2$ of ^{135}Ba. On the other hand, in Figure 5a the signals of ^{137}Ba and the $F_f = 1/2$ component of ^{135}Ba are produced by quasiresonant coherent two-photon absorption. Similarly, all signals displayed

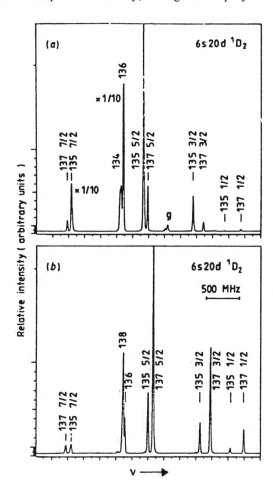

Figure 5. High-resolution spectra of the $6s20d$ $^1D_2$ state of Ba obtained by quasiresonant stepwise excitation and field ionization. For spectra (a) and (b) the first laser was tuned to the ^{135}Ba 3/2-5/2 and ^{137}Ba 3/2-3/2 hyperfine components of the 553.7 nm transition, respectively. The peaks belonging to the odd isotopes are labeled by the F quantum number of the final state. (Taken from Ref. 35.)

in Figure 5b are obtained in this way except for the hyperfine components $F_f = 1/2$, $3/2$, and $5/2$ of ^{137}Ba. Apart from a trivial shift in frequency, the two recordings differ in the relative intensities of the observed spectral lines, reflecting the strong enhancement obtained for stepwise (resonant) excitation.

2.1.4. Quasiresonant Two-Step Excitation and Collisional (Penning) Ionization

In an atomic beam collisions occur in the forward direction because of the distribution of thermal velocities along the beam axis. Such collisions can be exploited to ionize Rydberg atoms. In this section we describe an experimental setup, primarily designed to study highly excited Rydberg states in external electric and magnetic fields.[36] The technique is closely related to the experiment discussed in the previous section and can also be applied to measure hyperfine structures. The method takes advantage of the large cross sections of Rydberg states for collisional ionization as well as the high sensitivity of ion detection. The experimental setup is shown in Figure 6. Again, two stabilized cw dye laser beams intersect an atomic beam perpendicularly and excite the barium atoms to $6sns$ $^{1,3}S$ or $6snd$ $^{1,3}D$ Rydberg states via the $6s6p$ $^1P_1$ intermediate level. The Rydberg atoms are ionized by collisions with Ba atoms in the intermediate $6s6p$ $^1P_1$ or

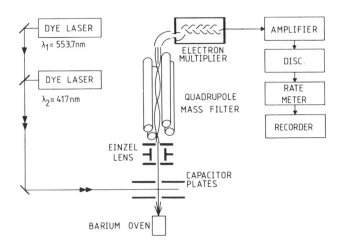

Figure 6. Experimental setup for two-step excitation of Ba Rydberg states followed by collisional ionization and mass selective detection of the Ba$^+$ ions produced. (Taken from Ref. 36.) The parallel plate capacitor served to compensate stray electric fields.

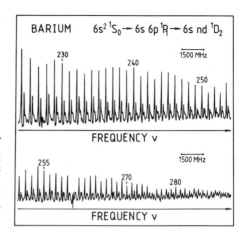

Figure 7. High-resolution spectrum of barium Rydberg states. The $6snd\ ^1D_2$ series can be followed up to 40 GHz ($n = 290$) below the ionization limit. Besides the $^1D_2$ series (strong signals), $^1P_1$ states appear as smaller lines. (Taken from Ref. 36.)

metastable $6s5d\ ^1D_2$ states. Such Penning ionization processes have been identified by high-resolution electron spectroscopy.[37] The Ba^+ ions are detected by means of a quadrupole mass spectrometer. As discussed in the previous section, resonant two-step excitation as well as quasiresonant two-photon transitions are involved, when studying the hyperfine structure (^{135}Ba, ^{137}Ba) of the Rydberg states. The spectrum shown in Figure 7 was obtained by tuning the first laser to the ^{138}Ba component of the resonance line and by scanning the frequency of the second laser across the upper atomic transitions. In this case the Rydberg states are reached by resonant two-step excitation, resulting in count rates up to 10^6/sec. The $6snd\ ^1D_2$ Rydberg series has been followed up to principal quantum numbers $n = 290$. Such states represent the largest atoms that have been produced under laboratory conditions up to now. Owing to Stark mixing caused by stray electric fields $^1P_1$ states were also excited and appear as smaller signals in Figure 7.

2.2. Experiments Using Vapor Cells

2.2.1. Doppler-Free Two-Photon Spectroscopy and Thermionic Detection

Doppler-free two-photon spectroscopy[6,32] offers several attractive features to study Rydberg series at high resolution.[38] Since two photons of equal frequency are used, only one laser is required, resulting in a simplified setup. Furthermore, the splittings and linewidths observed depend on the initial and final states only. Consequently, spectra obtained by this technique generally exhibit excellent resolution and can easily be analyzed.

The low oscillator strengths associated with off-resonant two-photon transitions can be compensated in a number of different ways. First of all, large power densities can be achieved using cw dye ring lasers with output powers of typically 0.1-1 W. Secondly, high atomic densities in vapor cells can be exploited favorably, since usually two-photon transitions cannot be saturated by cw dye laser radiation. Most decisive, however, is the use of thermionic diodes[39,40,41] to detect highly excited states. In a thermionic diode Rydberg states are ionized by collisions with, e.g., atoms in the atomic ground state. The ions partially neutralize the space charge surrounding a heated cathode, allowing electrons to flow to the anode. This current produces a voltage drop across a load resistor. The cross section for collisional ionization increases strongly with n and the detection efficiency of a thermionic diode approaches unity at high principal quantum numbers. Since the lifetime of the ions may be in the order of milliseconds, under favorable circumstances the diode may have an internal gain up to 10^5-10^7. Besides collisional ionization, photoionization may play a role under certain conditions.

In Figure 8 we present an instructive example for the sensitivity of two-photon spectroscopy combined with thermionic detection. Using the experimental setup shown in Figure 9, the $4s12d\ ^1D_2$ Rydberg state of Ca was populated by two-photon absorption, starting from the $4s^2\ ^1S_0$ ground state.[42] Only part of the recorded spectrum is displayed in Figure 8, revealing the isotope shift between the less abundant even isotopes ^{42}Ca (0.65%) and ^{44}Ca (2.09%) as well as the hyperfine structure of ^{43}Ca (0.14%, $I = 7/2$). All components with total angular momentum quantum numbers

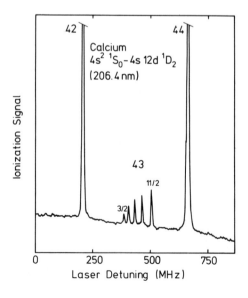

Figure 8. High-resolution two-photon spectrum of the $4s12d\ ^1D_2$ state of Ca. (Taken from Ref. 42.)

F ranging between 3/2 and 11/2 are clearly resolved. Compared to the $F = 7/2$ component of ^{43}Ca the signal of the most abundant isotope ^{40}Ca (not shown in Figure 8) is larger by a factor of 3500.

The ultimate limit for the sensitivity of a thermionic diode at very high n is set by the electric polarizability of Rydberg atoms, which increases as n^{*7} for states with low angular momentum quantum numbers. Small electric fields, present in the space charge of the diode, cause line broadenings and shifts due to the Stark effect. This seriously impedes the use of thermionic diodes as detectors for high-resolution studies of Rydberg states with principal quantum numbers $n > 100$. In Figure 10 two-photon spectra of $5s99d\ ^1D_2$ and $5s100s\ ^1S_0$ Rydberg states of Sr[25] are shown, demonstrating the considerable improvement in detector performance that can be achieved by carefully shielding the excitation region. In the lower part of Figure 10, a conventional diode was used that had been divided into two compartments by means of a fine mesh. Rydberg atoms excited in the lower section diffused to the detection region (upper part) containing the cathode. Only limited

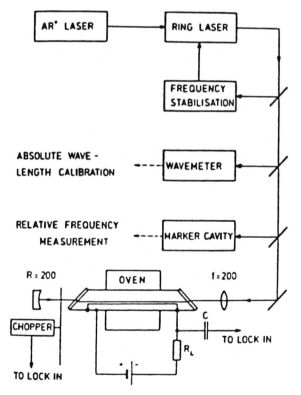

Figure 9. Experimental setup used for Doppler-free two-photon spectroscopy and thermionic detection. (Taken from Ref. 25.)

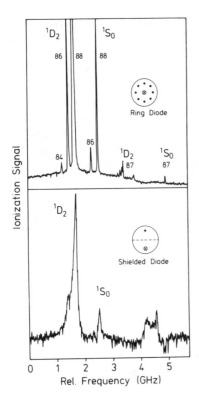

Figure 10. Two-photon excitation spectra of $5s99d\ ^1D_2$ and $5s100s\ ^1S_0$ Rydberg states of Sr. (Taken from Ref. 25.) The upper part was recorded using a ring diode, while the lower part was produced with a conventionally shielded diode. The spectra were obtained using the setup shown in Figure 9.

resolution was achieved, since at temperatures of 600–800°C necessary to produce a sufficient strontium vapor density, all metal parts, including the outer cylinder (anode) of the oven as well as the dividing mesh, are surrounded by space charges due to thermionic emission. To overcome this difficulty, Beigang *et al.*[41] constructed a ring diode where the cathode wires are arranged axially symmetric, thus creating a field-free space along the axis. The lower part of Figure 10 was obtained with the laser beams carefully aligned along the symmetry axis. Another important advantage of the annular arrangement of several cathodes is a correspondingly enhanced detection efficiency. Rydberg states up to $n = 200$ have been studied in this way.[41]

2.2.2. Resonant Two-Step Excitation and Thermionic Detection

Resonant two-step excitation by means of co- or counterpropagating cw dye laser beams results in Doppler-free spectra due to absorption line narrowing.[33,43] Starting from the atomic ground state $|g\rangle$, certain velocity ensembles, Doppler-tuned into resonance, are excited by one of the laser beams to the intermediate level $|i\rangle$. When the second laser is scanned across the upper atomic transition $|i\rangle \to |f\rangle$, the excited atoms are promoted further to the final (Rydberg) state $|f\rangle$. Although only a small fraction of the lower state population is utilized in this way, resonantly enhanced absorption cross sections more than compensate for the reduction in the effective particle density. In particular, for resonant dipole–dipole ($E1$–$E1$) transitions, only moderate laser output powers are required to achieve high transition rates, and purely optical techniques (see Section 2.2.4) may be sufficiently sensitive to monitor the excitation of the Rydberg states. When thermionic detection (cf. Section 2.2.1) is combined with resonant stepwise excitation, even forbidden ($M1$ or $E2$) atomic transitions can be exploited for high-resolution spectroscopy of Rydberg states. For example, Rydberg series with parity opposite to that of the atomic ground state can be reached by $E2$–$E1$ or $M1$–$E1$ transitions. Recently,[26,44] isotope shifts and hyperfine structures of $6snp\ ^{1,3}P$ and $6snf\ ^{1,3}F$ ($10 \leqslant n \leqslant 40$) Rydberg series of barium have been measured in this way, using the experimental setup shown in Figure 11.

In Figures 12 and 13 high-resolution spectra are shown for the $6s17p\ ^1P_1$ Rydberg state obtained for co- and counterpropagating laser beams, respectively. The wavelength $\lambda_1 = 2\pi c/\Omega_1$ of the first laser (see Figure 11) was kept fixed at an arbitrary position within the Doppler contour of the $6s^2\ ^1S_0 \to 5d^2\ ^1D_2$ quadrupole transition, while the frequency $\Omega_2/2\pi$ of the second laser was scanned across the upper atomic transition $5d^2\ ^1D_2 \to 6s17p\ ^1P_1$. The hyperfine components of ^{135}Ba, ^{137}Ba ($I = 3/2$) have been labeled F_i–F_f according to the hyperfine levels of the intermediate and final state, respectively. The total hyperfine splitting of the intermediate state is smaller than the Doppler width of the quadrupole transition. Hence all

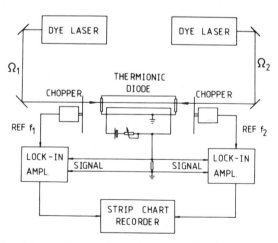

Figure 11. Experimental setup for stepwise excitation and thermionic detection. (Taken from Ref. 26.)

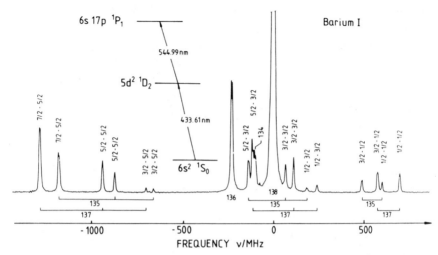

Figure 12. Two-step excitation ($E2$–$E1$) of the $6s17p\ ^1P_1$ Rydberg state of barium using copropagating laser beams. (Taken from Ref. 26.)

hyperfine components $6s^2\ ^1S_0\ F_g = 3/2 \rightarrow 5d^2\ ^1D_2\ F_i = 1/2, 3/2, 5/2$, and 7/2 are excited simultaneously by the first laser beam together with the even isotopes. Proceeding from the hyperfine levels of the intermediate state, different frequencies of the second laser are required to reach the same hyperfine component F_f of the final state. Hence the intermediate state contributes to the recorded hyperfine spectrum, in contrast to coherent two-photon absorption (cf. Section 2.2.1) and quasiresonant two-step excitation (cf. Section 2.1.3). It follows from a simple hole burning model[43] that

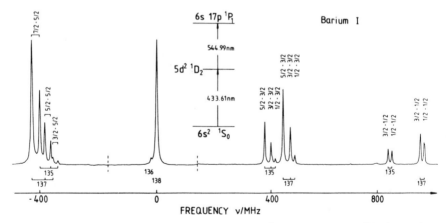

Figure 13. Two-step excitation ($E2$–$E1$) of the $6s17p\ ^1P_1$ Rydberg state of barium using counterpropagating laser beams. (Taken from Ref. 26.)

the splitting $\delta\omega_i^A(F_i - F_i')/2\pi \equiv \{\omega_i^A(F_i) - \omega_i^A(F_i')\}/2\pi$ $(A = 135, 137)$ between hyperfine components F_i and F_i' of the intermediate state (cf. Figure 14) appears enlarged by the factor $-(\omega_2 + \omega_1)/\omega_1$ for copropagating laser beams, but reduced by $(\omega_2 - \omega_1)/\omega_1$ when both laser beams are counterrunning. Also, the measured isotope shifts $\delta\Omega_2^{A-138}/2\pi \equiv (\Omega_2^A - \Omega_2^{138})/2\pi$ between the even isotopes $(A = 134, 136)$ and ^{138}Ba contain the isotope shifts $\delta\omega_1^{A-138} \equiv \omega_1^A - \omega_1^{138}$ and $\delta\omega_2^{A-138} \equiv \omega_2^A - \omega_2^{138}$ of the first and second transition, respectively. More specifically, the frequencies $\delta\Omega_2/2\pi$ of the recorded signals relative to ^{138}Ba are given by

$$\delta\Omega_2^{A-138}(F_i - F_f) = \delta\omega_2^{A-138}(F_f) - \varepsilon\frac{\omega_2}{\omega_1}\delta\omega_1^{A-138}$$

$$- \frac{\omega_1 + \varepsilon\omega_2}{\omega_1}\delta\omega_i^A(F_i\text{-c.g.}) \tag{2}$$

for the hyperfine component $F_i - F_f$ of the odd isotopes $(A = 135, 137)$, and by

$$\delta\Omega_2^{A-138} = \delta\omega_2^{A-138} - \varepsilon\frac{\omega_2}{\omega_1}\delta\omega_1^{A-138} \tag{3}$$

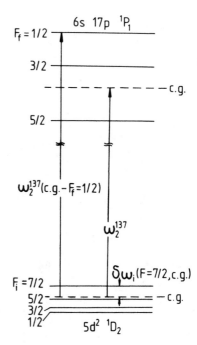

Figure 14. Illustration of transition frequencies and frequency separations used in Eqs. (2) and (3).

for even ($A = 134, 136$) isotopes, respectively. Equations (2) and (3) are valid for copropagating ($\varepsilon = +1$) as well as counterpropagating ($\varepsilon = -1$) laser beams. The symbol Ω refers to moving atoms, while ω applies to atoms at rest. In Eq. (2) the abbreviation $\delta\omega_2^{A-138}(F_f) \equiv \omega_2^A(F_f\text{-c.g.}) - \omega_2^{138}$ has been used where $\omega_2^A(F_f\text{-c.g.})/2\pi$ is the transition frequency between the center of gravity (c.g.) of the intermediate state and the hyperfine level F_f of the Rydberg state of the odd isotope A. In addition to the hyperfine splitting of the Rydberg state, $\delta\omega_2^{A-138}(F_f)$ contains the isotope shift $\delta\omega_2^{A-138}$ of the second transition. Furthermore $\delta\omega_i^A(F_i\text{-c.g.})/2\pi$ represents the frequency separation of the hyperfine component F_i of the intermediate state relative to its center of gravity. The meaning of the symbols appearing in Eqs. (2) and (3) is illustrated in Figure 14. Equations (2) and (3) can generally be used to analyze Doppler-free spectra obtained by resonant two-step excitation, starting from the $^1S_0$ ground state of two-electron systems.

Figures 12 and 13 provide instructive examples for the dependence of the recorded isotope shifts on the laser beam geometry. The hyperfine splittings corresponding to the intermediate ($5d^2\,^1D_2$) state are expanded in Figure 12 and compressed considerably in Figure 13. The total s-electron density at the nucleus decreases in the first, but increases in the second transition, and hence the isotope shifts of the two transitions are of opposite sign, i.e., $\delta\omega_1^{136-138} > 0$ and $\delta\omega_2^{136-138} < 0$. In Figure 12 a large separation $\delta\Omega_2^{136-138}$ is observed between the isotopes ^{136}Ba and ^{138}Ba since both isotope shifts ($\delta\omega_1^{136-138}$, $\delta\omega_2^{136-138}$) enter with the same sign. Likewise both isotope shifts nearly cancel for counterpropagating laser beams, resulting in a small splitting $\delta\Omega_2^{136-138}$ shown in Figure 13. In addition to the hyperfine structure of the Rydberg state, the hyperfine splitting of the intermediate state as well as the isotope shifts of both transitions can be inferred separately from spectra recorded for co- and counterpropagating laser beams.

At this point it is of interest to compare coherent two-photon with stepwise excitation, although the most suitable experimental scheme depends on the particular problem under study. In the case of Doppler-free coherent two-photon spectroscopy, only the initial and final state contribute to the recorded splittings and linewidths. Therefore, this technique offers better resolution, and generally simple spectra are obtained which are straightforward to interpret. On the contrary, the hyperfine structure of the intermediate state complicates the spectra recorded by resonant, two-step excitation, and a more involved analysis is required to deduce hyperfine structures and isotope shifts. Furthermore, the lifetime broadening of the intermediate state ($\Delta\nu_i$)

$$\Delta\nu_{tot} = \Delta\nu_f + \left| \frac{\omega_2 + \varepsilon\omega_1}{\omega_1} \right| \Delta\nu_i \tag{4}$$

contributes to the linewidth ($\Delta\nu_{tot}$), when two-step excitation is employed. Using counterrunning ($\varepsilon = -1$) rather than copropagating ($\varepsilon = +1$) laser beams, the additional broadening can be minimized. However, compared to off-resonant two-photon absorption, resonant two-step excitation greatly relaxes the power requirements of the laser radiations employed. Furthermore, the use of two lasers offers greater flexibility and allows a larger number of Rydberg series to be measured.

2.2.3. Two-Photon-Resonant Three-Photon Absorption and Thermionic Detection

Whereas Rydberg series of Ca, Sr, and Ba are easily accessible by coherent or stepwise two-photon absorption of cw dye laser radiation, starting from the atomic ground state, for elements with higher ionization limits, e.g., Mg or Yb, such techniques require at least one uv photon. Instead of using frequency-doubled laser radiation, higher excitation energies can be overcome by the absorption of three visible photons. Figure 15 displays the Doppler-free spectrum of the $6s15p\ {}^1P_1$ Rydberg state of Yb, obtained by two-photon-resonant three-photon absorption,[45] using the experimental setup shown in Figure 11. Starting from the $6s^2\ {}^1S_0$ ground state, velocity-selective, coherent two-photon absorption excited the atoms to the $6s7s\ {}^1S_0$ intermediate level. Contrary to conventional Doppler-free two-photon spectroscopy (cf. Section 2.2.1), two photons with identical wave vectors were absorbed from the first laser beam. The transition from the intermediate to the final ($6s15p\ {}^1P_1$) level was induced by the second (counterpropagating) laser beam, followed by thermionic detection of the Rydberg state. In Figure 15 the isotope shifts between the even Yb isotopes

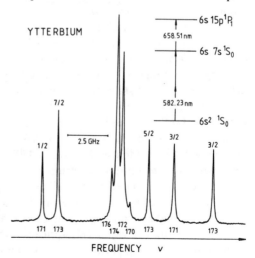

Figure 15. Doppler-free spectrum of the $6s15p\ {}^1P_1$ state of Yb obtained by two-photon-resonant three-photon absorption. (Taken from Ref. 45.)

($A = 170$–176) and the hyperfine structure of ^{171}Yb ($I = 1/2$) and ^{173}Yb
($I = 5/2$) are shown. Two-photon-resonant three-photon absorption is
closely related to resonant two-step excitation (cf. Section 2.2.2). Certain
velocity ensembles are excited to the intermediate state by a coherent
two-photon transition rather than by one-photon absorption. Again, because
of absorption line narrowing,[33,43] Doppler-free spectra are recorded when
scanning the second laser across the upper atomic transition. Rydberg states
$6snp\,^{1,3}P_1$ of Yb with principal quantum numbers up to $n = 50$ have beeen
studied in this way.[46] Two-photon resonant three-photon spectroscopy has
also been applied successfully to study $3snf\,^{1,3}F_3$ Rydberg states of Mg.[47]

2.2.4. Optical Detection of Resonant Two-Step Excitation (Optical–Optical Double Resonance)

Besides thermionic detection purely optical techniques have been
applied to monitor two-step excitation to Rydberg states. In the following
we discuss two examples, viz., fluorescence detection and absorption
measurements of the exciting laser radiations.

Fluorescence detection of Rydberg states is hampered by their long
lifetimes, by the many decay channels to lower levels, and by competing
collisional ionization. Therefore this technique is restricted to Rydberg states
with rather low ($n \leqslant 20$) principal quantum numbers. Some of these disad-
vantages can be overcome by monitoring the excitation to the Rydberg state
rather than its spontaneous decay. For two-step excitation via a short-lived
intermediate level the fluorescence emitted from that state may be used to
this end. When the intensity of the second laser beam, which excites the
atoms from the intermediate to the final Rydberg state, is modulated, the
population of the intermediate level is likewise modulated. This in turn
causes a periodic variation of the observed fluorescence intensity, which
can subsequently be detected by a lock-in amplifier. Doppler-free spectra
of $1snd\,^{1,3}D$ ($n = 12$–17) Rydberg states of ^{3}He were recorded in this way.[48]
Starting from the $1s2s\,^3S_1$ metastable state, populated by electron bombardment of the ground state, the frequency-doubled output of a cw dye laser
excited the atoms to the intermediate $1s5p\,^3P$ state. The second transition
$1s5p\,^3P \rightarrow 1snd\,^{1,3}D$ was induced by a counterpropagating color center laser
beam. The excitation of the He atoms to the final state was monitored by
observing the fluorescence decay of the $1s5p\,^3P$ state at $\lambda = 294.5$ nm. Figure
16 shows a high-resolution spectrum of the $1s13d\,^{1,3}D$ state. Signals marked
by an asterisk originated from collisional induced transitions between the
hyperfine components of the intermediate state.

At first sight, a measurement of the absorption of one of the laser
beams traversing the vapor cell appears to be too insensitive to detect the
excitation of Rydberg states. However, optical pumping may dramatically

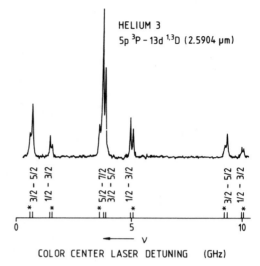

Figure 16. Doppler-free spectrum of the $1s2s\,^3S_1 \to 1s5p\,^3P \to 1s13d\,^{1,3}D$ two-step excitation of ^{3}He, monitored by the fluorescence of the $1s5p\,^3P$ intermediate level. (Taken from Ref. 48.)

increase the sensitivity, whereby this technique becomes suitable for the investigation of highly excited atomic states.[49] Figure 17 shows the experimental setup used to record high-resolution spectra of $6sns\,^{1,3}S$ and $6snd\,^{1,3}D$ Rydberg states with principal quantum numbers up to $n = 60$. Again two actively stabilized cw dye lasers were employed. The output of the first laser was split into a reference and a measuring beam, the latter one being crossed by the counterpropagating output of the second dye laser.

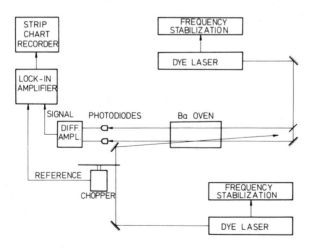

Figure 17. Experimental setup used for optical–optical double resonance of barium Rydberg states. (Taken from Ref. 49.)

While the frequency of the first laser was fixed at an arbitrary position within the Doppler contour of the barium resonance line (553.7 nm), the second laser was scanned across the upper atomic transition. The first laser sustains an optical pumping cycle, illustrated schematically in Figure 18. An atom excited to the intermediate state ($\tau = 8.4$ nsec) predominantly decays to the ground level by spontaneous emission. Since the transit time of an atom across the first laser beam amounts to several microseconds, it may absorb many photons, provided there is sufficient power density of the laser radiation (cf. Figure 18a). This optical pumping cycle is interrupted when the second laser excites the atom from the intermediate to the final state (see Figure 18b). Because of the long lifetimes of Rydberg states and the many possible decay channels, generally the atom does not return to the $6s6p\,^1P_1$ or to the ground state within its transit time. Hence the absorption of one photon from the second laser beam effectively suppresses the absorption of many photons from the first beam. The intensity of the first laser beam and the effective optical length of the vapor cell are chosen such that only a small fraction of the first laser radiation is transmitted in the absence of the second laser beam. Hence, when the second laser is at resonance with the upper atomic transition a dramatic increase in transmission of the first laser beam occurs. This signal appears on virtually zero background. Figure 19 illustrates the spectrum of the $6s14d\,^1D_2$ state of barium, obtained with counterpropagating laser beams. As discussed in Section 2.2.2, the hyperfine structure of the intermediate and final state appears in the spectrum, and Eqs. (2) and (3) should be applied to deduce isotope shifts and hyperfine structures. It is evident from Figures 2 and 19 that higher resolution and better signal-to-noise ratios have been achieved by optical–optical double resonance.

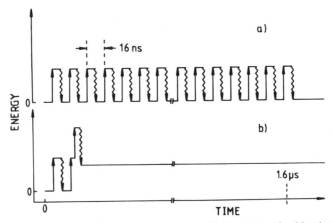

Figure 18. Schematic representation of the optical pumping cycle sustained by the first laser beam (a) without and (b) with the second laser being at resonance.

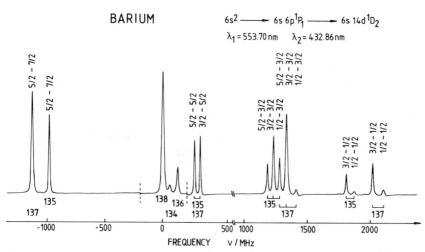

Figure 19. High-resolution spectrum of the $6s14d\ ^1D_2$ Rydberg state of barium recorded by optical–optical double resonance. (Taken from Ref. 50.)

3. Even-Parity Rydberg Series of Alkaline-Earth Elements

In this section we review isotope shift and hyperfine structure measurements concerning the even-parity Rydberg series $msns\ ^{1,3}S$ and $msnd\ ^{1,3}D$ of Ca, Sr, and Ba, which have been studied most extensively up to now. Hyperfine interactions occurring in $msns\ ^{1,3}S$ Rydberg states can be readily understood and will be considered first.

3.1. $msns\ ^{1,3}S$ Rydberg Series of Ca($m = 4$), Sr($m = 5$), and Ba($m = 6$)

3.1.1. Isotope Shifts of $6sns\ ^1S_0$ Rydberg States of Ba

The finite electronic charge density at the nucleus gives rise to an electrostatic interaction with the nuclear charge distribution which represents a correction of the Coulomb potential of the nucleus experienced by electrons. In an optical transition a change $\delta\rho_{el}$ of the average relativistic total charge density across the nucleus may occur that manifests itself in the field shift contribution

$$\delta\nu_{FS}^{A-A'} = \frac{2\pi}{3}\frac{Ze}{h}\,\delta\rho_{el}\,\delta\langle r_N^2\rangle^{A-A'} \tag{5}$$

to the total observed isotope shift

$$\delta\nu^{A-A'} = \delta\nu_{FS}^{A-A'} + \delta\nu_{NMS}^{A-A'} + \delta\nu_{SMS}^{A-A'} \tag{6}$$

Here, $\delta\langle r_N^2\rangle^{A-A'} = \langle r_N^2\rangle^A - \langle r_N^2\rangle^{A'}$ stands for the difference between the nuclear mean square charge radii of the isotopes with mass numbers A and A'. In Eq. (6) the second and third terms correspond to the normal and specific mass shift, respectively. Both terms are proportional to $(A - A')/A \cdot A'$. The normal mass shift takes the different reduced masses of the isotopes A and A' into account and can be evaluated easily.[4] On the contrary, the calculation of specific mass shifts, caused by the effect of correlations in the electronic motion on the recoil energy of the nucleus, represents a considerable theoretical challenge. Apart from a small relativistic share of $p_{1/2}$ electrons, the field shift originates from s electrons and can consequently be used to detect any change in the total s-electron density throughout a Rydberg series. In alkaline-earth atoms this concerns the ms valence electron of the ion core. At the origin, the electron density of the ns Rydberg electron decreases proportional to n^{*-3} and will henceforth be neglected. In order to deduce the change in s-electron density from experimental isotope shift data, the King plot procedure is usually applied. Modified isotope shifts ξ_1 and ξ_2 of two different optical transitions, ν_1 and ν_2, defined as

$$\xi_\alpha^{A-A'} = \frac{A \cdot A'}{A - A'} (\delta\nu_\alpha^{A-A'} - \delta\nu_{\alpha,\text{NME}}^{A-A'}) \qquad (\alpha = 1, 2) \qquad (7)$$

are plotted in a King diagram.[4] Since

$$\xi_1 = m \cdot \xi_2 + b \qquad (8)$$

holds independent of A, A', all pairs of isotopes fall on a straight line with slope m given by

$$m = \frac{\delta\rho_{\text{el},1}}{\delta\rho_{\text{el},2}} \qquad (9)$$

In this way the specific mass shifts and the changes in the mean square charge radii $\delta\langle r_N^2\rangle^{A-A'}$ are eliminated. However, only relative changes in the total s-electron density can be deduced.

In order to exploit field shifts for probing local admixtures of doubly excited configurations, the isotope shifts of $6sns\ ^1S_0$ (and $6snd\ ^{1,3}D_2$) Rydberg series of barium were studied.[51] It is known from MQDT analyses of these series[52,53] that the Rydberg states $6s18s\ ^1S_0$ and $6s14d\ ^1D_2$ are perturbed due to strongly localized configuration interactions with the $5d7d\ ^3P_0$ and $5d7d\ ^3F_2$ valence states, respectively. The admixture of $5d7d$ character reduces the total s-electron density at the nucleus and hence modifies the isotope shifts for these states. Figure 20 shows typical spectra for unperturbed (top row) and perturbed (bottom row) Rydberg states. The spectra were recorded by optical–optical double resonance (cf. Section 2.2.4), employing two-step excitation via the $6s6p\,^1P_1$ intermediate level. A comparison between corresponding upper and lower spectra immediately reveals the drastic influence of the $5d7d\,^3P_0$ and $5d7d\,^3F_2$ perturbers, leading to enlarged isotope shifts $^{136-138}$Ba and $^{134-138}$Ba for the $6s18s\,^1S_0$ and $6s14d\,^1D_2$ Rydberg states, respectively. Apart from configuration interactions, a systematic study of these isotope shifts with increasing principal quantum number discloses the contraction of the Ba$^+$ core that occurs when the outer electron is promoted to higher shells. For a quantitative analysis of these effects the modified isotope shifts $\xi(6s6p\,^1P_1 \rightarrow 6sns\,^1S_0)$ are compared with those of the Ba II resonance line in a King diagram (cf. Figure 21). As can be seen, identical King straight lines are found for all Rydberg states $11 \leq n \leq 23$ except for the perturbed state $6s18s\,^1S_0$. While the slopes corresponding to pure $6sns\,^1S_0$ Rydberg states are negative, both the $6s18s\,^1S_0$ level and its perturber $5d7d\,^3P_0$ yield King straight lines with positive slopes m. For unperturbed $6sns\,^1S_0$ Rydberg states, m can be written as

$$m(\overline{6sns}\,^1S_0) = \frac{|\Psi_{\overline{6sns}}(0)|^2 - |\Psi_{6s6p}(0)|^2}{|\Psi_{6p}(0)|^2 - |\Psi_{6s}(0)|^2} \tag{10}$$

In Eq. (10) a horizontal bar has been used to indicate a pure $6sns$ configuration. The denominator is known to be negative, since the total charge density at the nucleus decreases when the $6s$ electron of Ba$^+$ is excited to the $6p$ orbit. It follows from Eq. (10) and the negative slopes observed for pure Rydberg states of Ba that the $6s$ electron density at the origin is larger for $6sns\,^1S_0$ Rydberg states compared to the $6s6p\,^1P_1$ level. This can be

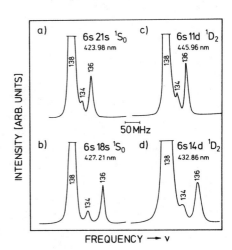

Figure 20. Isotope shifts of even barium isotopes recorded for unperturbed (top row) and perturbed (bottom row) Ba Rydberg states. (Taken from Ref. 51.)

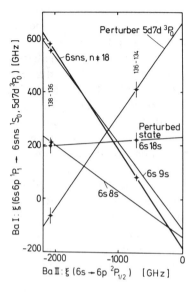

Figure 21. King diagram relating isotope shifts of transitions $6s6p\ ^1P_1 \to 6sns\ ^1S_0, 5d7d\ ^3P_0$ to the Ba II resonance line. (Taken from Ref. 51.)

explained by the screening of the $6s$ electron due to the penetrating $6p$ electron. A similar argument accounts for the difference of the slopes of the King straight lines corresponding to the $6s8s$ and $6s9s\ ^1S_0$ states. The screening of the $6s$ valence electron by the outer ns electron decreases with increasing n, resulting in a larger total s-electron density at the origin. For principal quantum numbers $n > 10$, the contraction of the Ba^+ core is completed and identical King straight lines with a slope $m_\infty = -0.37\,(5)$ are observed.

In order to derive the amount of $5d7d$ configuration contained in the $6s18s\ ^1S_0$ state quantitatively, its wave function is written as

$$|6s18s\ ^1S_0\rangle = \alpha\overline{|6s18s\ ^1S_0\rangle} + \varepsilon_1\overline{|5d7d\ ^1S_0\rangle} + \varepsilon_2\overline{|5d7d\ ^3P_0\rangle} \qquad (11)$$

with

$$\alpha^2 + \varepsilon_1^2 + \varepsilon_2^2 = 1$$

In Eq. (11) the state vector $|6s18s\ ^1S_0\rangle$ has been expanded into pure configuration, exactly SL-coupled basis vectors, denoted by the long horizontal bar. Whereas the total s-electron densities at the nucleus associated with pure $6sns$ and $5d7d$ configurations are sufficiently different to cause distinct isotope shifts, the fine structure components represented by the last two terms of Eq. (11) cannot be distinguished by isotope shift measurements. Hence, a simplified state vector

$$|6s18s\ ^1S_0\rangle = \alpha\overline{|6s18s\ ^1S_0\rangle} + \varepsilon\overline{|5d7d\ J = 0\rangle} \qquad (12)$$

will suffice to explain the isotope shift data, with $\varepsilon^2 = 1 - \alpha^2$ frequently denoted as perturber fraction. For the slope m of the perturbed $6s18s\,^1S_0$ state one obtains

$$m(6s18s\,^1S_0) = \frac{\alpha^2|\Psi_{\overline{6s18s}}(0)|^2 + \varepsilon^2|\Psi_{\overline{5d7d}}(0)|^2 - |\Psi_{6s6p}(0)|^2}{|\Psi_{6p}(0)|^2 - |\Psi_{6s}(0)|^2} \tag{13}$$

Equation (13) assumes the so-called "sharing rule" [4] to be applicable, i.e., matrix elements between $\Psi_{\overline{6s18s}}$ and $\Psi_{\overline{5d7d}}$ can be neglected. Because $|\Psi_{\overline{5d7d}}(0)|^2 < |\Psi_{\overline{6s18s}}(0)|^2$, the slightly positive slope of the King straight line of the perturbed state (see Figure 21) can be readily explained by a sufficient admixture of the doubly excited configuration.

Combining Eqs. (10) and (13), the admixture coefficient α^2 representing the total $6s$ character of the $6s18s\,^1S_0$ state can be expressed as

$$\alpha^2(6s18s\,^1S_0) = 1 - \frac{\{m(6s18s\,^1S_0) - m_\infty\}\{|\Psi_{6p}(0)|^2 - |\Psi_{6s}(0)|^2\}}{|\Psi_{\overline{5d7d}}(0)|^2 - |\Psi_{\overline{6s18s}}(0)|^2} \tag{14}$$

From relativistic HF calculations of Ba^+, $4\pi\{|\Psi_{6p}(0)|^2 - |\Psi_{6s}(0)|^2\}$ was found to be $-128a_0^{-3}$.[54] Furthermore, $|\Psi_{\overline{6s18s}}(0)|^2 = |\Psi_{6s}(0)|^2$ holds to a good approximation. The value for $|\Psi_{\overline{5d7d}}(0)|^2$ may be approximated by $|\Psi_{5d}(0)|^2$ (Ba II)[54] or derived from isotope shift data of the $6s6p\,^1P_1 \rightarrow 5d6d\,^3D_2$ transition. In this way $4\pi\{|\Psi_{\overline{5d7d}}(0)|^2 - |\Psi_{\overline{6s18s}}(0)|^2\} \cong -151a_0^{-3}$ was estimated.[51] Using $m(6s18s\,^1S_0) = +0.02(5)$ and $m_\infty = -0.37(5)$ the total Rydberg character of the perturbed $6s18s\,^1S_0$ state was found to be $\alpha^2(6s18s\,^1S_0) = 0.67(7)$. In a similar way the $6sns\,^1S_0$ character of the $5d7d\,^3P_0$ state was deduced to be $\alpha^2(5d7d\,^3P_0) = 0.40(7)$. Both results[51] are in excellent agreement with the s-characters $\alpha^2(6s18s\,^1S_0) = 0.71$ and $\alpha^2(5d7d\,^3P_0) = 0.35$, predicted by a MQDT analysis[52] of the $6sns\,^1S_0$ Rydberg series. Taking the ground state $6s^2\,^1S_0$ of Ba as reference, the change $4\pi\delta\rho_{el}$ in total s-electron density at the origin, associated with transitions to $6snd\,^1D_2$ and $6sns\,^1S_0$ states, has been plotted versus the principal quantum number n in Figure 22. The saturation behavior of $\delta\rho_{el}$ reached for $n > 10$ is clearly shown (cf. Figure 22b). The saturation limit corresponds to the difference in s-electron densities at the nucleus between the ground states of Ba and Ba^+. The deviation of the electron density of the perturbed states $6s14d\,^1D_2$ and $6s18s\,^1S_0$ from the regular behavior is conspicuous. Also, the larger electron density at the origin found for unperturbed $6sns\,^1S_0$ Rydberg states compared to the $6s6p\,^1P_1$ level is evident. The decrease in electron density $4\pi\delta\rho_{el} = -105ea_0^{-3}$ connected with the Ba I resonance line has been indicated by an arrow.

It follows from the preceding discussion that field shifts are well suited to probe the total s character of a Rydberg state, especially for heavy

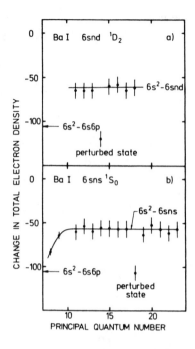

Figure 22. Change in total electron density $4\pi\,\delta\rho_{\mathrm{el}}(0)$ (a.u.) in barium, associated with transitions (a) $6s^2\,{}^1S_0 \to 6snd\,{}^1D_2$ and (b) $6sns\,{}^1S_0$. (Taken from Ref. 51.)

elements, for which the volume contribution $\delta\nu_{\mathrm{FS}}^{A-A'}$ generally represents a substantial fraction of the total observed isotope shifts. For sufficiently high n, the outer electron does not contribute directly or indirectly to the total s-electron density at the nucleus. Hence isotope shift measurements cannot discriminate between pure $msnl$ and $msnl'$ configurations. As a consequence, identical field shifts are observed for pure $6sns\,{}^1S_0$, $6sns\,{}^3S_1$, $6snd\,{}^1D_2$, and $6snd\,{}^3D_2$ Rydberg states of Ba, for example. Furthermore, only squares of amplitudes can be deduced and hence isotope shift measurements do not carry any information about the relative signs of the amplitudes representing the admixture of the different basis vectors.

3.1.2. Hyperfine Structure of Perturbed $6sns\,{}^3S_1$ Rydberg States of Ba

For isotopes with nonzero nuclear spin I [e.g., ^{43}Ca ($I = 7/2$), ^{87}Sr ($I = 9/2$), ^{135,137}Ba ($I = 3/2$)] one encounters a strong magnetic hyperfine coupling between the magnetic dipole moments of the nucleus and the ms valence electron of $msnl$ Rydberg states. This Fermi contact interaction is of the form

$$H_{\mathrm{hf}} = a_{ms}(n) \cdot \mathbf{I} \cdot \mathbf{s}_{ms} \tag{15}$$

$$H_{\mathrm{hf}} = \frac{a_{ms}(n)}{2}\,\mathbf{I} \cdot \mathbf{S} + \frac{a_{ms}(n)}{2}\,\mathbf{I} \cdot \boldsymbol{\sigma} \tag{16}$$

where $S = s_{ms} + s_{nl}$, $\sigma = s_{ms} - s_{nl}$, and s_{ms}, s_{nl} refer to the spin angular momenta of the ms and nl electrons, respectively. For $msns$ Rydberg states, the coupling constant $a_{ms}(n)$ can be written as

$$a_{ms}(n) = \frac{16\pi}{3} \mu_B \mu_N g_I \langle msns|\delta(r_{ms})|msns\rangle \tag{17}$$

Contrary to the preceding section (cf. 3.1.1), where $|\Psi_{msns}(0)|^2$ was used to denote the total charge density of s electrons at the nucleus, in Eq. (17) the matrix element represents the unpaired spin density of the ms valence electron at the origin. For low principal quantum numbers n, screening by the outer ns electron will affect the spin density of the ms valence electron at the nucleus. On the contrary, $a_{ms}(n)$ can be replaced by the free-ion value

$$a_{ms} = \frac{16\pi}{3} \mu_B \mu_N g_I \langle ms|\delta(r_{ms})|ms\rangle \tag{18}$$

for sufficiently high n, where screening effects can be neglected. The contribution of the outer ns electron being proportional to n^{*-3} has not been included in Eqs. (15) and (16) and will not be considered in the following.

When calculating magnetic hyperfine interaction energies E_F, the first and second term of Eq. (16) behave quite differently. Since the total angular momentum $F = I + S$ is conserved, both terms are diagonal in F. However, being proportional to the total spin angular momentum, the first term has nonvanishing matrix elements exclusively for the triplet states $|msns\ {}^3S_1, F\rangle$, $(F = I - 1, I, I + 1)$. On the contrary, the off-diagonal element $\overline{\langle msns\ {}^3S_1\ F = I|\mathbf{I}\cdot\boldsymbol{\sigma}|msns\ {}^1S_0\ F = I\rangle} = [I(I+1)]^{1/2}$ causes a mixing of singlet and triplet states. Again, the horizontal bar has been used to indicate pure-configuration (exactly SL-coupled) basis vectors. The hyperfine-induced singlet–triplet mixing leads to conspicuous shifts of odd versus even isotopes to be considered in detail in the next section (3.1.3).

For a discussion of the hyperfine splitting between the components $F = I + 1$ and $F = I - 1$, unaffected by singlet–triplet mixing, the $6sns\ {}^3S_1$ Rydberg series of ^{135,137}Ba is chosen.[34,55] As a typical example, Figure 23 shows the hyperfine structure of ^{135,137}Ba of the $6s15s\ {}^3S_1$ Rydberg state. From the matrix elements

$$E_{F=I+1} = \overline{\langle 6sns\ {}^3S_1\ F = I+1|H_{hf}|6sns\ {}^3S_1\ F = I+1\rangle} = \frac{a_{6s}}{2}I$$

$$\tag{19}$$

$$E_{F=I-1} = \overline{\langle 6sns\ {}^3S_1\ F = I-1|H_{hf}|6sns\ {}^3S_1\ F = I-1\rangle} = -\frac{a_{6s}}{2}(I+1)$$

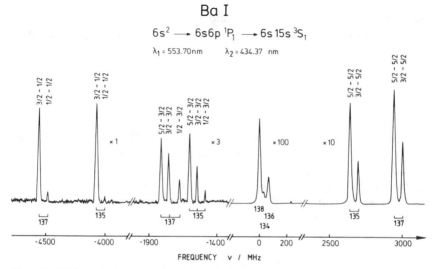

Figure 23. High-resolution spectrum of the $6s15s\,^3S_1$ Rydberg state of Ba, obtained by optical-optical double resonance (cf. Section 2.2.4). The signals of the odd isotopes have been labeled F_i-F_f according to the hyperfine levels of the intermediate and final state. (Taken from Ref. 55.)

one obtains for pure $6sns\,^3S_1$ Rydberg states

$$E_{F=I+1} - E_{F=I-1} = \frac{a_{6s}}{2}(2I+1) = A(2I+1) \tag{20}$$

where $A = a_{6s}/2$ is the usual hyperfine splitting factor. Apart from shielding effects, the hyperfine splitting factors $A(6sns\,^3S_1)$ are expected to be independent of n and equal or close to the saturation value $a_{6s}/2$ for unperturbed Rydberg states. Indeed, this is true for most hyperfine splitting factors ^{137}A, plotted versus the principal quantum number n in Figure 24. However, conspicuously reduced A factors are observed at $n = 15$, 19, and 20. This can readily be explained by configuration interaction of these Rydberg states with close-lying doubly excited perturbing states $5d7d\,^3S_1$ and $^1P_1$ (see Figure 24). Writing for the state vector at $n = 15$

$$|6s15s\,^3S_1\rangle = \alpha\overline{|6s15s\,^3S_1\rangle} + (1-\alpha^2)^{1/2}\overline{|5d7d\,J} = 1\rangle \tag{21}$$

the density of the $6s$ electron at the nucleus is seen to be scaled by the factor α^2. Compared to the strong Fermi-contact coupling of the $6s$ electron to the nuclear magnetic moment, the hyperfine interaction of a pure $5d7d$ configuration can be neglected.[55] Hence, with the aid of Eqs. (19)–(21), the

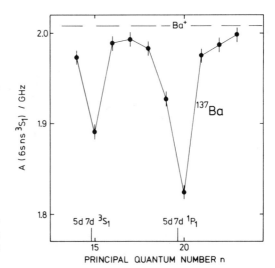

Figure 24. Hyperfine splitting factors ^{137}A of $6sns\,^3S_1$ Rydberg states of Ba. (Taken from Ref. 55.)

hyperfine splitting factor of the perturbed state $6s15s\,^1S_0$ is calculated to be

$$A(6s15s\,^3S_1) = \alpha^2\,\frac{a_{6s}}{2} \qquad (22)$$

and directly reflects the reduced density of the $6s$ valence electron at the origin. From measured splitting factors (cf. Figure 24) the Rydberg characters of perturbed $6s15s\,^3S_1$, $6s19s\,^3S_1$ and $6s20s\,^3S_1$ Rydberg states have been deduced to be 0.94 (1), 0.96 (1), and 0.91 (1), respectively, in excellent agreement with a MQDT analysis of this series.[56]

It is instructive to compare the suitability of isotope shift and hyperfine structure data to deduce the Rydberg character of (perturbed) $6sns\,^3S_1$ states. By means of a King diagram (cf. Section 3.1.1) isotope shifts $^{136-134}$Ba and $^{138-136}$Ba, observed for the $6s6p\,^1P_1 \to 6sns\,^3S_1$ transitions, were compared to the isotope shifts of the Ba II resonance line. In Figure 25 the slopes m of the corresponding King straight lines have been plotted versus the principal quantum number n. Figures 24 and 25 are strikingly similar. For the $6sns\,^3S_1$ series, both the hyperfine splitting factors and the field shifts can be analyzed to yield the total s character of the Rydberg states. However, the splitting factors provide the more direct and more sensitive way to probe configuration interactions.

3.1.3. Hyperfine-Induced Singlet–Triplet Mixing of Unperturbed $5sns\,^1S_0$ and $^3S_1$ Rydberg States of Sr

Hyperfine-induced mixing of fine structure components is well known and was first observed and explained in 1932 by Schüler and Jones[57] and

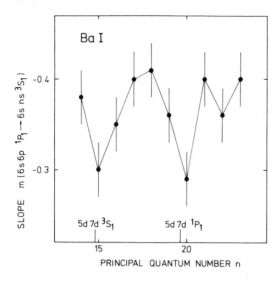

Figure 25. Change in total electron density for $6s6p\,^1P_1 \rightarrow 6sns\,^3S_1$ transitions, normalized to the Ba II resonance line. The data represent the slopes m of a corresponding King diagram. (Taken from Ref. 55.)

Casimir,[58] respectively. The recent interest in such effects has been stimulated by high-resolution laser spectroscopy of Rydberg series of alkaline-earth elements.[59] As discussed briefly in the Introduction, fine structure separations decrease proportionally to n^{*-3}, whereas the magnetic hyperfine interaction of the ion core remains virtually constant. Hence by increasing the principal quantum number the ratio between the hyperfine coupling constant a_{ms} and fine structure splittings, which determines the degree of mixing, can be varied between both extremes. For such investigations $msns\,^1S_0$ and $^3S_1$ Rydberg series are especially attractive, since mixing is restricted to the $F = I$ hyperfine levels. Experimental results will be presented first for the $5sns\,^1S_0$ and $^3S_1$ Rydberg series of ^{87}Sr, since these series are known not to be perturbed by configuration interactions and therefore illustrate the effects of hyperfine-induced singlet–triplet mixing most clearly.

For Rydberg states of pure $msns$ configuration, hyperfine-induced singlet–triplet mixing can be easily calculated.[60,61] Choosing the energy halfway between the $^1S_0$ and $^3S_1$ states as zero, the secular equation of the Hamiltonian $H = H_0 + H_{hf}$ [cf. Eq. (16)]

$$H = \frac{e^2}{r_{ms,ns}} + H_{hf} \tag{23}$$

is written as

$$\begin{vmatrix} -\dfrac{\Delta E_{ST}(n)}{2} - \dfrac{a_{ms}}{2} - E & \dfrac{a_{ms}}{2}[I(I+1)]^{1/2} \\[2ex] \dfrac{a_{ms}}{2}[I(I+1)]^{1/2} & \dfrac{\Delta E_{ST}}{2} - E \end{vmatrix} = 0 \tag{24}$$

The first and second row (column) of Eq. (24) refer to the basis vectors $|msns\,^3S_1\,F = I\rangle$ and $|msns\,^1S_0\,F = I\rangle$, respectively. The solutions of Eq. (24) are easily found to be

$$E_n^{(+)}(F = I) = -\frac{a_{ms}}{4} + \frac{1}{2}[a_{ms}^2(I + 1/2)^2$$

$$+ a_{ms} \cdot \Delta E_{ST}(n) + \Delta E_{ST}^2(n)]^{1/2} \qquad (25)$$

$$E_n^{(-)}(F = I) = -\frac{a_{ms}}{4} - \frac{1}{2}[a_{ms}^2(I + 1/2)^2$$

$$+ a_{ms} \cdot \Delta E_{ST}(n) + \Delta E_{ST}^2(n)]^{1/2} \qquad (26)$$

with $\Delta E_{ST}(n) = E_n(^1S_0, I = 0) - E_n(^3S_1, I = 0)$ denoting the singlet–triplet separation. For $|a_{ms}| \ll |\Delta E_{ST}(n)|$, expanding the square root up to second order, $E_n^{(+)}(F = I)$ and $E_n^{(-)}(F = I)$ can be approximated as

$$E_n^{(+)}(F = I) = \frac{\Delta E_{ST}(n)}{2} + \frac{a_{ms}^2 I(I + 1)}{4\Delta E_{ST}(n)} \qquad (27)$$

$$E_n^{(-)}(F = I) = -\frac{\Delta E_{ST}(n)}{2} - \frac{a_{ms}}{2} - \frac{a_{ms}^2 I(I + 1)}{4\Delta E_{ST}(n)} \qquad (28)$$

Neglecting the last term in Eqs. (27) and (28), the solutions $E_n^{(+)}(F = I)$ and $E_n^{(-)}(F = I)$ correspond to the energies expected for the $F = I$ components of the $msns\,^1S_0$ and $^3S_1$ Rydberg states, respectively, in the absence of any hyperfine-induced singlet–triplet mixing. As can be seen from Eqs. (25)–(28), the singlet–triplet mixing causes the $F = I$ levels to repel each other by equal but opposite amounts. For $msns\,^1S_0$ states this manifests itself as a shift of odd $(I \neq 0)$ versus even $(I = 0)$ isotopes of alkaline-earth elements, increasing in magnitude with increasing principal quantum number n. On the other hand, the hyperfine-induced isotope shifts cause deviations from the Landé interval rule for $msns\,^3S_1$ Rydberg states.[62] The pronounced shift of ^{87}Sr versus ^{86}Sr is illustrated in Figure 26 for $5sns\,^1S_0$ Rydberg states.[59] The spectra were recorded by Doppler-free two-photon spectroscopy. The shifts occur to higher frequencies since ΔE_{ST} is positive. In Figure 27 the experimentally observed isotope shifts $^{87-86}$Sr are plotted versus the principal quantum number n. The solid line has been calculated[60] using Eq. (25) together with the known hyperfine splitting constant of ^{87}Sr$^+$ ($a_{5s} = -1.0$ GHz) and singlet–triplet separations ΔE_{ST} derived from MQDT analyses of the $5sns\,^1S_0$ [63] and $^3S_1$ [64] Rydberg series. As can be seen from Figure 27, the agreement between experimental data and theoretical values is excellent.

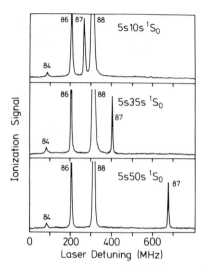

Figure 26. Doppler-free two-photon excitation spectra of $5sns\,^1S_0$ Rydberg states of Sr. (Taken from Ref. 59.)

Apart from shifts in energy, the hyperfine interaction modifies the wave functions of the $F = I$ components of $msns\,^1S_0$ and $^3S_1$ Rydberg states. The true wave functions, i.e., the eigenvectors of the secular equation [cf. Eq. (24)] can be expressed in terms of the $\overline{|msns\,^1S_0\,F = I\rangle}$ and $\overline{|msns\,^3S_1\,F = I\rangle}$ basis vectors

$$\Psi_n^{(+)}(F = I) \equiv |msns\,^1S_0\,F = I\rangle$$
$$= (1 - \Omega^2)^{1/2}\overline{|msns\,^1S_0\,F = I\rangle} + \Omega\overline{|msns\,^3S_1\,F = I\rangle} \tag{29}$$

$$\Psi_n^{(-)}(F = I) \equiv |msns\,^3S_1\,F = I\rangle$$
$$= -\Omega\overline{|msns\,^1S_0\,F = I\rangle} + (1 - \Omega^2)^{1/2}\overline{|msns\,^3S_1\,F = I\rangle} \tag{30}$$

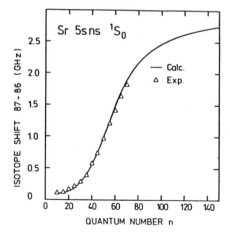

Figure 27. Experimental ($\triangle$) and calculated (solid curve) hyperfine-induced isotope shifts $^{87\text{-}86}$Sr. (Taken from Ref. 60.)

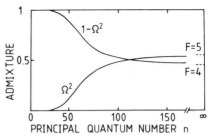

Figure 28. Calculated admixture coefficients Ω^2 and $1 - \Omega^2$ of $5sns$ $^{1,3}S$ Rydberg states of ^{87}Sr. On the right-hand side, the total angular momentum F^{ion} of ^{87}Sr$^+$ is indicated. (Taken from Ref. 65.)

In Figure 28 the singlet and triplet character of $^{1,3}S$ $F = I$ Rydberg states of ^{87}Sr have been plotted versus the principal quantum number n. The values shown were calculated[65] solving the secular equation [cf. Eq. (24)]. At low principal quantum numbers, the hyperfine interaction of the ms valence electron is small compared to the exchange interaction, i.e., the singlet–triplet splitting. Hence, both valence electrons form a total spin angular momentum $S = s_{ms} + s_{ns}$, which couples in turn with the nuclear spin I to yield the total angular momentum F. Hence the total spin angular momentum S remains a good quantum number and the wave functions $\Psi^{(+)}(F = I)$ and $\Psi^{(-)}(F = I)$ correspond to the $^1S_0$ and $^3S_1$ Rydberg states, respectively. With increasing n, however, the relative strength of both interactions reverses. Hence for sufficiently high principal quantum numbers the spin s_{ms} of the ms valence electron and the nuclear spin I add to form the total angular momentum of the ionic core $F^{ion} = I + s_{ms}$, which couples in turn with the spin s_{ns} of the Rydberg electron to the $F = I$ component of the total angular momentum F. Thus, due to hyperfine-induced singlet–triplet mixing, the total spin angular momentum S loses its meaning for the $F = I$ components of highly excited Rydberg states. As a consequence, the two-photon transition probabilities to these states may be different for odd and even isotopes. This is illustrated in Figure 29, which shows Doppler-free two-photon spectra corresponding to transitions to the $5s60d$ $^1D_2$ and $5s61s$ $^1S_0$ Rydberg states of Sr.[65] Signals of the latter Rydberg state appear on the right-hand side with the one corresponding to ^{87}Sr shifted towards higher frequencies. Since a $J = 1$ state cannot be populated by coherent two-photon absorption, starting from the $5s^2$ $^1S_0$ ground state, no lines related to the even isotopes were recorded for the $5s61s$ $^3S_1$ state. However, because of hyperfine-induced singlet–triplet mixing, the $F = I$ component of the $5s61s$ $^3S_1$ state contains some singlet character (cf. Figure 28) and hence the transition to the $5s61s$ $^3S_1$ $F = I$ level of ^{87}Sr appears on the left-hand side of Figure 29. The relative intensities of the $F = I$ components of the $^1S_0$ and $^3S_1$ Rydberg states qualitatively reflect the relative singlet character contained in the corresponding wave functions.

It is instructive to follow the $F = I$ hyperfine components of $msns$ $^{1,3}S$ Rydberg states up to the ionization limit. As can be seen from Eqs. (25)

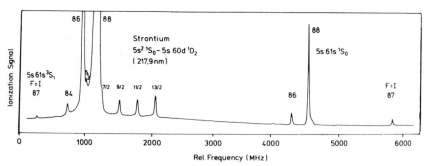

Figure 29. Doppler-free two-photon excitation spectrum of $5s60d$ $^1D_2$ and $5s61s$ $^1S_0$ Rydberg states of Sr. (Taken from Ref. 65.)

and (26), the solutions $E_n^{(\pm)}(F = I)$ approach the limits

$$E_{n=\infty}^{(\pm)}(F = I) = -\frac{a_{ms}}{4} \pm \frac{1}{2}|a_{ms}|\left(I + \frac{1}{2}\right) \tag{31}$$

for $\Delta E_{ST} \sim n^{*-3} \to 0$. Since $a_{5s}(^{87}Sr) < 0$, the $F = I$ component of the $5sns$ $^1S_0$ Rydberg series merges into the $F^{ion} = I - 1/2$ component

$$E_{F=I-1/2}^{ion} = -\frac{a_{5s}}{2}(I + 1) \tag{32}$$

of the $^2S_{1/2}$ ground state of $^{87}Sr^+$, as indicated in Figure 30. Likewise, the

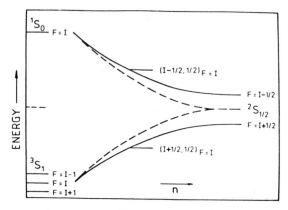

Figure 30. Schematic variation of the $5sns$ $^1S_0$ $F = I$ and $5sns$ $^3S_1$ $F = I$ hyperfine components of ^{87}Sr with principal quantum number n. The energies are not drawn to scale. For each value of n the origin of the energy scale has been taken to lie halfway between the corresponding $^1S_0$ and $^3S_1$ states of the even isotopes (dashed lines). The states of the free ion appear on the right-hand side. State mixing (cf. Section 3.1.4) has not been taken into account. (Taken from Ref. 65.)

$F = I$ component of the $5sns\,{}^3S_1$ Rydberg states finally coincides (cf. Figure 30) with the $F^{ion} = I + 1/2$ component of the ionic ground state

$$E^{ion}_{F=I+1/2} = \frac{a_{5s}}{2}\,I \tag{33}$$

Naturally, the $F = I + 1$ and $F = I - 1$ components of the ${}^3S_1$ states move towards the $F^{ion} = I + 1/2$ and $F^{ion} = I - 1/2$ hyperfine components of the ionic ground state, respectively, whereas the ${}^1S_0$ and ${}^3S_1$ states of the even Sr isotopes (dashed lines) merge at the ionization limit I_s into the unsplit ${}^2S_{1/2}$ ground state of the ion.

3.1.4. Hyperfine-Induced State Mixing (n-Mixing) at High Principal Quantum Numbers

In the preceding section hyperfine-induced singlet–triplet mixing has been considered between Rydberg states of the same principal quantum number only. At sufficiently high excitation of the outer electron, however, the hyperfine interaction of the ionic core becomes larger than its Coulomb interaction with the Rydberg electron. Consequently, the hyperfine-induced shift of the $5sns\,{}^1S_0\,F = I$ component of ^{87}Sr, for example, towards higher energies and the lowering of the adjacent $5s(n + 1)s\,{}^3S_1\,F = I$ level become comparable to the separation $\Delta E_{n,n+1}(I = 0) = E_{n+1}({}^3S_1, I = 0) - E_n({}^1S_0 I = 0)$ between the corresponding Rydberg states of the even isotopes. Hence both $F = I$ hyperfine components tend to cross as indicated by the dashed lines shown in Figure 31. However, when the energy separation $\Delta E_{n,n+1}(F = I)$ (cf. Figure 31) between both hyperfine levels is comparable to the hyperfine interaction constant a_{5s} of the inner valence electron, hyperfine-induced singlet–triplet mixing occurs again, but this time between Rydberg states of different principal quantum numbers (state-mixing). As a consequence, avoided crossings (cf. Figure 31) were observed by Beigang et al.,[66] corresponding to a repulsion of both hyperfine components. State mixing can be calculated[66] in a similar way as described for the hyperfine-induced singlet–triplet mixing in the previous section. Choosing the state vectors $\Psi^{(-)}_{n+1}(F = I) = |5s(n + 1)s\,{}^3S_1\,F = I\rangle$ and $\Psi^{(+)}_n(F = I) = |5sns\,{}^1S_0\,F = I\rangle$ [cf. Eqs. (29, 30)] as basis which include singlet–triplet mixing, and taking the energy halfway between singlet $[E_n({}^1S_0, I = 0)]$ and triplet $[E_{n+1}({}^3S_1, I = 0)]$ Rydberg states of the even isotopes as zero, the secular equation of the Hamiltonian $H = H_0 + H_{hf}$ [cf. Eqs. (15), (17), (23)] reads

$$\begin{vmatrix} \langle\Psi^{(-)}_{n+1}|H|\Psi^{(-)}_{n+1}\rangle - E & \langle\Psi^{(-)}_{n+1}|H|\Psi^{(+)}_n\rangle \\ \langle\Psi^{(+)}_n|H|\Psi^{(-)}_{n+1}\rangle & \langle\Psi^{(+)}_n|H|\Psi^{(+)}_n\rangle - E \end{vmatrix} = 0 \tag{34}$$

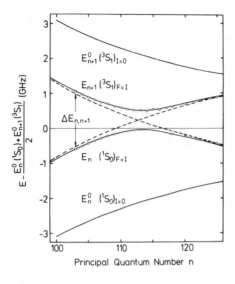

Figure 31. Avoided crossing between $5sns\ ^1S_0\ F = I$ and $5s(n+1)s\ ^3S_1\ F = I$ Rydberg states of ^{87}Sr. Open circles represent experimental data; solid lines have been calculated. Dashed lines show the expected trend of these hyperfine components, neglecting state mixing. The corresponding $5s(n+1)s\ ^3S_1\ (I = 0)$ and $5sns\ ^1S_0\ (I = 0)$ Rydberg states of the even isotopes are also included. (Taken from Ref. 66.)

The diagonal matrix elements contain the shifts of the $F = I$ components due to singlet–triplet mixing and can be expressed in terms of $E_n^{(+)}(F = I)$ and $E_{n+1}^{(-)}(F = I)$ [cf. Eqs. (25) and (26)] as

$$\langle \Psi_{n+1}^{(-)}|H|\Psi_{n+1}^{(-)}\rangle = \{E_{n+1}(^1S_0, I = 0) - E_n(^1S_0, I = 0)\}/2$$
$$+ E_{n+1}^{(-)}(F = I) \tag{35}$$

$$\langle \Psi_n^{(+)}|H|_n^{(+)}\rangle = -\{E_{n+1}(^3S_1, I = 0) - E_n(^3S_1, I = 0)\}/2$$
$$+ E_n^{(+)}(F = I) \tag{36}$$

The off-diagonal element is given by

$$\langle \Psi_n^{(+)}|H|\Psi_{n+1}^{(-)}\rangle = \frac{16\pi}{3}\mu_B\mu_N g_I\langle 5sns|\delta(r_{5s})|5s(n+1)s\rangle$$
$$\times \langle ^1S_0\,F = I|\mathbf{s}_{5s}\cdot\mathbf{I}|^3S_1\,F = I\rangle \tag{37}$$

In order to evaluate the off-diagonal matrix element, the radial overlap integral needs to be calculated. The overlap between the $5s$ and the Rydberg electron is proportional to n^{*-3} and can hence be neglected at high principal quantum numbers. On this assumption, one obtains

$$\langle \Psi_n^{(+)}|H|\Psi_{n+1}^{(-)}\rangle = a_{5s}\langle ^1S_0\,F = I|\mathbf{s}_{5s}\cdot\mathbf{I}|^3S_1\,F = I\rangle\langle ns|(n+1)s\rangle \tag{38}$$

In Eq. (38) the last factor would be zero if Rydberg electron wave functions of the even Sr isotopes were used. However, associated with the hyperfine-induced shift of the $F = I$ components, phase shifts in the radial part of the wave functions of the Rydberg electron occur, resulting in a nonvanishing overlap integral. It can be approximated as[66,67]

$$\langle ns|(n + 1)s \rangle = \langle n^*|(n + 1)^* \rangle$$
$$= -\frac{2[n^*(n + 1)^*]^{1/2}}{n^* + (n + 1)^*} \cdot \frac{\sin \pi\{(n + 1)^* - n^*\}}{\pi\{(n + 1)^* - n^*\}}, \quad (39)$$

where n^* and $(n + 1)^*$ are the effective principal quantum numbers of the $^1S_0\ F = I$ and $^3S_1\ F = I$ states, respectively. The effective principal quantum numbers

$$\nu^* = \left[\frac{R}{E^{\text{ion}} - E_\nu(F = I)}\right]^{1/2} \quad (40)$$

are readily obtained from the binding energies of these states. In Eq. (40) R stands for the mass-corrected Rydberg constant. E^{ion} corresponds to the ionization limits $I_s + E^{\text{ion}}_{F = I - 1/2}$ and $I_s + E^{\text{ion}}_{F = I + 1/2}$ of the $^1S_0\ F = I$ and $S_1\ F = I$ components with term values $E_n(F = I) = [E_n(^1S_0) + E_n(^3S_1)]/2 + E_n^{(+)}(F = I)$ and $E_{n+1}(F = I) = [E_{n+1}(^1S_0) + E_{n+1}(^3S_1)]/2 + E_{n+1}^{(-)}(F = I)$, respectively [cf. Eqs. (25), (26), (32), and (33)]. Solving the secular equation, the solid curves shown in Figure 31 were obtained. As can be seen, excellent agreement with experimental data was achieved.[66] At the crossing of the dashed lines, n^* and $(n + 1)^*$ are equal, resulting in an overlap integral $\langle n^*|(n + 1)^* \rangle = 1$, and consequently a large off-diagonal element. At even higher principal quantum numbers avoided crossings will occur between Rydberg states differing in their principal quantum numbers by more than 1 ($\Delta n \geq 2$). Similar effects have been observed for Rydberg states in external magnetic[68] and electric[69] fields.

In the region of the avoided crossing (cf. Figure 31), the wave functions $|5sns\ ^1S_0\ F = I\rangle$ and $|5s(n + 1)s\ ^3S_1\ F = I\rangle$ are strongly mixed, causing a characteristic intensity variation of the two-photon signals with increasing principal quantum numbers. As can be seen from Figure 28, the wave functions $|5sns\ ^1S_0\ F = I\rangle$ and $|5s(n + 1)s\ ^3S_1\ F = I\rangle$ contain about equal amounts of singlet and triplet character in the region of interest ($105 \leq n \leq 120$). Hence on either side of the avoided crossing, the two-photon signals of both $F = I$ components have approximately equal intensities. Owing to state mixing, symmetric and antisymmetric linear combinations of the wave functions $|5s(n + 1)s\ ^3S_1\ F = I\rangle$ and $|5sns\ ^1S_0\ F = I\rangle$ are formed, which correspond to the upper and lower branch (cf. Figure 31), respectively. On approaching the avoided crossing from below, the $^1S_0$ character first increases (decreases) and then decreases (increases) for states falling on the upper (lower) curve. The intensities of the observed two-photon signals directly reflect this behavior since only the $^1S_0$ part can be excited by a two-photon transition, starting from the $5s^2\,^1S_0$ ground state.

3.1.5. Hyperfine-Induced Singlet–Triplet Mixing of Perturbed $6sns\,^1S_0$ and $^3S_1$ Rydberg States of Ba

Besides strontium, hyperfine-induced singlet–triplet mixing of $msns\,^1S_0$ and $^3S_1$ Rydberg series has also been investigated in Ca[61] and Ba.[34,61,62] Contrary to the lighter alkaline earths Ca and Sr, however, the $6sns\,^1S_0$ and $^3S_1$ Rydberg series of Ba are locally perturbed, owing to configuration interaction with several low-lying members of $5dnd\,J = 0$ and $J = 1$ Rydberg series (cf. Sections 3.1.1 and 3.1.2). In the following we focus our attention on the singlet–triplet mixing of perturbed $^1S_0$ and $^3S_1$ Rydberg states of Ba.

Figure 32 compares displacements of the $F = I$ component of ^{137}Ba observed for $^1S_0$ and $^3S_1$ Rydberg states.[62] In the upper part (Figure 32a) open circles represent the experimental isotope shifts $^{137-138}$Ba of the transitions $6s^2\,^1S_0 \to 6sns\,^1S_0$, corrected for the normal mass contribution. The lower part (Figure 32b) displays experimental data (open circles) derived

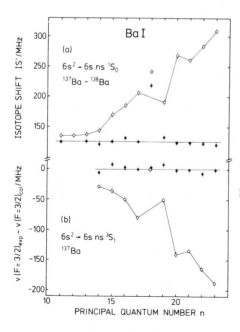

Figure 32. (a) Isotope shifts $^{137-138}$Ba of the transitions $6s^2\,^1S_0 \to 6sns\,^1S_0$, corrected for the normal mass shift. (b) Shift of the $6sns\,^3S_1$ $F = 3/2$ component of ^{137}Ba from the position predicted by the Landé interval rule. Open circles correspond to experimental data, solid ones to data corrected for hyperfine-induced singlet–triplet mixing. (Taken from Ref. 62.)

from the hyperfine structure of $6sns\,^3S_1$ Rydberg states. It has been discussed in Section 3.1.2 that the hyperfine splitting factor ^{137}A of $6sns\,^3S_1$ Rydberg states of ^{137}Ba varies with principal quantum number n (cf. Figure 24). This variation affects the position of the ^{137}Ba $F = I = 3/2$ hyperfine component relative to ^{138}Ba, even in the absence of any hyperfine-induced singlet–triplet mixing. Therefore the shift of the $F = 3/2$ hyperfine component was measured from the position $(E_{F=3/2}^{\text{calc}})$ predicted by the Landé interval rule (cf. Figure 32b). The overall splitting between the $6sns\,^3S_1\,F = 1/2$ and $F = 5/2$ hyperfine components [cf. Eq. (20)] was used to derive $E_{F=3/2}^{\text{calc}} = -a_{6s}(n)/2$. Apart from a (constant) isotope shift of ^{137}Ba versus ^{138}Ba, comprised of the volume and the specific mass shift (cf. Section 3.1.1), the upper and lower parts of Figure 32 essentially display the last term $[\pm a_{6s}^2(n)I(I+1)/4\Delta E_{\text{ST}}(n)]$ of Eqs. (27) and (28) and hence appear as mirror images of each other.

Comparing Figure 27 with Figure 32, the hyperfine-induced isotope shifts of ^{137}Ba are seen to vary no longer as smoothly as in the case of ^{87}Sr. For the most part, this is due to the rather irregular variation of the singlet–triplet splitting $\Delta E_{\text{ST}}(n) = E_n(^1S_0, I = 0) - E_n(^3S_1, I = 0)$ with principal quantum number n. Using experimental values[70] for $\Delta E_{\text{ST}}(n)$ to correct for the hyperfine-induced singlet–triplet mixing, the solid circles were derived. In Figure 32b the corrected data are seen to fall on a horizontal line through zero, confirming that the Landé interval rule applies to the hyperfine structure of $6sns\,^3S_1$ Rydberg states, provided the singlet–triplet mixing is taken into account. Likewise, the corrected isotope shifts $^{137-138}$Ba fall on a horizontal line (see Figure 32a), representing the volume and specific mass shifts, except for the Rydberg state $6s18s\,^1S_0$, perturbed by configuration interaction (cf. Section 3.1.1). Isotope shifts $^{137-138}$Ba and $^{135-138}$Ba obtained in this way are consistent with isotope shift data of the even isotopes discussed previously (see Figures 21 and 22). This is illustrated in Figure 33 by means of a King diagram. After taking the singlet–triplet mixing into account (solid symbols), isotope shifts of odd and even isotopes fall on King straight lines corresponding to unperturbed Rydberg states or to the perturbed $6s18s\,^1S_0$ level. The different slopes of both King straight lines immediately reveal the reduced s-electron density for the perturbed state (cf. Section 3.1.1).

Hyperfine-induced mixing of $^1S_0$ and $^3S_1$ Rydberg states of pure $msns$ configuration has been treated in detail in Section 3.1.3. However, when one or both fine structure components are perturbed by configuration interactions, the solutions given in Eqs. (25) and (26) or Eqs. (27) and (28) no longer apply. As an example the hyperfine-induced mixing of the $6s18s\,^1S_0$, $^3S_1$ and $5d7d\,^3P_0$ states will be discussed quantitatively. As before the mixing caused by the strong Fermi-contact interaction of the inner valence electron will be considered only.

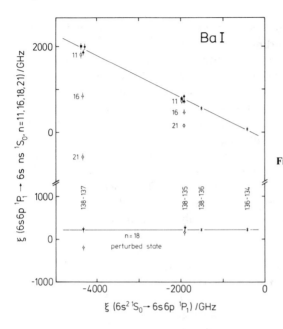

Figure 33. King diagram for unperturbed (upper part) and perturbed (lower part) $6sns\,^1S_0$ Rydberg states. Open symbols refer to experimental isotope shifts, solid ones to data corrected for hyperfine-induced singlet–triplet mixing. The modified isotope shifts shown in Figure 33 were obtained using the data plotted in Figure 32a. (Taken from Ref. 62.)

Hence the following simplified state vectors

$$|6s18s\,^1S_0\rangle = \alpha\overline{|6s18s\,^1S_0\rangle} + (1-\alpha^2)^{1/2}\overline{|5d7d\,J=0\rangle} \tag{12'}$$

$$|5d7d\,J=0\rangle = (1-\alpha^2)^{1/2}\overline{|6s18s\,^1S_0\rangle} - \alpha\overline{|5d7d\,J=0\rangle} \tag{41}$$

$$|6s18s\,^3S_1\rangle = \overline{|6s18s\,^3S_1\rangle} \tag{42}$$

which apply to the even barium isotopes, are used to set up the secular equation for the $F=I$ component of ^{135,137}Ba

$$\begin{vmatrix} -\dfrac{\Delta E_{ST}}{2} - \dfrac{a_{6s}}{2} - E & \dfrac{a_{6s}}{2}[I(I+1)]^{1/2}\alpha & \dfrac{a_{6s}}{2}[I(I+1)]^{1/2}(1-\alpha^2)^{1/2} \\[2ex] \dfrac{a_{6s}}{2}[I(I+1)]^{1/2}\alpha & \dfrac{\Delta E_{ST}}{2} - E & 0 \\[2ex] \dfrac{a_{6s}}{2}[I(I+1)]^{1/2}(1-\alpha^2)^{1/2} & 0 & \Delta E_{PT} - \dfrac{\Delta E_{ST}}{2} - E \end{vmatrix} = 0. \tag{43}$$

In Eq. (43) the energy halfway between the $6s18s\,^1S_0$ and $^3S_1$ states has been taken as zero. $\Delta E_{PT} = E(5d7d\,^3P_0,\ I=0) - E(6s18s\,^3S_1,\ I=0)$ denotes the perturber–triplet separation (cf. Figure 34). The present example requires the solution of a cubic secular equation, in contrast to the unperturbed case [cf. Eq. (24)]. Although general analytic expressions can be derived

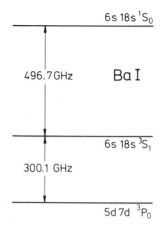

Figure 34. Relative positions of the $6s18s\ ^1S_0,\ ^3S_1$ Rydberg states and the $5d7d\ ^3P_0$ local perturber.

for the eigenvalues, it is sufficient to apply second-order perturbation theory, because $|\Delta E_{ST}|, |\Delta E_{PT}| \gg a_{6s}$. In this way one obtains[62]

$$E(^3S_1, F = I) = -\frac{\Delta E_{ST}}{2} - \frac{a_{6s}}{2} - \frac{a_{6s}^2 I(I + 1)}{4}\left(\frac{\alpha^2}{\Delta E_{ST}} + \frac{1 - \alpha^2}{\Delta E_{PT}}\right) \quad (44)$$

$$E(^1S_0, F = I) = \frac{\Delta E_{ST}}{2} + \frac{a_{6s}^2 I(I + 1)}{4\Delta E_{ST}}\alpha^2 \quad (45)$$

$$E(^3P_0, F = I) = \Delta E_{PT} - \frac{\Delta E_{ST}}{2} + \frac{a_{6s}^2 I(I + 1)}{4\Delta E_{PT}}(1 - \alpha^2) \quad (46)$$

Together with $\alpha^2(6s18s\ ^1S_0) = 0.67$ (cf. Section 3.1.1) Eqs. (44) and (45) have been used to derive the solid circles shown in Figures 32 and 33 at $n = 18$. As can be seen from Eqs. (45) and (46), the hyperfine-induced isotope shifts are a direct measure of the Rydberg character of the $^1S_0$ and $^3P_0$ states. The $F = I$ component of the $^3S_1$ level, however, experiences a total shift from the position predicted by the Landé interval rule caused by the interaction with both $J = 0$ states. Since the perturber $^3P_0$ lies below the $^3S_1$ state, both contributions are of opposite sign and fortuitously cancel. This is also evident from Figure 32b.

Whereas the hyperfine interaction of the $6s$ electron couples the $5d7d\ ^3P_0$ state, not normally affecting the $6sns\ ^3S_1$ Rydberg series, to the $6s18s\ ^3S_1(F = I)$ level of ^{135,137}Ba, the reverse situation can also be found, i.e., the strong perturbation of the $6s37s\ ^1S_0$ level[34] by the $5d7d\ ^3P_1$ perturber. This valence state profoundly influences the $6sns\ ^3S_1$ Rydberg series by configuration interaction.[56,70] Figure 35 compares experimental isotope shifts $^{138-137}$Ba of $6s^2\ ^1S_0 \rightarrow 6sns\ ^1S_0$ transitions, with theoretical predictions

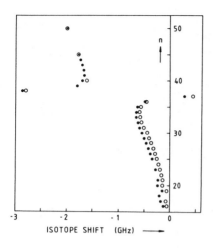

Figure 35. Isotope shifts $^{138-137}$Ba of $6s^2{}^1S_0 \rightarrow$ $6sns\,{}^1S_0$ transitions. Dots refer to experimental data; open circles were obtained using semiempirical (MQDT) wave functions. (Taken from Ref. 34.)

(open circles).$^{(34)}$ Wave functions derived from MQDT analyses of the $6sns\,{}^1S_0{}^{(52)}$ and $6sns\,{}^3S_1{}^{(56)}$ Rydberg series were used to account for the hyperfine-induced singlet–triplet mixing. Although the strong variation around $n = 37$ is well reproduced (cf. Figure 35), quantitative agreement between experimental and theoretical shifts could not be achieved at $n = 37$. A possible explanation for this discrepancy is given in Ref. 24. Figure 36 shows spectra$^{(34)}$ obtained by quasiresonant two-step excitation and field ionization (cf. Section 2.1.3), demonstrating the large variations of the hyperfine-induced isotope shifts around $n = 37$.

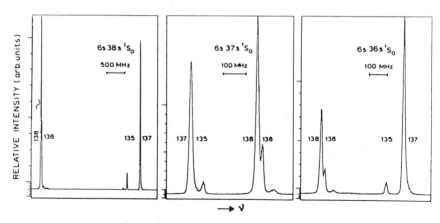

Figure 36. Excitation spectra $6s^2{}^1S_0 \rightarrow 6s6p\,{}^1P_1 \rightarrow 6sns\,{}^1S_0$ ($n = 36$–38) of barium. The first laser was locked to the ^{137}Ba $F = 3/2 \rightarrow 5/2$ component of the lower transition. The spectra show the strong variations of the hyperfine-induced isotope shifts $^{138-137}$Ba and $^{138-136}$Ba at $n = 37$. (Taken from Ref. 34.)

3.2. msnd $^{1,3}D$ Rydberg Series of Ca, Sr, and Ba

Compared to $msns$ $^{1,3}S$ Rydberg states, the hyperfine structure of $msnd$ $^{1,3}D$ Rydberg series is considerably more complicated. First of all, four fine structure components ($^1D_2$, $^3D_{1,2,3}$) rather than two can be mixed by the strong hyperfine interaction of the ms valence electron. Since the fine structure separations between ^3D states are generally smaller than the singlet–triplet splitting, strong mixing of the triplet states occurs at rather low principal quantum numbers where second-order perturbation theory might still be sufficient to account for hyperfine-induced shifts experienced by the different hyperfine levels of $msnd$ $^1D_2$ states. Secondly, whereas the total electronic spin angular momentum is a rigorous quantum number for pure $msns$ Rydberg states of even isotopes, this is no longer true for $msnd$ $^1D_2$ and $msnd$ $^3D_2$ levels. In particular for the heavier alkaline earths, spin–orbit interaction admixes some triplet character into the $^1D_2$ states (and vice versa), resulting in well-resolved hyperfine structures of $msnd$ ^1D states. Thirdly, $msnd$ $^{1,3}D$ Rydberg series are perturbed by a rather large number of valence states. Owing to configuration interaction, the singlet–triplet mixing can be altered dramatically, leading to conspicuous changes in the recorded hyperfine spectra. After correcting for hyperfine-induced mixing, valuable information can be derived from the observed hyperfine splittings concerning the electronic structure of these Rydberg states.

3.2.1. Spin–Orbit Coupling and Hyperfine Structure of Pure msnd Rydberg States

For Rydberg states of pure $msnd$ configuration, the Hamiltonian H_0, which takes Coulomb and spin–orbit interactions of the valence electrons into account, can be written as

$$H_0 = \frac{e^2}{r_{ms,nd}} + \xi_{nd}\, \mathbf{s}_{nd}\, \mathbf{l}_{nd} \tag{47}$$

where ξ_{nd} is the spin–orbit coupling constant of the nd Rydberg electron. In Eq. (47) spin–spin and spin–other-orbit interactions have been neglected. The eigenvalues and corresponding eigenvectors are known[71,72] to be

$$E(\overline{msnd}\ ^3D_3) = F^{(0)} - \tfrac{1}{5}G^{(2)} + \xi_{nd}$$

$$E(\overline{msnd}\ ^3D_1) = F^{(0)} - \tfrac{1}{5}G^{(2)} - \tfrac{3}{2}\xi_{nd} \tag{48}$$

$$E(\overline{msnd}\ ^3D_2) = F^{(0)} - \tfrac{1}{4}\xi_{nd} - [(\tfrac{1}{5}G^{(2)} + \tfrac{1}{4}\xi_{nd})^2 + \tfrac{3}{2}\xi_{nd}^2]^{1/2}$$

$$E(\overline{msnd}\ ^1D_2) = F^{(0)} - \tfrac{1}{4}\xi_{nd} + [(\tfrac{1}{5}G^{(2)} + \tfrac{1}{4}\xi_{nd})^2 + \tfrac{3}{2}\xi_{nd}^2]^{1/2}$$

$$\overline{|msnd\ {}^3D_3\rangle} = |\overline{msnd\ {}^3D_3}\rangle$$

$$\overline{|msnd\ {}^3D_1\rangle} = |\overline{msnd\ {}^3D_1}\rangle \tag{49}$$

$$\overline{|msnd\ {}^3D_2\rangle} = -\Omega_{so}|\overline{msnd\ {}^1D_2}\rangle + \Lambda_{so}|\overline{msnd\ {}^3D_2}\rangle$$

$$\overline{|msnd\ {}^1D_2\rangle} = \Lambda_{so}|\overline{msnd\ {}^1D_2}\rangle + \Omega_{so}|\overline{msnd\ {}^3D_2}\rangle$$

where $\Lambda_{so}^2 + \Omega_{so}^2 = 1$ and $F^{(0)}(msnd)$, $G^{(2)}(msnd)$ are the corresponding Slater integrals. In Eqs. (48) and (49) a short horizontal bar has been used to indicate that eigenvalues and eigenvectors refer to a pure $msnd$ configuration. The wave functions have been expanded in terms of pure-configuration, exactly SL-coupled basis vectors (intermediate coupling). The Slater integral $G^{(2)}(msnd)$, the spin–orbit coupling constant ξ_{nd}, and the eigenvectors [cf. Eq. (49)] can be derived from experimental term values of the (four) fine structure components. If the spin–orbit coupling constant ξ_{nd} is small compared to the exchange integral $G^{(2)}(msnd)$, second-order perturbation theory yields for the triplet (singlet) amplitude Ω_{so} of the ${}^1D_2$ (${}^3D_2$) state

$$\Omega_{so} = (3/2)^{1/2}\xi_{nd}/\Delta E_{ST}(n) \tag{50}$$

where the singlet–triplet separation is given by

$$\Delta E_{ST}(n) \equiv E(\overline{msnd\ {}^1D_2}) - E(\overline{msnd\ {}^3D_2})$$

$$= \tfrac{2}{5}G^{(2)}(msnd) + \tfrac{1}{2}\xi_{nd} \tag{51}$$

Since for Rydberg states of pure $msnd$ character both ξ_{nd} and $\Delta E_{ST}(n)$ decrease proportional to n^{*-3}, Ω_{so} is expected to be essentially independent of n.[19]

When hyperfine-induced mixing of fine structure components (cf. Section 3.1.3) can be neglected, the hyperfine structure of ${}^{1,3}D$ Rydberg states of pure $msnd$ configuration can easily be calculated using the wave functions given in Eq. (49). To this end the diagonal matrix element

$$E_F({}^1D_2) = \langle \overline{msnd\ {}^1D_2}, I, F|H_{hf}|\overline{msnd\ {}^1D_2}, I, F\rangle \tag{52}$$

and similar ones for the ${}^3D$ states need to be evaluated. Taking the Fermi contact interaction of the ms valence electron into account only [cf. Eq. (16)], one obtains from Table 1

$$E_F({}^3D_3) = \frac{a_{ms}}{12}[F(F+1) - I(I+1) - 12] \tag{53}$$

Table 1. Matrix Elements of the $\mathbf{S} \cdot \mathbf{I}$ and $\boldsymbol{\sigma} \cdot \mathbf{I}$ Operators between Pure-Configuration, Exactly SL-Coupled Basis Vectors $|msnl\,^{2S+1}L_J, I, F\rangle$ $(l = L > 0)$

$$\langle {}^1L_L, I, F | \boldsymbol{\sigma} \cdot \mathbf{I} | {}^3L_{L-1}, I, F \rangle = \frac{1}{2}\left[\frac{(F+I+L+1)(-F+I+L)(F-I+L)(F+I-L+1)}{L(2L+1)}\right]^{1/2}$$

$$\langle {}^1L_L, I, F | \boldsymbol{\sigma} \cdot \mathbf{I} | {}^3L_L, I, F \rangle = -\frac{1}{2}\frac{F(F+1) - I(I+1) - L(L+1)}{[L(L+1)]^{1/2}}$$

$$\langle {}^1L_L, I, F | \boldsymbol{\sigma} \cdot \mathbf{I} | {}^3L_{L+1}, I, F \rangle = -\frac{1}{2}\left[\frac{(F+I+L+2)(-F+I+L+1)(F-I+L+1)(F+I-L)}{(L+1)(2L+1)}\right]^{1/2}$$

$$\langle {}^1L_L, I, F | \mathbf{S} \cdot \mathbf{I} | {}^1L_L, I, F \rangle = 0$$

$$\langle {}^3L_J, I, F | \mathbf{S} \cdot \mathbf{I} | {}^3L_J, I, F \rangle = \frac{[F(F+1) - J(J+1) - I(I+1)][2 + J(J+1) - L(L+1)]}{4J(J+1)}$$

$$\langle {}^3L_L, I, F | \mathbf{S} \cdot \mathbf{I} | {}^3L_{L-1}, I, F \rangle = -\frac{1}{2L}\left[\frac{(L+1)(F+I+L+1)(-F+I+L)(F+I-L+1)(F-I+L)}{(2L+1)}\right]^{1/2}$$

$$\langle {}^3L_L, I, F | \mathbf{S} \cdot \mathbf{I} | {}^3L_{L+1}, I, F \rangle = -\frac{1}{(2L+1)}\left[\frac{L(F+I+L+2)(-F+I+L+1)(F-I+L+1)(F+I-L)}{(2L+1)}\right]^{1/2}$$

$$E_F({}^3D_1) = -\frac{a_{ms}}{8}[F(F+1) - I(I+1) - 2] \tag{54}$$

$$E_F({}^3D_2) = \frac{a_{ms}}{24}[\Lambda_{so}^2 + 2\sqrt{6}\Lambda_{so}\Omega_{so}][F(F+1) - I(I+1) - 6] \tag{55}$$

$$E_F({}^1D_2) = \frac{a_{ms}}{24}[\Omega_{so}^2 - 2\sqrt{6}\Lambda_{so}\Omega_{so}][F(F+1) - I(I+1) - 6] \tag{56}$$

Comparing Eqs. (53)–(56) with the usual expression[72] for the hyperfine components E_F in terms of the hyperfine splitting factor A

$$E_F = \frac{A}{2}[F(F+1) - I(I+1) - J(J+1)] \tag{57}$$

one finds

$$A({}^3D_3) = \frac{a_{ms}}{6} \tag{58}$$

$$A({}^3D_1) = -\frac{a_{ms}}{4} \tag{59}$$

$$A({}^3D_2) = \frac{a_{ms}}{12}(\Lambda_{so}^2 + 2\sqrt{6}\Lambda_{so}\Omega_{so}) \tag{60}$$

$$A(^1D_2) = \frac{a_{ms}}{12}(\Omega_{so}^2 - 2\sqrt{6}\Lambda_{so}\Omega_{so}) \qquad (61)$$

In Eqs. (58)–(61) the hyperfine coupling constant a_{ms} of the ms valence electron has been taken to be independent of the principal quantum number n. On the assumptions made, the hyperfine coupling constants $A(^3D_3)$ and $A(^3D_1)$ do not contain any mixing coefficients. On the contrary, from the measured splitting factors $A(^1D_2)$, $A(^3D_2)$ the amplitudes Ω_{so} and Λ_{so} can be derived. In Eqs. (60), (61) the first terms originate from the operator $a_{ms}\mathbf{S}\cdot\mathbf{I}/2$ [cf. Eq. (16)] and are a measure of the triplet character of the $^3D_2$ and $^1D_2$ states, respectively, while the interference term stems from the operator $a_{ms}\boldsymbol{\sigma}\cdot\mathbf{I}/2$ [see Eq. (16)].

The description of the hyperfine structure of $msnd$ Rydberg states by splitting factors A becomes inappropriate when hyperfine-induced mixing of the fine structure components cannot be ignored anymore. While this is the rule at high principal quantum numbers, dramatic effects may be observed even at low n when accidental degeneracies between fine structure components exist (cf. Section 3.2.6). Since hyperfine-induced mixing of $^1D_2$, $^3D_{1,2,3}$ states causes the corresponding hyperfine levels F to shift by different amounts, the energies E_F are no longer proportional to $[F(F+1) - I(I+1) - J(J+1)]$ and the Landé interval rule is violated. Hyperfine-induced mixing of $msnd$ fine structure components can be taken into account quantitatively in a similar way as described for $msns$ $^{1,3}S$ Rydberg states in Section 3.1.3. The energies $E(msnd, F)$ of the various hyperfine components F are obtained by diagonalizing the Hamiltonian $H = H_0 + H_{hf}$ [cf. Eqs. (47), (16)]. This can be done for each F value separately, since the total angular momentum is conserved. As basis, the pure-configuration, exactly SL-coupled vectors $|msnd\ ^{2S+1}L_J, I, F\rangle = |msnd\ S, L, J, I, F\rangle$ may be chosen with matrix elements of H_0 and H_{hf} given in Refs. 71 and 72, and Table 1, respectively. Alternatively, the wave functions $|\overline{msnd}\ ^{2S+1}L_J, I, F\rangle$ which correspond to the eigenvectors of the even isotopes given in Eq. (49) may be used to set up the secular equation. Whereas the matrix elements of H_{hf} depend on Λ_{so} and Ω_{so}, H_0 is diagonal with the corresponding matrix elements given in Eq. (48). The diagonal elements can also be derived from experimental term values. The latter basis is particularly well suited when singlet–triplet mixing is to be deduced from hyperfine structure data. Contrary to $msns$ $^{1,3}S$ Rydberg states [cf. Eq. (24)] generally fourth-order secular equations have to be solved to calculate hyperfine-induced mixing of $msnd$ fine structure components.

3.2.2. Hyperfine Structure of Perturbed msnd $^{1,3}D$ Rydberg States

As was mentioned above, the $msnd$ $^{1,3}D$ Rydberg series of Ca, Sr, and Ba are perturbed by configuration interaction with valence states of predominantly $(m - 1)dn's$ and $(m - 1)dn'd$ character. For our purposes, the wave functions corresponding to perturbed Rydberg states may be written as $(l = 0, 2)$

$$|msnd\ ^3D_3\rangle = \Lambda(^3D_3)|\overline{msnd\ ^3D_3}\rangle + \varepsilon(^3D_3)|\overline{(m-1)dn'l\ J = 3}\rangle \quad (62)$$

$$|msnd\ ^3D_1\rangle = \Lambda(^3D_1)|\overline{msnd\ ^3D_1}\rangle + \varepsilon(^3D_1)|\overline{(m-1)dn'l\ J = 1}\rangle \quad (63)$$

$$|msnd\ ^3D_2\rangle = \Omega(^3D_2)|\overline{msnd\ ^1D_2}\rangle + \Lambda(^3D_2)|\overline{msnd\ ^3D_2}\rangle$$
$$+ \varepsilon(^3D_2)|\overline{(m-1)dn'l\ J = 2}\rangle \quad (64)$$

$$|msnd\ ^1D_2\rangle = \Lambda(^1D_2)|\overline{msnd\ ^1D_2}\rangle + \Omega(^1D_2)|\overline{msnd\ ^3D_2}\rangle$$
$$+ \varepsilon(^1D_2)|\overline{(m-1)dn'l\ J = 2}\rangle \quad (65)$$

All wave functions are orthogonal and normalized. In Eqs. (62)–(65), the amplitude of the dominant $msnd$ character has been denoted by Λ. Conventionally, a Rydberg level is designated as $^1D_2$ state when its singlet character Λ^2 is larger than its triplet character Ω^2. Furthermore, the absolute phases of the wave functions have been chosen arbitrarily in such a way as to make Λ positive in each case. As discussed for perturbed $6sns\ ^3S_1$ Rydberg states of Ba (cf. Section 3.1.2), the Fermi contact interaction of the ms valence electron is generally much stronger than the hyperfine interaction due to core polarization of a pure $(m - 1)dn'l$ valence state. Hence, direct contributions of doubly excited configurations to the hyperfine structure of $msnd\ ^3D$ states may be neglected. The same holds true for $msnd\ ^1D_2$ Rydberg states, provided singlet–triplet mixing is strong enough to cause sufficiently large hyperfine splittings. If hyperfine-induced mixing of the $msnd$ fine structure components can be neglected, the hyperfine splitting factors of perturbed $msnd$ Rydberg states are found to be

$$A(^3D_3) = \frac{a_{ms}}{6}\,\Lambda^2(^3D_3) \quad (66)$$

$$A(^3D_1) = -\frac{a_{ms}}{4}\,\Lambda^2(^3D_1) \quad (67)$$

$$A(^3D_2) = \frac{a_{ms}}{12}[\Lambda^2(^3D_2) - 2\sqrt{6}\Omega(^3D_2)\Lambda(^3D_2)] \quad (68)$$

$$A(^1D_2) = \frac{a_{ms}}{12}[\Omega^2(^1D_2) - 2\sqrt{6}\Omega(^1D_2)\Lambda(^1D_2)] \quad (69)$$

The hyperfine splitting factors $A(^3D_3)$ and $A(^3D_1)$ [cf. Eqs. (66) and (67)] reflect the reduced *ms* electron density at the nucleus, and allow the *msnd* character of $^3D_3$ and $^3D_1$ Rydberg states to be deduced. The expressions for the splitting factors $A(^3D_2)$ and $A(^1D_2)$ are similar to the ones for unperturbed *msnd* Rydberg states. Indeed, setting $\Lambda(^1D_2) = \Lambda(^3D_2) = \Lambda_{so}$ and $\Omega(^1D_2) = -\Omega(^3D_2) = \Omega_{so}$, from Eqs. (68) and (69) the hyperfine splitting factors of unperturbed *msnd* $^{1,3}D_2$ Rydberg states [cf. Eqs. (60) and (61)] are obtained. However, the admixture of valence states into *msnd* $^1D_2$ and $^3D_2$ Rydberg states generally alters the corresponding amplitudes Λ and Ω and thus influences the hyperfine structure indirectly. Apart from a decrease in the total *msnd* character, i.e., $\Lambda^2 + \Omega^2 < 1$, the ratio between the corresponding amplitudes Λ and Ω may be changed by configuration interaction resulting in a complete breakdown of *SL*-coupling (cf. Section 3.2.3). Contrary to Rydberg states of pure *msnd* configuration, from measured splitting factors $A(^{1,3}D_2)$ of perturbed Rydberg states [cf. Eqs. (68) and (69)] the amplitudes Ω and Λ cannot be derived separately without additional information. The perturber character $\varepsilon^2 = 1 - (\Lambda^2 + \Omega^2)$, for example, may be used for this purpose. The perturber fraction can be obtained from measured lifetimes, isotope shifts, or from an analysis of term values within the framework of multichannel quantum defect theory (cf. Section 3.2.4). Expressions for the hyperfine splitting factors $A(^3L_l)$ and $A(^1L_l)$ [cf. Eqs. (70) and (71)] of perturbed *msnl* ($l > 0$) Rydberg states are readily obtained from Table 1, generalizing the results given in Eqs. (68) and (69). It should be noted, however, that direct contributions from the perturber part of the wave functions are not included.

$$A(^3L_l) = \frac{a_{ms}}{2l(l+1)} \{\Lambda^2(^3L_l) - 2[l(l+1)]^{1/2}\Lambda(^3L_l)\Omega(^3L_l)\} \qquad (70)$$

$$A(^1L_l) = \frac{a_{ms}}{2l(l+1)} \{\Omega^2(^1L_l) - 2[l(l+1)]^{1/2}\Omega(^1L_l)\Lambda(^1L_l)\} \qquad (71)$$

When hyperfine-induced mixing of the fine structure components cannot be neglected, the same procedure[50] as described for pure states at the end of the previous section can be followed to account for the hyperfine structure of perturbed *msnd* Rydberg states. Choosing the wave functions $|msnd \ ^{2S+1}L_J, I, F\rangle$ [cf. Eqs. (62)-(65)] as basis, the Hamiltonian H_0 is diagonal. If experimental term values are used for the (diagonal) matrix elements, an analysis of the fine structure of perturbed *msnd* Rydberg states is avoided. The matrix elements of H_{hf}, which can be easily calculated with the aid of Table 1, depend on $\Lambda(^1D_2, ^3D_{1,2,3})$ and $\Omega(^1D_2, ^3D_2)$. These amplitudes are essentially the only unknown parameters that appear in the matrices, representing the total Hamiltonian $H = H_0 + H_{hf}$, provided experimental term values are known with sufficient accuracy and any direct

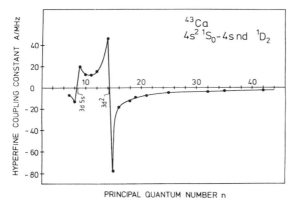

Figure 37. Hyperfine splitting factor A of $4snd$ $^1D_2$ Rydberg states of ^{43}Ca. The solid line connecting the data has been drawn to guide the eye. (Taken from Ref. 42.)

contributions of doubly excited configurations to the observed hyperfine structure can be neglected. A comparison between calculated and experimental hyperfine splittings makes it possible to derive the amplitudes Λ and Ω of the wave functions of *even* isotopes.

3.2.3. Resonance in Singlet-Triplet Mixing of $msnd$ $^{1,3}D$ Rydberg Series of Ca, Sr, and Ba, Detected by Hyperfine Structure

In Figures 37–39, the hyperfine splitting factors $A(^1D_2)$ of $msnd$ $^1D_2$ Rydberg series of Ca,[42] Sr,[73,74] and Ba,[17,18,35] have been plotted versus the principal quantum number n. Whereas the Landé interval rule is fulfilled for $4snd$ $^1D_2$ and $5snd$ $^1D_2$ ($n \neq 19$) Rydberg states of ^{43}Ca and ^{87}Sr, respectively, this is no longer true for perturbed $6snd$ $^1D_2$ states of ^{137}Ba.

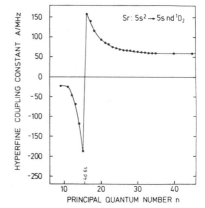

Figure 38. Hyperfine splitting factor A of $5snd$ $^1D_2$ Rydberg states of ^{87}Sr. The solid line connecting the data has been drawn to guide the eye. (Derived from Figure 3 of Ref. 74.)

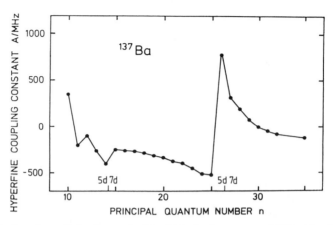

Figure 39. Hyperfine splitting factor A of $6snd$ $^1D_2$ Rydberg states of ^{137}Ba. The data points have been connected to guide the eye. (Taken from Ref. 17.)

In particular, for Rydberg states close to the $5d7d$ $^1D_2$ valence state located between $n = 26$ and 27 (cf. Figure 39), deviations from the interval rule are observed, caused by hyperfine-induced singlet–triplet mixing. In order to derive the approximate splitting factors $^{137}A(^1D_2)$ shown in Figure 39, the observed hyperfine structures were fitted to a power series in $C = F(F + 1) - I(I + 1) - J(J + 1)$, with the leading term given in Eq. (57). As common feature of Figures 37–39, resonant dispersionlike variations of the hyperfine splitting factors are observed, occurring at $n = 8$–9 and $n = 14$–15 in Ca, at $n = 15$–16 in Sr, and at $n = 25$–26 in Ba. These resonances are caused by configuration interaction between the $msnd$ $^1D_2$ Rydberg series and low-lying members of perturbing series, converging versus excited states of the corresponding alkaline-earth ion. At resonance, the recorded spectra immediately reveal the change in the sign of the hyperfine splitting factor A. This is illustrated in Figure 40, showing the hyperfine structure of $5snd$ $^1D_2$ states of ^{87}Sr$(I = 9/2)$ in the vicinity of the resonance. At $n = 13$ the $F = 13/2$ hyperfine component appears at low frequencies, corresponding to a negative splitting factor, while the opposite is true above $n = 15$. In addition, the spectra show the increase and decrease of the total splitting as the principal quantum number is swept through the resonance. In Figure 41, the frequencies of the $F_f = 1/2, 3/2, 5/2$, and $7/2$ hyperfine components of ^{137}Ba, measured relative to ^{138}Ba, are displayed for $6snd$ $^1D_2$ Rydberg states with principal quantum numbers between $n = 10$ and 35. More precisely, the quantity $\{\delta\omega_2^{137-138}(F_f) + \delta\omega_1^{137-138}\omega_2/\omega_1\}/2\pi$ (cf. Section 2.2.2) has been plotted versus the principal quantum number. It should be noted that the term $(\omega_2/\omega_1)\delta\omega_1^{137-138}/2\pi$, representing the contribution of the isotope shift of the first transition $6s^2$ $^1S_0 \to 6s6p$ $^1P_1$ of the two-step excitation, merely causes a vertical displacement, virtually independent of

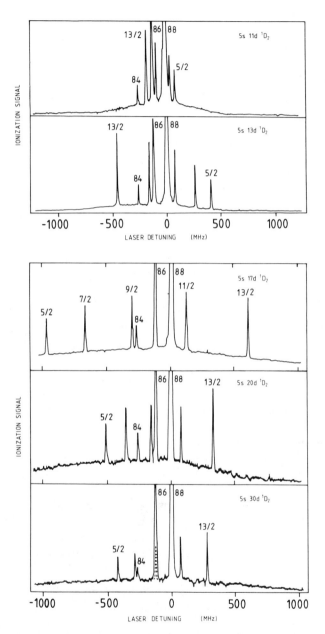

Figure 40. Doppler-free two-photon excitation spectra of $5snd\ ^1D_2$ Rydberg states of Sr[73,74] in the vicinity of the resonance in singlet–triplet mixing.

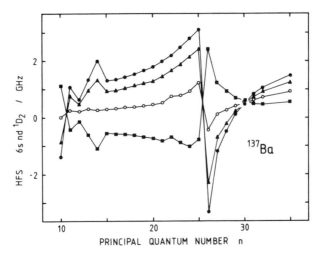

Figure 41. Position of the $F_f = 1/2$ (dots), 3/2 (triangles), 5/2 (open circles), and 7/2 (squares) hyperfine components of ^{137}Ba relative to ^{138}Ba for $6snd$ $^1D_2$ Rydberg states. (Taken from Ref. 18.)

n. All hyperfine components exhibit a dispersionlike variation as the principal quantum number increases from 15 to 35.

The triplet amplitudes $\Omega(^1D_2)$ [cf. Eq. (65)] are better suited for probing configuration interactions between $msnd$ $^1D_2$ Rydberg series and doubly excited valence states than the hyperfine splitting factors themselves, which depend on a_{ms}, i.e., on the sign and magnitude of the nuclear g factor and on the unpaired s-electron density at the nucleus. Since the hyperfine structure of $4snd$ $^1D_2$ and $5snd$ $^1D_2$ Rydberg states of ^{43}Ca and ^{87}Sr, respectively, obey the Landé interval rule, hyperfine-induced singlet–triplet mixing can be neglected and Eq. (69) can be employed to derive the triplet amplitudes $\Omega(^1D_2)$ from the measured hyperfine splitting factors. Using $a_{4s}(^{43}\text{Ca}) = -0.825$ GHz and $a_{5s}(^{87}\text{Sr}) = -0.99$ GHz and taking the total $msnd$ character from multichannel quantum defect analyses of $4snd$ $^1D_2$ and $5snd$ $^{1,3}D_2$ Rydberg series of Ca[75] and Sr,[63] respectively, the triplet amplitudes $\Omega(^1D_2)$ shown in Figure 42 were obtained. In the case of Ba, however, hyperfine-induced singlet–triplet mixing has to be taken into account to derive the triplet amplitudes quantitatively. This has been done[18] by following the procedure mentioned at the end of the last section. The total $6snd$ character $\Omega^2(^1D_2) + \Lambda^2(^1D_2)$ was taken from isotope shift data or lifetime measurements.[76]

Each of the dispersionlike resonances of the triplet amplitudes $\Omega(^1D_2)$ are caused by configuration interaction of the $msnd$ $^1D_2$ Rydberg series of Ca, Sr, and Ba with $(m-1)dn's$ or $(m-1)dn'd$ perturbing states. For example, configuration interaction of the $4snd$ $^1D_2$ Rydberg series of Ca with the $3d^2$ $^1D_2$ level leads to the resonance centered at $n = 14–15$, whereas

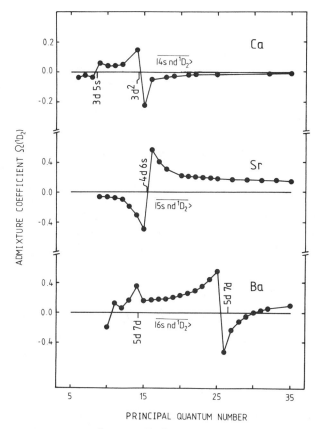

Figure 42. Triplet amplitudes $\Omega(^1D_2)$ of $msnd\ ^1D_2$ Rydberg states of Ca, Sr, and Ba, derived from hyperfine structure data.[42,74,18]

in Sr the resonance at $n = 15$–16 originates from the coupling of the $4d6s\ ^{1,3}D_2$ valence states to the $5snd\ ^{1,3}D_2$ Rydberg series. In either case the configuration interaction is spread out over a considerable number of Rydberg states. Since many of the $msnd\ ^1D_2$ Rydberg states contain some $3d^2\ ^1D_2$ (Ca) or $4d6s\ ^{1,3}D_2$ (Sr) character, the identification of a specific member as perturber state is no longer possible. In Figure 42, the approximate position of maximum $3d^2$ or $4d6s$ admixture has been indicated. In barium, the $5d7d\ ^3F_2$ valence state affects $6s14d\ ^{1,3}D_2$ states only, while the $5d7d\ ^1D_2$ perturber causes the pronounced resonance at $n = 25$–26 (see Figure 42). Qualitatively, the change in the coupling of both valence electrons can be understood in the following way. Whereas pure $msnd\ ^1D_2$ Rydberg states are predominantly SL-coupled, intermediate or even jj coupling are appropriate for the perturbing configurations because of the strong spin-orbit interaction of the $(m-1)d$ electron. It follows from an analysis[70] of the fine structure of the $5d7d$ configuration of barium, for

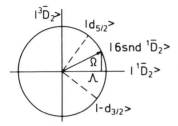

Figure 43. Illustration of the resonance in singlet-triplet mixing as rotation of the state vector $6snd\ ^1D_2$ of Ba. The horizontal and vertical axis correspond to the exactly SL-coupled basis vectors $|\overline{^1D_2}\rangle$ and $|^3D_2\rangle$, respectively.

example, that the $5d7d$ component of the perturber state called $5d7d\ ^1D_2$ is almost exactly jj coupled, corresponding to $|\overline{5d_{5/2}7d_{3/2}}\ J = 2\rangle$. Since the configuration interaction of $6snd\ ^1D_2$ Rydberg states with the $5d7d\ ^1D_2$ perturber is smaller than the fine structure splittings of the $5d7d$ configuration, for Rydberg states in the vicinity of the $5d7d\ ^1D_2$ perturber, configuration interaction admixes only the perturber component $|\overline{5d_{5/2}7d_{3/2}}\ J = 2\rangle$ into the wave functions. The degree of admixture depends on the matrix elements

$$\overline{\langle 6snd\ ^1D_2|(e^2/r_{12})|5d_{5/2}7d_{3/2}}\ J = 2\rangle$$

$$= -(\sqrt{3}/2)\overline{\langle 6snd\ ^1D_2|(e^2/r_{12})|5d7d\ ^1D_2\rangle}$$

and

$$\overline{\langle 6snd\ ^3D_2|(e^2/r_{12})|5d_{5/2}7d_{3/2}}\ J = 2\rangle$$

$$= (1/\sqrt{2})\overline{\langle 6snd\ ^3D_2|(e^2/r_{12})|(5d7d\ ^3D_2\rangle}$$

as well as the energy separation between the perturber and the particular Rydberg state. The admixture of the perturber component $|\overline{5d_{5/2}7d_{3/2}}\ J = 2\rangle$ into the wave functions of the Rydberg states becomes pronounced in the direct neighborhood of the $5d7d\ ^1D_2$ perturber and causes a recoupling of the $|6snd\ ^1D_2\rangle$ and $|6snd\ ^3D_2\rangle$ components, seen as a resonance in singlet-triplet mixing. Since the configuration interaction affects the energies of the Rydberg states besides their wave functions, the observed resonances can also be correlated to (avoided) crossings appearing in the corresponding Lu–Fano diagrams, to be discussed in the following section.

The resonant variation of the triplet amplitudes $\Omega(^1D_2)$ may be visualized as a rotation of the projection of the true state vector $|msnd\ ^1D_2\rangle$ onto the $|^1D_2\rangle$, $|^3D_2\rangle$ plane (cf. Figure 43). In barium, for example, the projection of the state vector is found to be close to the $|^1D_2\rangle$ axis at $n = 15$. Approaching the resonance from below, the vector rotates versus the jj-coupled state $|6s_{1/2}nd_{5/2}\rangle$. Between $n = 25$ and 26 the projection of the state vector suddenly turns toward the $-|6s_{1/2}nd_{3/2}\rangle$ basis vector. If n is increased still further, the $6snd$ component moves toward its original position, crossing the $|^1D_2\rangle$ axis at $n = 30$.

3.2.4. Three-Channel Quantum Defect Theory of 6snd $^{1,3}D_2$ Rydberg States of Ba

In the following the resonance in singlet–triplet mixing of $6snd$ $^{1,3}D_2$ Rydberg states of Ba will be described quantitatively within a three-channel quantum defect model. Multichannel quantum defect theory (MQDT)[11-13,27,77-80] provides a description of entire (interacting) Rydberg series in terms of a number of semiempirical parameters.

The $6snd$ $^{1,3}D_2$ Rydberg series of Ba are perturbed by a considerable number of doubly excited valence states.[53] These perturbing levels represent low-lying members of Rydberg series converging versus the second ($^2D_{3/2}$), third ($^2D_{5/2}$), and even higher ($^2P_{1/2}, ^2P_{3/2}$) ionization limits of neutral barium. Hence an analysis of the $6snd$ $^{1,3}D_2$ Rydberg series that takes most of the bound spectrum into account requires a large number of interacting channels. For example, nine channels have been used for the analysis[15,27,53] of the bound, even-parity, $J = 2$ states of barium with term values above $30\,000$ cm^{-1}. For a quantitative description of the resonance in singlet–triplet mixing, however, we restrict ourselves to $6snd$ $^{1,3}D_2$ Rydberg states ($18 \leq n \leq 35$) in the vicinity of the $5d7d$ $^1D_2$ perturber. Assuming again pure jj coupling for its $5d7d$ component (i.e., $|5d_{5/2}7d_{3/2} J = 2\rangle$, see Section 3.2.3) this level can be considered to be a low-lying member ($n' = 7$) of the Rydberg series $5d_{5/2} n'd_{3/2} J = 2$ converging versus the third ionization limit. Hence three channels are sufficient to analyze the resonance in singlet–triplet mixing. The influence of the remaining $5d7d$ fine structure components is neglected. Previously[76] this model has been employed successfully to explain the total perturber fraction of the $6snd$ $^1D_2$ and $^3D_2$ Rydberg states close to the $5d7d$ $^1D_2$ perturber.

In the following we outline the three-channel quantum defect model and list the pertinent equations. For a thorough discussion of multichannel quantum defect theory the reader is referred to the literature. Using the standard notation the jj-coupled collision channels $|i\rangle$ ($i = 1, 2, 3$) and the pure-configuration, exactly SL-coupled basis vectors $|\bar{\alpha}\rangle$, used throughout this chapter are given in Table 2. The recoupling matrix $U_{i\bar{\alpha}} = \langle jj|SL\rangle$[71,79] (cf. Table 2) transforms the collision channels $|i\rangle$ into the basis $|\bar{\alpha}\rangle$. The unitary matrix $U_{i\alpha}$ provides the transformation between the close-coupling channels $|\Psi_\alpha\rangle$ and collision channels $|\Psi_i\rangle$

$$|\Psi_i\rangle = \sum_{\alpha=1}^{3} U_{i\alpha}|\Psi_\alpha\rangle \tag{72}$$

where $|\Psi_\alpha\rangle$ is given by[13]

$$|\Psi_\alpha\rangle = \sum_i U_{i\alpha}|\Phi_i\rangle\{|f(\nu_i, r)\rangle \cos(\pi\mu_\alpha) - |g(\nu_i, r)\rangle \sin(\pi\mu_\alpha)\} \tag{73}$$

Here $|\Phi_i\rangle$ represents the wave function of the core including the spin and

Table 2. Results of the Three-Channel QDT Analysis of Ba $6snd$ $^{1,3}D_2$ Rydberg Series[18]

$i,a,\bar{a}$	1	2	3
$|i\rangle$	$6snd_{5/2}$	$6snd_{3/2}$	$5d\,7d$
$|\bar{a}\rangle$	$6snd\,^1D_2$	$6snd\,^3D_2$	$5d\,7d$
$U_{i\bar{a}}$	$(3/5)^{1/2}$	$(2/5)^{1/2}$	0
	$-(2/5)^{1/2}$	$(3/5)^{1/2}$	0
	0	0	1
I_i	42 035.02	42 035.02	46 908.89[a]
(cm^{-1})			47 709.86[b]
μ_i	0.694[a]	0.784[a]	0.345[a]
	0.708[b]	0.786[b]	0.658[b]
θ_{ij}	θ_{12}=0.193[a]	θ_{13}=±0.091[a]	θ_{23}=∓0.054[a]
	0.314[b]	±0.576[b]	∓0.172[b]

[a] $I_d=\,^2D_{3/2}(\text{Ba}^+)$.
[b] $I_d=\,^2D_{5/2}(\text{Ba}^+)$.

orbital angular momenta of the excited electron's wave function correspond-ing to the collisional channel i, and f, g are standard Coulomb wave functions. For each member of a Rydberg series the wave function $|\Psi^{(n)}\rangle$ is expressed in terms of the close-coupling channels

$$|\Psi^{(n)}\rangle = \sum_{\alpha=1}^{3} A_\alpha^{(n)}|\Psi_\alpha\rangle \tag{74}$$

For the corresponding term value E_n, a set of three effective quantum numbers $\nu_i^{(n)}$ is defined according to the following equations:

$$E_n = I_i - R/(\nu_i^{(n)})^2 \qquad (i = 1, 2, 3) \tag{75}$$

where I_i are the ionization limits associated with the chosen collision channels, and R stands for the mass-corrected Rydberg constant. Since our model involves two different ionization limits I_s and I_d only, i.e., $I_1 = I_2 = I_s$ $(^2S_{1/2})$, and $I_3 = I_d(^2D_{5/2})$, two different effective quantum numbers $\nu_1^{(n)} = \nu_2^{(n)} = \nu_s^{(n)}$, and $\nu_3^{(n)} = \nu_d^{(n)}$ are associated with each term value. The boun-dary conditions of the wave functions $|\Psi^{(n)}\rangle$ at $r \to \infty$ lead to the following equation:

$$\sum_{\alpha=1}^{3} A_\alpha F_{i\alpha} = \sum_{\alpha=1}^{3} A_\alpha U_{i\alpha} \sin[\pi(\nu_i + \mu_\alpha)] = 0 \tag{76}$$

where μ_α are the eigenquantum defects associated with the close-coupling channels $|\Psi_\alpha\rangle$. The (nontrivial) solutions of Eq. (76)

$$A_\alpha = C_{i\alpha}\bigg/\bigg(\sum_{\alpha'=1}^{3} C_{i\alpha'}^2\bigg)^{1/2} \qquad (\alpha = 1, 2, 3) \tag{77}$$

require the determinant $|F_{i\alpha}|$ to vanish. In Eq. (77), i may be taken arbitrarily

$(1 \leq i \leq 3)$ and $C_{i\alpha}$ is the corresponding cofactor of the determinant $|F_{i\alpha}|$. Using $|F_{i\alpha}| = 0$, ν_s can be written in terms of ν_d according to the following expression:

$$\nu_{s_{1,2}} = \nu_d + \frac{1}{\pi} \cot^{-1} \{-(M_{11} + M_{22})/2 \pm [(M_{11} - M_{22})^2/4 + M_{12}^2]^{1/2}\} \quad (78)$$

where M_{ij} is defined as[79]

$$M_{ij} = \sum_{\alpha=1}^{3} U_{i\alpha} U_{j\alpha} \cot [\pi(\nu_d + \mu_\alpha)] \quad (79)$$

Solutions (ν_s, ν_d) of Eq. (78) simultaneously satisfying

$$E_n = I_s - R/(\nu_s^{(n)})^2 = I_d - R/(\nu_d^{(n)})^2 \quad (75a)$$

correspond to theoretical term values E_n.

Apart from term values, MQDT makes it possible to calculate the wave functions of the Rydberg states and of the perturbing level. Choosing the collision channels $|i\rangle$ as basis (cf. Table 2) the expansion coefficients are given by

$$Z_i^{(n)} = (-1)^{l_i+1}(\nu_i^{(n)})^{3/2} \sum_{\alpha=1}^{3} U_{i\alpha} \cos [\pi(\nu_i^{(n)} + \mu_\alpha)] A_\alpha^{(n)} / N_n \quad (80)$$

where the normalization factor N_n is reported in Eq. (2.7) of Ref. 79. Since the $6snd$ Rydberg states are predominantly SL-coupled, it is preferable, though, to use the exactly SL-coupled, pure-configuration basis vectors $|\bar{\alpha}\rangle$ (cf. Table 2). The corresponding amplitudes are obtained by transforming the expansion coefficients $Z_i^{(n)}$ by means of the recoupling matrix $U_{i\bar{\alpha}}$

$$Z_{\bar{\alpha}}^{(n)} = \sum_{i=1}^{3} Z_i^{(n)} U_{i\bar{\alpha}} \quad (81)$$

In order to be consistent with the notation used in MQDT analyses, in Eq. (81) the amplitudes have been denoted as $Z_{\bar{\alpha}}^{(n)}$, rather than Ω, Λ, and ε. For $6snd$ $^1D_2$ Rydberg states Z_1, Z_2, and Z_3 correspond to $\Lambda(^1D_2)$, $\Omega(^1D_2)$, and $\varepsilon(^1D_2)$, respectively. Likewise, $Z_1(^3D_2) = \Omega(^3D_2)$, $Z_2(^3D_2) = \Lambda(^3D_2)$, and $Z_3(^3D_2) = \varepsilon(^3D_2)$.

Within a least-squares fit procedure, experimental energies E_n and triplet amplitudes $\Omega(6snd\ ^1D_2)$ $(20 \leq n \leq 35)$, derived from hyperfine structure data, were compared with theoretical values. In this way eigenquantum defects μ_α and the elements $U_{i\alpha}$ of the transformation matrix were derived. Following Lee and Lu,[79] the transformation matrix was decomposed into the recoupling matrix $U_{i\bar{\alpha}}$ and the matrix $V_{\bar{\alpha}\alpha}$ which describes the channel interactions and is generated by three successive rotations through generalized Eulerian angles θ_{12}, θ_{13}, and θ_{23}. The first angle (θ_{12}) leads to a mixing of the $|\overline{6snd\ ^1D_2}\rangle$ and $|\overline{6snd\ ^3D_2}\rangle$ basis vectors and thus takes the spin–orbit interaction of the nd-electron into account. Similarly, the configuration interaction of the Rydberg channels $|\overline{6snd\ ^1D_2}\rangle$, $|\overline{6snd\ ^3D_2}\rangle$, and the perturbing level $|\overline{5d_{5/2}7d_{3/2}}\ J = 2\rangle$ is described by the angles θ_{13} and θ_{23}, respectively. Table 2 lists the values obtained for μ_α and θ_{ij}, assuming the perturbing channel to converge either versus the $^2D_{3/2}$ (a) or $^2D_{5/2}$ (b) ionization limit. Actually both models provide an equally good description of term values E_n and triplet amplitudes $\Omega(6snd\ ^1D_2)$ within the limited energy range considered here. However, as was mentioned before, the analysis of the fine structure of the $5d7d$ configuration[70] shows the $5d7d$ component of the $5d7d\ ^1D_2$ perturber to be $|\overline{5d_{5/2}7d_{3/2}}\ J = 2\rangle$ to a good approximation.

Figure 44 displays the usual Lu–Fano diagram.[78] The solid lines represent the solutions ν_{s_1} and ν_{s_2} [cf. Eq. (78)] plotted (modulo 1) versus $\nu_d\,(^2D_{3/2})$. Open circles and squares correspond to ν_s^{exp} (modulo 1), obtained from experimental term values of $6snd\ ^1D_2$ and $6snd\ ^3D_2$ Rydberg states, respectively, with the aid of Eq. (75a). As can be seen from Figure 44, good agreement between experimental and theoretical term values was achieved. The avoided crossing at $n = 26, 27$ is clearly visible.

Besides term values, the three-channel quantum defect model makes it possible to describe the triplet amplitudes $\Omega(6snd\ ^1D_2)$ quantitatively, provided spin–orbit interaction of the Rydberg electron is taken into account ($\theta_{12} \neq 0$). This is illustrated in Figure 45a. On the contrary, excluding spin–orbit interaction by setting θ_{12} arbitrarily equal to zero, experimental and theoretical triplet amplitudes are seen (cf. Figure 45b) to differ by a constant amount, corresponding to Ω_{so}. It should be noted that term values are not affected, i.e., identical Lu–Fano diagrams belong to Figures 45a, 45b, and 45c. It follows that term values of $6snd\ ^1D_2$ and $^3D_2$ Rydberg states are not sufficient to deduce the amount of singlet–triplet mixing due to spin–orbit interaction. Additional experimental information, such as

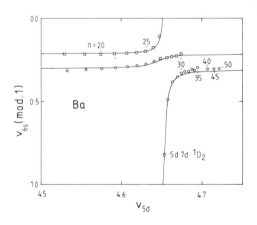

Figure 44. Lu–Fano diagram of the $6snd\ ^{1,3}D_2$ Rydberg series of Ba. The solid lines were calculated using the parameters listed in Table 2 ($I_d = {}^2D_{3/2}$). Open circles and squares correspond to experimental term values of $6snd\ ^1D_2$ and $6snd\ ^3D_2$ Rydberg states, respectively. (Taken from Ref. 18.)

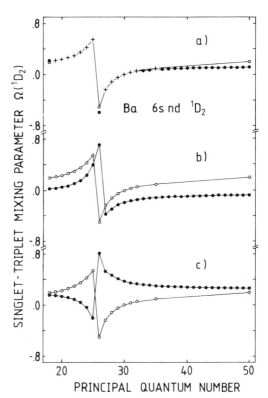

Figure 45. Experimental (○) and calculated (●) triplet amplitudes $\Omega(6snd\ ^1D_2)$, (+) indicates agreement within experimental accuracy. (a) Theoretical values were calculated using the values listed in Table 2 ($I_d = {}^2D_{3/2}$). (b) Spin–orbit coupling of the $6snd\ ^{1,3}D_2$ states was excluded by setting $\theta_{12} = 0$. (c) The relative sign of θ_{13} and θ_{23} was reversed. (Taken from Ref. 18.)

electronic g factors,[15] hyperfine structure data,[18] or fine structure splittings is needed. Since MQDT analyses refer to channels with the same total angular momentum J, the information about the spin–orbit interaction contained in the fine structure splittings is usually not fully exploited. The theoretical triplet amplitudes shown in Figure 45c were obtained by altering the relative signs of θ_{13} and θ_{23}, amounting to a change of the relative signs of the $U_{i\alpha}$ matrix elements. As can be seen from Figure 45c, experimental and theoretical triplet amplitudes appear approximately as mirror images of each other. It follows that hyperfine structure data make it possible to determine the relative sign of the $U_{i\alpha}$ matrix elements, in contrast to term values, lifetimes of Rydberg states, Stark shifts, and electronic g factors, which are sensitive to the squares of amplitudes.

Apart from the electronic structure of the Rydberg states the three-channel quantum defect analysis yields the wave function of the $5d7d\ ^1D_2$ perturber

$$|5d7d\ ^1D_2\rangle = \Lambda(5d7d)|\overline{6snd\ ^1D_2}\rangle + \Omega(5d7d)|\overline{6snd\ ^3D_2}\rangle$$
$$+ \varepsilon(5d7d)|\overline{5d_{5/2}7d_{3/2}}\ J = 2\rangle \tag{82}$$

Table 3. Composition of the $5d7d\ ^1D_2$ Perturbing State of Ba

	Λ	Ω	ε
MQDT[a]	0.69	−0.30	0.66
hfs[b]	0.71 (3)	−0.30 (3)	±0.64 (3)
Lifetime[c]	0.66	±0.48	±0.58
Stark shift[c]	0.46	±0.62	±0.64

[a] Reference 18.
[b] Reference 81.
[c] Reference 76.

with the amplitudes Λ, Ω, and ε given in Table 3. Because of configuration interaction this perturbing level gains a considerable amount of $6snd\ ^1D_2$ and $^3D_2$ character, resulting in well-resolved hyperfine structures of ^{135,137}Ba and isotope shifts between the even Ba isotopes (cf. Figure 46). Following the procedure described in Section 3.1.1 [cf. Eq. (14)], from the isotope shifts $^{134-136}$Ba and $^{136-138}$Ba of the transition $6s6p\ ^1P_1 \rightarrow 5d7d\ ^1D_2$ the total $6snd$ character $\Lambda^2 + \Omega^2 = 0.59(5)$ was derived. Since the neighboring $6s26d$ configuration is sufficiently well separated in energy from the $5d7d\ ^1D_2$ perturber, hyperfine-induced mixing is small and the hyperfine structure can be described by a splitting factor $^{137}A = 375(8)$ MHz. The amplitudes Λ, Ω, and ε (see Table 3) deduced with the aid of Eq. (69) fully corroborate

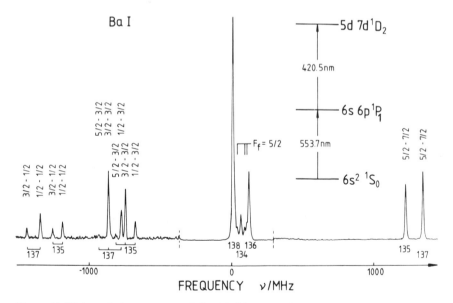

Figure 46. High-resolution spectrum of the $5d7d\ ^1D_2$ perturbing level obtained by optical-optical double resonance. (Taken from Ref. 81.)

the results of the quantum defect analysis. For comparison the amplitudes derived from lifetime and Stark shift measurements[76] are also included in Table 3. Whereas the total perturber fraction (ε^2) is well reproduced, apparently such measurements are less suited to derive the singlet and triplet amplitudes separately.

3.2.5. Singlet–Triplet Mixing and Hyperfine Structure of msnd $^3D_2$ Rydberg States of Sr and Ba

For the heavier alkaline earths Sr and Ba, the hyperfine structure of msnd $^3D_2$ Rydberg states has been investigated systematically as well. Naturally, from such measurements singlet–triplet mixing can also be inferred. However, owing to the proximity of the other ^3D fine structure components, hyperfine-induced mixing of msnd $^3D_2$ with corresponding $^3D_1$ and $^3D_3$ Rydberg states may lead to dramatic changes in the recorded hyperfine structures. These effects can be taken into account quantitatively, provided the fine structure separations $^3D_3$–$^3D_2$ and $^3D_2$–$^3D_1$ are known with sufficient accuracy. Compared to $^1D_2$ states, however, the singlet–triplet mixing is more difficult to derive from the hyperfine structure of $^3D_2$ states.

Figure 47 displays Doppler-free two-photon excitation spectra of 5snd $^3D_2$ Rydberg states of Sr just below the resonance in singlet–triplet

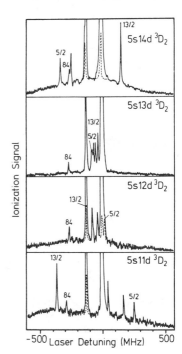

Figure 47. Two-photon excitation spectra of 5snd $^3D_2$ Rydberg states of Sr. (Taken from Ref. 82.)

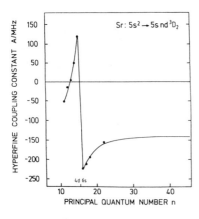

Figure 48. Hyperfine coupling constant $A(^3D_2)$ of $5snd\,^3D_2$ Rydberg states of ^{87}Sr. The solid line connecting the data has been drawn to guide the eye. (Derived from Figure 3 of Ref. 74.)

mixing (cf. Section 3.2.3). The hyperfine structure of the $5snd\,^3D_2$ Rydberg series has been followed up to $n = 22$.[74] Since the fine structure separations between $^3D_{1,2,3}$ states ($n \leq 22$) are still sufficiently large compared to $|a_{5s}|$, hyperfine-induced mixing of the ^3D fine structure components is small. Hence the recorded hyperfine spectra obey the Landé interval rule and can be described by a splitting factor $A(^3D_2)$, plotted versus the principal quantum number n in Figure 48. Analogous to $A(^1D_2)$ (cf. Figure 38) the splitting factor $A(^3D_2)$ shows a dispersionlike resonance in the region of strong singlet–triplet mixing, yet with the opposite phase. This is to be

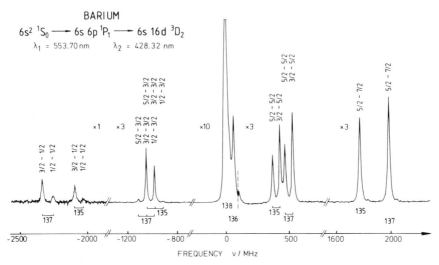

Figure 49. Doppler-free spectrum of the $6s16d\,^3D_2$ Rydberg state of Ba, obtained by optical-optical double resonance. The hyperfine components of ^{135,137}Ba have been labeled F_i–F_f according to the hyperfine levels of the intermediate and final state. (Taken from Ref. 83.)

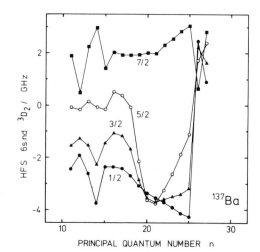

Figure 50. Hyperfine components of ^{137}Ba of $6snd\ ^3D_2$ Rydberg states, measured relative to ^{138}Ba. (Taken from Ref. 83.)

expected from Eqs. (68) and (69) since $\Omega(^3D_2) \cong -\Omega(^1D_2)$ generally holds to a good approximation.

The hyperfine structure of $6snd\ ^3D_2$ Rydberg states of Ba has been investigated[35,83] between $n = 11$ and $n = 30$. For illustration the spectrum of the $6s16d\ ^3D_2$ state is displayed in Figure 49. For the majority of the $6snd\ ^3D_2$ Rydberg states, the Landé interval rule is violated due to hyperfine-induced mixing of the fine structure components. In Figure 50, the frequencies of the $F_f = 1/2$, $3/2$, $5/2$, and $7/2$ hyperfine components of ^{137}Ba measured relative to ^{138}Ba have been plotted versus the principal quantum number n. The frequencies $\{\delta\omega_2^{137-138}(F_f) + \delta\omega_1^{137-138}\omega_2/\omega_1\}/2\pi$ (cf. Section 2.2.2) still contain the contribution of the isotope shift of the first transition of the cascade $6s^2\ ^1S_0 \rightarrow 6s6p\ ^1P_1 \rightarrow 6snd\ ^3D_2$. As mentioned before

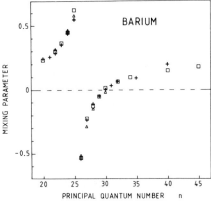

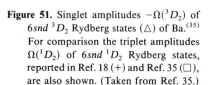

Figure 51. Singlet amplitudes $-\Omega(^3D_2)$ of $6snd\ ^3D_2$ Rydberg states ($\triangle$) of Ba.[35] For comparison the triplet amplitudes $\Omega(^1D_2)$ of $6snd\ ^1D_2$ Rydberg states, reported in Ref. 18 (+) and Ref. 35 ($\square$), are also shown. (Taken from Ref. 35.)

(Section 3.2.3), the term $\delta\omega_1^{137-138}\omega_2/\omega_1$ merely causes a vertical displacement, virtually independent of n. Contrary to the $^3D_2$ series of Sr, in the vicinity of the $5d7d$ $^1D_2$ perturber the hyperfine structure of $6snd$ $^3D_2$ states is dominated by strong fine structure perturbations, totally masking the resonance in singlet–triplet mixing. Using experimentally known fine structure separations, these perturbations have been taken into account quantitatively for Rydberg states up to $n = 25$.[35,83] In Figure 51, the singlet amplitude $-\Omega(^3D_2)$ [cf. Eq. (64)] has been plotted versus the principal quantum number n and compared to the corresponding triplet amplitude $\Omega(^1D_2)$ [cf. Eq. (65)]. As can be seen, $\Omega(^3D_2) \cong -\Omega(^1D_2)$ holds to a good approximation.

3.2.6. Fine Structure Perturbations of Hyperfine Structure of $msnd$ $^{1,3}D$ Rydberg States of Sr and Ba

At high principal quantum numbers, where fine structure separations are of the same order or even smaller than the hyperfine coupling constant $|a_{ms}|$, hyperfine-induced mixing of the $msnd$ fine structure components profoundly modifies the electronic wave functions of the odd isotopes. This situation may also be encountered at low principal quantum numbers, when accidental degeneracies of $msnd$ fine structure components exist. An instructive example[84] provides the hyperfine structure of the $5s19d$ $^1D_2$ Rydberg state of ^{87}Sr, shown in Figure 52. Below $n = 19$, the $5snd$ $^1D_2$ Rydberg states

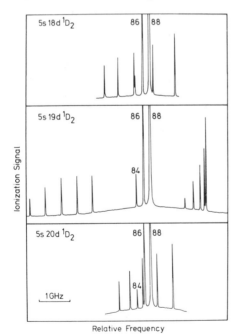

Figure 52. Doppler-free two-photon excitation spectra of $5snd$ $^1D_2$ ($n = 18-20$) Rydberg states of Sr. In addition, at $n = 19$ hyperfine components of ^{87}Sr belonging to the $^3D_3$ level are observed. (Taken from Ref. 84.)

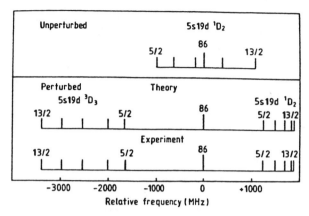

Figure 53. Hyperfine structure of ^{87}Sr predicted for the $5s19d\ ^1D_2$ Rydberg state, neglecting hyperfine-induced mixing of $^1D_2$ and $^3D_3$ fine structure components (upper part). Experimentally observed and theoretical line positions of the hyperfine components of the $5s19d\ ^1D_2$ and $^3D_3$ states are shown in the lower part. (Taken from Ref. 84.)

lie below the corresponding $^3D_3$ fine structure components, while the reverse ordering is observed for $n \geqslant 20$.[85] At $n = 19$, both fine structure components are nearly degenerate. Whereas the signals of the even isotopes ^{84,86,88}Sr are unaffected, hyperfine-induced mixing of the $^1D_2$ and $^3D_3$ fine structure levels manifests itself in a strong repulsion of corresponding hyperfine components of ^{87}Sr. This is illustrated in Figure 53. The upper part shows the hyperfine structure of ^{87}Sr, which would be expected for the $5s19d\ ^1D_2$ state if it were unperturbed by the proximity of the $^3D_3$ fine structure component. The hyperfine structure was obtained interpolating between measured hyperfine splitting factors $A(^1D_2)$ of the $5s18d$ and $5s20d$ Rydberg states (cf. Figure 38). The lower part compares experimentally observed with calculated line positions of the hyperfine components of the $^1D_2$ and $^3D_3$ Rydberg states. The hyperfine-induced mixing of the $^1D_2$ and $^3D_3$ levels was calculated[84] in a similar way as described for the $msns\ ^1S_0$ and $^3S_1$ Rydberg states (see Section 3.1.3). Since the remaining fine structure components $^3D_1$ and $^3D_2$ are sufficiently far removed in energy, hyperfine-induced mixing with these states can be neglected and hence a second-order rather than a fourth-order secular equation needs to be solved. The Hamiltonian $H = H_0 + H_{hf}$ [cf. Eqs. (47) and (16)] was set up using the wave functions $|5s19d\ ^3D_3\rangle$ [cf. Eq. (62)] and $|5s19d\ ^1D_2\rangle$ [cf. Eq. (65)] as basis. From MQDT analyses of the $5snd\ ^{1,3}D_2$ [63] and $5snd\ ^3D_3$ [85] Rydberg series, it is known that the admixture of the $4d6s$ perturbing configuration into the Rydberg states at $n = 19$ amounts to less than 5%. For the calculation of the hyperfine-induced mixing of the $^1D_2$ and $^3D_3$ states, the admixture of the $4d6s$ configuration was neglected, i.e., $\varepsilon(^3D_3) = \varepsilon(^1D_2) = 0$ [cf. Eqs. (62) and (65)]. However, the basis vector $|5s19d\ ^1D_2\rangle$ takes singlet–triplet mixing due to the configuration interaction (cf. Section 3.2.3) into account explicitly. With the aid of Eq. (69) the amplitudes $\Lambda(^1D_2)$ and $\Omega(^1D_2)$ were derived from the interpolated hyperfine splitting factor (cf. Figures 38 and 53). Consequently, the fine structure splitting $\Delta E = E(^1D_2) - E(^3D_3)$ was the only unknown parameter appearing in the secular equation. Excellent agreement between recorded and calculated hyperfine structures of the $^1D_2$ and $^3D_3$ states (see Figure 53) was achieved for $\Delta E = 1.78$ GHz. Because of the strong mixing, the angular momentum J loses its meaning for the odd isotope ^{87}Sr and only F remains a rigorous quantum number. Hence, the selection rule $\Delta J = 0, 2$ no longer applies to coherent two-photon excitation and both fine structure components can be reached from the $5s^2\,^1S_0$ $F = 9/2$ ground state. However, since two units of angular momentum can be transferred at most, the $F = 15/2$ ($^3D_3$) hyperfine level cannot be populated in this way.

In barium, the hyperfine structure of $6snd\,^3D_2$ Rydberg states frequently exhibits perturbations due to close-lying fine structure components. For example, $6snd\,^3D_1$ and $^3D_2$ Rydberg states of Ba are nearly degenerate for principal quantum numbers ranging between $n = 19$ and 24. Since hyperfine components with different total angular momentum quantum numbers F_f are orthogonal, the $^3D_1$ fine structure level does not affect the $F_f = 7/2$ hyperfine component of $6snd\,^3D_2$ Rydberg states, which is seen to vary smoothly between $n = 16$ and $n = 25$ (cf. Figure 50). In contrast, the $F_f = 5/2$ and $3/2$ components are strongly repelled by the corresponding hyperfine levels of the $^3D_1$ state, resulting in a dramatic variation of the hyperfine structure of $6snd\,^3D_2$ ($19 \leqslant n \leqslant 24$) Rydberg states (see Figure 50). As an example, the hyperfine structure of the $6s21d\,^3D_2$ and $^3D_1$

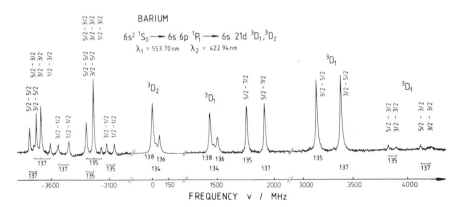

Figure 54. Doppler-free spectrum of the $6s21d\,^3D_1$ and $^3D_2$ Rydberg states of Ba, obtained by optical-optical double resonance. All frequencies have been measured relative to ^{138}Ba ($^3D_2$). The hyperfine components of ^{135,137}Ba have been labeled F_i-F_f according to the hyperfine levels of the intermediate and final state. (Taken from Ref. 83.)

Rydberg states are shown in Figure 54. The signals of the even isotopes ^{134,136,138}Ba corresponding to transitions to the $^3D_2$ and $^3D_1$ fine structure levels appear at the center. All frequencies have been measured relative to ^{138}Ba $(^3D_2)$. As before, the signals of the odd isotopes have been labeled F_i-F_f according to the hyperfine levels of the intermediate $(6s6p\ ^1P_1)$ and final state involved in the two-step excitation. The $F_f = 5/2$ components of both Rydberg states repel each other most strongly, resulting in a complete reordering of the hyperfine structure of the $6s21d\ ^3D_2$ Rydberg state.[83] The repulsion is caused by off-diagonal elements, such as $\langle 6s21d\ ^3D_2,$ $F_f|a_{6s}\mathbf{s}_{6s}\cdot\mathbf{I}|6s21d\ ^3D_1, F_f\rangle$. The interaction between corresponding hyperfine levels depends on the $|6s_{1/2}21d_{3/2}\rangle$ components of the wave functions of both Rydberg states. The strong repulsion observed (cf. Figures 50 and 54) immediately reveals the presence of a large fraction of the $d_{3/2}$ component in the state vector of the $6snd\ ^3D_2$ Rydberg states. Similar arguments can be used to qualitatively understand the strong hyperfine-induced mixing of $6s14d\ ^3D_2$ and $^3D_3$ Rydberg states of ^{135,137}Ba.[50]

4. Odd-Parity Rydberg Series of Alkaline-Earth Elements

Although equally interesting, odd-parity Rydberg series of alkaline-earth elements have been investigated less frequently by high-resolution laser spectroscopy than the even-parity ones discussed in the previous

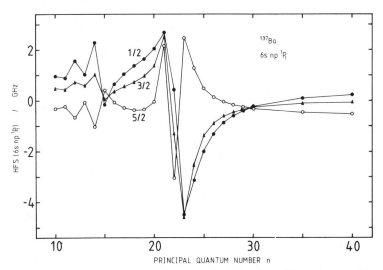

Figure 55. Relative frequencies of the hyperfine components F_f of ^{137}Ba corresponding to the one-photon transition $6s^2\ ^1S_0 \rightarrow 6snp\ ^1P_1$, and corrected for the normal mass shift $\delta\omega_{\mathrm{NME}}^{137-138}$. More specifically, along the vertical axis the frequencies $(\delta\omega_2^{137-138}(F_f) + \delta\omega_1^{137-138} - \delta\omega_{\mathrm{NME}}^{137-138})/2\pi$ have been plotted. Solid circles, triangles, and open circles correspond to the $F_f = 1/2$, 3/2, and 5/2 components, respectively. (Taken from Ref. 44.)

sections. Lacking narrow-band, tunable laser radiation of sufficiently short wavelengths to excite $msnp\ ^{1,3}P$ or $msnf\ ^{1,3}F$ Rydberg states directly from the ground state, the odd-parity Rydberg series are not as easily accessible as the even parity ones, which can be reached conveniently by Doppler-free two-photon absorption. Apart from a recent study of $3snf\ ^{1,3}F$ Rydberg states of Mg[47] populated by Doppler-free two-photon resonant three-photon absorption (cf. Section 2.2.3), extensive hyperfine structure data of odd-parity Rydberg series have been reported for barium only.[26,29,44]

Note added in proof: Recently term values of odd-parity Rydberg states of barium ($6snp\ ^{1}P_1$, $^{3}P_{0,1,2}$ and $6snf\ ^{1}F_3$, $^{3}F_{2,3,4}$) have been measured and analysed[95] within the framework of MQDT. Furthermore hyperfine structure data on $6snp\ ^{1,3}P_1$ and $6snf\ ^{1,3}F_3$ series have been reported and compared to MQDT calculations.[96]

4.1. Hyperfine Structure and Singlet–Triplet Mixing of 6snp $^1P_1$ Ba Rydberg States

Doppler-free spectra of $6snp\ ^{1}P_1$ ($10 \leqslant n \leqslant 40$) Rydberg states of barium (see Figures 12 and 13) have been recorded by employing thermionic detection after resonant two-step excitation via quadrupole ($6s^{2\ 1}S_0 \rightarrow 5d^{2\ 1}D_2$)– dipole ($5d^{2\ 1}D_2 \rightarrow 6snp\ ^{1}P_1$) transitions (cf. Section 2.2.2). In Figure 55 the frequencies of the hyperfine components $F_f = 1/2$, $3/2$, and $5/2$ of ^{137}Ba relative to ^{138}Ba have been plotted versus the principal quantum number n. Apart from the correction $\delta\omega_{\text{NME}}^{137-138}$ for the normal mass shift, the relative frequencies correspond to those that would be observed for the one-photon transition $6s^{2\ 1}S_0 \rightarrow 6snp\ ^{1}P_1$. The variation of the hyperfine splitting reflects changes in the coupling of the electronic angular momenta caused by configuration interaction or hyperfine-induced mixing with close-lying fine structure components. From the work of Garton and Tomkins[86] and Armstrong and co-workers,[87] it is known that valence states of configuration $5d4f$ and $5dn'p$ ($n' = 7, 8$) affect the $6snp\ ^{1}P_1$ series rather locally below $n = 16$. These perturbations manifest themselves in the irregular variation of the hyperfine splittings of the corresponding Rydberg states ($n < 16$). On the contrary, the $5d8p\ ^{1}P_1$ level, located just below the $6s$ ionization limit, interacts with many bound members of this series and causes the resonancelike change of the hyperfine splitting between $n = 16$ and $n = 30$. The resonance in singlet–triplet mixing is partially masked by the repulsion of the hyperfine components of the $6s21p$ and $6s22p\ ^{1}P_1$ states due to hyperfine-induced mixing with the corresponding $^{3}P_2$ Rydberg states.

For the wave functions of perturbed $6snp\ ^{1}P_1$ Rydberg states we write in complete analogy to Eq. (65)

$$|6snp\ ^{1}P_1\rangle = \Lambda(^{1}P_1)|\overline{6snp\ ^{1}P_1}\rangle + \Omega(^{1}P_1)|\overline{6snp\ ^{3}P_1}\rangle$$
$$+ \varepsilon(^{1}P_1)|\overline{5dn'l\ J = 1}\rangle \qquad (l = 1, 3) \qquad (83)$$

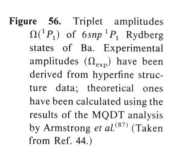

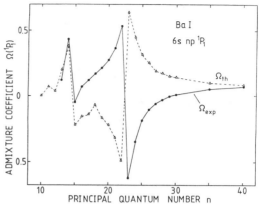

Figure 56. Triplet amplitudes $\Omega({}^1P_1)$ of $6snp\ {}^1P_1$ Rydberg states of Ba. Experimental amplitudes (Ω_{exp}) have been derived from hyperfine structure data; theoretical ones have been calculated using the results of the MQDT analysis by Armstrong et al.[87] (Taken from Ref. 44.)

In order to derive the triplet amplitudes $\Omega({}^1P_1)$ from the experimental hyperfine structure data, the Hamiltonian $H = H_0 + H_{\text{hf}}$ was diagonalized, using the wave function $|6snp\ {}^1P_1\rangle$ and similar ones for the $6snp\ {}^3P$ states as basis. Again, H_0 includes the Coulomb interaction of both valence electrons as well as spin–orbit effects. H_0 is diagonal in the basis chosen and experimental term values of the even isotope ^{138}Ba were used as diagonal elements. Furthermore, perturber amplitudes $\varepsilon({}^1P_1)$ predicted by the eight-channel quantum defect analysis by Armstrong et al.[87] were included in the calculation for some of the more strongly perturbed states. Figure 56 compares triplet amplitudes (Ω_{exp}) derived from hfs data with those (Ω_{th})

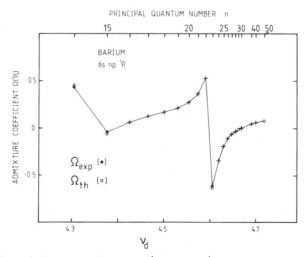

Figure 57. Theoretical triplet amplitudes $\Omega_{\text{th}}({}^1P_1)$ of $6snp\ {}^1P_1$ Rydberg states of Ba, obtained from a five-channel quantum defect analysis. For comparison, experimental amplitudes $\Omega_{\text{exp}}({}^1P_1)$ are included. A cross indicates agreement within error limits. The effective quantum number ν_d plotted along the horizontal axis refers to the second ionization limit ($I_{d_{3/2}} = 46\,908.89\ \text{cm}^{-1}$). (Taken from Ref. 44.)

predicted by the eight-channel quantum defect analysis. As before, the absolute phases of the $^1P_1$ wave functions have been chosen in such a way to make the singlet amplitude $\Lambda(^1P_1)$ positive in each case. Two distinct, dispersionlike resonances appear at $n = 14\text{-}15$ and $n = 22\text{-}23$. The latter resonance is related to an avoided crossing of corresponding branches in the Lu–Fano diagram of the $6snp\ ^{1,3}P_1$ series.[87] Also, the singlet–triplet separation reverses its sign between $n = 22$ and 23. The $5d8p\ ^1P_1$ perturber state, situated close to the ionization limit, interacts more strongly with the $6snp\ ^1P_1$ series compared to the triplet series. Hence for $n \geqslant 23$, $6snp\ ^1P_1$ states are pushed below the corresponding $^3P_1$ levels. As can be seen from Figure 56, the resonance at $n = 14\text{-}15$ is described reasonably well by the eight-channel quantum defect analysis, whereas the wrong phase is predicted for the pronounced resonance occurring at $n = 22\text{-}23$. This is not too surprising, since only term values were included in the analysis.[87] In order to remove this discrepancy, a five-channel quantum defect analysis has been carried out recently in our laboratory.[44] Restricting ourselves to $6snp\ ^{1,3}P_1$ states with $n \geqslant 14$, localized perturbations of these series by valence states of $5d4f$ configuration can be neglected. Besides term values, hyperfine structure data were included in the analysis, using $6snp\ ^1P_1$, $^3P_1$, $5dnp\ ^1P_1$, $^3P_1$, and $^3D_1$ as channels. As can be seen from Figure 57, the analysis describes well the resonances at $n = 14\text{-}15$ and $n = 22\text{-}23$.

4.2. Hyperfine Structure of $6snf\ ^3F$ Rydberg Series of ^{137}Ba

An exhaustive study of the hyperfine structure of $6snf\ ^3F$ Rydberg states $(20 \leqslant n \leqslant 56)$ of ^{137}Ba $(I = 3/2)$ has been reported recently by Eliel and Hogervorst.[29] Starting from the $5d6s\ ^3D_1$ and $^3D_2$ metastable states, one-step excitation and field ionization were employed to record Doppler-free spectra of odd-parity Rydberg levels. Figure 58 displays the energy of the hyperfine levels F_f of $6snf\ ^3F_2$, $^3F_3$, and $^3F_4$ Rydberg states measured relative to the $F_f = \frac{1}{2}(^3F_2)$ level as a function of the principal quantum number n (vertical axis). Although J loses its meaning at high principal quantum numbers due to hyperfine-induced mixing of the fine structure components, below $n = 30$ the well-separated groups of hyperfine levels correspond to the $^3F_2$, $^3F_3$, and $^3F_4$ fine structure components in the order of increasing energy. Above $n = 45$ hyperfine components of predominantly $^1F_3$ character appear on the lower right corner of Figure 58. Furthermore, the hyperfine levels below and above $F_f = \frac{1}{2}(^3F_2)$ begin to merge into the $F = 1$ and $F = 2$ hyperfine doublet of the Ba$^+$ ground state, respectively.

By diagonalizing the Hamiltonian

$$H = \frac{e^2}{r_{6s,nf}} + \xi_{nf}\,\mathbf{s}_{nf} \cdot \mathbf{l}_{nf} + a_{6s}(n) \cdot \mathbf{s}_{6s} \cdot \mathbf{I} \tag{84}$$

within the basis $\overline{|6snf\ ^1F_3,\ ^3F_{2,3,4}\rangle}$ Eliel and Hogervorst[29] analyzed the

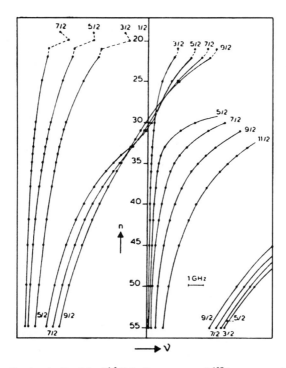

Figure 58. Hyperfine levels F_f of $6snf^{1,3}F$ Rydberg states of ^{137}Ba, measured relative to the $F_f = 1/2(^3F_2)$ component. (Taken from Ref. 29.)

measured hyperfine structure on a state-by-state basis to derive the Slater integrals $G^{(3)}(6snf)$, as well as the spin–orbit (ξ_{nf}) and hyperfine $[a_{6s}(n)]$ coupling constants. Contrary to the analysis described previously (cf. Sections 3.2.2 and 4.1), such an approach does not take perturbations caused by doubly excited configurations into account. Recently these authors have shown the $6snf\,^3F_2$ and $6snf\,^1F_3$ series to be perturbed below $n = 30$.[88,89] Recently Post *et al.*[96] showed that discrepancies between experimental and calculated hyperfine splittings encountered by Eliel and Hogervorst[29] are caused by not taking perturbations of $6snf$ series into account.

4.3. Hyperfine-Induced Singlet–Triplet Mixing of $3snf^{1,3}F$ Rydberg States of ^{25}Mg

The hyperfine structure of ^{25}Mg ($I = 5/2$) of $3snf^{1,3}F$ Rydberg states has been reported for principal quantum numbers between $n = 14$ and 84.[47] Starting from the atomic ground state, the Rydberg states were reached by two-photon resonant three-photon absorption via the $3s3d\,^1D_2$ intermediate state. Within the range of principal quantum numbers studied, the

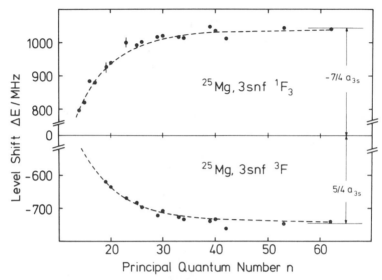

Figure 59. Repulsion of $3snf\ ^{1,3}F$ Rydberg states of ^{25}Mg. (Taken from Ref. 47.)

Rydberg states are not perturbed by configuration interactions, and spin-orbit coupling can be neglected. Employing experimental or extrapolated singlet–triplet separations $\Delta E_{ST} = E(3snf\ ^1F_3) - E(3snf\ ^3F_3)$ of the even isotopes (^{24,26}Mg), the Hamiltonian H_{hf} [cf. Eq. (16)] was diagonalized using the exactly SL-coupled vectors $|3snf\ ^1F_3\rangle$, $|3snf\ ^3F_{2,3,4}\rangle$ as basis. In this way the hyperfine-induced singlet–triplet mixing and the corresponding repulsion of the ^1F and ^3F levels of ^{25}Mg was inferred. Figure 59 compares experimental with calculated (dashed lines) level shifts. The unperturbed position ($\Delta E = 0$ MHz) corresponds to the true isotope shift of ^{25}Mg versus ^{24}Mg (1840 MHz) for the transition $3s^2\ ^1S_0 \to 3snf\ ^{1,3}F$.[47] From the data shown in Figure 59, the hyperfine coupling constant a_{3s} of the Mg$^+(3s\ ^2S_{1/2})$ ground state was inferred and good agreement with a previously published, more precise value[90] was observed.

5. Hyperfine Structure and Isotope Shifts of Rydberg States of Other Two-Electron Systems

Although we have been interested primarily in the investigation of Rydberg series of alkaline-earth elements by hyperfine structure measurements, in this section we briefly comment on similar studies performed on He and Yb.

5.1. 6snl Rydberg Series of Yb

The measurement of isotope shifts and hyperfine structures of $6snd\ ^3D_1$ Rydberg states of Yb (cf. Section 2.1.1), reported by Barbier and

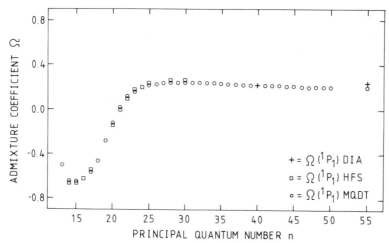

Figure 60. Triplet amplitude of $6snp\ ^1P_1$ Rydberg states of Yb. The amplitudes were derived from hyperfine structure splittings (open squares), a five-channel quantum defect analysis using term values and hyperfine structure data (open circles), and measurements of diamagnetic shifts (crosses). (Taken from Ref. 46.)

Champeau[20] was the first study of the electronic structure of an entire Rydberg series by hyperfine interactions. Within the range of principal quantum numbers investigated ($20 \leqslant n \leqslant 53$) the Rydberg states are unperturbed by configuration interactions. However, at $n = 26$ the hyperfine structure of ^{171}Yb and ^{173}Yb shows deviations from the expected trend due to hyperfine-induced mixing with a nearby $J = 2$ perturbing level of configuration $4f^{13}5d6s6p$. This is similar to the perturbation of the $6s18s\ ^3S_1$ Rydberg state of ^{135,137}Ba by the $5d7d\ ^3P_0$ valence state discussed in Section 3.1.5. With increasing principal quantum number n, the hyperfine levels of the odd isotopes were observed to converge toward the corresponding hyperfine components of the Yb II ground state. The isotope shifts between the even isotopes ^{172,174,176}Yb were found to be independent of n, a behavior expected for Rydberg states of pure configuration.

Recently,[45,46] hyperfine structures and isotope shifts of $6snp\ ^{1,3}P_1$ Rydberg states of Yb have been measured by Doppler-free, two-photon resonant three-photon absorption (cf. Section 2.2.3). Strong variations of the hyperfine structures have been observed, caused by configuration interactions as well as hyperfine-induced mixing with nearby $^3P_0$ and $^3P_2$ fine structure components. In Figure 60 the triplet amplitude $\Omega(^1P_1)$ of $6snp\ ^1P_1$ Rydberg states has been plotted versus the principal quantum number n. The pronounced variation at $n \sim 20$ is caused by a perturbing state of configuration $4f^{13}\ 5d^2\ 6s$.[46] Figure 60 compares amplitudes derived from hyperfine structure data, a five-channel quantum defect analysis,[46] as well as diamagnetic shifts of $6snp\ ^1P_1$ states.[91]

5.2. 1sns and 1snd Rydberg States of ^{3}He

Finally, we turn to the simplest two-electron system, i.e., ^{3}He. Since doubly excited configurations lie beyond the first ionization limit, Rydberg series are of pure $1snl$ character. Hyperfine-induced singlet–triplet mixing is pronounced, even at low principal quantum numbers, because fine structure splittings are small and the Fermi contact interaction of the $1s$ electron with the nuclear magnetic moment of ^{3}He $(I = 1/2)$ is strong $(a_{1s} = -8.67$ GHz). The hyperfine structure of $1sns\ ^3S_1$ $(n = 4$-$6)^{(92)}$ and $1snd\ ^{1,3}D$ $(n = 3$-$6, 12$-$17)^{(19,48,92,93)}$ Rydberg states of ^{3}He has been measured recently employing high-resolution laser spectroscopy. For the analysis of the recorded hyperfine structure, the Hamiltonian

$$H = \frac{e^2}{r_{1s,nl}} + H_{fs} + H_{hf} \qquad (85)$$

has been used. The fine structure Hamiltonian[94]

$$H_{fs} = A' \cdot \mathbf{L} \cdot \mathbf{S} + a' \cdot \mathbf{L} \cdot \boldsymbol{\sigma} + b' \frac{3(\mathbf{L} \cdot \mathbf{S})^2 + \frac{3}{2}\mathbf{L} \cdot \mathbf{S} - \mathbf{L}^2 \cdot \mathbf{S}^2}{2S(2S-1)L(2L-1)} \qquad (86)$$

takes spin–orbit, spin–other-orbit, and spin–spin interactions into account. The coupling constants A', a', and b' can be calculated, using hydrogenic wave functions. Alternatively, these constants as well as the singlet–triplet separation can be inferred from the spectrum of ^{4}He.[94] Using the hyperfine coupling constant a_{1s} of ^{3}He$^+$, generally excellent agreement between calculated and recorded hyperfine structures has been achieved in this way.[19,48]

6. Conclusion

In this chapter we have reviewed high-resolution laser spectroscopy of Rydberg states of two-electron systems. In particular, the even-parity Rydberg series $msns\ ^{1,3}S$ and $msnd\ ^{1,3}D$ of Ca, Sr, and Ba have been discussed in some detail. Apart from experimental techniques the main emphasis has been put on (i) hyperfine-induced mixing of fine structure components belonging to Rydberg states with identical or different principal quantum numbers and (ii) the interpretation of measured isotope shifts and hyperfine structures to infer the electronic structure of Rydberg states. In particular, mixing coefficients of the corresponding wave functions obtained in this way can be directly compared to theoretical amplitudes predicted by multichannel quantum defect theory. For such analyses, hyperfine structure data provide experimental information complementary to that obtainable from term values.

The electronic structure of Rydberg states can be probed by a large variety of atomic properties such as term values, lifetimes, oscillator strengths, Zeeman splitting factors, Stark and diamagnetic shifts, as well as fine and hyperfine structures. To some extent these observables are sensitive to different parts of the wave functions of Rydberg states. The volume effect, which represents a substantial fraction of the total isotope shift observed in heavy elements, probes changes in the s-electron density at the nucleus and hence allows the total Rydberg ($msnl$) character to be derived. This information can also be obtained from lifetimes or Stark shifts, for example. The coupling of the electronic angular momenta of both valence electrons strongly influences the hyperfine structure of Rydberg states. Therefore, hyperfine structure data are ideally suited to deduce the singlet–triplet mixing of the Rydberg ($msnl$) component of the state vector, including the relative sign of singlet and triplet amplitudes. Direct contributions of doubly excited configurations to the hyperfine structure of Rydberg states of two-electron systems perturbed by configuration interaction can often be neglected. Thus hyperfine structure data provide limited but specific information about the electron structure of Rydberg states. Furthermore, Zeeman splitting factors are sensitive to the Rydberg as well as perturber part of the wave function. For Rydberg states with large radial extent, i.e., high principal quantum numbers, diamagnetic shift can be used to infer singlet–triplet mixing as well as the total Rydberg character.

We hope that high-resolution laser spectroscopy of Rydberg states of two-electron systems, which has developed rapidly within the last three years and has contributed considerably to our knowledge of the electronic structure of such states, will continue to serve this purpose.

Acknowledgments

This work was supported by the Deutsche Forschungsgemeinschaft, Sonderforschungsbereich 161. I thank Professor E. Matthias for discussions and for his critical reading of parts of the manuscript, Dr. Beigang for allowing me to include unpublished results, and Dr. Neukammer for helping me with tables and figures. Last but not least I thank Mrs. Mallwitz for patiently typing the manuscript.

References

1. D. Kleppner, M. Littman, and M. Zimmerman, *Sci. Am.* **244**, 108 (1981).
2. S. Feneuille and P. Jacquinot, in *Atomic and Molecular Physics*, Vol. 17 (edited by D. Bates and B. Bederson), Academic, New York (1981), p. 99.
3. *Rydberg States of Atoms and Molecules* (edited by R. F. Stebbings and F. B. Dunning), Cambridge University Press, Cambridge (1983).
4. K. Heilig and A. Steudel, in *Progress in Atomic Spectroscopy*, Part A (edited by W. Hanle and H. Kleinpoppen), Plenum, New York (1978), p. 263.
5. L. Armstrong, *Theory of the Hyperfine Structure of Free Atoms*, Wiley-Interscience, New York (1971).
6. W. Demtröder, *Laser Spectroscopy*, Springer Series in Chemical Physics, Vol. 5, Springer, Berlin (1981).

7. C. Fabre and S. Haroche, Ref. 3, Chap. 4.
8. H. Kopfermann, *Nuclear Moments*, Academic, New York (1958).
9. S. Büttgenbach, *Hyperfine Structure in 4d and 5d Shell Atoms*, Springer Tracts on Modern Physics, Vol. 96, Springer, Berlin (1982).
10. A. Lurio, M. Mandel, and R. Novick, *Phys. Rev.* **126**, 1758 (1962).
11. M. J. Seaton, *Rep. Progr. Phys.* **46**, 167 (1983), and references therein.
12. U. Fano, *J. Opt. Soc. Am.* **65**, 979 (1975).
13. K. T. Lu, *Phys. Rev. A* **4**, 579 (1971).
14. J. J. Wynne, J. A. Armstrong, and P. Esherick, *Phys. Rev. Lett.* **39**, 1520 (1977).
15. P. Grafström, C. Levinson, H. Lundberg, S. Svanberg, P. Grundevik, L. Nilsson, and M. Aymar, *Z. Phys. A* **308**, 95 (1982).
16. U. Fano, private communication.
17. H. Rinneberg and J. Neukammer, *Phys. Rev. Lett.* **49**, 124 (1982).
18. H. Rinneberg and J. Neukammer, *Phys. Rev. A* **27**, 1779 (1983).
19. P. F. Liao, R. R. Freeman, R. Panock, and L. M. Humphrey, *Opt. Commun.* **34**, 195 (1980).
20. L. Barbier and R.-J. Champeau, *J. Phys. (Paris)* **41**, 947 (1980).
21. S. Svanberg, in *Laser Spectroscopy V* (edited by A. R. W. McKellar, T. Oka, and B. P. Stoicheff), Springer Series in Optical Sciences, Vol. 30, Springer, Berlin (1981), p. 301.
22. E. R. Eliel and W. Hogervorst, *J. Phys. (Paris)* **43**, C2-443 (1982).
23. E. Matthias, H. Rinneberg, R. Beigang, A. Timmermann, J. Neukammer, and K. Lücke, in *Atomic Physics 8* (edited by I. Lindgren and S. Svanberg), Plenum, New York (1983), p. 543.
24. W. Hogervorst, *Commun. At. Mol. Phys.* **13**, 69 (1983).
25. R. Beigang and A. Timmermann, *J. Phys. (Paris)* **44**, C7-137 (1983).
26. H. Rinneberg and J. Neukammer, *J. Phys. (Paris)* **44**, C7-177 (1983).
27. M. Aymar, *Phys. Rep.* **110**, 163 (1984).
28. P. Grafström, J. Zhan-Kui, G. Jönsson, S. Kröll, C. Levinson, H. Lundberg, and S. Svanberg, *Z. Phys. A* **306**, 281 (1982).
29. E. R. Eliel and W. Hogervorst, *Phys. Rev. A* **27**, 2995 (1983).
30. E. R. Eliel, thesis, University of Amsterdam, 1982 (unpublished).
31. T. F. Gallagher, Ref. 3, Chap. 5.
32. G. Grynberg and B. Cagnac, *Rep. Progr. Phys.* **40**, 791 (1977).
33. J. E. Bjorkholm and P. F. Liao, *Phys. Rev. A* **14**, 751 (1976).
34. W. Hogervorst and E. R. Eliel, *Z. Phys. A* **310**, 19 (1983).
35. E. R. Eliel and W. Hogervorst, *J. Phys. B* **16**, 1881 (1983)
36. H. Rinneberg, J. Neukammer, G. Jönsson, H. Hieronymus, A. König, and K. Vietzke, *Phys. Rev. Lett.* **55**, 382 (1985).
37. H. Rinneberg, J. Neukammer, U. Majewski, and G. Schönhense, *Phys. Rev. Lett.* **51**, 1546 (1983).
38. K. C. Harvey and B. P. Stoicheff, *Phys. Rev. Lett.* **38**, 537 (1977).
39. S. M. Curry, C. B. Collins, M. Y. Mirza, D. Popescu, and I. Popescu, *Opt. Commun.* **16**, 251 (1976).
40. K. C. Harvey, *Rev. Sci. Instrum.* **52**, 204 (1981).
41. R. Beigang, W. Makat, and A. Timmermann, *Opt. Commun.* **49**, 253 (1984).
42. R. Beigang, K. Lücke, and A. Timmermann, *Phys. Rev. A* **27**, 587 (1983).
43. C. Delsart and J.-C. Keller, *Opt. Commun.* **15**, 91 (1975).
44. J. Neukammer and H. Rinneberg, to be published.
45. U. Majewski, J. Neukammer, and H. Rinneberg, *Phys. Rev. Lett.* **51**, 1340 (1983).
46. U. Majewski, J. Neukammer, and H. Rinneberg, (unpublished).
47. R. Beigang, D. Schmidt, and A. Timmermann, *Phys. Rev. A* **29**, 2581 (1984).
48. L. A. Bloomfield, H. Gerhardt, and T. W. Hänsch, *Phys. Rev. A* **27**, 850 (1983).
49. J. Neukammer and H. Rinneberg, *J. Phys. B* **15**, 2899 (1982).

50. J. Neukammer and H. Rinneberg, *J. Phys. B* **15**, 3787 (1982).
51. J. Neukammer, E. Matthias, and H. Rinneberg, *Phys. Rev. A* **25**, 2426 (1982).
52. M. Aymar, P. Camus, M. Dieulin, and C. Morrilon, *Phys. Rev. A* **18**, 2173 (1978).
53. M. Aymar and O. Robaux, *J. Phys. B* **12**, 531 (1979).
54. M. Wilson, *Phys. Lett.* **65 A**, 213 (1978).
55. J. Neukammer and H. Rinneberg, *J. Phys. B* **15**, L425 (1982).
56. M. Aymar and P. Camus, *Phys. Rev. A* **28**, 850 (1983), and private communication.
57. H. Schüler and E. G. Jones, *Z. Phys.* **77**, 801 (1932).
58. H. Casimir, *Z. Phys.* **77**, 811 (1932).
59. R. Beigang, E. Matthias, and A. Timmermann, *Z. Phys. A* **301**, 93 (1981).
60. H. Rinneberg, *Z. Phys. A* **302**, 363 (1981).
61. R. Beigang and A. Timmermann, *Phys. Rev. A* **25**, 1496 (1982).
62. H. Rinneberg, J. Neukammer, and E. Matthias, *Z. Phys. A* **306**, 11 (1982).
63. P. Esherick, *Phys. Rev. A* **15**, 1920 (1977).
64. K. C. Pandey, Sudhanshu S. Jha, and J. A. Armstrong, *Phys. Rev. Lett.* **44**, 1583 (1980).
65. R. Beigang and A. Timmermann, *Phys. Rev. A* **26**, 2990 (1982).
66. R. Beigang, W. Makat, A. Timmermann, and P. J. West, *Phys. Rev. Lett.* **51**, 771 (1983).
67. S. A. Bhatti, C. L. Cromer, and W. E. Cooke, *Phys. Rev. A* **24**, 161 (1981).
68. M. L. Zimmerman, J. C. Castro, and D. Kleppner, *Phys. Rev. Lett.* **40**, 1083 (1978).
69. M. L. Zimmerman, M. G. Littman, M. M. Kash, and D. Kleppner, *Phys. Rev. A* **20**, 2251 (1979).
70. P. Camus, M. Dieulin, and A. El Himdy, *Phys. Rev. A* **26**, 379 (1982).
71. E. U. Condon and G. H. Shortley, *The Theory of Atomic Spectra*, Cambridge University Press, Cambridge (1979).
72. I. I. Sobelman, *Atomic Spectra and Radiative Transitions*, Springer, Berlin (1979).
73. R. Beigang, E. Matthias, and A. Timmermann, *Phys. Rev. Lett.* **47**, 326 (1981).
74. R. Beigang, E. Matthias, and A. Timmermann, *Phys. Rev. Lett.* **48**, 290(E) (1982).
75. J. A. Armstrong, P. Esherick, and J. J. Wynne, *Phys. Rev. A* **15**, 180 (1977).
76. T. F. Gallagher, W. Sandner, and K. A. Safinya, *Phys. Rev. A* **23**, 2969 (1981).
77. M. J. Seaton, *Proc. Phys. Soc. (London)* **88**, 801 (1966).
78. K. T. Lu and U. Fano, *Phys. Rev. A* **2**, 81 (1970).
79. C. M. Lee and K. T. Lu, *Phys. Rev. A* **8**, 1241 (1973).
80. J. J. Wynne and J. A. Armstrong, *Comments At. Mol. Phys.* **8**, 155 (1979).
81. H. Rinneberg and J. Neukammer, *J. Phys. B* **15**, L825 (1982).
82. R. Beigang and A. Timmermann, (unpublished).
83. J. Neukammer and H. Rinneberg, *J. Phys. B* **15**, L723 (1982).
84. R. Beigang, D. Schmidt, and A. Timmermann, *J. Phys. B* **15**, L201 (1982).
85. R. Beigang and D. Schmidt, *Phys. Scr.* **27**, 172 (1983).
86. W. R. S. Garton and F. S. Tomkins, *Astrophys. J.* **158**, 1219 (1969).
87. J. A. Armstrong, J. J. Wynne, and P. Esherick, *J. Opt. Soc. Am.* **69**, 211 (1979).
88. K. A. H. van Leeuwen, W. Hogervorst, and B. H. Post, *Phys. Rev. A* **28**, 1901 (1983).
89. B. H. Post, W. Hogervorst, and W. Vassen, *Phys. Rev. A* **29**, 2989 (1984).
90. W. M. Itano and D. J. Wineland, *Phys. Rev. A* **24**, 1364 (1981).
91. J. Neukammer, H. Rinneberg, and U. Majewski, *Phys. Rev. A* **30**, 1142 (1984).
92. E. de Clerq, F. Biraben, E. Giacobino, G. Grynberg, and J. Bauche, *J. Phys. B* **14**, L183 (1981).
93. J. E. Lawler, A. I. Ferguson, J. E. M. Goldsmith, D. J. Jackson, and A. L. Schawlow, *Phys. Rev. Lett.* **42**, 1046 (1979).
94. T. A. Miller and R. S. Freund, in *Advances in Magnetic Resonance*, Vol. 9 (edited by J. S. Waugh), Academic, New York (1977), p. 49.
95. B. H. Post, W. Vassen, W. Hogervorst, M. Aymar, and O. Robaux, *J. Phys. B* **18**, 187 (1985).
96. B. H. Post, W. Vassen, and W. Hogervorst, *J. Phys. B* **19**, 511 (1986).

5

Parity Nonconservation in Atoms

T. PETER EMMONS AND E. NORVAL FORTSON

1. Introduction

Atomic physics has played an important role in the advance of fundamental physics. The analysis of the spectral lines of hydrogen gave some of the important clues necessary for the development of quantum theory. Likewise the observation of the Lamb shift led to the modern theory of quantum electrodynamics. Although today most of the discoveries in fundamental physics come from high-energy accelerators, the field of atomic physics still has many contributions to make. A good example is the subject of this review, the atomic parity violation experiments.* These experiments have provided a direct test of the Weinberg–Salam–Glashow standard model of weak interactions, and should continue to test aspects of weak interaction theory. These experiments also possess an intrinsic interest within the domain of atomic physics, because they offer a striking example of the breakdown of left–right symmetry—a phenomenon not seen before in atoms.

Several other review articles[1-3] have been published about atomic parity violation experiments and about the connection with weak interaction theory. In this review we will not attempt a complete study of this field but will instead try to point out most of the salient points for the nonspecialist.

T. PETER EMMONS AND E. NORVAL FORTSON • Department of Physics, FM-15, University of Washington, Seattle, Washington 98195.
* Here, we will use the terms "parity violation" and "parity nonconservation" (PNC) interchangeably. Each refers to the violation of reflection symmetry, which implies that the parity operator is not conserved.

1.1. The Weak Interaction—Charged Currents and Neutral Currents

In the past, the weak interaction was not considered important in the study of atomic systems, not only because its effects were known to be intrinsically weak but also because originally the only known form of the interaction required a change or exchange of charge. For example, in beta decay, a neutron is converted into a proton while creating an electron and antineutrino. This process is said to be a "charged current" weak interaction. This is an interaction that in first order can cause the decay of the nucleus, but not a steady-state modification of an atom's electronic structure without changing the identity of the atom.

Many theories prior to the 1960s proposed that the weak interaction is carried by the $W^{\pm}$, a hypothetical charged boson of mass about 80 GeV. In the late 1960s, the Weinberg-Salam-Glashow theory[4-6] predicted that the weak force should also be carried by a neutrally charged particle given the name Z^0. In this case, "neutral current" weak processes should occur, i.e., those involving no exchange of charge. One could then have interactions between the electron of the atom and the nucleons in the nucleus that would not result in a change in the atom's identity but instead produce a slight modification in the electronic wave function. In principle, this modification can reflect parity violation inherent in this neutral current form of the weak interaction.

The predicted neutral current weak interaction was initially seen in neutrino-nucleon scattering in 1973.[7] There remained the question of the validity of the theory as regards parity violation in the electron-nucleon interaction. This question has been largely answered in high-energy electron-scattering experiments[8] and in the atomic parity violation experiments, the subject of this review.

1.2. Atomic Structure and Parity Nonconservation

Parity violation is the important key to observing the weak interaction in atoms. Since all the other forces in nature, so far as known, conserve parity, their effect is to put the atom in a state of definite parity. Any mixing of parity states would then be due to the weak interaction. Furthermore, as discussed above, these weak interactions must involve only the previously unseen neutral current form.

It turns out that the atom has built into it very good tests to determine if it is in a state of definite parity. A selection rule for electric dipole ($E1$) transitions is that the parity of the final state must be opposite from that of the initial state. Therefore, one cannot induce an $E1$ transition between two electronic energy levels of the same parity. But if these electronic levels have a small admixture of opposite parity state (due, for example, to the weak interaction) then there is a possibility for an $E1$ transition. The selection rules for two states of the same parity allow a magnetic dipole ($M1$) transition. One would then get an interference between the $E1$ and the $M1$ transitions. This interference will manifest itself in effects that distinguish between the left and the right such as circular dichroism and optical rotation.

1.3. Optical Rotation and Circular Dichroism

Optical rotation and circular dichroism are closely related. Optical rotation, the rotation of plane polarized light, is due to a difference in the index of refraction between left and right circularly polarized light. Plane polarized light can be broken into equal portions of left and right circularly polarized light. Since the phase velocity is inversely proportional to the index of refraction, one sense of circular polarization will be retarded with respect to the other and the plane of polarization will be rotated. Circular dichroism is the differential absorption between left and right circularly polarized light. Since the absorption and the index of refraction are related by the Kramers–Kronig dispersion relation, these two effects are integrally related. One cannot have one without the other. Optical rotation follows the real part of the index of refraction near an absorption line while circular dichroism follows the imaginary, i.e., absorptive, part. Circular dichroism is thus symmetrically peaked at an absorption line, while optical rotation is antisymmetrical and varies as a dispersion curve about the line. Figure 1 shows the two curves for a typical transition. A rigorous discussion of these effects appears later in this chapter.

Optical rotation is an effect usually associated with complicated molecules such as sugars. In fact, even today the quality of sugar is determined by looking at its optical rotation. The sugar molecule has a helical

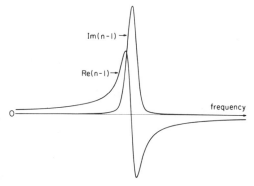

Figure 1. Real and imaginary parts of the index of refraction, n. The absorption follows the imaginary part while optical rotation follows the real part.

structure along which the valence electrons are constrained to move. Classically, light directed perpendicular to the helical axis with the optical electric field having a component along the axis will accelerate the valence electrons, producing an electric moment $E1$ along the axis. Since the electrons are constrained to move along the helix, they also generate a magnetic moment, $M1$, along the axis of the helix parallel to $E1$. The helical structure therefore creates both a magnetic and an electric dipole moment that are oscillating $\pi/2$ out of phase with each other, just as the charge and current are. More generally, the moments have to be orthogonal in phase in order for them not to violate time reversal symmetry. Since $M1$ and $E1$ moments are, respectively, odd and even under time reversal, their time-averaged product must be zero, which is the case for a $\pi/2$ phase difference. When the $M1$ and $E1$ moments both point in the same direction but differ in phase by $\pi/2$ their interaction with the magnetic and electric fields of a light wave will either add or subtract depending on the sense of circular polarization of the light. Thus sugar molecules cause circular dichroism and its companion, optical rotation.

This example, then, is like our parity-violating atom where there are both $M1$ and $E1$ moments. The helical structure within the atom is caused by the parity-violating term. Time reversal invariance guarantees the phase between the moments.

The question then arises whether the sugar molecule necessarily violates parity. The answer is no because the two states of the sugar are degenerate. Molecules with opposite helicity, but otherwise identical, can exist and produce opposite optical effects. In the case of a nondegenerate atom, the helicity can show up only if there is parity violation and will have the same sense for every atom.

We now investigate the theory somewhat further in order to understand the magnitude of the atomic parity violation.

2. Theory

Since the mass of the Z^0 is expected to be large, around 90 GeV, its interaction has a short range. Thus it is a very good approximation to take the weak interaction as a contact potential in an atom. This means that only those electronic states that have an overlap with the nucleus will be affected by the electron–nucleon weak force. Normally only s state electrons satisfy this condition, but relativistic corrections also cause a finite contribution for p electrons. This is important because the mixing of the p electronic state into the opposite parity s state or vice versa is what causes all the observable effects of the weak interactions in atoms.

The size of the weak interaction in the contact potential is governed by the Fermi coupling constant G_F, which is quite small in atomic units:

$$G_F = 89.6 \text{ eV fm}^3$$

$$= 3.29 \times 10^{-15} \text{ a.u.} \tag{1}$$

An important aspect of the Weinberg–Salam–Glashow (WSG) theory is a fundamental equivalence between the weak and electromagnetic interactions. The reason that the electromagnetic force has such a greater effect is that the photon, the carrier of the electromagnetic force, has no mass so the range of the force is unlimited rather than constrained to be the Compton wavelength of the Z^0 or $W^{\pm}$.

We now present a rather complete treatment of the neutral current weak interaction in atoms. We will start with the relativistic neutral current interaction between electrons and nucleons and use it with suitable approximations to discuss the amplitudes of parity mixing in atoms. Time reversal symmetry is assumed throughout. The PNC neutral current interaction between electrons makes only a small relative contribution in heavy atoms[9] and therefore will not be considered here.

In atoms, as noted already, it is adequate to replace the propagator for the massive bosons by a contact function, and the weak interaction reduces to a current–current interaction form. The electron current has vector and axial vector forms $\bar{\Psi}_e \gamma^\mu \Psi_e$ and $\bar{\Psi}_e \gamma^\mu \gamma^5 \Psi_e$, where Ψ_e is the electron wave function and γ^μ is the Dirac matrix 4-vector.[10] The nucleon current is similar, but can have an isospin dependence, i.e., the current may not be the same for a neutron and a proton. Parity nonconservation results when the Hamiltonian contains a product of vector and axial vector currents. The general form of the parity nonconserving Hamiltonian is thus given by[11]

$$H_{\text{PNC}} = G_F[\bar{\Psi}_n \gamma_\mu (v_0 + v_1 \tau_3) \Psi_n \bar{\Psi}_e \gamma^\mu \gamma^5 \Psi_e$$
$$+ \bar{\Psi}_n \gamma_\mu \gamma^5 (a_0 + a_1 \tau_3) \Psi_n \bar{\Psi}_e \gamma^\mu \Psi_e] \tag{2}$$

where τ_3 is the nuclear isospin operator.

In the Weinberg–Salam theory, the parameters of H_{PNC} are given in terms of a single parameter θ_w:

$$v_0 = -\sin^2 \theta_w, \qquad a_0 = 0$$

$$v_1 = 1/2(1 - 4\sin^2 \theta_w), \qquad a_1 = 1/2(1 - 4\sin^2 \theta_w)g_A \tag{3}$$

where $g_A \simeq 1.25$ is the axial vector coupling strength in the theory. The best current experimental value of θ_w is given by[12]

$$\sin^2 \theta_w = 0.23 \pm 0.02 \tag{4}$$

If radiative corrections are taken into account, a slightly different value of $\sin^2 \theta_w$ may be used[12]:

$$\sin^2 \theta_w = 0.219 \pm 0.014 \tag{5}$$

Frequently, one uses the vector and axial-vector coupling coefficients C_1 and C_2 for the protons and neutrons, where

$$C_{1P} = v_0 + v_1 = 1/2(1 - 4\sin^2 \theta_w)$$

$$C_{1N} = v_0 - v_1 = -1/2$$

$$\tag{6}$$

$$C_{2P} = a_0 + a_1 = 1/2(1 - 4\sin^2 \theta_w)g_A$$

$$C_{2N} = a_0 - a_1 = -1/2(1 - 4\sin^2 \theta_w)g_A$$

The second form in each case is that given by the Weinberg–Salam theory.

Independently of gauge theories, we could have written down Eq. (2) as a phenomenological current-current interaction with parameters to be determined by experiment.

In an atomic nucleus, the nucleons are nonrelativistic so we can use the nonrelativistic limits for the 4-vector operators[10]:

$$\gamma_0 \sim 1, \qquad \gamma\gamma^5 \sim \boldsymbol{\sigma}, \qquad \gamma^5 \sim \frac{\boldsymbol{\sigma} \cdot \mathbf{p}}{mc} \tag{7}$$

and apply them to the nucleon current in Eq. (2). We use $\bar{\Psi} = \Psi^+ \gamma^0$ and $\boldsymbol{\gamma} = \gamma^0 \boldsymbol{\alpha}$ to obtain

$$H_{\text{PNC}} = \frac{G_F}{\sqrt{2}} \sum_{en} \left[C_{1n} \int \Psi_n^+ \Psi_n \Psi_e^+ \gamma^5 \Psi_e \, d\mathbf{r} \right.$$

$$\left. + C_{2n} \int \Psi_n^+ \boldsymbol{\sigma}_n \Psi_n \Psi_e^+ \boldsymbol{\alpha} \Psi_e \, d\mathbf{r} \right] \tag{8}$$

where n stands for N or P, and the summation is over all electrons and nucleons in the atom.

The terms in C_{1n} are additive for all the nucleons because the electronic wave function has approximately the same phase over the entire nucleus.

The terms in C_{2n} depend upon nuclear spin and have a net contribution only from unpaired nucleons. Therefore, in heavy atoms the C_1 terms normally dominate and provide a total interaction for each electron of the form

$$H_{\mathrm{PNC}}^{(1)} = \frac{G_F}{\sqrt{2}} \frac{Q}{2} \int \rho_n \Psi_e^+ \gamma^5 \Psi_e \, d\mathbf{r} \tag{9}$$

where $Q = 2(ZC_{1P} + NC_{1N})$. Z and N are the numbers of protons and neutrons in the nucleus, and ρ_n is the nuclear density. In the Weinberg–Salam theory,

$$Q = Z(1 - 4\sin^2 \theta_w) - N \tag{10}$$

There are some cases[13] in heavy atoms where the interaction containing the C_{2n} terms might be measurable, even though it is much smaller than $H_{\mathrm{PNC}}^{(1)}$ in Eq. (9).

In hydrogen the experiments in principle can be made equally sensitive to either C_{1P} or C_{2P}. Experiments in both hydrogen and deuterium could be combined to yield C_{1N} and C_{2N} as well.

Thus far we have retained the form of H_{PNC} which is correct for arbitrary electron energy, since in heavy atoms the electrons are indeed fully relativistic near the nucleus where they contribute to H_{PNC}. However, the nonrelativistic form of H_{PNC} is useful for calculations in light atoms and also has some conceptual value in heavy atoms as well. Making the nonrelativistic approximations of Eq. (7) for the Dirac operators in Eq. (8) we obtain the nonrelativistic potential for a single electron interacting with a point nucleus at $r = 0$:

$$V_{\mathrm{PNC}}^{\mathrm{Nr}} = \frac{G_F}{m_e c \sqrt{2}} \sum_n \frac{\delta(\mathbf{r})}{2} [C_{1n} \boldsymbol{\sigma}_e \cdot \mathbf{p}_e$$

$$- C_{2n}(\boldsymbol{\sigma}_n \cdot \mathbf{p}_e + i\boldsymbol{\sigma}_n \cdot \boldsymbol{\sigma}_e \times \mathbf{p}_e)] + \mathrm{H.C.} \tag{11}$$

The parity nonconserving aspect of the potential is evident from the pseudo-scalar combinations of the vectors in Eq. (11). This interaction will mix electronic states of opposite parity. As pointed out earlier, only mixing between $s_{1/2}$ and $p_{1/2}$ states need be considered usually, because other states make a negligible contribution at the origin.

2.1. A Simple Calculation

For illustration we show the calculation of the C_1 (electronic spin) term for the simplest case, the hydrogen ground state. Since this is a light element, Eq. (11) is a very good approximation.

In order to calculate the mixing fraction ε_{PNC}, one resorts to first-order perturbation theory. We will proceed with a first-order calculation for the ground $(1s)$ state of hydrogen. Although all p states are mixed into the ground state by H_{PNC}, we will calculate only the contribution of the nearest opposite parity state, the $2p$. First-order perturbation theory then takes the form

$$\varepsilon_{PNC}^{(1s,2p)} = \frac{\langle 1s|H_{PNC}^{(1)}|2p\rangle}{E_{1s} - E_{2p}} \tag{12}$$

We substitute V_{PNC} from Eq. (11), and integrate out the delta functions, leaving

$$\varepsilon_{PNC}^{(1s,2p)} = \frac{G_F}{4m_e c\sqrt{2}} \frac{|\langle 1s_{1/2}|\boldsymbol{\sigma} \cdot \mathbf{p}|2p_{1/2}\rangle|}{E_{1s} - E_{2p}} r = 0 \tag{13}$$

The values for the wave functions at the origin are known exactly for hydrogen[14] and can be substituted in. The $2p$ state can have total angular momentum $J = 3/2$ and $J = 1/2$. Only the $J = 1/2$ term contributes, as can be seen by evaluating directly the matrix elements of V_{PNC}. This, of course, is expected because our potential is a (pseudo)-scalar.

$$\varepsilon_{PNC}(1s, 2p) = i1.3 \times 10^{-16}(Q)$$

$$Q = 0.08 \qquad \text{for } \sin^2 \theta_w = 0.23 \tag{14}$$

Note the factor of i in this expression. It will always appear in matrix elements of H_{PNC} when the interaction is of the form of Eq. (2) or its special cases, Eq. (8) or Eq. (11), and when the conventional choice of wave function phase is made as was done here. The relative phase of matrix elements is independent of phase convention, and plays an important role in the PNC interference effects to be discussed later. This point was illustrated in the qualitative discussion in Section 1.3.

In Eq. (14), we are left with a value for the mixing that is too small to measure. An enhancement can be obtained by looking at the $2s$ state in hydrogen where the energy denominator of Eq. (12) is only the Lamb shift. The energy denominator can be further reduced by the addition of a magnetic field. Several experiments to look for PNC in the hydrogen $2s$ state are currently underway.[15-17]

2.2. Effect in Heavy Atoms

An extremely important enhancement occurs in heavy atoms as first pointed out by Bouchiat and Bouchiat.[18] The essential points can be understood by applying the nonrelativistic formalism to heavy atoms. As in the above discussion of hydrogen, the matrix element must be of the form $\langle n'p_{1/2}| V_{\text{PNC}}^{nr}|ns_{1/2}\rangle$, because the two states must be opposite in parity but have the same total angular momentum, J. The momentum operator acting on the $p_{1/2}$ state may be approximated at the origin by

$$\left|\frac{h}{i}\frac{d}{dr}\Psi_p(r)\right|_0 \cong Zh/a_0|\Psi_s(0)| \tag{15}$$

One would expect the momentum to increase as Z near the nucleus because of the increased nuclear charge attracting the electron. Another factor of Z enhancement can be seen by using the expression Fermi and Segre[19] derived for $|\Psi_s(0)|^2$ in order to calculate the isotope shift in alkali-like atoms:

$$|\Psi_s(0)|^2 = \frac{ZZ_0^2}{\pi(n-\sigma)^3 a_0^3}\left(1 - \frac{d\sigma}{dn}\right) \tag{16}$$

where $Z_0 \sim 1$ is the outer charge seen by the electron, σ is the quantum defect, and n is the principal quantum number. (The formula was originally derived for alkali metals, but the qualitative behavior is more general.) Note the factor of Z, which indicates how much the wave function is "pulled in" where the nuclear charge is incompletely shielded. The coherence over the nucleus contributes the factor Q of Eq. (10), which adds roughly another factor of Z. The nonrelativistic result is a Z^3 dependence of $\langle n'p_{1/2}| V_{\text{PNC}}^{nr}|ns_{1/2}\rangle$. A somewhat similar enhancement in heavy atoms of the effect of an electron's electric dipole moment was demonstrated long ago by Sandars.[20]

Using Eqs. (11), (15), and (16), we arrive at the expression

$$\langle n'p_{1/2}| V_{\text{PNC}}^{nr}|ns_{1/2}\rangle \sim \frac{G_F}{4\pi}\frac{Z^2 QhK_r}{Ca_0^4(n'-\sigma')^{3/2}(n-\sigma)^{3/2}}$$

$$\sim 10^{-16}Z^2 QK_r \tag{17}$$

We have added K_r, the relativistic correction factor which may be as large as 10 for heavy atoms. Equation (17) was first derived by Bouchiat and Bouchiat[18] and is useful for estimating the order of magnitude of PNC effects in atoms.

An accurate calculation in heavy atoms requires a thoroughgoing relativistic treatment. The central field-independent particle approximation gives a first-order answer, but corrections involving electron exchange, noncentral effects, and other electron–electron correlations are needed in many cases. Details of the calculations have been reviewed elsewhere[1,2] and will not be presented here. Theoretical values of ε_{PNC} for transitions of interest are about 10^{-10}. This is about 10^6 larger than the above calculation for hydrogen, in rough agreement with the expected Z^3 dependence.

The values of $E1_{PNC}$ are still small, but the atomic system gives an added enhancement. Allowed $M1$ moments are also small, typically of order v/c times an allowed $E1$ moment. For allowed $M1$ transitions, the measured ratio $E1_{PNC}/M1$ is at the 10^{-7} level. For extremely weak $M1$ transitions, this ratio can be even larger, as discussed later.

To show the basis of more thorough calculations, we generalize the simple treatment above. The initial definite parity state $|i\rangle$ has states of opposite parity, denoted here by $|n\rangle$, mixed in to form the state $|i_{PNC}\rangle$:

$$|i_{PNC}\rangle = |i\rangle + \sum_n \varepsilon_{PNC}(i, n)|n\rangle \tag{18}$$

Here, $\varepsilon_{PNC}(i, n)$ is defined as in Eq. (12). The same is true with the final state $|f_{PNC}\rangle$. Since the $E1$ matrix element links only states of opposite parity, we can write the complete $E1$ matrix element as

$$E1_{PNC} = \langle f_{PNC}|E1|i_{PNC}\rangle$$
$$= \sum_n \left\{ \frac{\langle f|H_{PNC}|n\rangle\langle n|E1|i\rangle}{E_f - E_n} + \frac{\langle f|E1|n\rangle\langle n|H_{PNC}|i\rangle}{E_i - E_n} \right\} \tag{19}$$

The two terms are just the PNC perturbation acting, respectively, on the final and initial states.

The magnetic dipole matrix element between the same two states,

$$\langle f_{PNC}|M1|i_{PNC}\rangle = \langle f|M1|i\rangle \tag{20}$$

has the same dependence on magnetic quantum numbers as $E1_{PNC}$. Hence the ratio $E1_{PNC}/M1$ is independent of orientation of the atom.

2.3. Transitions of Interest

We now survey those transitions in heavy atoms that have been picked for experimental study.

The primary selection rule for allowed $M1$ transitions is that the electron configuration not change. Since we are only interested in those configurations that have valence electrons with finite probability of being at the nucleus, we are limited to p electrons; there is always one configuration for purely s valence electrons. The stable elements with the largest Z^3 enhancement that also have p valence electronic configurations are thallium

$(Z = 81, \lambda = 1.283 \ \mu m)$, lead $(Z = 82, \lambda = 1.279 \ \mu m)$, and bismuth $(Z = 83, \lambda = 0.876, 0.648, 0.462, 0.302 \ \mu m)$. The values of λ shown are the allowed $M1$ transition wavelengths. These elements turn out to be well suited for optical rotation experiments.

The fractional effect $E1_{PNC}/M1$ can be made larger by looking at very weak $M1$ transitions. Bouchiat and Bouchiat[9,18,21] were the first to call attention to this possibility. They suggested looking at transitions such as $6s$-$7s$ in cesium, or $6p$-$7p$ in thallium. Since the electronic configuration changes, these transitions are forbidden except for a relativistic correction, which gives them an amplitude about 10^{-4} (and intensity about 10^{-8}) times that of a normal $M1$ transition. In practice, these $M1$ transitions are so weak that an electric field must be applied in order to see the transition through Stark mixing. These experiments thus do not actually measure circular dichroism or optical rotation, but instead measure an interference between PNC and Stark amplitudes, as will be discussed in some detail in Sections 3.1 and 5.

3. Circular Dichroism and Optical Rotation— Rigorous Discussion

We now derive expressions for the circular dichroism and optical rotation that occur when a PNC-induced $E1$ transition between two states interferes with an $M1$ transition between the same two states.

A circularly polarized electromagnetic wave traveling in the $\hat{z}$ direction has electric and magnetic field vectors:

$$\mathbf{E}_\eta = \frac{E_0}{\sqrt{2}}(\hat{x} - i\eta\hat{y}), \qquad \mathbf{B}_\eta = \frac{B_0}{\sqrt{2}}(i\eta\hat{x} + \hat{y}) \tag{21}$$

where $\eta = \pm 1$ gives the two states of circular polarization, the positive sign denoting right-hand circularly polarized light by the standard optical convention (which corresponds to negative helicity). Including both electric and magnetic dipole radiation between initial and final atomic states i and f, the transition amplitude is proportional to

$$A_\eta(f_i) = \langle f_{PNC} | \mathbf{E}1 \cdot \mathbf{E}_\eta + \mathbf{M}1 \cdot \mathbf{B}_\eta | i_{PNC} \rangle \tag{22}$$

The circular polarization in emission is then given by

$$P_c \equiv \sum_{m_f m_i} \frac{(|A_+|^2 - |A_-|^2)}{(|A_+|^2 + |A_-|^2)} = \frac{2 \ \text{Im} \ E1_{PNC} M1}{|E1_{PNC}|^2 + |M1|^2}$$

$$\sim 2 \ \text{Im}(E1_{PNC}/M1) \tag{23}$$

where we have assumed $E1_{PNC} \ll M1$ in the last expression. The dependence of $E1_{PNC}$ and $M1$ on directional quantum numbers is the same, and hence cancels out in Eq. (23).

Similarly, in the case of absorption, the absorptivity K will change between right and left circularly polarized light by a fractional amount 2δ, where

$$2\delta = \frac{K_+ - K_-}{K} = 4 \, \text{Im} \left(\frac{E1_{PNC}}{M1} \right) \tag{24}$$

δ is called the circular dichroism, and we have assumed $K \sim K_+ \sim K_-$, implying $\delta \ll 1$.

Any substance which exhibits circular dichroism must also demonstrate optical activity. This feature is a result of the difference in the real part of the refractive index n for right and left circularly polarized light. From dispersion relations we have

$$\frac{n_+ - n_-}{n - 1} = \frac{K_+ - K_-}{K} \tag{25}$$

where $n \cong n_+ \cong n_-$ for small effects. Because of the differing phase velocities $c/n_\pm$, the polarization plane of linearly polarized light of wavelength λ will be rotated in a distance l by an angle

$$\phi = \frac{\pi l}{\lambda}(n_+ - n_-) = \frac{4\pi l}{\lambda}(n - 1) \, \text{Im} \left(\frac{E1_{PNC}}{M1} \right) = \frac{4\pi l}{\lambda}(n - 1)R \tag{26}$$

$$R \equiv \text{Im} \, (E1_{PNC}/M1)$$

where we have used the optical convention in which the rotation is considered positive when it appears clockwise to an observer looking into the source.

Note that the angle ϕ follows the dispersive line shape of $(n - 1)$, which is always a useful discriminant against spurious rotations. This shape is shown in Figure 1. Such PNC optical rotation is the object of experiments made by groups at Seattle, Oxford, Novosibirsk, and Moscow.

Whenever the PNC interaction has time-reversal symmetry, as it does in the forms of interest here given in Eqs. (8) or (11), then $E1_{PNC}$ and $M1$ will have a relative phase of $\pi/2$.[9] This phase maximizes the circular dichroism and optical rotation shown in Eqs. (24) and (26). We have already mentioned the importance of this phase in the qualitative discussion in Section 1.3, and we noted the factor of i in the result of the simple calculation of PNC mixing in Eq. (14).

3.1. PNC-Stark Interference

In the second class of experiments the PNC-induced $E1$ moment interferes with a Stark-induced $E1$ moment instead of an $M1$ moment. When an external static electric field E_s is applied to the atom, a Stark $E1$ amplitude $E1_S$ appears with a form similar to the PNC amplitude in Eq. (19):

$$E1_S = \langle f_S | E1 | i_S \rangle = e \sum_n \left(\frac{\langle f | E1 | n \rangle \langle n | \mathbf{E}_s \cdot \mathbf{r} | i \rangle}{E_f - E_n} + \frac{\langle f | \mathbf{E}_s \cdot \mathbf{r} | n \rangle \langle n | E1 | i \rangle}{E_i - E_n} \right) \quad (27)$$

where the Stark perturbed states are given a subscript S, and we use i and f for initial and final states.

The Stark–PNC form of interference is utilized in experiments with heavy atoms at Berkeley and Paris, and in many experiments with hydrogen. Here we present a qualitative sketch of the scheme used at Paris and Berkeley. The basic idea, originally pointed out by Bouchiat and Bouchiat,[21] is that an electronic polarization (i.e., a nonzero expectation value of the electronic angular momentum $\langle J \rangle$) in the excited state of the atom is induced by absorption of a circularly polarized photon directed perpendicular to an applied static electric field.

For definiteness, let the static field be in the $\hat{x}$ direction, $\mathbf{E}_s = E_s \hat{x}$, and let the incident photon have its momentum $\mathbf{k}$ in the $\hat{y}$ direction, so that its circularly polarized electric vector is given [cf. (21)] by $\hat{E}_\eta = E_0(\hat{z} - i\eta\hat{x})/\sqrt{2}$. We look for electronic polarization in the z direction given by

$$(P_e)_z = \langle J_z^f \rangle = \frac{\sum_{m_f m_i} m_f |\langle f_s | \mathbf{E}_\eta \cdot E1 | i_s \rangle + \langle f_{PNC} | \mathbf{E}_\eta \cdot E1 | i_{PNC} \rangle|^2}{\sum_{m_f m_i} |\langle f_s | \mathbf{E}_\eta \cdot E1 | i_s \rangle + \langle f_{PNC} | \mathbf{E}_\eta \cdot E1 | i_{PNC} \rangle|^2} \quad (28)$$

where i and f denote initial and final states. Using the matrix element for vector operators[14] in this expression, the sums over the initial, intermediate, and final values of the magnetic quantum number, m, give in general a nonvanishing result proportional to η. Thus, a measurement of J_z^f provides the pseudoscalar $\mathbf{J}^f \cdot \eta\mathbf{k} \times \mathbf{E}_s$, which reveals the PNC interaction of the atom. More details of the calculation of Eq. (28) are given in Refs. 1 and 21.

4. Optical Rotation Experiments

Experiments designed to look for optical rotation in heavy metal vapors began in 1974 at Seattle,[22] Oxford,[23] and Novosibirsk[24] and improved experiments are continuing at all these places. A fourth experiment began a year later at Moscow.[25] All four groups chose initially to look for optical activity in atomic bismuth because bismuth has allowed $M1$ absorption lines (see Section 2.3) that are accessible with tunable lasers available at the time the experiments were begun. Our group in Seattle is now also measuring optical rotations in lead and thallium, using lasers that have only recently become available.

For illustration, we show the bismuth energy levels and the transitions studied by the different groups in Figure 2. The Seattle experiment uses the $^4S_{3/2} \rightarrow {}^2D_{3/2}$ line at 876 nm. All of the other experiments use the $^4S_{3/2} \rightarrow {}^2D_{5/2}$ line at 648 nm, although more recently the Oxford group began a second experiment at the 876-nm line as well.

Comparisons between theoretical predictions and experimental results are usually made in terms of the quantity $R \equiv \mathrm{Im}(E1_{\mathrm{PNC}}/M1)$, which is related to the rotation ϕ_{PNC} in Eq. (26). In that equation we see that the rotation ϕ_{PNC} follows a dispersion curve $(n-1)$, with the maximum change in ϕ_{PNC} occurring across the absorption line center from one dispersion peak to the other, with magnitude

$$\Delta\phi_{\mathrm{PNC}} = lK_0 R \tag{29}$$

where an isolated Lorentz-shaped line is assumed. This expression serves as a useful estimate even when the line shape is not Lorentzian. Predicted magnitudes of ϕ_{PNC} are of order 10^{-7} rad per absorption length lK_0. The efforts of all experimenters are devoted to assessing the value of this asymmetric rotation associated with PNC, while discriminating against other rotations that are not related to PNC.

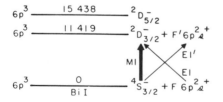

Figure 2. Electronic energy levels for bismuth. The absorption lines from the ground state to the two lowest excited states at 876 and 648 nm have been studied for PNC.

4.1. Angle Resolution

Shot noise in the rotation experiments is very small. The rms fluctuation in angle due to shot noise in the detected photons is

$$\Delta\phi_{shot} = \left[\frac{q\Delta\nu}{\varepsilon I}\right]^{1/2} \tag{30}$$

where $\Delta\nu$ is the observation bandwidth, q is the electron charge, ε is the detector efficiency in A/W, and I is the laser power in watts.

The lasers used in these experiments produce typical powers greater than 2 mW. The detectors often are of nearly unity quantum efficiency. Thus, in one second, $\Delta\phi_{shot} < 10^{-8}$ rad. Even with the decreased efficiency of photomultiplier tubes, and less light intensity in experiments which use many absorption lengths of vapor, one finds $\Delta\phi_{shot} < 10^{-7}$ rad in an observation time $(\Delta\nu)^{-1} = 1$ sec. In practice, the noise in the measured angles often turns out to be larger than the shot noise limit, owing to a combination of mechanical and optical instabilities, and detector noise in some cases. In the most recent experiments, 10^{-8} rad can be resolved in less than 1 min of averaging time.

4.2. Faraday Rotation

There is another rotation associated with the absorption line called Faraday rotation. This is only present when there is a magnetic field parallel to the light. Faraday rotation in a known magnetic field is used to calibrate rotation measurements, but Faraday rotation due to residual magnetic fields is a potential source of systematic error that is carefully guarded against.

The origin of Faraday rotation may be understood as follows: The addition of a magnetic field causes a splitting of the electronic states of different magnetic quantum number m_f. This is the well-known Zeeman effect. Since the two circular polarizations cause transitions having opposite signs of Δm_f, these transitions will be displaced from one another in frequency ν because of the Zeeman effect. Therefore, at a given wavelength, one circular polarization will be more absorbing than the other, and will also have a different refractive index. From Eq. (26), we see that optical rotation will occur, which in this case is called Faraday rotation. Since this rotation is proportional to the derivative $dn/d\nu$ of the index of refraction, it is symmetric in shape about line center (for an isolated absorption line).

Faraday rotation also arises from another effect of a magnetic field, the mixing of hyperfine states of different angular momentum F. This second form of Faraday rotation is antisymmetric in shape about an absorption line. Both the size and sign of the antisymmetric portion vary for each hyperfine component. The sum over all hyperfine components, however, is identically zero. Therefore, the parity curve and the Faraday curve are nearly orthogonal when compared over all the hfs components of a transition, and conventional curve-fitting routines differentiate clearly between the two effects.

The calculation of both the symmetric and antisymmetric effect is straightforward.[26,27] Later on, the actual Faraday rotation for lead is displayed (Figure 6b) as is the Faraday rotation for bismuth (Figure 4b) with its more complicated hyperfine structure. Included in these curves are actual experimental data and the theoretical predictions for their respective magnetic fields.

4.3. The Seattle Optical Rotation Experiments

To illustrate the procedures and problems encountered in optical rotation PNC experiments, we will discuss the University of Washington experiments in some detail. These experiments have yielded measurements of PNC in atomic bismuth and, more recently, in atomic lead.

Of great importance in the optical rotation experiments is the laser source. The laser must be tunable, and should be highly monochromatic, stable in intensity, and emit a beam of good optical quality. Tunable dye lasers satisfying all these criteria operated well at the 648-nm bismuth line in 1974 when work began in several places. The groups at Oxford and Novosibirsk and later at Moscow have all used such lasers to measure the effect on this line.

The group at the University of Washington elected to concentrate on the 876-nm bismuth infrared line because that line is not overlaid by the strong absorption due to Bi_2 molecules that is found at 648 nm. However, initially the group had to use a Nd:YAG pumped optical parametric laser with a large (30-GHz) linewidth, making spectroscopy difficult.[22] Unfortunately, this laser never gave good consistent results.[28,29] In 1977, a semiconductor laser diode was adopted instead, and has proven to be satisfactory in most respects. Since diode lasers are relatively new in this kind of application, we feel some discussion of them here would be appropriate.

Diode lasers generate light by the recombination of electrons with holes as they cross a $p-n$ junction. The band gap determines the lasing wavelength. Since these lasers are small, about 100 μm in length, the corresponding Fabry-Perot cavity length causes a mode separation on the order of 100 GHz (3 cm^{-1}). Generally, these lasers are not totally single mode but have anywhere from 10% to 96% of the total light on a single longitudinal mode. Certain structures, such as channeled substrate planar[30] and transverse junction stripe,[31] have been found to enhance the fraction of light on a single mode. Fortunately, they run on a single transverse mode so the beam shape is good.

The linewidth of an individual mode varies dramatically between lasers, but the best appear to have about 20 MHz width. There is also fine structure on either side of the main mode about 1 to 2 GHz away. These fine structure modes can have as little as 0.1%, and up to 10% of the total light on them depending on the type of laser. These side modes can be quite troublesome. Luckily, at the 0.1% level, they do not add significantly to errors.

Diode lasers are tuned in wavelength by changing either their operating current or temperature. Both of these change the Fabry-Perot cavity length and therefore tune the individual mode. They also change the band gap energy, causing mode hops. In general, one can tune an individual mode about 50 GHz before the mode hops to a frequency over 100 GHz away. Unfortunately, this causes gaps in wavelength which one cannot tune through with a single diode.

Currently there are GaAlAs lasers operating in the wavelength region from 790 to 920 nm, and InGaASP lasers[32] in the 1.1–1.6-μm region. While diode lasers have some drawbacks as spectral sources, their simplicity and low cost make them quite useful. Much work has been done also in putting these lasers in external cavities to decrease their linewidth[33] and eliminate all extraneous mode structures.

Another important part of optical rotation experiments is the plane polarizer and analyzer. We have used birefringent polarizing prisms of both the Glan Thompson and Nicol variety for the experiment at the University of Washington. Although they differ in details of their geometry, both of these polarizers are made of two calcite wedges glued together so only one polarization is transmitted. We have found that the best extinction ratio under the conditions of our experiment is about 10^7 for both types of polarizers. This means that when two polarizers are perfectly crossed, 10^{-7} of the light still gets through.

When the two polarizers are uncrossed slightly, the transmission T grows as the square of the angle of uncrossing ϕ:

$$T = T_e + \phi^2 \tag{31}$$

Here, T_e is the transmission at maximum extinction. The light doubles when the polarizers are uncrossed by an angle $\phi_0 = (T_e)^{-1/2}$, which equals about 3×10^{-4} for our polarizers. One obtains maximum sensitivity at ϕ_0, i.e., the largest fractional change in T for a given rotation $\delta\phi$ between the polarizers. By operating at this point with our polarizers, rotations of 10^{-7} rad cause a fractional change in T of 3×10^{-4}.

In our experiment the polarization is sinusoidally rocked between $\pm\phi_0$ at a 1-kHz rate. Any additional rotation $\delta\phi$ then leads to a 1-kHz cross term in Eq. (31) that is linear in $\delta\phi$, but insensitive to absorption and other changes in intensity. The rocking of the polarization is done by using the

Faraday rotation in some suitable substance placed between the polarizers. Water works well in the visible while toluene can be used beyond 1 μm. A solenoid wound around a cell containing the substance generates an oscillating magnetic field. Usually a field of about 100 G is sufficient to cause a rotation of 10^{-3} rad in a cell length of 10 cm.

Figure 3 shows a schematic outline of the Seattle experiment. As mentioned, the light source is a diode laser. The light from the laser comes out at a diffraction-limited 30° angle. The light then is collimated by using an ordinary 20× microscope objective. We then use a telescope for further control over the beam shape. The light goes through the initial polarizer and then passes through a well-shielded Faraday rotator and into an oven. For the metals of interest, Bi, Tl, and Pb, temperatures up to 1300 K are necessary in order to obtain about 1 torr of vapor and about one absorption length on the $M1$ transition used. Care therefore has to be taken in order to assure that no thermal stresses in the oven windows destroy the polarization. After the oven, the light passes through an analyzing polarizer and into a detector.

The detected light intensity varies greatly as the laser wavelength is swept across an absorption line. The Faraday modulation serves to subtract out these variations, but we choose to divide them out further in order to maximize rejection of the absorption pattern from the angle signal. This is done by taking a reflection off the front face of the analyzing polarizer and feeding the light into a detector. We then divide the two signals before detecting at the 1-kHz modulation frequency. The resultant signal is proportional to angle. It is calibrated in two different ways, firstly by use of levers

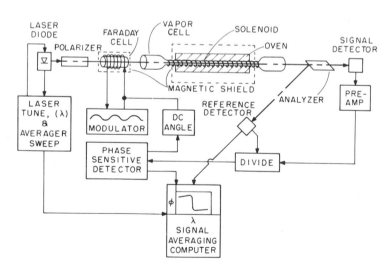

Figure 3. Schematic view of the Seattle apparatus.

and micrometers to accurately rotate our polarizer by 10^{-4} rad, and secondly by the Faraday rotation associated with the $M1$ transition in the metal vapor. We are able to get agreement to within 1% between the two different calibrations. Inside the oven, the vapor cell is wound with a solenoid, which, by passing a known dc current producing a known magnetic field, provides the desired Faraday rotation calibration. Figure 4b shows this Faraday rotation for bismuth. A 30-mG field produces a rotation of about 10^{-4} rad, which can be accurately calculated once the absorption (Figure 4a) is known.

Perhaps it should be emphasized that at the wavelengths used in our experiments, the Faraday rotation in water and toluene is practically independent of wavelength (unlike Figure 4b), and therefore these substances can be used as polarization modulators.

We sweep the laser across the $M1$ absorption and record both the angle and the absorption. This is initially done with a large (30-mG) field which generates the large Faraday rotation for calibration. The direction of the sweep is reversed and the same data independently stored as we sweep from the opposite direction. We then turn off the field and again sweep in both directions across the line to search for PNC rotations. This procedure takes about 6 sec and is repeated 150 times for a given data run.

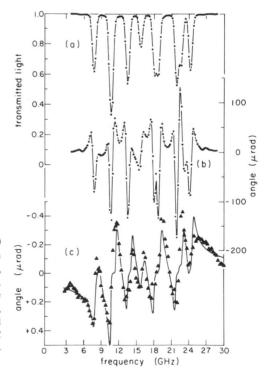

Figure 4. Theoretical curves (lines) fitted to experimental Bi 876-nm data (points) for (a) absorption, (b) Faraday rotation at 0.18 G, and (c) PNC rotation at relative optical depths of 1, 1, and 10, respectively. The theoretical curves for Faraday and PNC take into account dilution by off-mode laser light.

The data are stored in 128 separate bins for each sweep. The noise per bin corresponds to about 3×10^{-7} rad. Since about one second is spent on each bin, this is essentially the measured angle uncertainty per $Hz^{1/2}$.

From these data, we use standard iterative least-squares curve-fitting techniques to generate values for the absorption length, angle sensitivity, residual magnetic field, and finally PNC rotation. A several-parameter curve is used to fit the angle data, which has, besides the theoretical PNC curve, a quadratic polynomial to remove background rotations, the experimental Faraday curve to remove residual field effects, and an absorption-shaped curve to remove any feedthrough due to common mode rejection failure of the divider and phase sensitive detector. Extensive correlation tests assure us that none of these parameters projects onto our PNC curve.

Figure 4c shows the sum of about 50 of these runs added together for the bismuth infrared (876-nm) $M1$ line along with the associated theoretically expected PNC pattern. This theoretical PNC curve, along with theory curves for the Faraday rotation and absorption, also depend on experimentally derived parameters such as hyperfine splitting, Lorentz width, and Doppler width. These parameters have uncertainties that contribute to errors in the desired quantity R [see Eq. (26)] at less than the 1% level. Further confirmation comes from the agreement between the two techniques for calibrating the angle as discussed above.

The statistical variation among individual runs is about 5×10^{-8} for R. This variation is larger than expected purely on the basis of the measured fluctuations in angle at a given wavelength. It is due to wavelength-dependent angles in the polarizers as discussed below. Assuming random variations from run to run, the statistical error for the 50 runs shown should be about 10^{-8}. In addition there are systematic uncertainties to be described.

The largest error in these experiments is due to the polarizers. There are wavelength-dependent rotations associated with them which only slowly average out in time if the polarizers are not moved. It turns out that one limit on the extinction of polarizers is that the orientation of the crystal is not the same throughout the polarizer. One therefore registers different angles as one moves the beam about on the surface. The best extinction is obtained by decreasing the laser spot size on the polarizers. Any wavelength-dependent steering of the laser beam by any optical element or by the vapor itself will then cause a rotation.

Furthermore, there is an observed variation of rotation angle with wavelength of the period associated with the length of the polarizers. This suggests some etalonlike effect in the polarizer. This can be as large as 10^{-4} rad, but by tilting the polarizers, this effect can be reduced to the order of 10^{-7} rad for wavelength changes of interest. Both of these effects limit the sensitivity of our experiments and limit our ability to measure angles.

These uncertainties due to the polarizers, together with other possible sources of systematic error, are estimated and controlled in the operation of the experiment and in the analysis of the data.[34-36] The total uncertainty in the measured value of R is estimated at about $\pm 1 \times 10^8$ in our bismuth experiment.

4.4. Results of the Seattle Bismuth Experiment

A list of all bismuth PNC measurements carried out in Seattle is given in Table 1. The table includes the first three measurements, made without the systematic checks and controls developed later. These measurements are not included in computing the final result. The remaining six measurements, each containing between 50 and 100 data sets and each taken under a broad range of conditions, are seen to have a common value of R to within the quoted uncertainties. These six measurements yield the resultant experimental value[35,36]

$$R = \text{Im}\,(E1_{\text{PNC}}/M1) = -10.5 \pm 1.2 \times 10^{-8} \qquad (32)$$

where the quoted error is dominantly systematic.

In Table 2 we present recent results of atomic calculations for the bismuth 876-nm line based on the Weinberg–Salam theory. The earliest published central-field values[11,37] of R were about twice the central field values shown in the first entry.[11,38,39] The change is due to the decrease in the experimental value of $\sin^2 \theta_W$ from 0.35 to below 0.23, and to improved calculations. The semiempirical value[40] shown in the table utilizes empirical

Table 1. PNC Optical Rotation Measured on the Bi 876-nm Line[a]

Data period	Ref.	Laser	$R(10^{-8})$
Sp 1976	28	OPO	(-8 ± 3)
Wi 1977	29	OPO	(-0.7 ± 3.2)
Wi 1978	1, 34, 36	TJS 1	(-2.4 ± 1.4)
Au 1978	1, 34, 36	TJS 2	-10.2 ± 3.1
Wi 1979	1, 34, 36	TJS 3	-11.8 ± 3.9
Su 1979	1, 34, 36	TJS 3	-9.8 ± 2.4
Sp 1980	35, 36	CSP 1	-9.7 ± 2.5
Su 1980	35, 36	CSP 1	-10.8 ± 1.9
Wi 1981	35, 36	CSP 1	-10.7 ± 3.0
			$-10.5 \pm 1.2 \times 10^{-8}$
Weighted average (last 6 entries only)			

[a] Errors include systematic uncertainties except when numbers are in parentheses.

Table 2. Calculation of R for Bi 876-nm Line Using
Weinberg–Salam–Glashow Theory ($\sin^2 \theta_w = 0.219$)[a]

Calculation	Ref.	$R(10^{-8})$
Central field IPM	11, 37, 39	-17
Hartree–Fock IPM	38	-16
Semiempirical	40	-13
Parametric potential with shielding and first order	39	-11
Hartree–Fock with shielding and first order	41	-8

[a] The last three entries go beyond the independent particle model (IPM).

data to include effects not contained in the central field independent-particle
(IPM) model. The last two entries[39,41] represent the most complete calculations made to date. If we take the mean of the three calculations (the last
three entries in the table) that go beyond the IPM model, we have the
theoretical value

$$R = -10.7 \pm 2.5 \times 10^{-8} \tag{33}$$

where $\sin^2 \theta_W = 0.219$, and the quoted uncertainty is made large enough to
overlap the three calculations.

The Seattle result in Eq. (32) and the theoretical value in Eq. (33) are
in good agreement with each other.

4.5. Measurements on the 1.28-μm Line of Atomic Lead

It has long been known that lead and thallium should have a PNC
optical rotation at a level comparable to that in bismuth, yet the allowed
magnetic dipole transitions in both elements fall near 1.28 μm, as shown
in the case of lead in Figure 5. Only recently[32,42] have lasers been developed
that can reach this wavelength. Diode lasers now work handily at 1.28 μm.
The InGaAsP/InP diode lasers at 1.28 μm have remarkably similar characteristics to the GaAlAs lasers used in the Bi experiment. Therefore the
Seattle bismuth apparatus described above has been adapted to the longer
wavelength with only minimal changes.

The first measurements with this modified apparatus have been made
on the $J = 0 \rightarrow J = 1$ Pb transition at 1.2788 μm. Naturally occuring lead is
used, which has 78% even isotopes with no hfs structure and 22% Pb^{207}
having a two-component hfs structure. The observed absorption pattern
and Faraday rotation are shown in Figure 6, together with the expected
theoretical pattern.

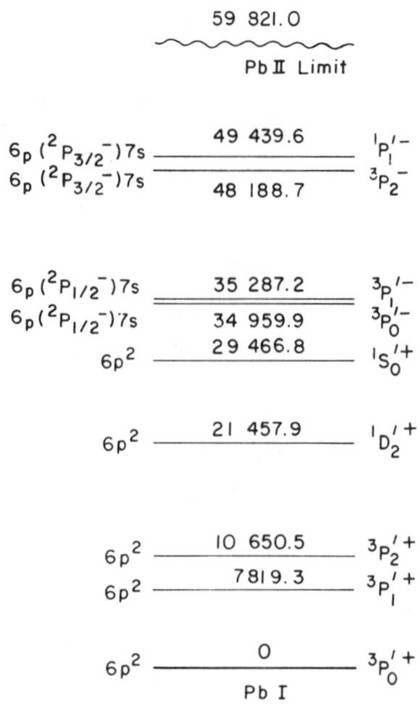

59 821.0

Pb II Limit

| $6p\,(^2P_{3/2}^-)7s$ | 49 439.6 | $^1P_1'^-$ |
| $6p\,(^2P_{3/2}^-)7s$ | 48 188.7 | $^3P_2^-$ |

$6p\,(^2P_{1/2}^-)7s$	35 287.2	$^3P_1'^-$
$6p\,(^2P_{1/2}^-)7s$	34 959.9	$^3P_0'^-$
$6p^2$	29 466.8	$^1S_0'^+$

| $6p^2$ | 21 457.9 | $^1D_2'^+$ |

| $6p^2$ | 10 650.5 | $^3P_2'^+$ |
| $6p^2$ | 7819.3 | $^3P_1'^+$ |

Figure 5. Electronic energy levels for lead. The absorption from the ground state to the first excited state has been studied for PNC.

| $6p^2$ | 0 | $^3P_0'^+$ |
| | Pb I | |

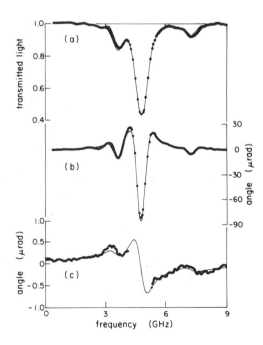

Figure 6. Theoretical curves (lines) fitted to experimental lead data (points) for (a) absorption, (b) Faraday rotation at 0.03 G, and (c) PNC rotation at relative optical depths of 1, 1, and 8, respectively. The curves for Faraday and PNC take into account dilution by off-mode laser light.

At comparable temperatures, the $M1$ absorption on the even isotope line in Pb is 10 times greater than on the largest hfs component at 876 nm in Bi. This allows operation at lower temperatures where the system is more stable, or alternatively at high absorption lengths where the PNC rotation is larger.

We have taken PNC data at several temperatures corresponding to a range of from 10 to 50 absorption lengths on the even isotope line. Each oven setting was used with both Glan–Thompson and Nicol polarizers. There is agreement among all measurements within their uncertainties. Figure 6c shows the PNC data summed separately from some 20% of these measurements and reveals clearly the expected PNC dispersion shape. All the systematic tests and controls developed for Bi have been utilized with these data as well.

The Pb data thus far yield the value

$$R = -9.2 \pm 2.5 \times 10^{-8} \qquad (34)$$

where the quoted error is mainly a composite of residual systematic uncertainties. There is good agreement with the only theoretical value calculated so far[2,38]:

$$R = -11 \times 10^{-8} \qquad (35)$$

where $\sin^2 \theta_W = 0.219$. The authors state a 20% uncertainty. This calculation uses the semiempirical technique similar to that given for bismuth in Table 2.

5. Stark–PNC Experiments; Cesium and Thallium

As mentioned already, experiments using highly forbidden $M1$ transitions in Cs and Tl were first suggested by Bouchiat and Bouchiat,[9,18,21] and PNC measurements have been underway since then with Cs at Paris,[43] and with Tl at Berkeley.[44] Observations of PNC have been reported in both experiments. Other experiments with Cs began more recently at Zurich[45] and at Michigan,[46] but they have not yet reported results accurate enough to observe PNC. (See note added at end of this article.)

In order to increase the transition rate for these forbidden transitions, all groups use an electric field E. In Section 3.1, we pointed out that the combination $(\boldsymbol{\sigma}_i \times \hat{E}) \cdot \mathbf{J}_f$ is the observable pseudoscalar, which is odd under parity. Here $\boldsymbol{\sigma}_i = \eta \hat{k}$ [see Eq. (21)] is the circular polarization of the pump laser that excites the atom using the forbidden transitions enhanced by Stark mixing due to E. $\mathbf{J}_f$ is the angular momentum polarization of the excited state. This polarization is revealed, for example, by circular polarization of the fluorescent light from the decay of the excited state to some intermediate state. Figure 7 shows the electronic structures of cesium and thallium, and the transitions of interest.

Let us denote the polarization of the fluorescence by Δ:

$$\Delta = (n_+ - n)/(n_+ + n_-) \qquad (36)$$

where $n_\pm$ are the counting rates for the two different circular polarizations of the fluorescence. In general, this polarization of the fluorescence depends upon the initial and final hyperfine states F and F' involved in the absorption of the pump laser light. For the $F = 0$ to $F' = 1$ transition in Tl used at Berkeley, the polarization is[47,48]

$$\Delta(F = 0 \rightarrow F = 1) = \Delta_m \pm \Delta_{\text{PNC}} = 2(M1 + \eta E1_{\text{PNC}})/\beta E \qquad (37)$$

while for the $F = F'$ transition in Cs used at Paris, the polarization is[43]

$$\Delta(F \rightarrow F) = \Delta_m \pm \Delta_{\text{PNC}} = \frac{8}{3} \frac{F(F + 1)}{(2I + 1)^2} (M1 + \eta E1_{\text{PNC}})/\alpha E \qquad (38)$$

where α and β are, respectively, the scalar and vector polarizabilities[21] due to the electric field E. Δ_m and Δ_{PNC} arise from the interference of the Stark amplitude of the transition with the $M1$ and $E1_{\text{PNC}}$ amplitudes. The $\pm$ signs signify that Δ_{PNC} is dependent on the circular polarization $\hat{\sigma}_i = \eta \hat{k}$ of the pump beam. One tells the difference between Δ_{PNC} and Δ_m by switching the sense of the circular polarization of the pump beam. Because $M1$ is about 10^3 (thallium) to 10^4 (cesium) times larger than the parity-violating induced electric dipole, $E1_{\text{PNC}}$, one has to be extremely careful how one reverses the sense of the pump beam's polarization.

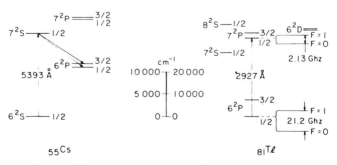

Figure 7. Electronic energy levels for cesium and thallium.

An additional distinction between Δ_{PNC} and Δ_m may be made by reversing the direction k of the pump beam. Δ_m is odd and Δ_{PNC} is even under k reversal. All the distinctions can be understood by noting that Δ_m measures the scalar $(\hat{k} \times \hat{E}) \cdot \mathbf{J}_f$, while Δ_{PNC} measures the pseudoscalar $(\boldsymbol{\sigma}_i \times \hat{E}) \cdot \mathbf{J}_f$.

As an example of these principles, the group in Paris working on cesium boosts the pump rate while simultaneously decreasing the size of Δ_m by using a multipass system for their pump beam. They put the cesium cell inside a cavity in which the pump beam makes about 50–70 double passes. The net direction of the pump approaches zero, reducing Δ_m by about a factor of 180 with no reduction in Δ_{PNC}. This system also increases the net power in the pump beam therefore increasing the signal by that same factor.

A second class of systematic effects can come from imperfect electric field reversal. For example, if there is some stray (nonreversing) electric field acting on the atoms, with a component parallel to the pump laser beam, then a portion of the detected polarization Δ can reverse with σ_i and E just like the PNC portion. Fortunately, this effect can be measured independently, as can other spurious electric field effects.[49,50]

Figure 8 shows the experimental setup for the Paris group working on cesium.[43] The pump wavelength is at 539.4 nm, easily in the reach of conventional dye lasers. The pump beam passes through a longitudinal Pockels cell sandwiched between two half wave plates rotating at different frequencies, to produce a superposition of a linear polarization rotating at frequency ω_b and a circular polarization. Both polarizations are amplitude modulated at frequency ω_e.

The laser is then fed into the multipass system with the cesium cell in the center. The fluorescence at 1.36 μm is measured in an orthogonal direction. This is passed through a circular polarization analyzer rotating at ω_f and then an interference filter to reject oven light and into a Ge photodiode.

By using phase sensitive detection at different combinations of ω_e, ω_b, and ω_f, specific labels of circular polarization and linear polarization effects are obtained with negligible cross modulation. The PNC effect, which changes with incident circular polarization, is proportional to $\sigma_i[\sin(\omega_e + \omega_f)t - \sin(\omega_e - \omega_f)t]$, and has continuous sinusoidal modulations as well

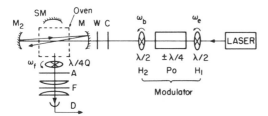

Figure 8. Schematic view of the Paris cesium apparatus.

as discrete modulations due to σ_i reversal by the Pockels cell. Important systematic effects due to asymmetry in polarization reversal can be estimated by the dependence on the linear polarization modulation.

The group at Berkeley chose thallium to work on.[44,50] The size of $E1_{PNC}$ is about 10 times greater than in cesium, but the pump transition is in a more difficult part of the spectrum.

Figure 7 shows the atomic structure for thallium. The pump transition is at 293 nm, which is reached by doubling a flash lamp pumped pulsed dye laser using an ammonium dihydrogen arsenate (ADA) doubling crystal. This light is circularly polarized by a Pockels cell and fed into the thallium cell. The light can be reflected back once through the thallium cell to reduce Δ_m.

The overall experimental arrangement at Berkeley is shown in Figure 9. The polarization of the $7P$ state is measured not by its fluorescence, which is far in the infrared, but by using another laser at 2.18 μm which is circularly polarized to sweep all the $7P$ atoms of the proper polarization up to the $8s$ state. One then observes the fluorescence at 323 nm from the $8s$ state as a function of the polarization of the 2.18-μm beam, and thereby measures the polarization J_f of the $7P$ state. Since one does not need to know the polarization of this fluorescence, one can detect a greater solid angle and improve the counting rate. One defines Δ similarly to Eq. (36), except n_+ and n_- refer to the numbers of fluorescent photons for the two states of circular polarization of the 2.18-μm laser beam.

Both the Berkeley and the Paris experiments are sensitive to the systematics discussed above and great care is taken to account for them. In each

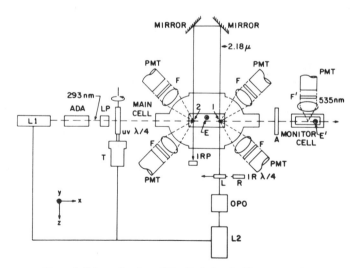

Figure 9. Schematic view of the Berkeley thallium apparatus.

experiment, the data were corrected for measured systematic shifts; although in the case of the Paris experiment, only a very small (5%) correction was needed. The results of these experiments,[43,50] along with the results of all the optical rotation experiments,[35,51-54] are discussed in the next section.

6. Discussion of Results; Conclusions

As we have seen there now exist significant measurements of PNC in four different atoms, carried out on five distinct magnetic-dipole absorption lines. In Table 3 we show the most recent result of each experiment together with the value predicted by the Weinberg–Salam weak interaction theory. The predicted values come from atomic calculations[39-41,55,56] taking into account the major corrections to the central field approximation. We also show the values of Q defined in Eqs. (9) and (10) that may be inferred in each case from the experimental result and the atomic calculations.

First of all, these results leave little doubt that parity conservation is violated at the level predicted by the Weinberg–Salam theory. To go farther, and discuss the collection of atomic results quantitatively, we suppose that each measured absorption line presents us with an independent determination of Q (and therefore of $\sin^2 \theta_W$). This assumption is clearly justified as regards the experimental results, because the errors in the different experiments should not be correlated with each other. As regards the atomic calculations, however, the predicted values for the two Bi lines and to some extent for the Pb line could in principle have common errors. However,

Table 3. Comparison of the Predicted and Measured Results of the Atomic Experiments*

Atomic transition (nm)	Measured quantity	Measured value	Theoretical value[a]	$\left(\dfrac{Q^{\text{meas}}}{Q^{\text{WS}}}\right)^b$
2Bi 876	$E1_{\text{PNC}}/M1$	$(-10.5 \pm 1.3 \times 10^{-8})^{(35,36)}$	$(-11 \times 10^{-8})^{(39-41)}$	0.95
Bi 648[c]	$E1_{\text{PNC}}/M1$	$(-9 \pm 2 \times 10^{-8})^{(51,53)}$	$(-13 \times 10^{-8})^{(39-41)}$	0.69
Pb 1279	$E1_{\text{PNC}}/M1$	$(-9.2 \pm 2.5 \times 10^{-8})^{(54)}$	$(-11 \times 10^{-8})^{(40)}$	0.85
Tl 293	$E1_{\text{PNC}}/M1$	$(2.8^{+1.0}_{-0.9} \times 10^{-3})^{(50)}$	$(1.7 \times 10^{-3})^{(56)}$	1.63
Cs 539[d]	$E1_{\text{PNC}}/\beta$	$(-1.34 \pm 0.22 \pm 0.11 \text{ mV/cm})^{(43)}$	$(-1.73 \text{ mV/cm})^{(55)}$	0.73

Average $Q^{\text{meas}}/Q^{\text{WS}} = 0.97 \pm 0.17$

* As of 1983. For more recent results, see Note added at the end of this article.
[a] Theoretical values are averaged from those atomic calculations that include the major corrections to the central field independent particle model (IPM), with $\sin^2 \theta_W = 0.219$.
[b] For the weak charge Q, see Eqs. (9), (10), and (17). The ratio here comes from the previous two columns.
[c] For the 648-nm line, we use the Oxford result. As of this writing, the various 648-nm experiments remain mutually inconsistent, although the average agrees with the value taken here.
[d] The measured value shows statistical and systematic errors in that order.

there are significant differences even between the two bismuth lines in those portions of the atomic calculations that involve the major uncertainties, namely, in the noncentral electron–electron interactions. With some justification, therefore, we simply take an average of the values in Table 3 and obtain the mean value

$$Q_{\text{meas}}/Q^{\text{WS}} = 0.97 \pm 0.17 \tag{39}$$

where we have used the value $\sin^2 \theta_w = 0.219 \pm 0.014$. It is clear that, using a value of θ_w consistent with high energy neutral current results, the Weinberg–Salam theory correctly predicts the atomic PNC results to well within the atomic uncertainties.

Alternatively, the atomic experiments could be interpreted as measuring θ_w, yielding a mean value

$$\sin^2 \theta_w = 0.21 \pm 0.05 \tag{40}$$

If one adopts a model-independent approach to the weak neutral current interaction, the atomic experiments supply certain information about the coupling constants not provided by high-energy experiments. Atomic experiments are necessary, as we will now show, in order to set stringent limits on the neutral current parameters and test theories (including, but not restricted to Weinberg–Salam) that predict values of these parameters.

Using the simplest model-independent assumptions, Hung and Sakurai[57] characterize the theory by ten coupling constants which must be determined by experiment. Of these all but four $(\tilde{\alpha}, \tilde{\beta}, \tilde{\gamma}, \tilde{\delta})$ are determined by neutrino experiments. These remaining four are defined in terms of the C parameters of Eq. (6):

$$\tilde{\alpha} \equiv \tfrac{1}{2}(C_{1p} - C_{1N})$$

$$\tilde{\beta} \equiv \tfrac{1}{2}(C_{2p} - C_{2N})$$

$$\tilde{\gamma} \equiv \tfrac{1}{2}(C_{1p} + C_{1N}) \tag{41}$$

$$\tilde{\delta} \equiv \tfrac{1}{2}(C_{2p} + C_{2N})$$

The polarized electron scattering experiment[8] yields

$$\tilde{\alpha} + \tfrac{1}{3}\tilde{\gamma} = -0.60 \pm 0.16$$

$$\tilde{\beta} + \tfrac{1}{3}\tilde{\delta} = 0.31 \pm 0.51 \tag{42}$$

The heavy atom experiments are sensitive to an almost orthogonal linear combination of $\tilde{\alpha}$ and $\tilde{\gamma}$:

$$Q(Z, N) = -(Z - N)\tilde{a} - 3(Z + N)\tilde{\gamma}$$
$$= 43\tilde{\alpha} - 627\tilde{\gamma} \quad \text{(for Bi)}$$

(43)

In Figure 10 we make a two-dimensional plot of $\tilde{\alpha}$ and $\tilde{\gamma}$, and show the area allowed by Eq. (43) and all atomic results from Table 3. Also shown in Figure 10 is the area allowed by the polarized electron scattering results [Eq. (42)], plus further restrictions imposed on $\tilde{\alpha}$, $\tilde{\gamma}$ by neutrino experiments if the "factorization" hypothesis[57] is included. One can see that the atomic PNC results significantly exclude values of the parameters that would otherwise be allowed. It is noteworthy, of course, that the Weinberg–Salam theory falls within all allowed regions.

Using all data, the region allowed in the diagram yields

$$\tilde{\alpha} = -0.68 \pm 0.15$$

(44)

$$\tilde{\gamma} = +0.18 \pm 0.3$$

Various theoretical models other than Weinberg–Salam can be tested against Eq. (44). For illustration we mention one other class of models, namely, those that include two or more neutral Z bosons. As pointed out

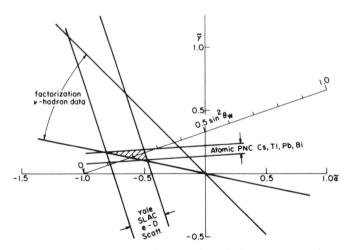

Figure 10. Restrictions placed on $\tilde{\gamma}$ and $\tilde{\alpha}$ (see text) by the heavy atom atomic experiments, polarized electron scattering, and neutrino experiments. The Weinberg–Salam theory falls well within the intersection of all experiments.

by Robinett and Rosner[58] atomic experiments set the most severe constraints on such models. For example, in the two-Z model considered by Robinett and Rosner the limits in Eq. (44) imply that the mass of the two Z bosons must satisfy

$$M_1 \cong M_0$$
$$M_0 \cong 89 \text{ Gev} \qquad (45)$$
$$M_2 \gtrsim 3M_0$$

the dominant restrictions coming from the atomic results.

6.1. Future Prospects

In the case of heavy atoms improved experimental accuracy will pay off in all elements, particularly in Cs and to a somewhat lesser extent in Tl because of the accuracy of the atomic theory and the degree of understanding of the levels involved in these two atoms. Because the transition of interest involves a $J = 0$ state, Pb also offers some advantages for advanced atomic calculations.[59] Improved measurements of Q_W will yield more refined comparisons of the type shown in Eqs. (43)–(45) above. It remains to be seen whether improved experimental accuracy will allow resolution of the C_2 terms of Eqs. (8) and (11), which should show up as small fractional differences in PNC between hyperfine components of a transition. The ongoing experiments in atomic hydrogen offer the hope of determining C_1 and C_2 with no uncertainties in atomic theory, and furthermore are sensitive to radiative corrections that test the essence of gauge theories at ~ 100 Gev/c momentum transfer.[60]

In this article we have discussed experiments and theories that are consistent with the principle of time-reversal symmetry. There are other experiments[23,61–63] searching for a permanent electric dipole moment of an atom (similar to the searches for a neutron electric dipole moment[64]) which provide all-important tests of both parity and time-reversal symmetry violation.

It is clear there is still much to be done in this very young field of atomic PNC. The mysteries of elementary particle interactions still have great relevance for atomic physics, and in turn, atomic physics continues to offer possibilities of new insights into these interactions.

Note added in proof: Since this article was completed, a number of new experimental results have been published. New measurements have been carried out on the Cs 539 nm line at Paris[65] and Boulder,[66] on the Tl 293 nm line at Berkeley,[67] and on the Bi 648 nm line at Moscow.[68] The overall conclusions given in connection with Table 3 remain unchanged.

References

1. E. N. Fortson and L. Wilets, in *Advances in Atomic and Molecular Physics* (edited by B. Bederson and D. R. Bates), Vol. 16, p. 319, Academic, New York (1980).
2. L. M. Barkov, I. B. Khriplovich, and M. S. Zolotorev, *Phys. Rev. (Sov. Sci. Rev., Sec. A)* **3**, 1 (1981).
3. E. D. Commins and P. H. Bucksbaum, *Ann. Rev. Nucl. Part. Sci.* **30**, 1 (1980).
4. S. Weinberg, *Phys. Rev. Lett.* **19**, 1264 (1967).
5. A. Salam, *Elem. Part. Theory, Proc. Nobel Symp. 8th* 367 (1968).
6. S. L. Glashow, *Nucl. Phys.* **22**, 579 (1961).
7. F. J. Hasert *et al.*, *Phys. Lett. B* **46**, 138 (1973); A. Benvenuti *et al.*, *Phys. Rev. Lett.* **32**, 800 (1974); S. J. Barish *et al.*, *Phys. Rev. Lett.* **33**, 448 (1974),
8. C. Y. Prescott *et al.*, *Phys. Lett. B* **77**, 347 (1978).
9. M. A. Bouchiat and C. C. Bouchiat, *J. Phys.* **35**, 899 (1974).
10. J. D. Bjorken and S. D. Drell, *Relativistic Quantum Mechanics*, McGraw-Hill, New York (1969).
11. E. M. Henley and L. Wilets, *Phys. Rev. A* **14**, 1411 (1976); E. M. Henley, L. Wilets, and M. Klapisch, *Phys. Rev. Lett.* **39**, 994 (1977).
12. J. F. Wheater, *Phys. Lett.* **105B**, 483 (1981).
13. V. N. Novikov, O. P. Sushkov, V. V. Flambaum, and I. B. Khriplovich, *Sov. JETP* **46**, 420 (1977).
14. A. Messiah, *Quantum Mechanics*, Wiley, New York (1968).
15. E. G. Addelberger, T. A. Trainor, and E. N. Fortson, *Bull. Am. Phys. Soc.* **23**, 546 (1978); E. G. Addelberger *et al.*, *Nucl. Instrum. Methods* **179**, 181 (1981).
16. R. W. Dunford, R. R. Lewis, and W. L. Williams, *Phys. Rev. A* **18**, 2421 (1978).
17. E. A. Hinds and V. W. Hughes. *Phys. Lett.* **67B**, 468 (1977).
18. M. A. Bouchiat and C. C. Bouchiat, *Phys. Lett.* **48B**, 111 (1974).
19. E. Fermi and E. Segre, *Z. Phys.* **82**, 729 (1933).
20. P. G. H. Sandars, *Phys. Lett.* **14**, 194 (1965).
21. M. A. Bouchiat and C. C. Bouchiat, *J. Phys.* **36**, 493 (1975).
22. D. C. Soriede and E. N. Fortson, *Bull. Am. Phys. Soc.* **20**, 491 (1975); D. C. Soreide, D. E. Roberts, E. G. Lindahl, L. L. Lewis, G. R. Apperson, and E. N. Fortson, *Phys. Rev. Lett.* **36**, 352 (1976).
23. P. G. H. Sandars, *Atomic Physics* (edited by G. zu Putlitz, E. W. Weber, and A. Winnacker), p. 71, Plenum Press, New York (1975).
24. I. B. Khriplovich, *JETP Lett.* **20**, 315 (1974).
25. D. V. Saskyan, I. I. Sobel'man, and E. A. Yukov, *JETP Lett.* **29**, 258 (1979).
26. G. J. Roberts, P. E. G. Baird, M. W. S. M. Brimicombe, P. G. H. Sandars, D. R. Selby, and D. N. Stacey, *J. Phys. B.* **13**, 1389 (1980).
27. V. N. Novikov, O. P. Sushkov, and I. B. Khriplovich, *Opt. Spectrosk.* **43**, 370 (1977).
28. P. E. G. Baird, M. W. S. M. Brimicombe, G. J. Roberts, P. G. H. Sandars, D. Stacey, E. N. Fortson, L. L. Lewis, E. G. Lindahl, and D. C. Soreide, *Nature* **264**, 528 (1976).
29. L. L. Lewis, J. H. Hollister, D. C. Soreide, E. G. Lindahl, and E. N. Fortson, *Phys. Rev. Lett.* **39**, 795 (1977).
30. A. Aiki *et al.*, *Appl. Phys. Lett.* **30**, 649 (1977).
31. H. Namizake, *IEEE J. Quantum Electron*, **QE-11**, 427 (1975).
32. J. J. Hsieh, *Appl. Phys. Lett.* **38**, 238 (1976).
33. M. W. Fleming and A. Mooradian, *IEEE J. Quantum Electron* **QE-17**, 44 (1981).
34. G. A. Apperson, thesis, University of Washington, Seattle, 1979.
35. J. H. Hollister, thesis, University of Washington, Seattle, 1982.

36. J. H. Hollister, G. R. Apperson, L. L. Lewis, T. P. Emmons, T. G. Vold, and E. N. Fortson, *Phys. Rev. Lett.*, **46**, 643 (1981).
37. M. W. S. M. Brimicombe, C. E. Loving, and P. G. H. Sandars, *J. Phys. B* **41** (1976).
38. S. L. Carter and H. P. Kelly, *Phys. Rev. Lett.* **42**, 966 (1979).
39. M. J. Harris, C. E. Loving, and P. G. H. Sandars, *J. Phys. B* **11**, L749 (1980); P. G. H. Sandars, *Phys. Scr.* **21**, 284 (1980).
40. V. N. Novikov, D. P. Sushkov, and I. B. Khriplovich, *Sov. JETP* **46**, 420 (1976).
41. A. M. Martensson, E. M. Henley, and L. Wilets, *Phys. Rev. A* **24**, 305 (1981).
42. H. Nishi, M. Yano, Y. Nishitani, Y. Akita, and M. Takusagawa, *Appl. Phys. Lett.* **35**, 232 (1979).
43. M. A. Bouchiat, J. Guena', L. Hunter, and L. Pottier, *Phys. Lett.* **117B**, 358 (1982).
44. E. D. Commins, P. H. Bucksbaum, and L. R. Hunter, *Phys. Rev. Lett.* **46**, 640 (1981).
45. J. Hoffnagel, V. L. Telegdi, and A. Weis, *Phys. Lett.* **86A**, 451 (1981).
46. S. Gilbert, R. Watts, and C. Wieman, *Bull. Am. Phys. Soc.* **28**, 780 (1983).
47. D. V. Neuffer and E. D. Commins, *Phys. Rev. A* **16**, 844 (1977).
48. R. Conti *et al.*, *Phys. Rev. Lett.* **42**, 343 (1979).
49. M. A. Bouchiat, J. Guena', and L. Potter, *J. Phys. Lett.* (*Paris*) **41**, L-299 (1980).
50. P. H. Bucksbaum, E. D. Commins, and L. R. Hunter, *Phys. Rev. D* **24**, 1134 (1981).
51. P. E. G. Baird *et al.*, *Proceedings of the Seventh Vavilov Conference* (edited by S. G. Rautian) (Novosibirsk), p. 22 (1981).
52. L. M. Barkov and M. S. Zolotorev, *JETP Lett.* **27**, 357 (1978); L. M. Barkov *et al.*, *Comments At. Mol. Phys.* **8**, 79 (1979).
53. V. V. Bogdanov, I. I. Sobel'man, V. N. Sorokin, and I. I. Struck, *JETP Lett.* **31**, 214 and 522 (1980).
54. T. P. Emmons, J. M. Reeves, and E. N. Fortson, to be published (1983).
55. C. Bouchiat, in *Atomic Physics 7* (edited by D. Kleppner and F. Pipkin), Plenum Press, New York (1981); C. Bouchiat, D. Pignon and C. Piketty, to be published (1983).
56. B. P. Das *et al.*, *Phys. Rev. Lett.* **49**, 32 (1982).
57. P. Q. Hung and J. J. Sakurai, *Phys. Lett.* **88B**, 91 (1979).
58. R. W. Robinett and J. L. Rosner, *Phys. Rev. D* **25**, 3036 (1982).
59. P. G. H. Sandars, private communication (1982).
60. W. J. Marciano and A. Sirlin, *Phys. Rev. D* **27**, 552 (1983).
61. E. A. Hinds and P. G. H. Sandars, *Phys. Rev. A* **21**, 471 and 480 (1980).
62. D. Wilkening, thesis, Harvard University, Cambridge, 1981.
63. F. J. Raab, T. G. Vold, and E. N. Fortson, *Bull. Am. Phys. Soc.* **27**, 37 (1982).
64. N. F. Ramsey, in *Atomic Physics 7* (edited by D. Kleppner and F. Pipkin), Plenum Press, New York (1981); I. S. Altarev *et al.*, *Phys. Lett.* **102B**, 13 (1981).
65. M. A. Bouchiat, in *Atomic Physics 9* (edited by R. S. van Dyck, Jr., and E. N. Fortson), World Scientific, Singapore (1985).
66. S. L. Gilbert and C. E. Wieman, *Phys. Rev. A* **34**, 792 (1986).
67. P. S. Drell and E. D. Commins, *Phys. Rev. A* **32**, 2196 (1985).
68. G. N. Birich, Y. V. Bogdanov, I. I. Sobel'man, V. N. Sorokin, I. I. Struk, and E. A. Yukov, *JETP* **60**, 442 (1984).

6

Energy Structure of Highly Ionized Atoms

BENGT EDLÉN

1. Introduction

Early studies of the spectra of highly ionized atoms culminated in the 1930s when the sequence of sodiumlike spectra was extended to Cu XIX, the copper sequence to Sb XXIII, and the resonance lines of cobaltlike spectra were traced to Sn XXIV.[1] The spectra were produced in a high-voltage vacuum spark. Since at that time it seemed very improbable that such highly charged ions would ever be found anywhere else, on earth or in the heavens, there was no strong motivation for pursuing this line of research. The reactivation of this field that began in the 1960s was largely a consequence of the discovery that the solar spectrum in the extreme ultraviolet observed from space vehicles was dominated by emission lines from highly ionized atoms, a large part of which required further laboratory studies for their explanation. An additional impetus has later come from the need for diagnostic tools in the study of the hot plasmas in fusion research.

The renewed activity during the last decades has led to great progress in the observation and analysis of spectra of highly charged ions. The light sources that have proved most successful in this work are the low-inductance spark and the laser-produced plasma. The former was developed from the old vacuum spark by minimizing the inductance in the discharge circuit,

BENGT EDLÉN • Department of Physics, University of Lund, Sölvegatan 14, S-223 62, Lund, Sweden.

thereby increasing the peak current and the ionizing power. The laser-produced plasma is obtained by focusing a high-power laser pulse on a solid target, the ionization produced in the plasma depending on the applied power density. The two light sources are capable of reaching equivalent ionization stages, and both have a rather high plasma density with consequent line broadening.

The spectrum of the solar corona and solar flares has been a rich source of accurate information on the energy structure of the ground configurations through the forbidden transitions which appear in these low-density sources.[2] The information is limited, however, to elements of high cosmic abundance. This limitation does not exist for the tokamak plasma, which is similar to that of flares in temperature and density and where a large number of forbidden lines in elements up to $Z = 42$ have been discovered recently.

Spectroscopic work on highly ionized atoms has been reviewed by Fawcett[3,4] in two papers with comprehensive listings of references up to 1980. In the present review we will discuss the observed energy structure in representative sequences of isoelectronic ions. In some of these sequences the observations have now been extended to ions up to 50 times ionized which provides excellent illustration of the Z dependence of the energy level structure. Many of these data have become available since 1980.

The recent remarkable progress in the analysis of the spectra of highly ionized atoms would not have been possible without the parallel progress in the theoretical methods of interpreting the experimental data. The theoretical treatment of atomic structure has now advanced to a stage where it can reproduce with fair approximation the level systems encountered in the study of highly charged ions. In the mostly used approach the relativistic effects are added as corrections to the nonrelativistic Hartree–Fock calculations, and the correlation is accounted for either by empirical scaling of the energy parameters (Cowan and others) or by including a large number of configurations (Froese Fischer and others). Empirical parameters may also be introduced via a model potential (Klapisch and others). In the Dirac–Hartree–Fock approach the relativistic effects are more fully considered as the main part is included from the start (Desclaux, Grant, *et al.*) but the computational complexity increases rapidly with the number of electrons. The perturbation method in which the coefficients in the $1/Z$ expansion of the energy are derived (Safronova *et al.*) is particularly suited to describe isoelectronic ions but has been applied only to systems with relatively few electrons. The experimental analyses are usually checked and completed by a least-squares fitting of the energy parameters and subsequent diagonalization of the energy matrix. For a comprehensive discussion of the theoretical methods we refer to *The Theory of Atomic Structure and Spectra* by Cowan.[5]

2. General Energy Relations in Isoelectronic Ions

The Coulomb energy of an orbital nl increases as the square of the screened nuclear charge, $(Z - s)^2$, and since the screening s depends on l but is approximately independent of Z, the energy difference between orbitals with the same n but different l will increase linearly with Z. This means that with increasing Z in an isoelectronic sequence the energy order of configurations will ultimately be determined by the n values of the orbitals. The change from $(n + l)$ to n ordering is typically illustrated by elements of the iron group where the ground configuration has changed from $4s^2 3d^{k-2}$ in the neutral atom to $3d^k$ already in the doubly charged ion. A more far-reaching transformation will be found in the sequence of silverlike ions.

In a given configuration, the electron interaction that is caused by electrostatic forces and represented by the integrals F^k and G^k increases linearly with Z, while the spin–orbit interaction and other relativistic effects increase as $(Z - s)^4$. The latter effects will, therefore, ultimately dominate the structure, causing a transformation from LS to jj coupling. In the limit their influence will become even larger than that of l, and the structure of a set of configurations having the same n but different l values will approach the hydrogenic structure as is shown by the theoretical diagram[6] in Figure 1 of the configurations $2s^2 p^4$, $2s2p^5$, and $2p^6$ in the oxygen sequence.

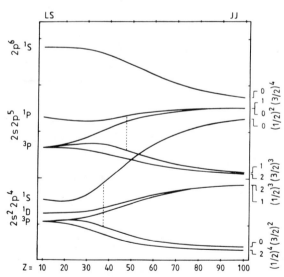

Figure 1. Theoretical energy structure of the set of $n = 2$ configurations in oxygenlike ions according to Kononov and Safronova. The levels are referred to the center of gravity and scaled (approximately) by the total width of the set. "Avoided crossings" are marked by dotted lines.

3. Survey of the Low Configurations in Isoelectronic Sequences

The character of the spectrum of an ion is to a large extent determined by the structure of its lowest configurations as most of the strong lines are due to transitions into these configurations. They also form the series limits which determine the structure of the higher part of the term system in the next lower stage of ionization. For convenient reference we have collected in Table 1 the ground configurations in the first 54 isoelectronic sequences. The sequences are designated by their total number of electrons N and grouped according to the n and l values of the asymptotically lowest configuration. The simplest structure is found near the closed n shells ($N = 2$, 10, and 28), and the corresponding sequences are, therefore, the most extensively observed ones.

In the helium sequence ($N = 2$) theoretical calculations have reached a very high accuracy, and the aim of recent beam–foil observations of highly charged heliumlike ions has mainly been to determine QED effects from the difference between measured and calculated level intervals.

Table 1. The Low Configurations in Ions with Total Numbers of Electrons $N = 1$–54

Group	N	Configuration	Sequence
1	1	$1s$	H
	2	$1s^2$	He
2	3	$(1s^2)2s$	Li
	4	$2s^2$	Be
	5–10	$2s^2 2p^k$	B–Ne
3a	11	$(2s^2 2p^6)3s$	Na
	12	$3s^2$	Mg
	13–18˅	$3s^2 3p^k$	Al–Ar
3b	19	$(3s^2 3p^6)4s \to 3d$	K
	20	$4s^2 \to 3d^2$	Ca
	21–28	$4s^2 3d^{k-2} \to 3d^k$	Sc–Ni
4a	29	$(3d^{10})4s$	Cu
	30	$4s^2$	Zn
	31–36	$4s^2 4p^k$	Ga–Kr
4b	37	$(4s^2 4p^6)5s \to 4d$	Rb
	38	$5s^2 \to 4d^2$	Sr
	39–46	$5s^2 4d^{k-2} \to 4d^k$	Y–Pd
4c	47	$(4d^{10})5s \to 4f$	Ag
	48	$5s^2 \to 4f^2$	Cd
	49–54	$5s^2 5p^k \to 4f^{k+2}$	In–Xe

In group 2 of Table 1 the $n = 2$ complex contains at most three configurations, each with a relatively simple structure, whereas in group 3a the 3d orbits introduce an increasing complexity in the set of asymptotically low configurations. In group 3b excitation of core electrons adds configurations such as $3p^5 3d^{k+1}$ to the $n = 3$ set. No such interference occurs in group 4a since the $n = 3$ shell is closed with $N = 28$. The situation in group 4b is analogous to that in 3b. In group 4c, finally, considerable rearrangements take place with increasing Z until the 4f orbits reach the lowest position in accordance with their n value. The $n = 4$ shell will be complete in the asymptotic ground configuration, $(4d^{10})4f^{14}$, of the sequence $N = 60$, but it is questionable whether this situation will be reached before the end of the periodic system.

4. The $n = 2$ Configurations

The low configurations in the sequences $N = 3$–9 have been extensively studied both experimentally and theoretically. Since the set of $n = 2$ configurations is well separated from the rest of the term system and therefore free from accidental perturbations, all level intervals must show a perfectly smooth Z dependence, which permits a check on the observational data. In two recent papers[7,8] we have studied the level intervals in ions isoelectronic with Li, Be, B, O, and F ($N = 3, 4, 5, 8, 9$) by comparing reported experimental values with the values from the Dirac–Fock calculations of Cheng et al.[9] and expressing the difference by a simple function of Z which corrects the theoretical values for residual correlation and relativistic effects. By this procedure, using the theoretical values as a frame of reference, it has been possible to smooth out the errors in the observations to a large extent and to supplement the observed data by inter- and extrapolation. As a result we have obtained complete sets of recommended level values up to $Z = 36$ for all $n = 2$ configurations. For the carbon and nitrogen sequences, only the ground configurations, $2s^2 2p^2$ and $2s^2 2p^3$, have so far been analyzed in this way.[10]

To illustrate the typical structure of the $n = 2$ set we show in Figures 2 and 3 the level diagrams for the beryllium sequence ($N = 4$) and the conjugate oxygen sequence ($N = 8$) which contain the same ten levels, arranged in roughly opposite order. The diagrams show the gradual transformation of the LS structure with increasing Z and the trend towards a mixing of the three configurations and the ultimate formation of three groups in the hydrogenic limit (see Figure 1).

The ground configurations $2s^2 2p^k$ ($k = 1$–5) contain metastable levels which may give rise to forbidden lines. The ionization stages around $Z = 26$ (Fe XVIII–Fe XXII) are produced in the temperature range of solar flares

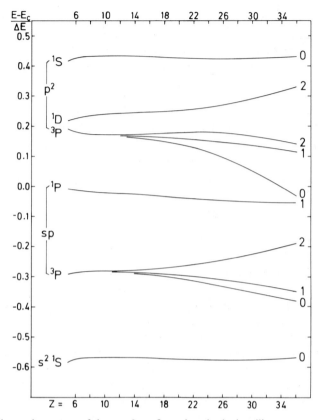

Figure 2. Observed structure of the $n = 2$ configurations in the beryllium sequence. E_c is the center of gravity and ΔE is the total width of the set.

and tokamak plasmas where these lines provide important diagnostic tools. Figure 4 shows the structure of $2s^2 2p^2$ and the extent to which the forbidden transitions have been observed in tokamaks.[11] Corresponding diagrams for $2s^2 2p^3$ and $2s^2 2p^4$ may be found in Ref. 10. The forbidden transitions, which in general fall at much longer wavelengths than the permitted lines, give accurate information on the level structure of these configurations. Also, they tie together different multiplicities (singlets–triplets, doublets–quartets) which in the $n = 2$ period are difficult to connect by other means.

5. The Neon Sequence ($N = 10$)

With $N = 10$ the $n = 2$ shell is complete and gives the single level $2s^2 2p^6\,^1S_0$. Since its combinations with the next higher configurations,

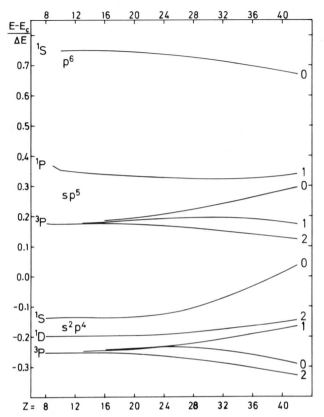

Figure 3. Observed structure of the $n = 2$ configurations in the oxygen sequence (cf. Figure 1).

$2s^2 2p^5(^2P)3s$, $3d$, and $2s2p^6(^2S)3p$ are restricted to $J = 1$ levels, the 2–3 transition group consists of at most 7 lines. These few lines stand out very clearly in the spectrum and could be identified quite early[12] up to Co XVIII. In a series of recent papers[13–18] observations are reported which extend the sequence to Xe 44^+ where the lines fall in the 2.5–3.0 Å range. The xenon spectrum was produced by laser-induced implosion of microballoons filled with xenon.[18]

The diagram in Figure 5 gives a survey of the observations in the upper part of the sequence. It shows the relative positions of the $n = 3$, $J = 1$ levels, scaled by the splitting of the limit $2s^2 2p^5$ ^2P (see Table 2). This makes the intervals in each l-group nearly independent of Z. On this scale the interval $3p(^1P_1 - ^3P_1)$ remains close to the hydrogenic ratio $(2/3)^3 = 0.30$ of the spin–orbit splittings of $3p$ and $2p$. We further note the increasing dominance of the $2s^2 2p^5$ ^2P splitting which has reached the same size as

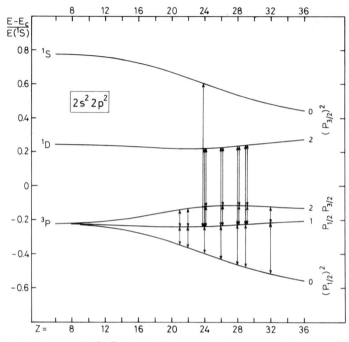

Figure 4. Level diagram of $2s^2 2p^2$ with forbidden transitions observed in tokamaks indicated.

the $3s$–$3d$ energy difference at $Z = 54$. The theoretical analysis[18] shows that $3s(1/2\ 1/2)_1$ and $3d(3/2\ 5/2)_1$ are now strongly mixed.

The strongest lines of the corresponding 2–3 transition group of the fluorine and oxygen sequences have also been identified up to xenon, the mean wavelength decreasing and the complexity of the group increasing with the ion charge. In addition, at slightly longer wavelengths are observed some faint lines (satellites) which represent the same 2–3 transition in the core-excited sodiumlike ion.

Table 2. The Splitting of $2s^2 2p^5\ ^2P$ (in units of 10^3 cm^{-1})^a

Z	$\Delta\,^2P$	Z	$\Delta\,^2P$	Z	$\Delta\,^2P$
43	984.1	47	1459.9	51	2097.5
44	1089.7	48	1602.7	52	2286.5
45	1203.9	49	1756.2	53	2488.5
46	1327.1	50	1921.0	54	2704.2

^{a} Extrapolation of data from Ref. 7.

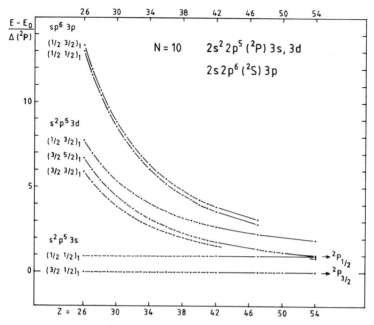

Figure 5. Relative positions of the $n = 3$ levels that combine with $2s^2 2p^6\ {}^1S_0$ in the upper part of the neon sequence. The intervals are scaled by the splitting of $2s^2 2p^5\ {}^2P$.

6. Ions with Ground Configurations of 3s and 3p Electrons

For the sodiumlike ions ($N = 11$) we refer to the paper by Edlén,[19] which gives accurate isoelectronic relations for the $3s-3p$ and $3p-3d$ transitions. As a model for ions with more than one valence electron we take the sequence of magnesiumlike ions ($N = 12$) which is observed up to high Z and illustrates the main features of the spectra in group 3a of Table 1. In Figure 6 are plotted the excitation energies of the low levels as far as they have been observed, scaled by the energy of $3s3p\ {}^1P$. We have included the lowest level, $3s4s\ {}^3S$, of the $n > 3$ configurations in order to show how the $n = 3$ configurations eventually form a group well separated from the higher part of the level system. Any interactions of levels with the same symmetry will take the form of "avoided crossings" in which a smooth Z dependence of the level intervals is conserved. An example of such a "crossing" is shown by the levels ${}^3P_2$ and ${}^1D_2$ of $3p^2$. The overlapping of the two configurations $3p^2$ and $3s3d$ gives rise to a strong mixing of the two ${}^1D$ levels for which the configuration assignment is arbitrary. Of the data plotted in Figure 6 those for $Z = 29-34$ are from Fawcett and Hayes[23] and for $Z = 38-45$

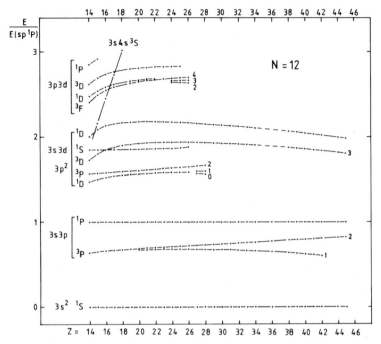

Figure 6. Observed levels of the $n = 3$ configurations in the magnesium sequence.

from Reader.[24] For Mo^{30+} tentative identifications of additional 3–3 transitions are given by Mansfield *et al.*[72] Values for the intersystem transition $3s^2\,{}^1S_0-3s3p\,{}^3P_1$ given by Finkenthal *et al.*[70,71] up to $Z = 42$ have recently been improved by Peacock and Stamp.[25]

The addition of $3d$ to the $n = 3$ set has greatly complicated the situation as compared to the analogous beryllium sequence. The set now contains six different configurations, including $3d^2$, which is not shown in the diagram but has been observed for P IV [20] and S V.[21] In the next sequence ($N = 13$) there are nine configurations in the set, the number increasing by three for each added electron.[22]

Metastable levels in the ground configurations $3s^23p^k$ ($k = 1$–5) give rise to numerous forbidden lines, which are of great interest for the study of hot plasmas. They include all the strong lines in the solar corona ($Z = 26$, 28) and some 60 lines that have recently been observed in tokamaks[26,27] for elements from $Z = 29$ to 42. A survey of the forbidden lines observed in $3s^23p^2$ and $3s^22p^4$ is given in Figures 7 and 8. A similar diagram for $3s^23p^3$ may be found in Ref. 27.

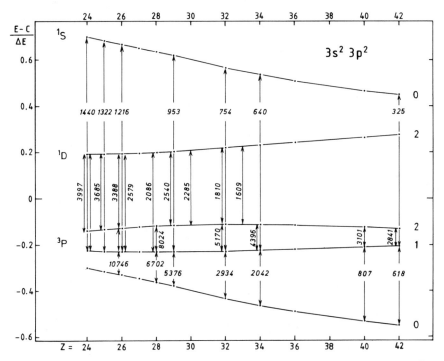

Figure 7. Levels of the ground configuration in the silicon sequence with the wavelengths (Å) of forbidden transitions observed in the solar corona ($Z = 24–28$) and in tokamaks ($Z = 29–42$) indicated.

7. The Configurations $3d^k$

7.1. The Potassium Sequence (N = 19)

In the sequence starting with argon the ground configuration $3s^2 3p^6$ does not have the unique position that $2s^2 2p^6$ has in the neon sequence since the excitation of a core electron to $3d$ still gives an $n = 3$ configuration only moderately higher than the ground term. Thus no abrupt change in the asymptotic structure is caused by the closure of the $3p^6$ shell.

The characteristic arrangement of configurations in the group of $3d^k$ ions is exhibited already by the first sequence ($N = 19$). Figure 9 shows how the one-electron level system is invaded by $3p^5 3d^2$, which soon becomes the lowest excited configuration on account of its lower n values. The perturbations arising when it crosses $3p^6 4f$ and $3p^6 4p$ are clearly visible in the diagram. The sequence has been followed to $Z = 29$, but for the last four spectra it is only transitions to the ground term $3d\,^2D$ that have been

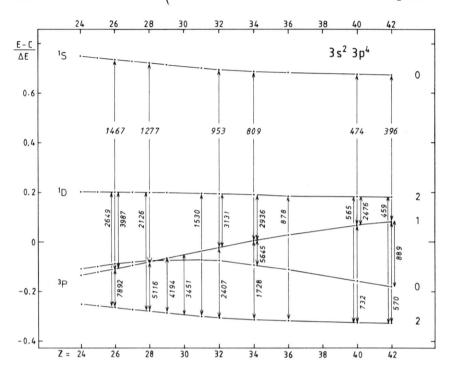

Figure 8. The ground configuration in the sulfur sequence with forbidden transitions observed in the corona ($Z = 26, 28$) and tokamaks indicated.

observed. This means that $4s\,^2S$, which in the beginning of the sequence is the ground term and then is metastable up to $Z = 26$, has not been found beyond $Z = 25$. Transitions are observed also from $3p^53d4s$, which shows the same asymptotic trend in the energy diagram as the other $n = 4$ levels. Each of the configurations $3p^53d^2$ and $3p^53d4s$ contains a large number of levels, of which only those combining with $3d\,^2D$ are observed, and only part of the observed ones are plotted in Figure 9.

Higher core-excited configurations may be expected, in the first place $3s3p^63d^2$ which will run parallel to $3s^23p^53d^2$ and eventually become the lowest excited even-parity configuration. Combinations between the numerous levels of these two configurations may give rise to a large number of lines spread over a wide wavelength range. The original one-electron spectrum is thus gradually transformed into a quite different, much more complicated type of spectrum.

Most of the data for $Z = 25$–29 are from Ramonas and Ryabtsev[28] and for $Z = 24$ from Ekberg.[29] Other data were taken from the NBS compilations.[30-33] The curve for $(^1G)^2F$ confirms the identity of the $^2F_{5/2}$ level found in Ti IV.[34]

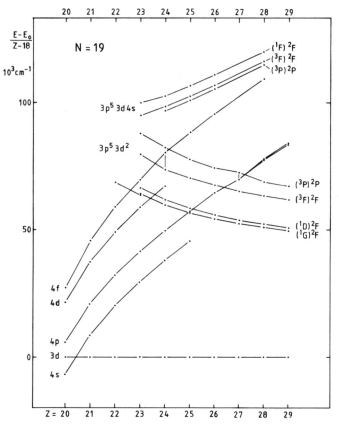

Figure 9. The lower part of the term systems in the potassium sequence. Only a few of the levels of $3p^53d4s$ and $3p^53d^2$ are plotted. Note the perturbations at the crossing of $4p$ and $4f$ by $3p^53d^2$.

7.2. The Iron Sequence (N = 26)

The rapidly increasing complexity of the level systems as further $3d$ electrons are added to the low configurations makes a detailed analysis increasingly difficult for high stages of ionization. In the calcium sequence ($N = 20$), Fawcett et al.[35] were still able to identify some of the strongest lines of the transition $3p^63d^2-3p^53d^3$ in Ni IX and Cu X. However, when the $3d$ shell begins to fill up the level structure is determined by the number of electrons missing from a full shell and becomes again amenable to analysis. The manganese sequence ($N = 25$) was recently studied by Wyart et al.[36] By support of parametric calculations they managed to identify a number of lines of the transition $3p^63d^7-3p^53d^8$ in eight elements from yttrium ($Z = 39$) to silver ($Z = 47$), from which they could derive partial level systems for Y 14^+, Zr 15^+, Nb 16^+, and Mo 17^+. The analysis was aided by their previous study of the corresponding transitions in the iron sequence[44] which provided scaling factors for the parameters.

In the iron sequence ($N = 26$) $3p^6 3d^8$ soon becomes the ground configuration and $3p^5 3d^9$ will ultimately become the first excited configuration. Figure 10 shows the level structure of $3d^8$ from Cu IV to Ag XXII. The data derive from Meinders[37] (Cu IV), van Kleef et al.[38] (Zn V), Podobedova et al.[39] (Ge VII), Uylings et al.[40] (As VIII), van Kleef et al.[41] (Se IX), Ryabtsev[42] ($Z = 37, 38$), Reader et al.[43] ($Z = 39$–42), and Wyart et al.[44] ($Z = 39$–47). Some values of ^1S in Ref. 43 were corrected

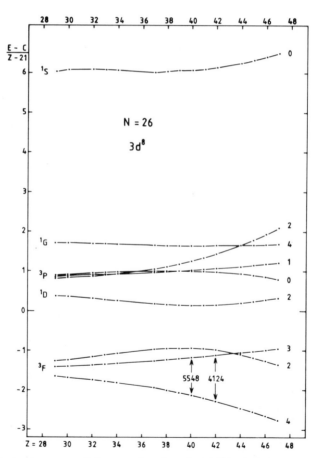

Figure 10. The level structure of the ground configuration $3d^8$ in the iron sequence. C is the center of gravity. The wavelengths (Å) of forbidden transitions observed in the Princeton tokamak are indicated.

in Refs. 44 and 69, and the diagram suggests that the value for $Z = 37$ may need further checking. The levels of $3d^8$ were derived from fairly complete analyses of the transition group $3p^63d^8$-$3p^53d^9$. Four additional configurations in the $n = 3$ set are possible, of which the lowest is $3s3p^63d^9$.

Figure 10 illustrates the considerable changes in relative level positions that take place over the observed range, especially through the mutual repulsion of the three $J = 2$ levels, whose properties become thoroughly mixed.[43]

7.3. The Cobalt Sequence (N = 27)

In the cobalt sequence the asymptotically lowest configurations are reduced to $3p^63d^9\,{}^2D$ and $3p^53d^{10}\,{}^2P$, which give rise to a resonance multiplet of three outstanding lines which were early identified high up in the sequence.[45] Recently, the observations of this multiplet have been extended by Reader[46] to $Z = 70$ (Yb 43^+). Also, the group of lines of $3d^9\,{}^2D$-$3d^84p$ has been observed and analyzed by Wyart *et al.*[47] from Y 12^+ to Ag 20^+ and by Ryabtsev and Reader[48] from Sr 11^+ to Mo 15^+. Finally, the group of $3d$-$4p$ transitions emitted by ions of Tm to Pt ($Z = 69$-78) in a laser-produced plasma in the range 9-6 Å have been studied by Mandelbaum *et*

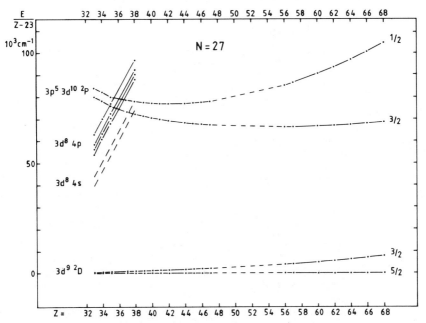

Figure 11. The low configurations in the cobalt sequence. Only a few of the levels of $3d^84s$ and $3d^84p$ are plotted.

$al.$[49] They used the relativistic model potential method to identify the strongest lines representing $3d$-$4p$ transitions in the cobalt, nickel, copper, and zinc sequences, all falling in about the same wavelength range. The groups are very narrow, e.g., the 16 identified features of the $3d^9$-$3d^84p$ transition in Hf 45^+ lie between 7.56 and 7.08 Å.

In the diagram of the cobalt sequence in Figure 11 the data for $Z = 33$-35 are from Refs. 51-52. The configuration $3d^84p$ is represented in the diagram by a few selected levels and the position of $3d^84s$ is indicated by some extrapolated values. This configuration is metastable up to $Z = 38$.

The $n = 3$ complex in the cobalt sequence is completed by $3s3p^63d^{10}\,^2S$, which might be found from combinations with $3p^53d^{10}\,^2P$. The structure of the group of low configurations is thus reduced to the same terms, ^2S, ^2P, and ^2D, as in the sodiumlike ions where the building up of the $n = 3$ shell started.

It is of interest to note that forbidden transitions between the fine-structure levels of the ground terms $3d\,^2D$, $3d^2\,^3F$, $3d^8\,^3F$, and $3d^9\,^2D$ have

Table 3. Interpolation Formula for the Splitting of $3d^9\,^2D$

Z	$\sigma(\text{calc})^a$	$\sigma(\text{obs})^b$	obs-calc
38	14 658	14 660 ± 60	+2
39	17 233	17 230 ± 65	−3
40	20 126	20 126 ± 1.2	0
41	23 360	23 380 ± 80	+20
42	26 960	26 960 ± 1.5	0
44	35 367	35 380 ± 60	+13
45	40 229	40 240 ± 70	+11
46	45 568	45 520 ± 70	−48
47	51 415	51 320 ± 80	−95
48	57 801	57 750 ± 80	−51
49	64 757	64 740 ± 90	−17
50	72 317	72 400 ± 100	+83
51	80 514	80 640 ± 100	+126
52	89 383		
53	98 960		
54	109 283		
55	120 387		
56	132 314	132 230 ± 240	−84
57	145 102	144 940 ± 260	−162
60	189 044	189 070 ± 300	+26
62	223 421	223 630 ± 340	+209
64	262 282	262 420 ± 380	+138
74	537 781		

$^a \sigma(\text{calc}) = 0.03607\ (Z - s)^4;\ s = 11.7623 + 25.51\ (Z - s)^{-1} - 1.30 \times 10^{-6}$ $(Z - s)^3;\ 0.03607 = R\alpha^2/n^3 l(l + 1)$ for $3d$.
$^b \sigma(\text{obs})$ for $Z = 44$-51 from J. Reader and N. Acquista (private communication).

been found in tokamak spectra of zirconium and molybdenum ($Z = 40$, 42).[53] The intervals have by then grown to a size that makes the transitions fall in the visible region. The identifications are supported by comparison with isoelectronic data.[27] The interval $3d^9(^2D_{3/2}-^2D_{5/2})$ can be expressed by a simple semiempirical formula as shown in Table 3.

7.4. The Nickel Sequence ($N = 28$)

In the nickel sequence the $n = 3$ complex is reduced to $3d^{10}\,^1S_0$ and the resonance lines consist of transitions from the $J = 1$ levels of $3p^63d^9(^2D)4p$, $4f$ and $3p^53d^{10}(^2P)4s$, $4d$. These few strong lines stand out against similar transitions in neighboring ions and have in fact been seen up to Pt 50$^+$ where they fall between 6 and 5 Å.[54] The upper part of the sequence is illustrated by the diagram in Figure 12, where the relative level positions are scaled by the splitting of $3d^9\,^2D$ (see Table 3). This splitting is closely reproduced by the interval $(3/2\,3/2)_1-(5/2\,3/2)_1$ of $3d^94p$, and the large splitting of the second limit, $3p^53d^{10}\,^2P$, is reflected in the interval $(1/2\,1/2)_1-(3/2\,1/2)_1$ of $3p^53d^{10}(^2P)4s$. The third conceivable limit may give rise to two more levels, belonging to $3s3p^63d^{10}(^2S)4p$, that could combine with $3d^{10}\,^1S$.

The data shown in Figure 12 derive from Zigler et al.[54] ($Z = 72-78$), Klapisch et al.[55] ($Z = 69-70$), and Burkhalter et al.[56] ($Z = 62-66$). Further references to this sequence are Schweitzer et al.[57] ($Z = 39-47$) and Burkhalter et al.[58] ($Z = 50$). To appreciate the scale of the diagram we note that

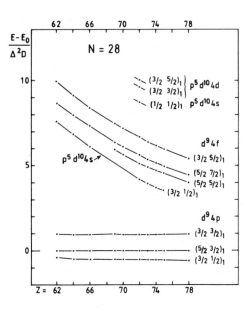

Figure 12. Relative positions of the $n = 4$ levels combining with $3d^{10}\,^1S_0$ in the uppermost part of the nickel sequence. The level intervals are scaled by the splitting of $3d^9\,^2D$. $E_0 = 3d^94p(5/2\,3/2)_1$.

the deviation from a smooth curve for $Z = 62$ in the lowest curve corresponds to 0.02 Å.

The lowest excited configuration $3d^9 4s$ is metastable since it has the same parity as the ground configuration. However, in Mo XV the electric-quadrupole transitions to $3d^{10}\,^1S_0$ from the two $J = 2$ levels of $3d^9 4s$ were found by Klapisch et al.[59] in the spectrum of the tokamak at Fontenay-aux-roses and by Mansfield et al.[72] in DITE. The two lines, which fall around 58 Å, do not appear in the low-inductance spark, but in the low-density tokamak plasma they are actually stronger than the allowed transitions from $3d^9 4p$ owing to a higher population rate. The intensity ratio is predicted[59] to stay approximately constant up to an electron density of about 10^{15} cm^{-3}. Also, five lines of the analogous transition $3d^9$–$3d^8 4s$ in Mo XVI were found in the DITE tokamak.[72]

8. The Copper Sequence (N = 29)

After the closing of the $n = 3$ shell the building up of the $4s$, $4p$, and $4d$ shells leads to configurational structures in the sequences $N = 29$–46 which are essentially similar to those of the sequences $N = 11$–28, the main difference being the increased importance of the spin–orbit interaction. We will, therefore, limit the discussion to the first and simplest case, viz., the copper sequence. Here the one-electron structure is conserved throughout the sequence, and the $n = 4$ configurations form an asymptotically isolated and unperturbed group. This is shown by Figure 13, where the excitation energies are scaled by the mean energy of $4p\,^2P$.

The transitions $4s$–$4p$, $4p$–$4d$, and $4d$–$4f$ were observed in a laser-produced plasma by Reader and Luther[60] for ten elements from $Z = 56$ to $Z = 74$ (W 45$^+$), the identifications being aided by calculations of Cheng and Kim.[61] At the upper end of the sequence the spin–orbit splitting of $4p\,^2P$ is seen to be larger than $4s\,^2S_{1/2}$–$4p\,^2P_{1/2}$, making the wavelength of this transition more than twice as long as that of $4s\,^2S_{1/2}$–$4p\,^2P_{3/2}$. This reduces its relative intensity by a large factor and accounts for its not having been observed. The wave number of $4s\,^2S_{1/2}$–$4p\,^2P_{1/2}$ is found to be approximately linear in Z, which means that the relativistic effects are nearly equal in the two levels. For the ions up to Dy 37$^+$ ($Z = 66$) many of the $n = 4$–5 transitions as well as $5g$–$6h$ were also observed. Observations of corresponding transitions in the ions from Ru 15$^+$ ($Z = 44$) to Sn 21$^+$ ($Z = 50$) were recently published by Reader et al.,[62] who give references to their earlier papers on $Z = 37$–42. Data for $Z = 35$ and 36 are from Livingston et al.[63,64]

The simple and unperturbed structure of the copperlike spectra has made it possible to determine not only the $n = 4$ levels but also a good deal of the higher levels, including such terms as $5g$ and $6h$. This has permitted

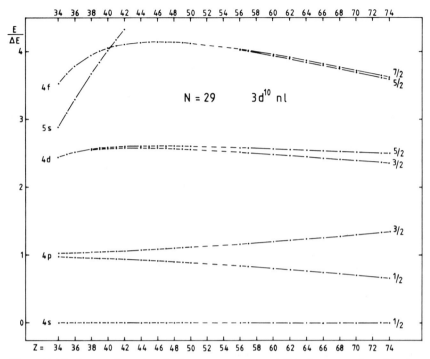

Figure 13. The low configurations in the copper sequence. ΔE is the mean value of $4s-4p$.

an accurate derivation of ionization energies up to higher ionization stages than in any other sequence. The ionization energy for W 45$^+$ was thus determined[60] to be 2412.7 ± 1.9 eV, which may be compared with the value 2415 eV calculated by Cowan.[65] It should be noted, however, that some core-excited configurations, e.g., $3d^94s^2$ and $3d^94s4p$, will remain below the ionization limit and may interfere with the uppermost part of the one-electron system.

9. The Silver Sequence (N = 47)

We pass over the rest of group 4a and 4b (see Table 1), where the main features are similar to those in 3a and 3b, and take the silver sequence as the last example in this review. The structure is illustrated in Figure 14, where the observed levels are plotted relative to $5s$ and scaled by the mean value of $5s-5p$. The data for I VII to Dy XX ($Z = 53-66$) are from Sugar and Kaufman.[66,67] Their observations include also a few lines of Ho XXI.

This sequence provides a most striking example of transition to n-ordering with increasing Z. From a high position in the beginning of the

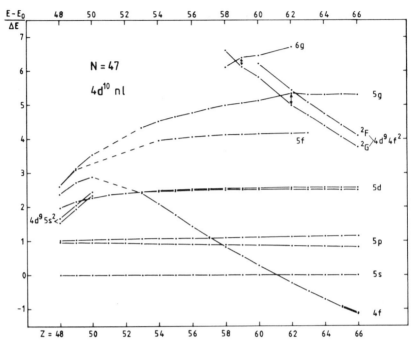

Figure 14. Configurations in the silver sequence. $E_0 = 5s\,^2S$ and ΔE is the mean value of $5s$-$5p$. Note the perturbations as $4d^9 4f^2$ is crossing $6g$ and $5g$.

sequence the $4f$ term, by virtue of its low n value, dives down through the level system, crossing $5d$ at I VII, $5p$ at La XI, and finally $5s$ to become the ground term from Sm XVI on. A similar dive is made by the configuration $4d^9 4f^2$, and the transitions from $4d^9 4f^2$ to $4d^{10} 4f$ will eventually form the resonance lines. Sugar and Kaufman[68] were able to identify this group of lines in tungsten, W 27^+, in the spectrum of the Oak Ridge tokamak. Thus we see how the simple one-electron system is gradually transformed into a quite different type of structure. The complexity of this structure will increase very rapidly with increasing number of $4f$ electrons in the following sequences.

Note Added in December 1986

This note gives references to important work published after the manuscript was first submitted.

The critical examinations of the observations of $2s$-$2p$ transitions were concluded by Edlén.[73-75] New observations of $2s$-$2p$ transitions by Hinnov et al.[76] include the Li- and Be-like resonance lines in Ge as well as the resonance lines in Zn^{21+}–Zn^{23+}. Feldman *et al.*[77] observed oxygen- and fluorine-like $2s$–$2p$ transitions for $Z = 33$–35, 37, and Dietrich *et al.*[78] for $Z = 36$. The oxygen-like transitions for Y, Zr, and Nb are reported by Behring *et al.*[79] Finally, the resonance transitions $2s^2 2p^5$–$2s 2p^6$ have been measured by Reader *et al.*[80] in eight fluorine-like ions from Zr^{31+} to Sn^{41+} in good agreement with the semiempirical extrapolations.[7]

In the sequence of Na-like ions we note an accurate and comprehensive study of Sr^{27+} by Reader.[81] Also, new measurements of the principal transitions $3s$–$3p$, $3p$–$3d$, and $3d$–$4f$ have been made by Reader *et al.*[82] for a number of ions from Cu^{18+} to Sn^{39+}. By fitting the differences between observed and calculated wavenumbers to simple expressions in Z they obtained least-squares fitted wavelengths for all Na-like ions from Ar^{7+} to Xe^{43+} with an accuracy that should make them useful as reference lines.

For the Mg-like ions Fe^{14+}, Co^{15+}, and Ni^{16+} significant additions and revisions to the $n = 3$ configurations up to $3p3d$ have been published by Churilov *et al.*[83] The intercombination line $3s^2 \, {}^1S_0$–$3s3p \, {}^3P_1$ has been measured by Seely *et al.*[84] in spectra from the PLT tokamak for the ions Zn^{18+}, Ge^{20+}, Se^{22+}, and Mo^{30+}.

The transitions $3p^6 3d^9$–$3p^5 3d^{10}$ and $3p^6 3d^8$–$3p^5 3d^9$ of the Co- and Fe-like ions of Ag, Cd, In, and Sn have been identified in a laser plasma by Kononov *et al.*[85] Wyart *et al.*[86] observed in the TFR tokamak numerous lines belonging to the Cu- to Cr-like ions Xe^{25+} to Xe^{30+}. Finally, Seely *et al.*[87] have produced spectra of highly charged ions of Au, Pb, Bi, Th, and U in plasmas generated by the 24-beam OMEGA laser at Rochester. Lines were identified as belonging to the Fe, Co, Cu, and Zn isoelectronic sequences, including the persistent transition $3p^6 3d^9 \, {}^2D_{5/2}$–$3p^5 3d^{10} \, {}^2P_{3/2}$ all the way up to U^{65+}.

References

1. B. Edlén, *Phys. Scr.* **T3**, 5 (1983).
2. C. Jordan, in *Progress in Atomic Spectroscopy*, Part B (edited by W. Hanle and H. Kleinpoppen), Plenum Press, New York (1979), p. 1453.
3. B. C. Fawcett, in *Advances in Atomic and Molecular Physics*, Vol. 10, p. 223, Academic Press, New York (1974).
4. B. C. Fawcett, *Phys. Scr.* **24**, 663 (1981).
5. R. D. Cowan, *The Theory of Atomic Structure and Spectra*, University of California Press (1981).
6. E. Ya. Kononov and U. I. Safronova, *Opt. Spectrosc.* **43**, 3 (1977).
7. B. Edlén, *Phys. Scr.* **28**, 51 (1983).
8. B. Edlén, *Phys. Scr.* **28**, 483 (1983).
9. K. T. Cheng, Y.-K. Kim, and J. P. Desclaux, *At. Data Nucl. Data Tables* **24**, 111 (1979).
10. B. Edlén, *Phys. Scr.* **26**, 71 (1982).
11. E. Hinnov, S. Suckewer, S. Cohen, and K. Sato, *Phys. Rev. A* **25**, 2293 (1982).

12. F. Tyrén, *Z. Phys.* **111**, 314 (1938).

13. H. Gordon, M. G. Hobby, N. J. Peacock, and R. D. Cowan, *J. Phys. B* **12**, 881 (1979).

14. V. A. Boiko, A. Ya. Faenov, and S. A. Pikuz, *J. Quant. Spectrosc. Radiat. Transfer* **19**, 11 (1978).

15. H. Gordon, M. G. Hobby, and N. J. Peacock, *J. Phys. B* **13**, 1985 (1980).

16. R. J. Hutcheon, L. Cooke, M. H. Key, C. L. Lewis, and G. E. Bromage, *Phys. Scr.* **21**, 89 (1980).

17. E. V. Aglitsky, E. Ya. Golts, Yu. A. Levikin, and A. M. Livshits, *J. Phys. B* **14**, 1549 (1981).

18. Y. Conturie, B. Yaakobi, U. Feldman, G. A. Doschek, and R. D. Cowan, *J. Opt. Soc. Am.* **71**, 1309 (1981).

19. B. Edlén, *Phys. Scr.* **17**, 565 (1978).

20. P. O. Zetterberg and C. E. Magnusson, *Phys. Scr.* **15**, 189 (1977).

21. I. Joelsson, P. O. Zetterberg, and C. E. Magnusson, *Phys. Scr.* **23**, 1087 (1981).

22. B. Edlén, in *Beamfoil Spectroscopy*, Vol. 1 (edited by I. A. Sellin and D. J. Pegg), Plenum Press, New York (1976), p. 1.

23. B. C. Fawcett and R. W. Hayes, *J. Opt. Soc. Am.* **65**, 623 (1975).

24. J. Reader, *J. Opt. Soc. Am.* **73**, 796 (1983).

25. N. J. Peacock, M. F. Stamp, and J. D. Silver, *Phys. Scr.* **T8**, 10 (1984).

26. B. Denne, E. Hinnov, S. Suckewer, and S. Cohen, *Phys. Rev. A* **28**, 206 (1983).

27. B. Edlén, *Phys. Scr.* **T8**, 5 (1984).

28. A. A. Ramonas and A. N. Ryabtsev, *Opt. Spectrosc.* **48**, 348 (1980).

29. J. O. Ekberg, *Phys. Scr.* **8**, 35 (1973).

30. J. Sugar and C. Corliss, *J. Phys. Chem. Ref. Data* **9**, 473 (1980).

31. J. Sugar and C. Corliss, *J. Phys. Chem. Ref. Data* **7**, 1191 (1978).

32. C. Corliss and J. Sugar, *J. Phys. Chem. Ref. Data* **6**, 1253 (1977).

33. C. Corliss and J. Sugar, *J. Phys. Chem. Ref. Data* **10**, 197 (1981).

34. J. W. Swensson and B. Edlén, *Phys. Scr.* **9**, 335 (1974).

35. B. C. Fawcett, A. Ridgeley, and J. O. Ekberg, *Phys. Scr.* **21**, 155 (1980).

36. J.-F. Wyart, M. Klapisch, J.-L. Schwob, and P. Mandelbaum, *Phys. Scr.* **28**, 381 (1983).

37. E. Meinders, *Physica* **84C**, 117 (1976).

38. Th. A. M. van Kleef, L. I. Podobedova, A. N. Ryabtsev, and Y. N. Joshi, *Phys. Rev. A* **25**, 2017 (1982).

39. L. I. Podobedova, A. A. Ramonas, and A. N. Ryabtsev, *Opt. Spectrosc.* **49**, 247 (1980).

40. P. Uylings, Th. A. M. van Kleef, Y. N. Joshi, A. N. Ryabtsev, and L. I. Podobedova, *Phys. Scr.* **29**, 330 (1984).

41. Th. A. M. van Kleef, P. Uylings, Y. N. Joshi, L. I. Podobedova, and A. N. Ryabtsev, *Physica C* **124**, 67 (1984).

42. A. N. Ryabtsev, *Phys. Scr.* **28**, 176 (1983).

43. J. Reader and A. Ryabtsev, *J. Opt. Soc. Am.* **71**, 231 (1981).

44. J.-F. Wyart, M. Klapisch, J.-L. Schwob, N. Schweitzer, and P. Mandelbaum, *Phys. Scr.* **27**, 275 (1983).

45. B. Edlén, *Physica* **13**, 545 (1947).

46. J. Reader, *J. Opt. Soc. Am.* **73**, 63 (1983).

47. J.-F. Wyart, M. Klapisch, J.-L. Schwob, and N. Schweitzer, *Phys. Scr.* **26**, 141 (1982).

48. A. N. Ryabtsev and J. Reader, *J. Opt. Soc. Am.* **72**, 710 (1982).

49. P. Mandelbaum, M. Klapisch, A. Bar-Schalom, J.-L. Schwob, and A. Zigler, *Phys. Scr.* **27**, 39 (1983).

50. Th. A. M. van Kleef and Y. N. Joshi, *J. Opt. Soc. Am.* **70**, 491 (1980).

51. Y. N. Joshi, Th. A. M. van Kleef, and H. Benschop, *Can. J. Phys.* **54**, 1545 (1976).

52. Y. N. Joshi and Th. A. M. van Kleef, *Phys. Scr.* **23**, 249 (1981).

53. S. Suckewer, E. Hinnov, S. Cohen, M. Finkenthal, and K. Sato, *Phys. Rev. A* **26**, 1161 (1982).

54. A. Zigler, H. Zmora, N. Spector, M. Klapisch, J. L. Schwob, and A. Bar-Shalom, *Phys. Lett.* **75A**, 343 (1980); *J. Opt. Soc. Am.* **70**, 129 (1980).
55. M. Klapisch, A. Bar-Shalom, P. Mandelbaum, J. L. Schwob, A. Zigler, H. Zmora, and S. Jackel, *Phys. Lett.* **79A**, 67 (1980).
56. P. G. Burkhalter, D. J. Nagel, and R. R. Whitlock, *Phys. Rev. A* **9**, 2331 (1974).
57. N. Schweitzer, M. Klapisch, J. L. Schwob, M. Finkenthal, A. Bar-Shalom, P. Mandelbaum, and B. S. Fraenkel, *J. Opt. Soc. Am.* **71**, 219 (1981).
58. P. G. Burkhalter, U. Feldman, and R. D. Cowan, *J. Opt. Soc. Am.* **64**, 1058 (1974).
59. M. Klapisch, J. L. Schwob, M. Finkenthal, B. S. Fraenkel, S. Egert, A. Bar-Shalom, C. Breton, C. DeMichelis, and M. Mattioli, *Phys. Rev. Lett.* **41**, 403 (1978).
60. J. Reader and G. Luther, *Phys. Scr.* **24**, 732 (1981).
61. K. T. Cheng and Y.-K. Kim, *At. Data Nucl. Data Tables* **22**, 547 (1978).
62. J. Reader, N. Acquista, and D. Cooper, *J. Opt. Soc. Am.* **73**, 1765 (1983).
63. A. E. Livingston, L. J. Curtis, R. M. Schectman, and H. G. Berry, *Phys. Rev. A* **21**, 771 (1980).
64. A. E. Livingston and S. J. Hinterlong, *Phys. Rev. A* **23**, 758 (1981).
65. R. D. Cowan, Los Alamos report No. LA-6679-MS (1977).
66. V. Kaufman and J. Sugar, *Phys. Scr.* **24**, 738 (1981).
67. J. Sugar and V. Kaufman, *Phys. Scr.* **24**, 742 (1981); **26**, 419 (1982).
68. J. Sugar and V. Kaufman, *Phys. Rev. A* **21**, 2096 (1980).
69. J. Reader and A. Ryabtsev, *J. Opt. Soc. Am.* **73**, 1207 (1983).
70. M. Finkenthal, R. E. Bell, H. W. Moos, and TFR Group, *Phys. Lett.* **88A**, 165 (1982).
71. M. Finkenthal, E. Hinnov, S. Cohen, and S. Suckewer, *Phys. Lett.* **91A**, 284 (1982).
72. M. W. D. Mansfield, N. J. Peacock, C. C. Smith, M. G. Hobby, and R. D. Cowan, *J. Phys. B* **11**, 1521 (1978).
73. B. Edlén, *Phys. Scr.* **30**, 135 (1984).
74. B. Edlén, *Phys. Scr.* **31**, 345 (1985).
75. B. Edlén, *Phys. Scr.* **32**, 86 (1985).
76. E. Hinnov, F. Boody, S. Cohen, U. Feldman, J. Hosea, K. Sato, J. L. Schwob, S. Suckewer, and A. Wouters, *J. Opt. Soc. Am. B* **3**, 1288 (1986).
77. U. Feldman, J. F. Seely, W. E. Behring, M. C. Richardson, and S. Goldsmith, *J. Opt. Soc. Am. B* **2**, 1658 (1985).
78. D. D. Dietrich, R. E. Stewart, R. J. Fortner, and R. J. Dukart, *Phys. Rev. A* **34**, 1912 (1986).
79. W. E. Behring, J. F. Seely, S. Goldsmith, L. Cohen, M. Richardson, and U. Feldman, *J. Opt. Soc. Am. B* **2**, 886 (1985).
80. J. Reader, C. M. Brown, J. O. Ekberg, U. Feldman, J. F. Seely, W. E. Behring, *J. Opt. Soc. Am. B* **3**, 1609 (1986).
81. J. Reader, *J. Opt. Soc. Am. B* **3**, 870 (1986).
82. J. Reader, V. Kaufman, J. Sugar, J. O. Ekberg, U. Feldman, C. M. Brown, J. F. Seely, W. L. Rowan, *J. Opt. Soc. Am. B*, submitted November 1986.
83. S. S. Churilov, E. Ya. Kononov, A. N. Ryabtsev, and Yu. F. Zayikin, *Phys. Scr.* **32**, 501 (1985).
84. J. F. Seely, J. O. Ekberg, U. Feldman, J. L. Schwob, S. Suckewer, and A. Wouters, *J. Opt. Soc. Am. B*, submitted November 1986.
85. E. Ya. Kononov, L. I. Podobedova, and S. S. Churilov, *Opt. Spectrosc.* **57**, 26 (1984).
86. J. F. Wyart, C. Bauche-Arnoult, E. Luc-Koenig, and TFR Group, *Phys. Scr.* **32**, 103 (1985).
87. J. F. Seely, J. O. Ekberg, C. M. Brown, U. Feldman, W. E. Behring, J. Reader, M. C. Richardson, *Phys. Rev. Lett.* **56**, 2924 (1986).

7

Inner-Shell Spectroscopy with Hard Synchrotron Radiation

WOLFGANG JITSCHIN

1. Introduction

The interaction between photons and matter is one of the most fundamental processes. The interaction mechanism is well understood in the usual case of nonrelativistic photon energies and nonexcessive intensities. Studies of photon absorption and photon-induced processes thus provide a powerful tool to investigate properties of atoms, molecules, condensed matter, and surfaces. For example, cross sections for photon absorption probe the wave function overlap of electronic states; spectroscopy of absorbed or emitted radiation makes it possible to determine energy and width of levels. Atomic levels span an energy range of a few eV up to about 100 keV. Therefore, experimental studies have to employ photons in a correspondingly broad energy range (Figure 1).[1] Feasibility and quality of experimental photoabsorption and photon-induced emission studies are governed by the available photon sources and related instruments, which are quite different in the various ranges of photon energies.

The advent of the dye-laser photon source about 15 years ago has revolutionized high-resolution *outer-shell* spectroscopy. The outstanding features of laser radiation are its small spectral bandwidth, tunability, high output power, and achievable short output pulses. Synchrotron radiation emitted by orbiting electrons exhibits similar outstanding features and

---

WOLFGANG JITSCHIN • Fakultät für Physik, Universität Bielefeld, D-4800 Bielefeld 1, West Germany. New address: Physikalisch-Technische Bundesanstalt, Institut Berlin, D-1000 Berlin 10, Germany.

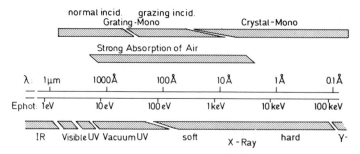

Figure 1. Spectral regions of synchrotron radiation from 1 eV to 100 keV and appropriate monochromators. (From Ref. 1.)

extends toward higher energies. In the last years a vastly increasing number of experiments has made use of the intense ultraviolet and soft x-ray synchrotron radiation. These investigations have made large impact on a variety of fields as discussed in comprehensive review books.[2-4] An impression of the variety of ongoing studies may be obtained from the annual research reports issued by the synchrotron radiation laboratories.

For *inner-shell* studies hard x-rays have to be employed. The continuous spectrum of synchrotron radiation has a sharp intensity decrease toward high photon energies whose position depends strongly on the energy of the orbiting electrons. Therefore, intense radiation in the hard x-ray regime is only delivered by the larger electron accelerators and storage rings. In the last few years electron storage rings have become available as (partly) dedicated x-ray sources; monochromators for hard x-rays have been installed, and progress has been made to increase the primary intensity by inserting special magnetic devices in the electron orbit. Now monochromatized hard x radiation of considerable intensity is available as a new tool for inner-shell experiments.

One essential advantage of synchrotron radiation is the high intensity. Comparing the intensity of synchrotron radiation sources with standard x-ray tubes or γ-ray nuclear sources one has to take into account the different spectral and spatial emission characteristics. Any general comparison must be very crude since the result of a comparison depends on the specific requirements of the application, i.e., how well an experimental setup can accept the delivered radiation.[1,5] Synchrotron radiation extends over a broad spectral range; however, its small azimuthal angular spread is favorable for the use of monochromators. X-ray tubes emit intense characteristic lines of the anode element superimposed on a continuous background. If one is interested in the x-ray intensity at energies coinciding with characteristic x-ray lines only, monochromatized synchrotron radiation may be more intense than radiation from x-ray tubes by one or two orders of magnitude only. In contrast, if one is interested in arbitrary (tunable) x-ray energies, synchrotron radiation is superior in intensity by several orders of magnitude. The available high intensity of synchrotron radiation after monochromatization not only drastically reduces measuring times (absorption edge spectra of solid probes may be recorded in less than 1 sec[6]), it also makes new kinds of studies feasible, as, e.g., studies of adsorbate atoms on surfaces by x-ray absorption spectroscopy, studies of induced fluorescence or electron emission with high resolution but low efficiency detectors, and studies of higher-order processes in photon scattering.

The present article concentrates on atomic physics experiments involving hard synchrotron radiation. Selected experiments are reviewed in order to elucidate basic principles as well as experimental possibilities and achievements. No attempt is made to cover the whole rapidly expanding field or to speculate on future developments. The discussion starts with instrumental details, e.g., the properties of synchrotron radiation, and a brief review of x-ray monochromators and detectors. Thereafter x-ray absorption studies are described that are experimentally very simple (at least in principle). More detailed information on atomic structure may be obtained if one observes not only the x-ray absorption, but additionally the induced fluorescence or emitted electrons. Finally, studies of photon scattering are sketched.

2. Instrumental Details

2.1. Characteristics of Synchrotron Radiation

Electromagnetic radiation originating from fast orbiting electrons has first been observed in an electron synchrotron,[7] yielding the name Synchrotron Radiation (SR) for radiation emitted by orbiting electrons. Nowadays preferentially electron storage rings are used as sources of intense synchrotron radiation. The physical process responsible for radiation emission is the elementary interaction between an electron and the electromagnetic field in which the electron moves. The field affects the electron motion through the Lorentz force causing a change of the electron trajectory. Any change of the absolute value or of the direction of the electron velocity may result in an emission of radiation. The emission process has been treated theoretically by Ivanenko and Pomeranchuk[8] and Schwinger.[9,10] The theoretical predictions for intensity, spectral and angular distribution, and polarization of the radiation are now confirmed by measurements with a high accuracy.[11] Thus, a storage ring is a well-understood and calculable radiation source.

The theoretical results will be reviewed for a simple case. Consider a single electron moving in a homogeneous magnetic field with field direction perpendicular to the electron velocity (Figure 2).[12] The trajectory is a circular orbit, and for the intensity radiated from a monoenergetic relativistic electron one obtains

$$I(\lambda, \psi) = \frac{27}{32\pi^3} \frac{e^2 c}{R^3} \left(\frac{\lambda_c}{\lambda}\right)^4 \gamma^8 [1 + (\gamma\psi)^2]^2 \left[K_{2/3}^2(\xi) + \frac{(\gamma\psi)^2}{1 + (\gamma\psi)^2} K_{1/3}^2(\xi) \right] \quad (1)$$

where λ denotes the wavelength of the emitted radiation, ψ the azimuthal angle, R the bending radius of the orbit, and $K_{2/3}(\xi)$ and $K_{1/3}(\xi)$ are modified Bessel functions of the second kind. In Eq. (1) some abbreviations have been used:

$$\gamma = E/m_e c^2 \quad (2)$$

is the ratio of electron energy to rest mass energy.

$$\lambda_c = 4\pi R/3\gamma^3 \quad (3)$$

denotes the so-called cutoff or characteristic wavelength and

$$\xi = (\lambda_c/2\lambda)[1 + (\gamma\psi)^2]^{3/2} \quad (4)$$

is a parameter containing the azimuthal dependence. For hard x-rays frequently the photon energy $\varepsilon = hc/\lambda$ instead of the wavelength λ is used. The characteristic energy ε_c is

$$\varepsilon_c = 3hc\gamma^3/4\pi R \quad (5)$$

A closer inspection of Eq. (1) reveals the following behavior of the intensity

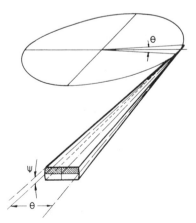

Figure 2. Geometry of synchrotron radiation emission. (From Ref. 12.)

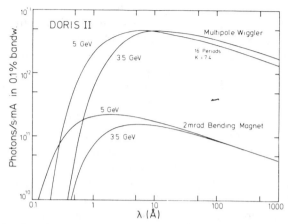

Figure 3. Intensity of synchrotron radiation emitted from the storage ring DORIS II for electron energies of 3.5 and 5 GeV. (From Ref. 13.)

distribution (Figure 3): The spectral intensity varies little with photon energy ε (or wavelength λ) for photon energies smaller than the characteristic energy, i.e., $\varepsilon < \varepsilon_c$. In contrast, for $\varepsilon > \varepsilon_c$ the intensity diminishes steeply with increasing ε. The spatial distribution of the radiation emission is concentrated to the plane of the electron orbit (Figure 4). The azimuthal width depends on the photon energy and at the characteristic energy ε_c the half width $\langle \psi \rangle$ is $1/\gamma$. For example, in the case of 5-GeV electron energy the emission width at ε_c is as small as 0.1 mrad (0.006°). The two Bessel function terms in the intensity formula [Eq. (1)] are associated with the two polarization components $I_\parallel$ and $I_\perp$ with electric vector parallel and

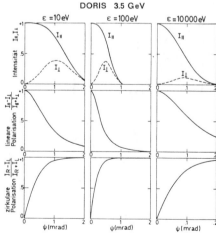

Figure 4. Angular distribution of intensity components with electric vector parallel ($I_\parallel$) and normal ($I_\perp$) to the electron orbital plane; linear and circular polarization calculated for the three photon energies $\varepsilon = 10$, 100, and 10 000 eV. ψ is the elevation angle perpendicular to the orbital plane. (From Ref. 12.)

perpendicular to the plane of the electron orbit, respectively. The radiation emission in both polarization directions is correlated, resulting in elliptically polarized radiation. The total intensity radiated from an electron storage ring is obtained by multiplying the single electron intensity $I(\lambda, \psi)$ [Eq. (1)] with the number of orbiting electrons.

In order to obtain a reasonable intensity at a specific x-ray energy ε, the cutoff energy ε_c which depends on electron energy and curvature radius of the orbit is a crucial parameter. Magnetic fields commonly achieved in bending magnets are 1.0–1.5 T (10–15 kG). At a field strength of 1.0 T a characteristic x-ray energy of, say, 10 keV corresponds to an electron energy of 3.9 GeV. This high electron energy can be achieved in rather large electron storage rings. The requirement for high electron energy is somewhat reduced for most experiments since the radiation intensity may be still sufficient for photon energies up to about five times the characteristic energy.

A shift of the emission spectrum toward higher photon energies and a significant increase of the radiation intensity can be obtained by special magnetic devices. These are conveniently inserted into a rectilinear part of the electron orbit between two bending magnets. A *wavelength shifter* is a dipole magnet with strong magnetic field, thereby causing a small bending radius of the electron orbit and enhanced radiation emission; with superconducting coils, field strengths of about 5 T (50 kG) can be achieved.[14] Enhancement of the radiation intensity can be obtained by periodic magnetic structures in which the electron beam is deflected back and forth (Figure 5).[16] The radiation produced in the different periods of such a structure

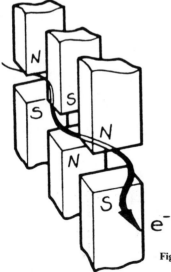

Figure 5. Electron trajectory in a multiperiod wiggler magnet. (From Ref. 15.)

adds up coherently or incoherently depending on both the properties of the magnet and the wavelength of interest. In a *wiggler*, the angular deflection of the electron trajectory is larger than the angular width of radiation emission, resulting in incoherent superposition. In contrast, coherent super-position in an *undulator* gives high emission intensity with intensity peaks at one or a few photon energies.

Synchrotron radiation exhibits time structure due to the operation requirements of storage rings: the energy loss of the orbiting electrons caused by the radiation emission is compensated by accelerating radio-frequency fields. Only electrons passing the radiofrequency cavities within a time interval close to a maximum of the accelerating field strength will remain on the orbit. Thus the acceleration method results in the formation of electron bunches where the bunch length depends on the operating conditions and is of the order of 20 psec to 1 nsec. Synchrotron radiation is delivered to an experiment only during the passage of a bunch. The time interval between bunches depends on the circumference of the storage ring and the mode of operation, typically 2–1000 nsec. In the present article only processes where the time structure is not resolved and which depend linearly on the photon intensity are discussed. Therefore, the time structure of the radiation will be neglected in the following.

2.2. X-Ray Monochromators

The broad spectral bandwidth of synchrotron radiation makes it necessary to employ monochromating devices in most spectroscopic studies which transmit only a small part ΔE, $\Delta \lambda$ of the radiation spectrum. The spectral resolution required for inner-shell spectroscopy or Compton scatter-ing studies is $\Delta E/E = -\Delta \lambda/\lambda = 10^{-3}-10^{-4}$ as determined by the ratio of width to energy of inner-shell levels. However, a resolution better than 10^{-6} is necessary to detect an energy transfer between an x-ray photon and crystal phonons and a resolution of about 10^{-12} is required to observe Mössbauer nuclear-resonant absorption.[17] Monochromators for the regime of hard x-rays generally make use of the wavelength-selective x-ray diffraction by crystals. The high demands on synchrotron radiation monochromators as well as the available high primary intensity stimulated intensive instrumental research resulting in improved knowledge of crystal properties and in advances in instrument building (see, e.g., Ref. 18).

X-ray crystal monochromators may be operated in the reflection mode (Bragg case) or in the transmission mode (Laue case). The condition for diffraction to occur is given by the familiar Bragg's law

$$2d \sin \theta = \lambda \qquad (6)$$

where d is the spacing of crystal lattice planes [for higher orders of diffraction the spacing d in Eq. (6) is the corresponding fraction of the actual lattice spacing], λ the x-ray wavelength, and θ the angle that the incident radiation makes to the reflection plane. The energy resolution for Bragg reflection is obtained from the derivative of Eq. (6) with respect to λ:

$$|\Delta E / E| = |\Delta\lambda / \lambda| = \Delta\theta \cot\theta \qquad (7)$$

The resolution is thus determined by the angular collimation $\Delta\theta$ as well as the Bragg angle θ. Two factors contribute to $\Delta\theta$: angular spread of the incoming radiation (which depends on the experimental geometry) and the reflection width of the crystal. Both factors will now be discussed in more detail.

The azimuthal emission width of synchrotron radiation (see above) is small for hard x-rays, say, e.g., 0.1 mrad. This value corresponds to the collimation obtained by two slit systems, each having 1 mm vertical aperture and separated by a distance of 20 m, which is typical for synchrotron monochromators.[19] Further reduction of the angular spread may be obtained by smaller apertures with some loss of intensity. One notes that for such small apertures problems associated with the size and stability of the electron beam spot may arise. In principle, the intensity loss caused by a single slit can be avoided by a parallel-plate collimation system (Soller slits)[20] as has proved to be useful in investigations of fluorescence spectra (see, e.g., Ref. 21). However, the angular collimation achievable with Soller collimators is limited by total reflection at the collimator plates; e.g., for 8-keV x-rays the glancing angle for an Au surface is as large as 10 mrad.[18]

The reflection width of a crystal depends on the type of x-ray diffraction, i.e., kinematical or dynamical, or an intermediate of the two. For kinematical reflection, e.g., in imperfect crystals, the incoming photons are scattered only once in the crystal and the reflection width depends on the crystal structure (Darwin width). For dynamical reflection multiple scattering at the crystal lattice is responsible, resulting in a small intrinsic width. Commercially available high-quality crystals of silicon, germanium, quartz, and some other materials are (practically) perfect, and their diffraction properties are well described by the dynamical theory.[22] Some properties of perfect crystals are listed in Table 1. The reflection width decreases with increasing order of reflection; furthermore, for higher orders the Bragg angle θ becomes larger, which is also in favor of improved resolution [Eq. (7)]. X-rays are reflected by a crystal in a small interval of diffraction angles θ close to the Bragg angle. Although the peak value of the reflectivity $R(\theta)$ may have values close to unity (Figure 6) the integral reflecting power P of a crystal

$$P = \int d\theta\, R(\theta) \qquad (8)$$

is typically as small as 10^{-5} due to the small width of the reflection interval (Table 1). One notes that the Bragg angle becomes rather small for high x-ray energies; e.g., at 10 keV photon energy the Bragg angle for the 111 reflection of Si and Ge amounts to about 11°.

An x-ray monochromator may be built simply by using a single plane crystal. Such an instrument gives reasonable performance due to the high intensity and good collimation of synchrotron radiation. However, significant improvements are achieved by using more complex arrangements. Intensity increase may be obtained by focusing devices as, e.g., focusing mirrors and singly and doubly bent crystals.[24] Monochromator designs with two and more crystals give smaller spectral bandwidth and permit a suppression of unwanted harmonics.[25,22,26] A monochromator design frequently employed for hard x-rays uses two plane crystals in the nondispersive parallel setting (Figure 7). This arrangement offers practical advantages: since the primary and diffracted beam are parallel, access to the diffracted beam is easy even for small Bragg angles.

The two crystals in a double-crystal arrangement may be made very similar by using a monolith which is grooved or channel cut. However, employing two individual crystals it becomes possible to suppress unwanted orders of reflection since the reflection curves for the various harmonics differ (Figure 6). The suppression may be achieved by slightly detuning the second crystal with respect to the first one, by making use of asymmetric

Table 1. Intrinsic Bragg Reflection Width $\Delta\theta_s$, Energy Resolution $\Delta E/E$ and Integral Reflecting Power P of Perfect Crystals at 8.05 keV (1.54 Å)[a]

Crystal	2d (Å)	hkl	$\Delta\theta_s$ (arcsec)	$\Delta E/E$ (10^{-5})	P (10^{-6})
Silicon	6.271	111	7.395	14.10	39.9
	3.840	220	5.459	6.04	29.7
	3.275	311	3.192	2.90	16.5
	2.715	400	3.603	2.53	19.3
	2.492	331	2.336	1.44	11.8
	2.217	422	2.925	1.47	15.5
	2.090	333	1.989	0.88	9.9
	1.920	440	2.675	0.96	14.0
Germanium	6.533	111	16.338	32.64	85.9
	4.000	220	12.449	14.46	67.4
	3.412	311	7.230	6.92	37.1
	2.829	400	7.951	5.94	42.3
	2.596	331	5.076	3.34	25.4
	2.310	422	6.178	3.34	32.4
	2.178	333	4.127	2.00	20.2
	2.000	440	5.339	2.14	27.5

[a] Reference 18.

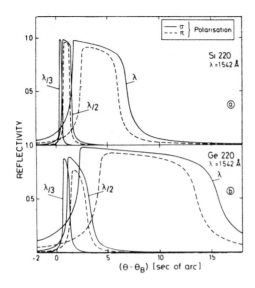

Figure 6. Single-crystal reflection curve for the symmetrical Bragg case 220 (a) silicon crystal, (b) germanium crystal. For comparison, the curves for $\lambda = 1.54$ Å (8.05 keV), $\lambda/2 = 0.77$ Å (16.1 keV), and $\lambda/3 = 0.51$ Å (24.1 keV) are plotted. (From Ref. 23.)

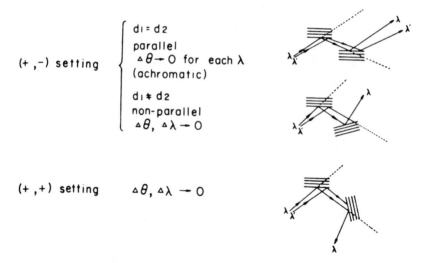

Figure 7. Double-crystal arrangements. In the parallel $(+, -)$ setting, where both lattice spacings of the first and the second crystal parts are equal, all the wavelengths components from the first crystal are diffracted by the second crystal as well. When the lattice spacing of the first is different from the second crystal, it is called the nonparallel $(+, -)$ setting. The scheme as shown in the bottom is called the $(+, +)$ setting. In the nonparallel $(+, -)$ and the $(+, +)$ setting only small angular and wavelength spreads can be diffracted by the second crystal. (From Ref. 22.)

reflection by crystals where the surface is inclined by an angle to the crystal axis, or by using two crystals with different lattice spacing (e.g., Si and Ge).[22,23]

For the choice of a proper diffraction crystal one has to take several aspects into account: e.g., high reflected intensity, small intrinsic reflection width, and small content of heavy atoms which might fluoresce. Also the thermal conductivity of the crystal may be important since for synchrotron radiation of high intensity the heat load of the first optical element is large.

Crystal spectrometers affect the polarization of the diffracted beam due to the polarization dependence of Bragg reflection.[27] At Bragg angles θ in the vicinity of 45° the reflection is strongly polarization dependent, whereas at small angles the polarization dependence is rather small and thus the polarization of the incoming radiation is only slightly changed by the reflection.

Several spectroscopic studies require precise absolute values of the x-ray energy. According to Eq. (6) the x-ray energy (wavelength) transmitted by a monochromator is determined by the Bragg diffraction angle and the crystal lattice spacing. In a precise calibration procedure corrections for refraction of x-rays at the crystal surfaces have to be applied. By a careful calibration of the monochromator absolute accuracies of the photon energy of 10^{-5} can be achieved.[19] Table 2 compiles some properties of the radiation available at the exit of synchrotron x-radiation monochromators.

2.3. X-Ray Detectors

As detectors for hard synchrotron radiation in principle all kinds of x-ray detectors may be employed.[29] Most detectors make use of the photo-effect, in which the absorption of a photon creates an electron–ion pair. The number of photons N absorbed in the active material of a detector is

Table 2. Some Properties of Monochromatized Synchrotron Radiation Available at the *Stanford Synchrotron Radiation Laboratory* (SSRL)[a]

	A	B
Photon energy	7.1 keV	7.1 keV
Accuracy of absolute energy	<0.1 eV	
Spectral bandwidth	<0.5 eV	±2.5 eV
Photon flux	$2 \cdot 10^{10} \, \text{sec}^{-1}$	$10^{12} \, \text{sec}^{-1}$
Spot size	$1 \times 20 \, \text{mm}^2$	$2 \times 4 \, \text{mm}^2$

[a] Monochromators: (A) double-crystal arrangement, (B) with additional toroidal double-focusing mirror.[19,28]

given by the Lambert–Beer law:

$$N = N_0[1 - \exp(-\sigma\rho L)] \tag{9}$$

where N_0 is the number of the incoming photons, σ the absorption cross section, ρ the density, and L the thickness of the material.

For monitoring high total intensities one may conveniently use ionization chambers.[30] In ionization chambers the x-rays pass through a gas cell (Figure 8); the number of electrons and ions produced by the absorption of one photon, i.e., the photoionization yield, depends on the x-ray energy and the gas. The electrons and ions produced are accelerated to electric field plates; the voltages applied to the plates are generally kept small to avoid secondary ionization processes. A measurement of the induced current makes it possible to determine the x-ray flux. The detection efficiency of an ionization chamber can be adjusted to the experimental requirements by proper choice of chamber gas (Ar, Kr, etc.) and chamber length.

For an energy-dispersive detection of x-rays or a detection of single photons other types of detectors have to be used, such as, e.g., proportional counters (gas ionization or gas scintillation counters) or semiconductor detectors [Si(Li), Ge(Li)].[32,33]

For certain applications the polarization of the x radiation is important. As follows from theory, synchrotron radiation is perfectly linearly polarized in the plane of the electron orbit and elliptically polarized outside the plane (Figure 4). However, in practice one has to take into account the finite size and position stability of the radiation source as well as the polarization-changing properties of monochromators. Therefore it is desirable to determine the actual polarization experimentally. The linear polarization can be measured by diffraction methods, such as, e.g., by Bragg reflection at $2\theta = 90°$ or by observing the high Laue transmission for the polarization component with electric vector parallel to the crystal lattice (Borrmann effect).[34] By employing multiple reflection arrangements polarization ratios can be determined even at the level of 10^{-5}.[35] A simple and fast method for a polarization measurement of hard x-rays makes use of x-ray diffraction by powder.[36] Nonresonant scattering at 90° only occurs for the x-ray

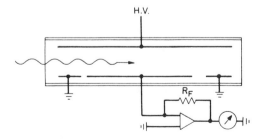

Figure 8. Cross section of a parallel-plate ionization chamber. (From Ref. 31.)

component with electric vector perpendicular to the plane of scattering. Thus the linear polarization can be determined by measuring the intensity of the scattered radiation at various angles. Experimental methods for determining the circular polarization of synchrotron radiation are also available. For example, one can make use of magnetic x-ray scattering by a sample of known magnetization.[37]

3. X-Ray Absorption by Free Atoms

3.1. X-Ray Attenuation

When a photon beam passes through matter the number of photons diminishes. For the systems discussed in the present article, the dominant absorption mechanism is the photoeffect. In a photoionization process the photon energy is used to excite an atom from an initial state ψ_i to a final state ψ_f with a free electron (photoelectron). In the simplest case only one electron changes its state by the photon absorption, whereas all other electrons remain in their previous states. The total absorption cross section contains contributions from all atomic electrons with binding energy less than the photon energy. The cross section varies smoothly with photon energy except at some energies where abrupt changes occur, called absorption edges. These edges correspond to the energies that are required to excite additional occupied electron levels of an atom. For x-ray absorption the main contribution to the total cross section comes from inner shells.

The cross section for the absorption of a photon with momentum $\mathbf{k}_\omega$ and polarization vector ε by an atom is given in the single-electron, single-photon approach by

$$\sigma = (4\pi^2 \alpha^5/k_\omega) \sum_{i,f} \left| \int \psi_f^* \boldsymbol{\alpha}\varepsilon \, e^{i\mathbf{k}_\omega \mathbf{r}} \psi_i \right|^2 \delta(E_f - E_i + \hbar\omega) \qquad (10)$$

where $\boldsymbol{\alpha}$ denotes a 2×2 matrix with Pauli spin matrices as off-diagonal elements; the summation extends over all initial states ψ_i and final states ψ_f.[38] Comprehensive calculations for the photoionization of individual subshells of neutral atoms with $Z = 1$ to 101 for photon energies ranging from 1 to 1500 keV have been performed.[39] In these calculations the electrons are treated relativistically and are assumed to be moving in the same Hartree–Slater central potential both before and after the absorption of the photon.

Experimentally the total x-ray attenuation cross section (summed over the contributions of all electron shells) can be obtained from transmission measurements (Figure 9).[40] In a comparison of experimental total x-ray

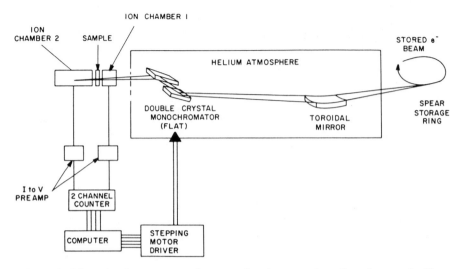

Figure 9. Schematic of an apparatus for measuring the x-ray absorption of a sample. The incoming synchrotron radiation is monochromatized by a two-crystal x-ray spectrometer. The toroidal focusing mirror in front of the monochromator gives strong enhancement of the photon flux to the debit of some increase of the bandwidth of the transmitted radiation. (From Ref. 40.)

absorption cross sections with calculated photoionization cross sections, the small contribution by coherent and incoherent scattering has to be taken into account. Comprehensive calculations of scattering cross sections are available.[41,42] Theoretical calculations are generally performed for free atoms whereas most x-ray attenuation measurements employ solid samples. It is a reasonable assumption that the phase state has only minor influence on the gross features of x-ray absorption (fine details of the absorption spectrum caused by the atomic environment will be discussed in Section 4). Comparison of theoretical and experimental x-ray photoelectric cross sections gives good agreement in general; differences typically amount to 5%–10% from 1.0 to 5.0 keV x-ray energy and 1%–5% from 5.0-keV energy up to energies above which the photoeffect cross section becomes fractionally too small to be accurately determined from the total attenuation coefficient.[43]

The ionization cross sections of the individual L subshells may provide a sensitive test for details of the ionization process as has been demonstrated for ion impact ionization.[44,45] The binding energies of the three L subshells are comparable, but the corresponding wave functions differ significantly. Photoionization occurs dominantly by electric dipole transitions ($\Delta l = \pm 1$), so that only a few final states are occupied. For ionization of the individual

L subshells the following electron transitions are allowed:

$$2s_{1/2} \rightarrow \varepsilon p_{1/2}, \varepsilon p_{3/2}$$

$$2p_{1/2} \rightarrow \varepsilon s_{1/2}, \varepsilon d_{3/2} \tag{11}$$

$$2p_{3/2} \rightarrow \varepsilon s_{1/2}, \varepsilon d_{3/2}, \varepsilon d_{5/2}$$

In a recent photoionization measurement of Au L-subshell cross sections in the regime of the L edges[46] agreement between experiment and theory[39] was found within 2%, which is less than the expected experimental uncertainty (Figure 10). The experimental uncertainty partly arises from the use of a solid Au sample; the absorption spectrum of solid Au exhibits fine structure oscillations[47] which have not been taken into account in the data evaluation. The energy dependences of the L_3- and L_2-subshell cross sections as predicted by theory[39] are rather similar. For the L_3 cross section the experimental slope is in agreement with theory, whereas there is indication that for the L_2 cross section the experimental slope differs from the theoretical prediction. A possible explanation for the observed deviation is

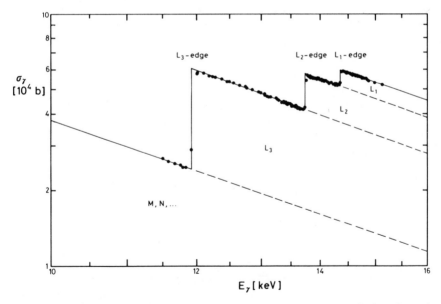

Figure 10. Mass absorption coefficient of gold in the regime of the L-edges. ●, Experimental points. Curves are fits to the experimental data assuming power laws with exponents as suggested by theory.[39] In the fits the regions of at least 100 eV above the edges have been omitted in order to avoid disturbances by EXAFS oscillations. The absolute cross-section values have been obtained by normalizing the relative experimental cross section between the L_2 and L_3 edges to theory. (From Ref. 46.)

a failure of the single-electron assumption in the theory. Recent calculations performed in the independent particle model as well as in the linear response model[48] show that the linear response model reproduces the experimentally observed L_3 and L_2 slopes better than the independent particle model.

To summarize, the gross features of inner-shell ionization by photons, i.e., the total photoelectric cross section as well as the contributions by individual subshells, can be rather well described by theory; the agreement with experiment is typically a few percent. Photoionization is thus more precisely known as ionization by charged particle impact.

3.2. Energies and Widths of Inner-Shell Levels

A precise experimental determination of inner-shell binding energies provides a delicate test for state-of-the-art atomic structure theories.[49,50] Energies are large and so relativistic and quantum electrodynamic effects become strong. Isolated inner-shell vacancies have pronounced single-particle character with correlations generally contributing only approximately one eV to the binding energies of the $1s$ and $2p$ levels. Experimentally the atomic energy levels may be determined by x-ray attenuation studies in the vicinity of absorption edges. The edge corresponds to the onset of photoexcitation from a specific level; for photon energies below the edge, no absorption from electrons in this level occurs, whereas for energies above the edge absorption sets in. The detailed edge shape of the absorption spectrum will now be analyzed for the case of a free atom.[51,52]

In a photoionization process an electron is promoted from a bound state to the continuum. For a free atom the corresponding cross section exhibits only a small dependence on the photon energy ε and may be regarded as constant in a very small energy regime above threshold. The finite lifetime of the final vacancy state results in a broadening of the jump of the absorption cross section at the edge. By integration over the Lorentzian level profile one obtains for the energy dependence of the ionization cross section $\sigma_{ion}(\varepsilon)$

$$\sigma_{ion}(\varepsilon) \propto (1/2) + (1/\pi) \arctan [2(\varepsilon - E_i - E)/\Gamma] \qquad (12)$$

In this formula both inner-shell binding energy E_i and levelwidth Γ enter; E denotes the energy of the photoelectron.

So far, we have neglected photoabsorption due to excitation of an inner-shell electron to bound unoccupied levels. Photoexcitation processes occur already for photon energies below the ionization threshold. The corresponding excitation cross section $\sigma_{exc}(\varepsilon)$ is a sum of Lorentzian profiles:

$$\sigma_{exc}(\varepsilon) \propto \sum_f c_i \frac{\Gamma/2\pi}{(\varepsilon - E_i + E_f)^2 + (\Gamma/2)^2} \qquad (13)$$

where the summation extends over all final excited levels; c_i is a weight factor; E_i and E_f denote the binding energies of the initial and final state, respectively. The widths of all individual excitation curves are (approximately) the same, and they are equal to the width Γ of the final vacancy state.

The photoelectric cross section at an absorption edge is determined by the sum of the two contributions given by Eqs. (12) and (13). The energy separation of outer shell levels to which excitation can occur is of the order of a few eV for all atoms; in contrast, the levelwidth may be less than 0.1 eV for light atoms and larger than 10 eV for heavy atoms. Thus only for light atoms does the photoabsorption spectrum show a clear separation between the two contributions of ionization and excitation. The absorption cross section for excitation to the lowest-lying level may be rather strong, which gives a significant peak in the absorption spectrum called a white line (Figure 11). For heavy atoms the detailed edge structure is blurred by the comparatively large levelwidth. Figures 11–14 show some recorded absorption spectra with fitted decompositions into the different contributions. The L-shell absorption spectra of the neighboring elements Xe and Ba exhibit different edge structure which is due to the different outer-shell configurations.

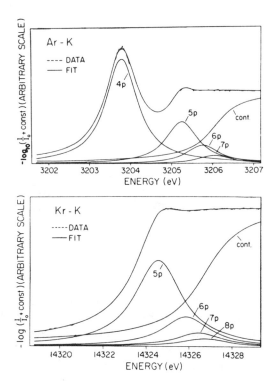

Figure 11. Measured argon and krypton absorption spectra near the K-edge and partition into their components. (From Ref. 19.)

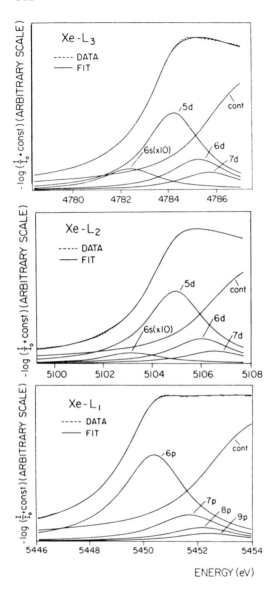

Figure 12. Xenon absorption spectra at the L-edges and fitted decomposition. (From Ref. 19.)

For an extraction of precise energy values from measured absorption edges additional information is required: one may, for example, take the energy differences between the distinct bound and continuum levels from optical spectroscopy data of the next element (optical match, $Z + 1$ rule).[52] With this method inner-shell binding energies have been determined with absolute accuracies of about 10^{-4}.[19,53] The experimental binding energies

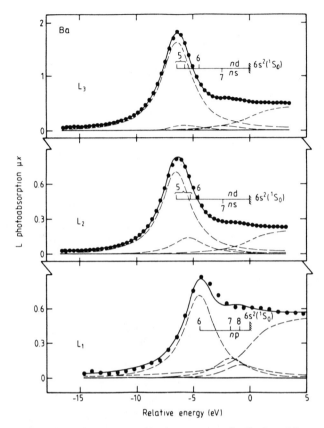

Figure 13. The photoabsorption spectra of barium vapor at the L-edges. Measured data are shown by dots, total fitted spectra by solid curves and their components by broken curves. (From Ref. 53.)

for the K, L_2, and L_3 subshells are reproduced within the error limits by advanced theories.[54–57] In contrast, a clear discrepancy is observed for the L_1 binding energy. This discrepancy is of systematic nature, being common to elements throughout the Periodic Table. It can be surmised that the failure of theory in case of L_1 levels is traceable to the independent-particle approximation.[19,49] In detail, the interaction between an inner-shell vacancy and the Coster–Kronig continuum (dynamic relaxation) is strong especially for the L_1 subshell, where it amounts to a few eV.[58] A major source for the discrepancies in the L_1 binding energies is presumably the quantum-electrodynamic correction as calculated in the one-electron approximation.[53]

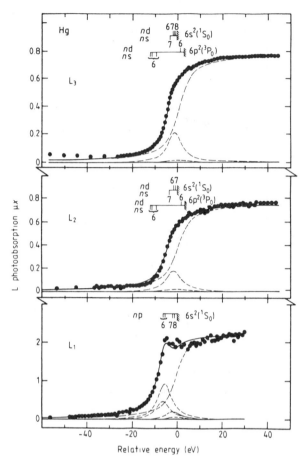

Figure 14. As Figure 13, but for mercury vapor. Owing to ground-state configuration interaction the L_2 and L_3 spectra converge to two separate continua $6s^2\,{}^1S_0$ and $6p^2\,{}^3P_0$. Thus the broken curve for the continuum contribution is also a sum of two arctan curves. (From Ref. 53.)

3.3. Correlation Effects in X-Ray Absorption

Inner-shell electrons of heavy atoms are subjected to the strong nuclear potential; the interaction with other electrons is weak as compared to the interaction with the nucleus. Therefore, electron correlations are expected to become more important for outer shells and lighter atoms. One correlation effect is a two-electron transition induced by single-photon absorption. This process might heuristically be described by the shake model[59]: By absorption of the photon, one (inner-shell) electron is transferred to an unoccupied excited or continuum state thereby causing a change of the effective potential

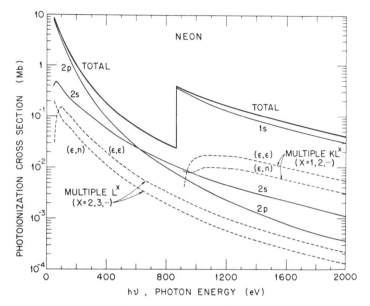

Figure 15. Partition of the photoionization cross section of neon. The one-photon two-electron processes of shake-off and shake-up are denoted by $(\varepsilon, \varepsilon)$ and (ε, n), respectively. (From Ref. 60.)

seen by outer-shell electrons. The outer electrons may react to the changed potential by adjustments of their wave functions and binding energies while evolving to the final "relaxed" state. Alternatively, there is a finite chance that one of the outer electrons is transferred to an excited state ("shake-up") or to a continuum state ("shake-off"). The probabilities for shake-up and shake-off in K-shell photoionization of light elements may be as high as about 10% (Figure 15).[60] The simple take model does not predict a dependence of shake probabilities on the photon energy. A more advanced description of electron correlation effects has to take three mechanisms of configuration interaction into account: initial-state and final-ionic-state configuration interaction (which are photon-energy independent) and continuum-state configuration interaction (which is highly energy dependent and more important near threshold).

K-shell photoionization of argon close to threshold has recently been studied in great detail.[61-63] The high intensity of synchrotron radiation made it possible to record the x-ray absorption spectrum with high quality, as well as to analyze the induced fluorescence and the photoelectron spectrum with high resolution. The gross features of the recorded absorption spectrum (Figure 16) are due to one-electron processes. The pronounced peak close to threshold (white line) corresponds to the $1s$–$4p$ excitation (compare Figure 11). The total absorption cross section agrees well even

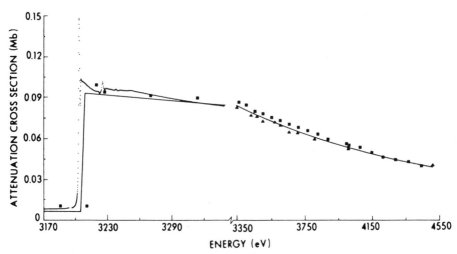

Figure 16. Argon x-ray absorption spectrum scanned (dots) and discrete data (triangles, squares). Solid line is a calculation with hydrogenic wavefunctions. (From Ref. 62.)

quantitatively with calculations in the hydrogenic approximation.[64] The agreement possibly is fortuitous, as, e.g., the hydrogenic calculations predict a smooth energy dependence, whereas the experimental absorption spectrum reveals pronounced features. For energies immediately above the white line the experimental cross section shows significantly greater slope than the hydrogenic calculations. Calculations with more realistic wave functions such as Hermann–Skillman[65] or Hartree–Fock[66,67] give more structure in the energy dependence than the hydrogenic calculations, which is due to the energy dependence of the continuum amplitude in the vicinity of the nucleus arising in the normalization process (Figure 17).

An important point in a treatment of inner-shell photoionization is the rearrangement of electron shells in the residual ion after the creation of an inner-shell vacancy. The rearrangement results from the response of the wave functions of all electrons to the creation of a vacancy and its subsequent decay. This leads to the changes in the mean field in which the ejected photoelectron moves and, thus, affects the probability of electron ejection. The approximation of a frozen core results in a significant overestimate of the photoionization cross section while the approximation of complete statical rearrangement leads to a prominent underestimate. Best agreement is obtained by assuming a dynamical rearrangement in which the electron shells only partly relax during the finite lifetime of the 1s vacancy.

For photon energies ca. 15–40 eV above the threshold the experimental Ar absorption spectrum shows a lot of structure (Figure 18) which can be attributed to two-electron excitation processes. To identify the observed structures, energy positions for various final states and configurations have

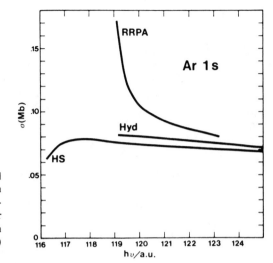

Figure 17. Theoretical *K*-shell photoabsorption cross section of Argon showing the hydrogenic (Hyd), Hartree–Slater (HS) and relativistic random phase approximation (RRPA) results. (From Ref. 69.)

been calculated[68]; any multiplet splitting of doubly excited states may be suppressed as it is expected to be smaller than the *K* levelwidth. Comparison of calculated energy levels and the measured spectra (Figure 18) clearly shows the onset of various two-electron processes when energetically allowed. For example, the peak B in Figure 18 corresponds to the excitation of two electrons initially in the $1s3p$ state to the final $4s^2$ and $4p^2$ states.

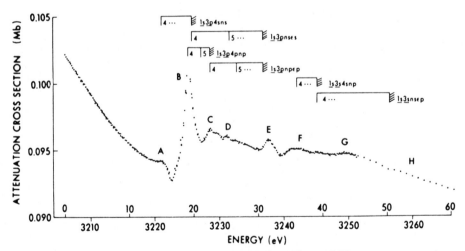

Figure 18. Expansion of the suprathreshold structure from Figure 16. The upper energy scale starting from zero gives the energy above the onset of the one-electron continuum at 3206.0 eV. Energy positions of several configurations obtained by Hartree–Fock calculations are indicated; underlined electron states in the designation of configurations denote vacancies. (From Ref. 62.)

The experimental absorption spectrum in general does not allow a reliable separation of the various two-electron processes and the one-electron process in photoabsorption. In order to perform a separation one may additionally investigate the induced fluorescence and photoelectron spectra. In case of K-shell ionization of Ar a one-electron process leads to a final $1s^{-1}$ ionic state, whereas two-electron processes may lead, e.g., to a $1s^{-1}3p^{-1}4p$ ionic state. Since the energies of both final states differ, the energies of the corresponding fluorescence and photoelectron lines are different. The intensity ratio of satellite line to diagram line reflects the contribution of one-electron and two-electron processes. Measurements of fluorescence spectra[62] have the disadvantage that they are influenced by an inneratomic vacancy rearrangement (e.g., by Coster–Kronig transitions) within the time interval between ionization and fluorescence emission. In contrast, measurements of the photoelectron spectra give direct information on the contributions of the various processes to the total photoionization cross section (Figure 19).[63,70]

4. X-Ray Absorption by Bound Atoms

The x-ray absorption spectra of an element recorded for free atoms, atoms in molecules, or atoms in various chemical compounds exhibit small but significant differences. Apparently the photoabsorption by an atom depends on its chemical environment. Much progress has been made in the interpretation of the detailed features of absorption spectra and nowadays absorption measurements provide a powerful tool to extract information on a variety of materials including three-atom molecules and small clusters. Principles and results of experimental investigations have been compiled by several authors.[71-75]

Basically, the near edge structure of the absorption spectrum makes it possible to determine the chemical binding of an atom, whereas the structure extending above the edges allows determination of the local atomic environment, i.e., kinds and distances of surrounding atoms.

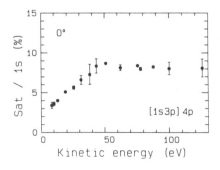

Figure 19. Intensity of photoelectrons from the argon K-shell. The intensity of the photoelectron satellite line for an excited final $1s^{-1}3p^{-1}4p$ Ar$^+$ state relative to the main line for a final $1s^{-1}$ Ar$^+$ state is plotted. Although the data scatter, the intensity of the satellite line apparently increases with increasing kinetic energy of the photoelectrons, i.e., with increasing energy of the primary photons. (From Ref. 70.)

4.1. X-Ray Absorption near Edge Structure (XANES)

The structure of edges in an absorption spectrum allows a precise determination of inner-shell binding energies (see Section 3.2). Inner-shell energies are mainly determined by the strong nuclear potential, but to some extent they also depend on the electron density in outer shells, which, in turn, depends on the specific chemical state. Interesting cases are the transition and rare-earth elements with partly unfilled $3d$ and $4f$ states, respectively. We discuss some studies of the occupancy of the $4f$ valence band.

The L-absorption edges of the rare-earth elements cerium, samarium, gadolinium, and erbium have been studied for free atoms and for solids.[76] The recorded absorption spectra at the L_2 and L_3 edges are rather similar. A significant feature in the spectra (Figure 20) is the white line whose position allows a precise determination of the corresponding L-shell binding energies. The observed energy shift of the white line for atoms either in the gaseous or in the solid phase is as small as 1–2 eV for Ce and Gd, but as large as about 8 eV for Sm and Er. The origin of the large differences in energy shift is the different chemical behavior of the individual elements: Ce and Gd have the same $4f^{n}$ configuration in the atom and in the solid (Ce $4f^{1}$, Gd $4f^{7}$). In contrast to this, Sm and Er undergo a change of the $4f^{n}$ configuration on solidification (Sm $4f^{6} \rightarrow 4f^{5}$; Er $4f^{12} \rightarrow 4f^{11}$).

The valence change on solidification can be investigated in detail by embedding the atoms in a noble gas matrix at different concentrations. For example, the valence of samarium is 2 for the free atom and 3 for the solid.

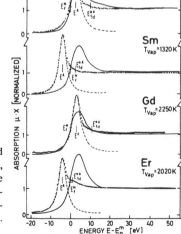

Figure 20. L_3 absorption spectra of atomic (dotted lines) and metallic (solid lines) cerium, samarium, gadolinium, and erbium. The positions of the Fermi levels E_F^M, the atomic $5d$ excitation energies E^A and the ionization energies E_{5d}^{*A} are indicated. The dashed lines are Lorentzian white lines. (From Ref. 76.)

The experimental data for Sm in an Ar matrix (Figure 21) show the expected Sm valence change with increasing concentration and the formation of Sm clusters.[77] The valence of transition elements even in the solid state may be sensitive to the physical conditions: if, e.g., solid cerium is subjected to high pressures in the order of 100 kbar, it may undergo structural transitions thereby changing its valence state.[78,79]

The experimental technique of x-ray absorption studies can be applied also to crystalline or amorphous alloys. XANES studies thus make it possible to probe the nature of chemical binding and to investigate changes in valence orbitals and relaxation of electron configurations.

4.2. Extended X-Ray Absorption Fine Structure (EXAFS)

So far, we have concentrated on the edge and near-edge structure of absorption spectra. Significant features are observed in this regime. For

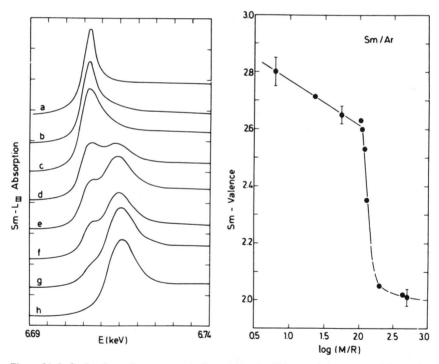

Figure 21. Left: L_3-absorption spectrum of samarium in different environments: (a) atomic Sm, (h) solid Sm; in Ar matrix at various concentrations (radical:matrix), (b) 1:440, (c) 1:200, (d) 1:130, (e) 1:120, (f) 1:55, (g) 1:6. Right: Sm valence as function of the concentration M/R (matrix:radical) in an Ar matrix at a temperature of 5 K. (From Ref. 77.)

photon energies of a few 10 eV above threshold the spectra exhibit rather smooth behavior in the case of free atoms. The situation changes if one investigates the absorption spectra of atoms in condensed matter. Here the absorption of x-rays on the high-energy side of absorption edges does not vary monotonically but exhibits an oscillatory structure that extends above the edge by an amount typically of the order of 1 keV (Figure 22). This structure has received the name *extended x-ray absorption fine structure* (EXAFS), and it only occurs when atoms are surrounded by other atoms and not for isolated atoms.

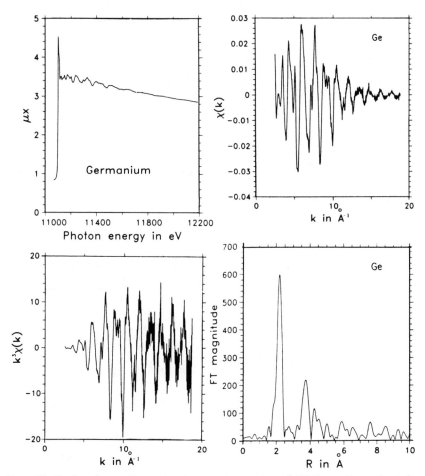

Figure 22. K-edge of crystalline germanium at a temperature of 100 K. μx: X-ray absorption spectrum, the sharp rise near 11 keV is the K-edge and the modulation above the edge is the EXAFS; $\chi(K)$: derived EXAFS spectrum; $K^3\chi(K)$: EXAFS spectrum multiplied by K^3; FT: Magnitude of the Fourier transform of the $K^3\chi(K)$ curve. (From Ref. 71.)

Quantitatively, the EXAFS χ is defined as the normalized oscillatory part of the x-ray absorption μ:

$$\chi = (\mu - \mu_0)/\mu_0 \qquad (14)$$

where μ_0 is the smoothly varying portion of χ past the edge and physically corresponds to the absorption coefficient of an isolated atom. χ depends on the photon energy $\hbar\omega$; however, it is convenient to regard χ as function of the wave vector K of the emitted photoelectron

$$K = [2m_e(\hbar\omega - E_0)]^{1/2}/\hbar \qquad (15)$$

where E_0 denotes the initial binding energy of the electron and $\hbar$ is the Planck constant.

What is the explanation for the occurrence of the EXAFS oscillations? In photoabsorption above threshold an electron is transferred to a continuum state. For a free atom the photoelectron wave function depends only weakly on the wave vector resulting in a smooth absorption spectrum (see Section 3). However, when an atom is in condensed matter and surrounded by other atoms, the outgoing photoelectron may be reflected by the surrounding atoms (Figure 23). Thus its wave function is a sum of an outgoing and a scattered portion. Constructive or destructive interference between outgoing and backscattered waves results in a decrease or increase of the photoelectron wave function in the region of the initial state, thus decreasing or increasing the photoabsorption cross section.[80] Since the phase between the waves depends on the photoelectron energy an oscillatory behavior of the photoabsorption cross section as a function of photon energy may occur.

In the interpretation of EXAFS data multiple scattering of the ejected photoelectron may be neglected in a good approximation. Thus it is sufficient

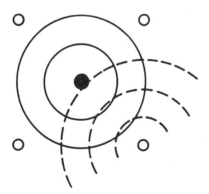

Figure 23. Mechanism responsible for EXAFS: photoelectron waves (solid lines) originating from the absorbing atom (full circle) propagate to neighboring atoms (open circles). The backscattered waves (dashed lines) modify the wave function at the absorbing atom. (From Ref. 71.)

to treat scattering from each of the surrounding atoms just once, and one does not have to worry about scattered waves bouncing off other atoms in between. Assuming single scattering of the photoelectron by a single atom (or by several atoms having all the same distance to the central absorbing atom) one finds the theoretical expression for the EXAFS[72,81,82]:

$$\chi(K) = [F_j(K)/KR_j^2]\, e^{-2R_j/\lambda}\, e^{-2K^2\sigma_j^2} \sin[2KR_j + \phi_j] \qquad (16)$$

In this expression, R_j is the distance between the central absorbing atom and the backscattering atom(s), and $\phi_j(K)$ is the scattering phase shift. $F_j(K)$ is the amplitude for electron backscattering, its value depends on the kind of atom(s) responsible for scattering as well as on the energy of the photoelectron (Figure 24).[83] The first exponential term $\exp(-2R_j/\lambda)$ in Eq. (16) describes the damping due to inelastic scattering of photoelectrons by a mean free path λ, the second exponential term $\exp(-2K^2\sigma_j^2)$ is a Debye–Waller-type term which accounts for a blurring of the EXAFS oscillations by the thermal vibrations of the scattering atoms. Equation (16) applies only in the case where all final photoelectrons have the same angular momentum 1, e.g., for photoionization of the K and L_1 shell; for L_2- and L_3-shell photoionization the formula is still a good approximation since

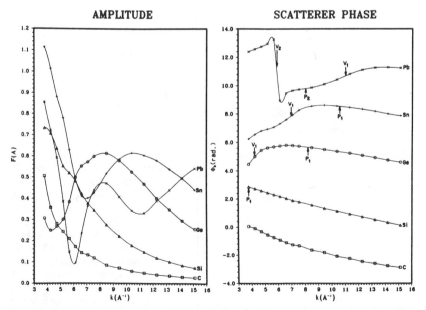

Figure 24. Backscattering amplitude $F(K)$ and phase $\phi(K)$ versus photoelectron momentum K for a number of elements. (From Ref. 83.)

the cross section for final d states is more than one order of magnitude larger than for final s states.[84,85]

EXAFS is caused by the interference of the photoelectron state with itself. Since the amplitude of the electron wave decreases with $1/R^2$ and the mean free path λ of electrons amounts to only a few angstroms, only neighboring atoms contribute to the interference. Thus EXAFS probes the distance between the central absorbing atom and its nearest neighbors, i.e., the short-range order (SRO) of atoms in condensed matter.

In experimental EXAFS studies one measures the x-ray absorption; the extraction of the EXAFS spectrum $\chi(K)$ from the absorption spectrum is straightforward.[86] In the further data evaluation the Fourier transformation of the measured EXAFS weighted by a factor K^n is performed; in practice, $n = 1$ and $n = 3$ are used routinely. The Fourier transform, which is a function of the length R, peaks approximately at those distances that correspond to the positions of the scattering atoms (Figure 22). EXAFS measurements make it possible to determine distances between neighboring atoms with an accuracy of up to 0.01–0.03 Å, if calibration

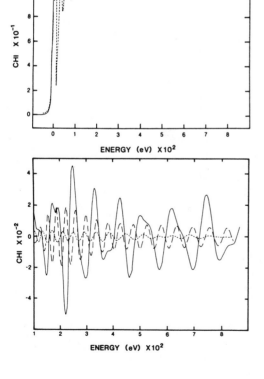

Figure 25. L_3 absorption edge of metallic platinum. The upper figure shows experimental data (solid line) and calculated spectrum (dashed line). The figure at bottom shows the calculated partition of the EXAFS: single scattering contribution (solid line), double scattering contribution (short dashed line), and the triple scattering contribution (long dashed line.) (From Ref. 88.)

measurements on reference materials with known distances (e.g., crystals) are performed.[82]

Multiple scattering of the photoelectron may also contribute to EXAFS[87]; here the photoelectron may return to its original atom via several paths. In consequence, the actual EXAFS is a sum over many terms corresponding to the different paths. In general, the lengths of scattering paths increase with increasing number of scattering processes. Owing to the $1/R^2$ dependence of the backscattered electron wave function the contributions from multiple scattering are expected to be small. The experimental EXAFS data for platinum metal have been compared to various calculations (Figure 25).[88] As expected, only minor contributions of multiple scattering processes are observed except for one triple scattering path. The agreement between measured and calculated EXAFS is good in general; deviations at 0–50 eV energy above threshold can be traced to disturbances of the photoelectron by the central absorbing atom which become large at small photoelectron energies.

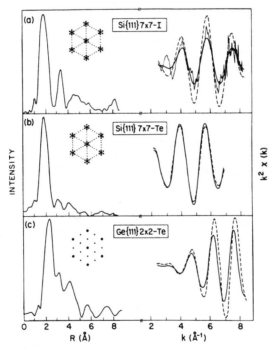

Figure 26. Fourier-transformed SEXAFS data from the iodine and tellurium L_3-edges (left); the iodine and tellurium atoms are adsorbed on silicon and germanium crystals. The inset shows the corresponding LEED patterns. Filtered EXAFS data of the first nearest neighbors (right); solid and dashed curves correspond to different photon polarizations with respect to the surface normal. (From Ref. 90.)

The EXAFS method can not only be applied to investigate the arrangements of atoms in the bulk, but also to investigate the arrangement of atoms on a surface, e.g., to determine the positions of adsorbate atoms. Studies of the surface extended x-ray absorption fine-structure (SEXAFS) offer the same advantages as the EXAFS studies: SEXAFS probes the short-range order of adsorbates at a surface; the accuracy for determining lengths is unbalanced by other methods; SEXAFS is complementary to other methods for surface analysis (e.g., low-energy electron diffraction, LEED), which probe the long-range order within the overlayer.[89] The absorption of iodine and tellurium on clean silicon (111) and germanium (111) crystals was recently investigated by SEXAFS studies.[90] The SEXAFS data (Figure 26) allowed extraction of bond lengths as well as site geometries of the adsorbates. Comparing the recorded EXAFS data of the different investigated systems, it is obvious that the adsorbate I occupies the same sites on Si and Ge, but the adsorbate Te behaves differently from I and differently for each substrate. In the data analysis, the nearest-neighbor and second-nearest-neighbor distances were derived with an accuracy of up to 0.02 Å and also the site geometries of the adsorbates could be determined.

5. Induced X-Ray Fluorescence and Auger-Electron Emission

5.1. Application for Chemical Analysis

The fluorescence spectrum (and also the Auger electron spectrum) of a sample exhibits characteristic lines whose energies depend on the species of the emitting atom. A spectral analysis of the fluorescence makes it possible to identify specific elements and to determine their content in the sample, i.e., an analysis of the chemical composition. The fluorescence may be excited by charged particle impact, where the method is known as particle induced x-ray emission (PIXE).[91] Alternatively, the fluorescence may be excited by x-radiation, which offers several advantages: The fluorescence cross sections for x-ray excitation are 10–1000 times larger than for particle impact. By using polarized monochromatized synchrotron radiation, background in the fluorescence spectra by photon scattering can be kept small, resulting in a high fluorescence-to-background ratio (Figure 27).[92,93] For the same fluorescence signal excitation by photons deposits order of magnitudes less energy in a sample as excitation by charged particles. Accordingly by photoexcitation heat load and radiation damage of the sample are strongly reduced, which opens new fields for nondestructive analyses as, e.g., analysis of biological materials.[94]

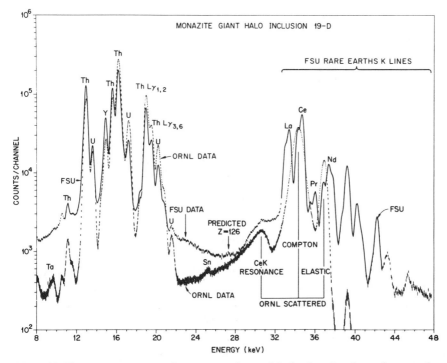

Figure 27. Fluorescence spectrum from a monazite particle (native phosphate of rare-earth elements) excited with 37-keV synchrotron radiation (ORNL data) shows an improved signal-to-background ratio over that excited with 5.7-MeV protons (FSU data). (From Ref. 93.)

For an elemental analysis at different spots in a sample, spatial resolution is required. In so-called microprobes the primary exciting radiation is focused or collimated to a small spot of the sample. Achievable resolutions both for charged particle impact and x-ray photoionization are in the μm range.[93]

5.2. Decay Channels of Inner-Shell Vacancies

Inner-shell vacancies are rapidly filled by electrons from higher shells. The time for vacancy filling, i.e., the vacancy lifetime, can be determined from levelwidth measurements (see Section 3.2). In the filling of a vacancy three different processes compete in general (Figure 28):

Radiative decay: An inner-shell vacancy is filled by an electron from an intermediate or outer shell; this process transfers the vacancy from an inner shell to a higher shell. The energy difference between the initial and the final state is carried away by a photon resulting in the emission of the

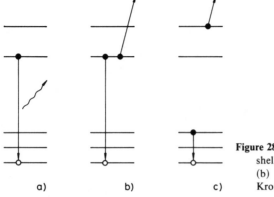

Figure 28. Various decay modes of inner-shell vacancies. (a) Radiative decay, (b) Auger decay, and (c) Coster–Kronig decay.

characteristic fluorescence. Radiative decay is (essentially) a one-electron process.

Auger decay: The interaction of two electrons in higher shells by the electrostatic repulsion forces one of the two electrons to fill the inner-shell vacancy while the other electron is emitted into the continuum, thereby carrying the excess energy as kinetic energy.

Coster–Kronig decay: This decay process is a special Auger decay in which the original vacancy is transferred to a higher subshell of the *same* shell. Auger and Coster–Kronig transitions are (essentially) two-electron processes and cause the emission of electrons with characteristic energies.

An accurate knowledge of the individual rates of vacancy decay is interesting in several fields. Firstly, transition rates present a sensitive tool to investigate details of atomic structure since they probe static properties (atomic wave functions) as well as dynamic properties (electron correlation and relaxation).[95] Secondly, an accurate knowledge of relative decay rates is important in practical applications: In experimental studies of ion–atom collisions either fluorescence or electron emission is detected and the ionization cross sections are derived. In the L-shell case uncertainties of fluorescence and Coster–Kronig yields are a limiting factor upon deriving ionization cross sections.[45]

Comprehensive theoretical calculations of radiative transition rates[96] as well as Auger and Coster–Kronig transition rates[97] are available. However, uncertainties are large for Coster–Kronig transitions with small excess energy due to the strong influence of several effects: (i) many-body interactions in the initial and final atomic systems, (ii) relaxation in the final ionic state, and (iii) exchange interaction between the continuum electron and the final bound-state electrons.[98,99] For an experimental determination of decay rates, various techniques have been employed, e.g., the use of radioactive sources or coincidence techniques.[95] Most techniques are limited to the determination of only a few decay rates. For those reasons, adopted decay rates still bear uncertainties up to several 10%.[100] An appropriate technique for a reliable experimental measurement is to employ a calibration system in which the distribution of primary vacancies in the individual subshells can be varied and is always known *a priori*.

These requirements can be fulfilled by photoionization studies. Cross sections for photoionization are rather accurately known (see Section 3.1); furthermore the ionization of the individual subshells can be switched "on" and "off" by varying the photon energy in the regime of the L edges. In a recent experiment on vacancy decay rates[46] the mass attenuation coefficient and the induced x-ray fluorescence were measured simultaneously (Figure 29). The fluorescence spectra (Figure 30) depend strongly on the energy E_γ of the primary radiation. For E_γ smaller than the L_3 edge of Au, only coherently and incoherently scattered primary radiation but no Au L radiation is observed. For E_γ between the L_3 and L_2 edges of Au only those characteristic Au lines are emitted that originate from the L_3 subshell. If the energy of the primary photons is increased above the L_2 edge, additionally L_2 vacancies are produced resulting in the emission of those characteristic lines that originate from the L_2 subshell. Furthermore, at the L_2 edge the intensity of lines originating from the L_3 subshell increases (Figure 31). This increase is due to a Coster–Kronig vacancy transfer from the L_2 to the L_3 subshell and directly allows a determination of the f_{23} vacancy transfer parameter. At the L_1 edge an intensity increase for lines originating from the L_2 and L_3 subshell is observed which corresponds to a vacancy transfer from L_1 to L_2 (parameter f_{12}) and from L_2 to L_3 (parameter $f_{13} + f_{13}f_{23}$). From the total intensity of all lines originating from one subshell i the fluorescence yield ω_i can be derived.

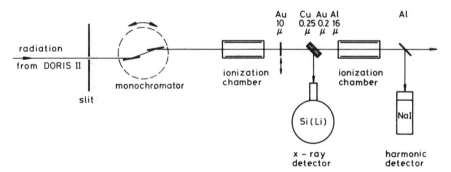

Figure 29. Sketch of an experimental setup for measuring the mass attenuation and the induced x-ray fluorescence; in practice, the Si(Li) detector is mounted perpendicular to the drawing plane, i.e., in the polarization direction of the primary photons in order to minimize the scattering of primary radiation into the Si(Li) detector. (From Ref. 46.)

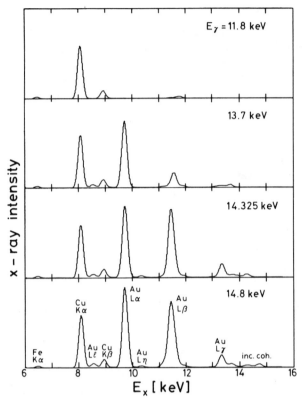

Figure 30. X-ray spectra of a gold–copper sandwich sample excited by primary radiation at different energies. From top to bottom: energy below the Au L edges, between the L_3 and L_2 edges, between the L_2 and L_1 edges, and above the L_1 edge. "Inc." and "coh." denote incoherently and coherently scattered primary radiation. (From Ref. 46.)

The experimental vacancy decay rates may be compared to theoretical predictions. A comparison of relative rates normalized to unity for the total decay rate (i.e., fluorescence and Coster–Kronig yields) may be misleading: For example, assume that one decay channel is strongly dominant and that the transition rate only of this channel is predicted incorrectly by theory. In such a case the dominant channel has a yield close to unity both experimentally and theoretically and thus only little deviation will be observed for this channel. However, experimental and theoretical yield of a weak decay channel depend strongly on the rate of the strong channel due to the normalization, and thus for the weak channel a significant deviation arises. For this reason, a conclusive comparison between experiment and theory has to be made for absolute rates. In order to convert the experimentally determined relative rates to absolute values, a normalization

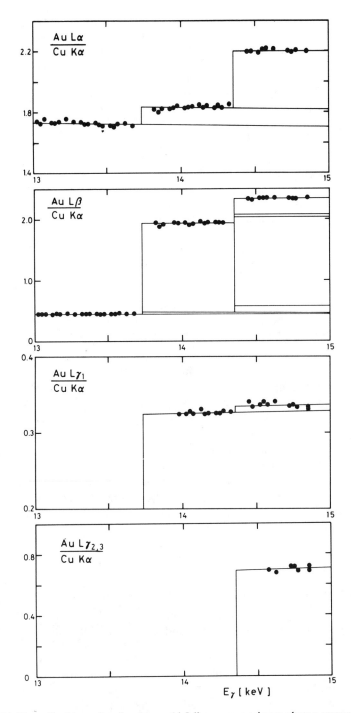

Figure 31. Normalized intensity of various gold *L* lines versus primary photon energy. Curves represent fits to the line intensities using the experimental subshell photoionization cross sections. Jumps at the *L* edges are due either to the onset of photoionization of the corresponding subshell or to the onset of Coster–Kronig vacancy transfer from another subshell. (From Ref. 46.)

Table 3. Compilation of Radiative (ω), Auger (a), and Coster-Kronig (f) Decay Rates for L-Subshell Vacancies in $_{79}$Au or $_{80}$Hg^a

Sub-shell	Decay mode	Experiment on Au,[46] relative rates	Experiment on Au,[46] absolute rates	Theory for Hg,[97] absolute rates	Experiment for Hg,[53] total levelwidth
L_3	ω_3	0.320 (12)	*1.72 eV		
	a_3	0.680 (12)	3.65 eV	3.88 eV	
	Total	1.000	5.37 eV		5.8 (0.8) eV
L_2	ω_2	0.401 (20)	*1.96 eV		
	a_2	0.499 (22)	2.44 eV	2.88 eV	
	f_{23}	0.100 (9)	0.49 eV	0.72 eV	
	Total	1.000	4.89 eV		6.0 (1.0) eV
L_1	ω_1	0.135 (9)	*1.12 eV		
	a_1	0.228 (24)	1.89 eV	2.09 eV	
	f_{12}	0.047 (10)	0.39 eV	0.99 eV	
	f_{13}	0.590 (20)	4.88 eV	10.20 eV	
	Total	1.000	8.28 eV		6.4 (2.0) eV

^{a} An asterisk * denotes a normalization to the radiative rate predicted by theory.[96]

procedure is required. One procedure is to adopt the total vacancy decay rate from measurements of the levelwidth (see Section 3.2).[101] Alternatively, one may adopt one partial decay rate for each subshell from theory; best suited is the radiative rate, which presumably is the most reliably predicted rate. For the L subshell of atoms with nuclear charge $Z = 79, 80$, both procedures lead to the same absolute decay rates within the uncertainties. Comparison between theory and experiment (Table 3) gives good agreement for Auger rates, but for the Coster-Kronig rates large discrepancies are observed amounting to a factor of 2 for f_{12} and f_{13}. Apparently theory fails to predict those rates in which the kinetic energy of the ejected electron is comparatively low.

5.3. Resonant Raman Auger (RRA) Process

So far we have regarded the creation of an inner-shell hole and its decay as independent succeeding processes. In this picture, the energy spectrum of vacancy decay products, e.g., Auger electrons, does not depend on the process of vacancy creation. The separation of excitation and decay breaks down in the case of photoexcitation and ionization close to threshold where new phenomena show up, such as, for example, the resonant Raman Auger (RRA) effect. The RRA process was first observed in the xenon $L_3-M_{4,5}^2$ Auger line.[102] In the experiment the energy of the primary ionizing photons was scanned in the region of the xenon L_3 edge. The spectrum of the induced Auger electrons was found to undergo a dramatic change when excited near threshold (Figure 32).[103] Two lines with different behavior behavior can be identified in the L_3-$M_{4,5}^2$ Auger electron spectrum: The lower energy line essentially stays at the same energy independent of the photon energy; it can be identified with the diagram Auger line. In contrast, the higher energy line exhibits interesting features (Figure 33); its intensity has a sharp maximum for a photon energy about 2.5 eV below the ionization threshold, and its energy is linearly dispersed with the primary photon energy. This line can be interpreted as a satellite line corresponding to the same Auger transition lines with a spectator electron in the $5d$ shell: by photoabsorption below threshold the initial L_3 electron is not promoted into the continuum but to the open $5d$ shell with fixed binding energy. The observed energy shift of the satellite line as function of photon energy is linear with a slope of 1. This slope is an unambiguous indication of an RRA process in which energy is conserved in the excitation–deexcitation sequence.

A particularly interesting Auger process is the xenon L_2-L_3N_4 Coster-Kronig decay excited by photons close to threshold. The emitted L_2-L_3N_4 Coster-Kronig electron has a low electron energy of 228.4 eV, which makes it sensitive to electron–electron interactions. Close to threshold the photoelectron recedes only slowly from the atom and is still in the vicinity of the

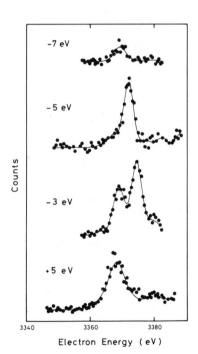

Figure 32. Xenon L_3-$M_{4,5}^2$ ^1G Auger electron spectra (normalized to the photoelectron-peak intensity) excited by x-rays with energies from -7 to $+5$ eV with respect to the Xe L_3 binding energy. (From Ref. 103.)

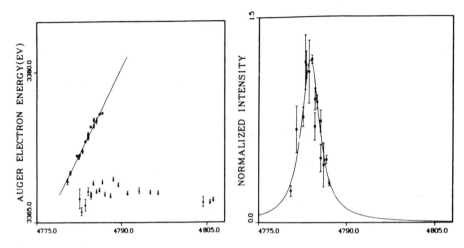

Figure 33. Properties of the xenon $L_3-M_{4,5}^2\,{}^1G_4$ Coster–Kronig line as function of the exciting x-ray energy (in eV). Left: energies of $5d$-spectator satellite line and diagram line; right: normalized intensity of the $5d$-spectator satellite line. (From Ref. 103.)

atom when the Coster–Kronig electron is emitted. The Coulomb interaction of both electrons is rather strong and causes a significant change of the energies of the two electrons. Thus the common neglect of interaction between the two continuum electrons is dissolved here. The experimentally observed energy of the Coster–Kronig L_2–L_3N_4 diagram line clearly is shifted as a function of the primary photon energy (Figure 34). The observed shift is due to the *post*collision *i*nteraction (PCI) of the two electrons and can be well described even quantitatively in a semiclassical treatment.[104]

6. Scattering of X Rays

6.1. Rayleigh and Compton Scattering

When a photon beam moves through matter, some of the photons are scattered out of the beam. Depending on the energy of the scattered photons, the scattering can be divided into the two contributions of elastic and inelastic scattering. *Elastic or coherent* (Rayleigh) scattering may be regarded as an emission of radiation from a (bound) electron, which is forced to oscillations by the primary photon beam; the state of the electron is not changed by the elastic photon scattering. In the process of *t*hermal *d*iffuse *s*cattering (TDS), which occurs in condensed matter, the electron also remains in its state, but the lattice is excited. Owing to the smallness of phonon energies, the energy of the scattered photon is very close to the energy of the primary photon. *Inelastic or incoherent* (Compton) scattering

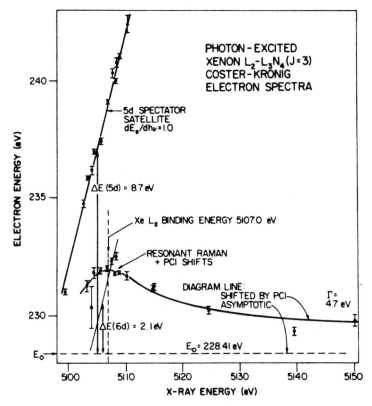

Figure 34. Energies of xenon L_2-L_3N_4 Coster–Kronig electrons as a function of the exciting x-ray energy. The solid curve for the diagram line indicates the prediction of the semiclassical Niehaus theory above threshold. (From Ref. 103.)

with significant energy loss of the photon occurs when the scattering electron carries part of the photon momentum away. Inelastic scattering may be regarded as a binary encounter between a photon and an electron. In the case of free electrons initially at rest, the energy of the scattered photon is uniquely determined by the initial photon energy and the photon scattering angle. Inelastic photon scattering by electrons initially bound in an atom may be treated in the so-called impulse approximation.[105] In this approximation the electron photon interaction again is treated as a binary encounter, but the electron is assumed to have an initial momentum as given by its momentum wave function. Owing to the spread of the momentum distribution the energy spectrum of the scattered photon is not sharp but exhibits some broadening. Photon scattering by a specific electron with definite initial momentum is only allowed if the final electron state is not occupied, i.e., if excitation occurs to high-lying or continuum levels. The energy

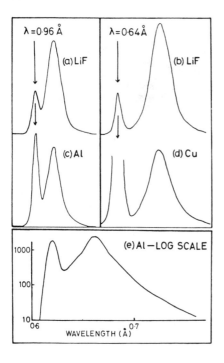

Figure 35. Spectra of Compton scattered photons from LiF, Al, and Cu samples as recorded by a Si(Li) detector. In (a) and (c), the primary photon energy is 13.0 keV (0.96 Å) and the scattering angle 155°; in (b) and (d) 19.9 keV (0.64 Å) and 145°. The two peaks in the spectra correspond to elastic (left peak) and Compton (right peak) scattering. The excellent background ratio in this raw data is illustrated by the logarithmic plot (e) for the Al sample. (From Ref. 106.)

distribution of the scattered photons (Compton profile) reflects the electron-momentum distribution. For atoms with several electrons one has to sum over the contributions from all individual electrons in order to obtain the atomic Compton profile.

Scattering studies using synchrotron radiation have started later than spectroscopic studies. The delay is directly related to the fact that for many scattering experiments the absolute value and the tunability of the primary photon energy (where the monochromatic synchrotron radiation is superior to the radiation from x-ray tubes or γ-ray sources) are not critical points. An exploratory experiment, however, has demonstrated the advantages of using synchrotron radiation in x-ray scattering studies.[106] In these Compton-profile measurements far higher signal count rates and better signal-to-background ratios were achieved as compared to studies employing conventional sources (Figure 35).

6.2. Resonant Raman Scattering (RRS)

Higher-order effects in photon scattering are small in general, but cross sections may be strongly enhanced in the vicinity of resonances. The tunability of monochromatized synchrotron radiation makes it possible to investigatate the higher-order effects in the resonance regime and to perform experiments that are virtually impossible with conventional photon sources.

A second-order photon scattering process is the *resonant Raman scattering* (RRS), which was first observed experimentally by Sparks[107] and found theoretically by Gavrila and Tugulea.[108] Resonant Raman scattering has similarities to the resonant Raman Auger process discussed in Section 5.3. To illustrate the process of resonant Raman scattering, we consider an example and artificially divide the RRS into two steps (Figure 36).[109] Firstly, an incoming photon excites a K-shell electron to an unoccupied bound state or to the continuum. Secondly, the K-shell vacancy is filled by an L- (or higher shell) electron. This process is called K–L–RRS in accord with the electrons involved. As the K vacancy is only an intermediate state the constraint of energy conservation does not apply to the individual steps but only to the transition from initial to final state. Therefore, K–L–RRS can also occur for photon energies below the threshold for K-shell excitation. The double-differential cross section for the K–L–RRS is given by[110,111]

$$\frac{d^2\sigma}{d\Omega\, d\omega_f} = \frac{r_0^2}{\pi m_e^2 \hbar} \frac{\omega_f}{\omega_i} \frac{|\langle f|p_z|K\rangle\langle K|p_z|L\rangle|^2}{(E_K - E_L - \hbar\omega_f)^2} \tag{17}$$

where, because of energy conservation,

$$\hbar\omega_i - \hbar\omega_f = E_L + E_f \tag{18}$$

Here E_K, E_L denote the binding energies of the K and L shells, respectively,

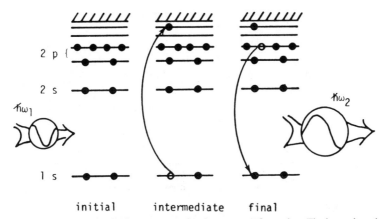

initial   intermediate   final

Figure 36. Inelastic x-ray scattering processes for the neon configuration. The incoming photon with energy $\hbar\omega_1$ excites an $1s$ electron into an intermediate state whereby the energy may not be conserved in a higher-order process. In the final state the $1s$ hole is filled by a $2p$ electron and a photon with energy $\hbar\omega_2$ is emitted. (From Ref. 109.)

and $E_f \geq 0$ the kinetic energy of the photoelectron. $\hbar\omega_i$ is the energy of the primary photon; $\hbar\omega_f$ is the energy and $d\Omega$ the solid angle of the scattered photon. As usual, m_e denotes the electron rest mass and r_0 the classical electron radius.

Second-order Raman scattering leads to the same final electronic states as first-order Compton scattering; for example, K–L–RRS and L-shell Compton scattering both create a final L vacancy. These two processes, however, are experimentally distinguishable by specific features: energy and angular distribution of the scattered photons are different. The Raman cross section is in general much smaller than the Compton cross section but becomes strongly enhanced when the energy of the incoming photon approaches a resonance, i.e., when the denominator in Eq. (17) becomes small. The expected behavior of the cross sections can clearly be seen from

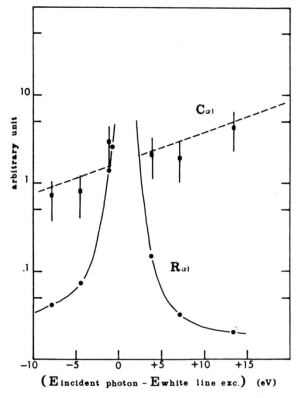

Figure 37. Observed intensity of the Raman line $R\alpha_1$ and the Compton line $C\alpha_1$ of manganese versus incident photon energy; the index α_1 indicates that the scattering process creates a final vacancy in the L_3 subshell. Error bars associated with the Compton intensity account for the small observed fraction of the whole spectrum whose shape is not known. Lines through the points are to guide the eye only. (From Ref. 112.)

measurements on manganese where the scattered radiation was recorded in the vicinity of the K edge with a high resolution of 5×10^{-5} (Figure 37).[112]

For heavy elements the energy separation of the higher electron shells and subshells is sufficiently large to be resolved by semiconductor x-ray detectors. Recently, the individual resonant Raman scattering processes resulting in final vacancies in the L_2, L_3, M, and N-shell have been studied for neodymium.[113] The spectra of the scattered photons were taken at three different primary photon energies close below the K-shell binding energy of 43.57 keV (Figure 38). By choosing the direction of observation close to the polarization direction of the incoming polarized radiation, the intensities of the Rayleigh and Compton peaks in the spectra were largely reduced. The experimental results for K-L-RRS and K-N-RRS were compared to calculations using the ansatz (17). The calculated cross sections were found to agree with the experimental data on an absolute scale within the error limits of a few percent. The same good agreement also holds for the angular and polarization dependences of RRS.[114]

So far, we have discussed K-shell RRS below the ionization threshold. When the energy of the primary photons is tuned toward the K edge, the cross section for RRS increases, attains a maximum at threshold, and decreases again for energies above threshold. On the other hand, at threshold the usual photoexcitation accompanied by fluorescence emission sets in. Thus at a resonance the photon scattering evolves continuously into fluorescence. This behavior can be described by a single formula for fluorescence and scattering.[115,116] Calculated x-ray emission spectra are shown in Figure 39.[117] The emission line with constant energy (independent from the energy of the primary photons) can be attributed to fluorescence, whereas the emission line which is linearly dispersed with the primary energy can be attributed to resonant scattering. A detailed investigation of the emission spectra exhibits an interesting feature: slightly below the threshold the inelastic line is narrowed. For example, the K-L_3-RRS line attains a linewidth that is not determined by the K levelwidth (0.68 eV for Ar) but which may approach the smaller L_3 levelwidth (0.13 eV for Ar). Owing to the reduction of the linewidth, high-resolution Raman spectroscopy provides more accurate information on the discrete part of the threshold regime than the corresponding absorption spectrum. Experimentally, the line narrowing in the transition regime from resonant scattering to fluorescence has been confirmed.[118] However, a detailed investigation of the line profile still remains a challenge for the experimentalist. The demands on small spectral bandwidth both for the primary photons and the induced x-radiation are high. Nevertheless, corresponding experiments seem feasible in the near future using an intense synchrotron radiation source and high-resolution x-ray monochromators.

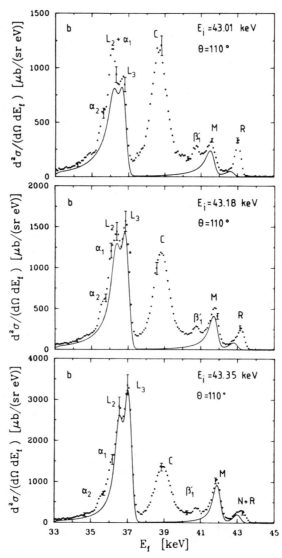

Figure 38. Double-differential scattering cross section for neodymium at 110° photon scattering angle. Spectra were taken for primary energies of 43.01, 43.18, and 43.35 keV and a bandwidth of 40 eV. Points are experimental data as recorded by an intrinsic Ge planar detector. Curves are theoretical spectra. *L*: *K*-*L*-RRS; *M*: *K*-*M*-RRS; *R*: Rayleigh scattering; α, β: *K*α, *K*β-fluorescence of a praseodymium impurity in the sample. (From Ref. 113.)

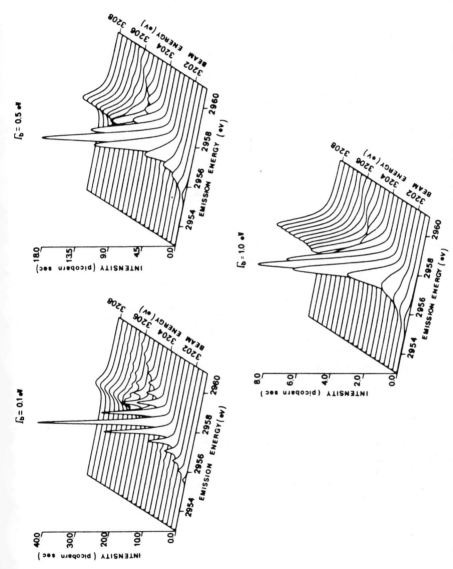

Figure 39. Calculated differential cross sections for the induced argon x-ray emission originating from resonant Raman scattering and fluorescence. The energy bandwidths Γ_b of the primary photon beams are 0.1, 0.5, and 1.0 eV, respectively. Only the spectrum corresponding to final states with a $2p_{3/2}$ vacancy is shown. (From Ref. 117.)

Acknowledgments

The author thanks Professor H. Kleinpoppen and Professor H. O. Lutz for numerous discussions which stimulated his interest in synchrotron-based experiments. Special thanks go to Dr. G. Materlik, who made valuable comments on the manuscript. The manuscript was completed in March 1985.

References

1. C. Kunz, in *Synchrotron Radiation, Techniques and Applications* (edited by C. Kunz), Springer, Berlin (1979), pp. 1-23.
2. C. Kunz (ed.), *Synchrotron Radiation, Techniques and Applications*, Springer, Berlin (1979).
3. H. Winick and S. Doniach (eds.), *Synchrotron Radiation Research*, Plenum Press, New York (1980).
4. E.-E. Koch (ed.), *Handbook on Synchrotron Radiation 1*, North-Holland, Amsterdam (1983).
5. E.-E. Koch, D. E. Eastman, and Y. Farge, in Ref. 4, pp. 1-63.
6. R. P. Phizackerley, Z. U. Rek, G. B. Stephenson, S. D. Conradson, K. O. Hodgson, T. Matsushita, and H. Oyanagi, *J. Appl. Crystallogr.* **16**, 220-232 (1983).
7. F. R. Elder, A. M. Gurewitsch, R. V. Langmuir, and H. C. Pollock, *Phys. Rev.* **71**, 829-830 (1947).
8. D. Ivanenko and J. Pomeranchuk, *Phys. Rev.* **65**, 343 (1944).
9. J. Schwinger, *Phys. Rev.* **70**, 798-799 (1946).
10. J. Schwinger, *Phys. Rev.* **75**, 1912-1915 (1949).
11. F. Riehle and B. Wende, *Opt. Lett.* **10**, 365-367 (1985).
12. C. Kunz, *Phys. Bl.* **32**, 9-21 (1976).
13. HASYLAB Jahresbericht 1982 (1982).
14. D. E. Baynham, P. T. M. Clee, and D. J. Thompson, *Nucl. Instrum. Methods* **152**, 31-35 (1978).
15. DARESBURY Annual Report 1983/4 (1984).
16. J. E. Spencer and H. Winick, in Ref. 3, pp. 663-716.
17. E. Gerdau and R. Rüffer, in HASYLAB Jahresbericht 1984, pp. 305-307.
18. T. Matsushita and H. Hashizume, in Ref. 4, pp. 261-314.
19. M. Breinig, M. H. Chen, G. E. Ice, F. Parente, B. Crasemann, and G. S. Brown, *Phys. Rev. A* **22**, 520-528 (1980).
20. W. Soller, *Phys. Rev.* **24**, 158-167 (1924).
21. W. Jitschin, B. Wisotzki, U. Werner, and H. O. Lutz, *J. Phys. E* **17**, 137-140 (1984).
22. K. Kohra and M. Ando, *Nucl. Instrum. Methods* **117**, 117-126 (1980).
23. G. Materlik and V. O. Kostroun, *Rev. Sci. Instrum.* **51**, 86-94 (1980).
24. R. D. Deslattes, *Nucl. Instrum. Methods* **117**, 147-151 (1980).
25. U. Bonse, G. Materlik, and W. Schröder, *J. Appl. Crystallogr.* **9**, 223-230 (1976).
26. H. Hashizume, *J. Appl. Cryst.* **16**, 420-427 (1983).
27. B. Cleff, *Acta Phys. Pol. A* **61**, 285-319 (1982).
28. H. Winick in Ref. 3, pp. 27-60.
29. J. G. Timothy and R. P. Madden in Ref. 4, pp. 315-366.
30. J. A. R. Samson, *Techniques of Vacuum Ultraviolet Spectroscopy*, Wiley, New York (1967).
31. G. S. Brown and S. Doniach, in Ref. 3, pp. 353-385.
32. G. Bertolini and G. Restelli, in *Atomic Inner Shell Processes II* (edited by B. Crasemann), Academic, New York (1975), pp. 123-167.

33. R. W. Fink, in *Atomic Inner Shell Processes II* (edited by B. Crasemann), Academic, New York (1975), pp. 169-186.
34. H. Cole, F. S. Chambers, and C. G. Wood, *J. Appl. Phys.* **32**, 1942-1945 (1961).
35. M. Hart and A. R. D. Rodrigues, *Philos. Mag. B* **40**, 149-157 (1979).
36. G. Materlik and P. Suortti, *J. Appl. Crystallogr.* **17**, 7-12 (1984).
37. M. Brunel, G. Patrat, F. DeBergevin, F. Rousseaux, and M. Lemonnier, *Acta Crystallogr. A* **39**, 84-88 (1983).
38. J. W. Cooper, in *Atomic Inner-Shell Processes I* (edited by B. Crasemann), Academic, New York (1975), pp. 159-199.
39. J. H. Scofield, Lawrence Livermore Rad. Lab. Report No. UCRL-51326 (1973).
40. J. B. Hastings, B. M. Kincaid, and P. Eisenberger, *Nucl. Instrum. Methods* **152**, 167-171 (1981).
41. Wm. J. Veigele, *At. Data Tables* **5**, 51-111 (1973).
42. J. H. Hubbell, Wm. J. Veigele, E. A. Briggs, R. T. Brown, D. T. Cromer, and R. J. Hoverton, *J. Phys. Chem. Ref. Data* **4**, 471-538 (1975).
43. J. H. Hubbell and Wm. J. Veigele, Technical Note 901, Nat. Bureau of Standards, Washington, DC (1976).
44. L. Sarkadi, *Nucl. Instrum. Methods* **214**, 43-48 (1983).
45. W. Jitschin, *Nucl. Instrum. Methods B* **4**, 292-295 (1984).
46. W. Jitschin, G. Materlik, U. Werner, and P. Funke, *J. Phys. B* **18**, 1139-1153 (1985).
47. P. Rabe, G. Tolkiehn, and A. Werner, *J. Phys. C* **12**, 899-905 (1979).
48. G. Doolen, private communication (1984).
49. B. Crasemann, in *X84 X-Ray and Inner-Shell Processes in Atoms, Molecules and Solids* (edited by A. Meisel and J. Finster), Karl-Marx-Universität, Leipzig (1984), pp. 51-60.
50. R. D. Deslattes and E. G. Kessler, in *X84 X-Ray and Inner-Shell Processes in Atoms, Molecules and Solids* (edited by A. Meisel and J. Finster), Karl-Marx-Universität, Leipzig (1984), pp. 165-175.
51. W. Kossel, *Z. Phys.* **1**, 119-134 (1920).
52. L. G. Parratt, *Phys. Rev.* **56**, 295-297 (1939).
53. O. Keski-Rahkonen, G. Materlik, B. Sonntag, and J. Tulkki, *J. Phys. B* **17**, L121-126 (1984).
54. N. Beatham, I. P. Grant, B. J. McKenzie, and S. J. Rose, *Phys. Scr.* **21**, 423-431 (1980).
55. J. P. Desclaux, *Phys. Scr.* **21**, 436-442 (1980).
56. M. H. Chen, B. Crasemann, M. Aoyagi, K.-N. Huang, and H. Mark, *At. Data Nucl. Data Tables* **26**, 561-574 (1981).
57. M. H. Chen, B. Crasemann, N. Mårtensson, and B. Johansson, *Phys. Rev. A* **31**, 556-563 (1985).
58. M. H. Chen, B. Crasemann, and H. Mark, *Phys. Rev. A* **24**, 1158-1161 (1981).
59. T. Åberg, *Phys. Rev.* **156**, 35-41 (1967).
60. F. Wuilleumier and M. O. Krause, *Phys. Rev. A* **10**, 242-258 (1974).
61. R. Frahm, R. Haensel, W. Malzfeldt, W. Niemann, and P. Rabe in HASYLAB Jahresbericht 1982, pp. 121-122.
62. R. D. Deslattes, R. E. LaVilla, P. L. Cowan, and A. Henins, *Phys. Rev. A* **27**, 923-933 (1983).
63. P. H. Kobrin, S. Southworth, C. M. Truesdale, D. W. Lindle, U. Becker, and D. A. Shirley, *Phys. Rev. A* **29**, 194-199 (1984).
64. D. F. Jackson and D. J. Hawkes, *Phys. Rep.* **70**, 169-223 (1981).
65. S. T. Manson and M. Inokuti, *J. Phys. B* **13**, L323-L326 (1980).
66. M. Ya. Amusia, V. K. Ivanov, and V. A. Kupchenko, *J. Phys. B* **14**, L667-L671 (1981).
67. S. Younger, unpublished, cited in Ref. 62.
68. C. Froese Fischer, *J. Comput. Phys.* **27**, 221-240 (1978).
69. S. T. Manson, in *X-Ray and Atomic Inner Shell Physics—1982* (edited by B. Crasemann), American Institute of Physics, New York (1982), pp. 321-330.

70. P. A. Heimann, T. A. Ferrett, D. A. Shirley, U. Becker, and H. G. Kerkhoff, Stanford Synchrotron Radiation Laboratory, Activity Report for 1985, edited by K. Cantwell (1985), p. IX-9.

71. P. A. Lee, P. H. Citrin, P. Eisenberger, and B. M. Kincaid, *Rev. Mod. Phys.* **53**, 769-806 (1981).

72. E. A. Stern and S. M. Heald, in Ref. 4, pp. 955-1014.

73. A. Bianconi, L. Incoccia, and S. Stipcich (eds.), *EXAFS and Near Edge Structures*, Springer Series in Chem. Phys., Vol. 27, Springer, Berlin (1983).

74. K. O. Hodgson, B. Hedman, and J. E. Penner-Hahn (eds.), *EXAFS and Near Edge Structure III*, Springer Proc. Phys. Vol. 2, Springer, Berlin (1984).

75. D. Koningsberger and R. Prins (eds.), *Principles, Techniques and Applications of EXAFS, SEXAFS and XANES*, Wiley, New York (1984).

76. G. Materlik, B. Sonntag, and M. Tausch, *Phys. Rev. Lett.* **51**, 1300-1303 (1983).

77. W. Niemann, M. Lübcke, W. Malzfeldt, P. Rabe, and R. Haensel, in HASYLAB Jahresbericht 1984, pp. 222-223.

78. D. Wohlleben and J. Röhler, *J. Appl. Phys.* **55**, 1904-1909 (1984).

79. J. Röhler, D. Wohlleben, J. P. Kappler, and G. Krill, *Phys. Lett.* **103A**, 220-224 (1984).

80. R. de L. Kronig, *Z. Phys.* **75**, 468-475 (1932).

81. D. R. Sandstrom and F. W. Lytle, *Ann. Rev. Phys. Chem.* **30**, 215-238 (1979).

82. F. W. Lytle, G. H. Via, and J. H. Sinfelt, in Ref. 3, pp. 401-424.

83. B. K. Teo and P. A. Lee, *J. Am. Chem. Soc.* **101**, 2815-2832 (1979).

84. S. T. Manson and J. W. Cooper, *Phys. Rev.* **165**, 126-138 (1968).

85. S. M. Heald and E. A. Stern, *Phys. Rev. B* **15**, 5549-5559 (1977).

86. F. W. Lytle, D. E. Sayers, and E. A. Stern, *Phys. Rev. B* **11**, 4825-4835 (1975).

87. B. K. Teo, *J. Am. Chem. Soc.* **103**, 3990-4001 (1981).

88. V. A. Biebesheimer, E. C. Marques, D. R. Sandstrom, F. W. Lytle, and R. B. Greegor, *J. Chem. Phys.* **81**, 2599-2604 (1984).

89. D. Norman, in *X-Ray and Inner Shell Physics—1982* (edited by B. Crasemann), American Institute of Physics, New York (1982), pp. 745-758.

90. P. H. Citrin, P. Eisenberger, and J. E. Rowe, *Phys. Rev. Lett.* **48**, 802-805 (1982).

91. S. A. E. Johansson and T. B. Johansson, *Nucl. Instrum. Methods* **137**, 473-516 (1976).

92. C. J. Sparks, Jr., S. Raman, E. Ricci, R. V. Gentry, and M. O. Krause, *Phys. Rev. Lett.* **40**, 507-511 (1978).

93. C. J. Sparks, Jr., in Ref. 3, pp. 459-512.

94. A. Knöchel, W. Petersen, and G. Tolkiehn, *Nucl. Instrum. Methods* **208**, 659-663 (1983).

95. W. Bambynek, B. Crasemann, R. W. Fink, H.-U. Freund, H. Mark, C. D. Swift, R. E. Price, and P. Venugopala Rao, *Rev. Mod. Phys.* **44**, 716-813 (1972).

96. J. H. Scofield, *At. Data Nucl. Data Tables* **14**, 121-137 (1974).

97. M. H. Chen, B. Crasemann, and H. Mark, *At. Data Nucl. Data Tables* **24**, 13-37 (1979).

98. Kh. R. Karim, M. H. Chen, and B. Crasemann, *Phys. Rev. A* **29**, 2605-2610 (1984).

99. Kh. R. Karim and B. Crasemann, *Phys. Rev. A* **31**, 709-713 (1985).

100. M. O. Krause, *J. Phys. Chem. Ref. Data* **8**, 307-327 (1979).

101. W. Jitschin, U. Werner, K. Finck, and H. O. Lutz, in *High-Energy Ion-Atom Collisions II* (edited by D. Berényi and G. Hock), Elsevier, Amsterdam (1985), pp. 79-95.

102. G. S. Brown, M. H. Chen, B. Crasemann, and G. E. Ice, *Phys. Rev. Lett.* **45**, 1937-1940 (1980).

103. G. E. Ice, G. S. Brown, G. B. Armen, M. H. Chen, B. Crasemann, J. Levin, and D. Mitchell, in *X-Ray and Inner-Shell Physics—1982* (edited by B. Crasemann), American Institute of Physics, New York (1982), pp. 105-113.

104. A. Niehaus, *J. Phys. B* **10**, 1845-1857 (1977).

105. B. Williams (ed.), *Compton Scattering*, McGraw-Hill, New York (1977).

106. M. Cooper, R. Holt, P. Pattison, and K. R. Lea, *Commun. Phys.* **1**, 159–165 (1976).
107. C. J. Sparks, Jr., *Phys. Rev. Lett.* **33**, 262–265 (1974).
108. M. Gavrila and M. N. Tugulea, *Rev. Roum. Phys.* **20**, 209–230 (1975).
109. J. Tulkki, in "X-84 Conference Book of Abstracts," Karl-Marx-Universität, Leipzig (1984), pp. 474–475.
110. J. M. Hall, K. A. Jamison, O. L. Weaver, P. Richard, and T. Åberg, *Phys. Rev. A* **19**, 568–578 (1979).
111. T. Åberg and J. Tulkki, in *Atomic Inner-Shell Physics* (edited by B. Crasemann), Plenum Press, New York (1985), pp. 419–463.
112. J. P. Briand, D. Girard, V. O. Kostroun, P. Chevalier, K. Wohrer, and J. P. Mossé, *Phys. Rev. Lett.* **46**, 1625–1628 (1981).
113. D. Schaupp, H. Czerwinski, F. Smend, R. Wenskus, M. Schumacher, A. H. Millhouse, and H. Schenk-Strauss, *Z. Phys. A* **319**, 1–7 (1984).
114. H. Czerwinski, F. Smend, D. Schaupp, M. Schumacher, A. H. Millhouse, and H. Schenk-Strauss, *Z. Phys. A* **322**, 183–189 (1985).
115. Y. R. Shen, *Phys. Rev. B* **9**, 622–626 (1974).
116. J. R. Solin and H. Markelo, *Phys. Rev. B* **12**, 624–629 (1975).
117. J. Tulkki, *Phys. Rev. A* **27**, 3375–3378 (1983).
118. P. Eisenberger, P. M. Platzman, and H. Winick, *Phys. Rev. Lett.* **36**, 623–626 (1976).

8

Analysis and Spectroscopy of Collisionally Induced Autoionization Processes

R. MORGENSTERN

1. Introduction

When light with variable wavelength is shined on atoms, sharp variations of the absorption cross section in the ionization continuum can indicate the existence of stationary atomic states in this energy range. An absorption spectrum of Xe atoms as measured by Beutler[1] in 1935 is shown in Figure 1. It exhibits characteristic structures at photon energies that are sufficient for the ionization of one electron, but not for the excitation or ionization of a second one. These structures are due to quasistationary states, which are embedded in the ionization continuum. Such states can also be excited by other methods such as electron or ion impact. They can decay in two competing ways: by emission of either a photon or an autoionization electron. The autoionization process is induced by the Coulomb interaction between two electrons, which causes an exchange of energy such that one of them is ejected. This process can be described by the matrix element

$$A^\rho = \langle \mathbf{k} \cdot \psi_{\text{ion}} | V_c | \psi^\rho \rangle \tag{1}$$

Here $V_c = \sum_{ij} e^2/(r_i - r_j)$ is the autoionization operator and ψ^ρ, ψ_{ion}, and

R. MORGENSTERN • Kernfysisch Versneller Instituut, Rijksuniversiteit Groningen, Zernikelaan 25, 9747 AA Groningen, The Netherlands.

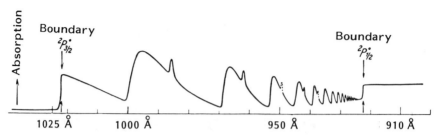

Figure 1. Absorption spectrum of Xe as measured by Beutler[1] at photon energies above the first ionization limit. (From Ref. 1.)

k describe the excited atom, the remaining ion, and the emitted electron, respectively. The transition matrix element (1) corresponds to typical autoionization times of 10^{-12}–10^{-15} sec. This is very short compared with typical decay times for photon emission of $\sim 10^{-9}$ sec. Therefore the decay by autoionization is in most cases strongly favored over the photon decay mode. This implies that electron spectroscopy is often the best suited method to investigate the decay of such highly excited atoms.

Distributions of ejected electrons on the one hand and of ejected photons on the other hand contain similar information, and in this sense electron spectroscopy of autoionizing states can be regarded as analogous to photon spectroscopy of "normal" excited atomic states. However, there is more about electron spectroscopy: (i) owing to the short lifetimes the ejected electrons often carry information about a short-lived complex as, e.g., a quasimolecule, formed during a collision, whereas photon emission mostly occurs from the separated collision partners long after the collision; (ii) the de Broglie wavelength of the ejected electrons is much shorter than typical photon wavelengths. Therefore electron distributions are more sensitive to structural details of such collision complexes than photons; (iii) angular distributions of ejected electrons often contain more structure than those of ejected photons, since the angular momentum of the emitted electrons is not limited to $l = 1$, as is the case for photons due to dipole radiation.

In the last few years a large number of experimental and theoretical investigations has been devoted to the study of autoionizing atoms. First of all a proper description of the correlated motion of the excited electrons in the quasistationary states is interesting in itself. Secondly one can obtain information about dynamical excitation mechanisms during electronic or atomic collision processes, if one succeeds in identifying the collisionally excited states and the magnetic substates. Also charge transfer processes in collisions between highly charged ions and atoms can be investigated by observing the population of autoionizing states. Charge transfer into such states is in fact a process that occurs with a large cross section and which is to a large extent responsible for the formation of a charge equilibrium in a plasma. However, to obtain dependable information about the population of autoionizing states one needs in the first place a detailed understanding of the decay process, which is often strongly influenced by the so called postcollision interaction (PCI). Without such an understanding it is not possible to deduce the full information about the excited atoms that is contained in the distributions of ejected electrons or scattered projectiles. Therefore PCI effects have been studied intensively during recent years.

In this report autoionization processes and various aspects of electron spectroscopy will be discussed. It will especially concentrate on autoionization processes induced by ion or atom collisions. A more general discussion of atomic compound states was given earlier by Williams in Chapter 23 of Part B of *Progress in Atomic Spectroscopy*.

2. Description of Autoionizing States

2.1. The Independent Particle Model

In a straightforward approach to describe an autoionizing atom one can use an independent particle model, in which each electron moves in the average potential created by the other electrons. In this approach the electron–electron interaction is averaged and the wave functions do not depend on the instantaneous positions of the electrons, i.e., electron correlations are not taken into account. In the Hartree–Fock self-consistent field method[2] this approach is used for the determination of electronic wave functions. For autoionizing states with only one excited electron—e.g., $Ar(3s3p^64s)$—this approach is certainly as justified as for other singly excited states with energies below the ionization limit. The situation is different for doubly excited states of the type (core) $(nl, n'l')$, e.g., $He(2s2p)$ in which the correlated motion of the two excited electrons has a large influence on the state. Although there are much more refined descriptions of autoionizing states that take correlation effects into account, it is common practice in the literature to use the independent particle model to identify autoionization states.

Read has shown[3] that the energies of such autoionization states can be described to a good approximation by a modified Rydberg formula, in which electron correlation is reflected in a phenomenological way. In this approach the energies are given by

$$E_{nlnl'} = I - \frac{R(Z - \sigma)^2}{(n - \delta_{nl})^2} - \frac{R(Z - \sigma)^2}{(n - \delta_{nl'})^2} \tag{2}$$

where I is the energy of the core configuration and Z is the charge of the core. The quantum defect δ_{nl} can to a good approximation be taken from the singly excited parent configuration (core) (nl). The screening σ depends on the amount of correlation between the two excited electrons. For the case of a state (core)$(ns^2\,{}^1S)$, for example, the screening is exactly $\sigma = 0.25$ if the two electrons behave completely correlated, being always on opposite sides of the core, and $\sigma = 0.5$, if they behave completely uncorrelated. Read[3] has shown that states of this type can for a large number of atoms be described by Eq. (2) with values for σ between 0.23 and 0.27. This corresponds to maximum correlation and has to be compared with the values to be expected from "Slater's rules",[4] namely, $\sigma \approx 0.35$ for $1s$ electrons and $\sigma \approx 0.30$ for other values of n. For other states with the configuration (core)$(nlnl')$ the different couplings of the electrons yield a larger spread of σ, which then lies between 0.22 and 0.37. However, these numbers clearly indicate that the motion of the two excited electrons occurs in a strongly correlated way. This also implies that the coupling of the excited electrons to each other is more important than their coupling to the core. For the description of doubly excited states, therefore, an outer coupling scheme is often appropriate, which can be written as (core)$L'[(nl, n'l')L'']^{2S+1}L$. The so called "aufbau" coupling scheme, written as (core)$L'(nl)L''(n'l')^{2S+1}L$, is the more appropriate one for cases in which the two excited electrons have very different binding energies and therefore are less correlated.

2.2. Correlated Electron Motion

For the simplest autoionizing system, i.e., the doubly excited He atom, refined descriptions of the excited states have been developed. Since the first observation of He autoionization states in photon absorption spectra by Madden and Codling[5,6] a lot of experimental investigations have been performed in which He autoionization states were excited by fast[7-10] or slow[11-13] electrons or by ions or atoms.[14-22] Therefore detailed empirical data are available. The first attempt to take electron correlation into account in the description of these states was made by Cooper et al.[23] They realized that the near degeneracy of the levels $(2snp)$ and $(2pns)$ causes a strong configuration interaction leading to the formation of composite levels of the type $(2snp \pm 2pns)$. These states are noted as $(sp2n+)$ and $(sp2n-)$. Although certain aspects—e.g., the different decay widths of these states—were properly described, Macek[24] showed that such a simple configuration mixing did not fully account for the properties of the observed series of autoionizing states. Therefore Macek introduced the concept of hyperspherical coordinates,[25] which was subsequently extended[26-31] especially by Fano and co-workers. This concept proved to be very successful in the description of doubly excited states. The hyperspherical coordinates are especially suited to visualize the electron correlation: the radial correlation of the two electrons with position vectors $\mathbf{r}_1$ and $\mathbf{r}_2$ is described by a variable $\alpha = \arctan(r_1/r_2)$, the angular correlation is described by the angle θ_{12} formed by the position vectors, and a coordinate $R = (r_1^2 + r_2^2)^{1/2}$ describes the overall size of the electron cloud. Macek[25] showed that the Schrödinger equation is approximately separable in these coordinates. The coordinate R only varies slowly in time and therefore resembles to some extent the internuclear distance in the description of diatomic molecules. By using the hyperspherical coordinates it is possible to determine correlated wave functions for the doubly excited states. To illustrate the pattern of these correlated wave functions Lin[32] has drawn instructive plots of the corresponding charge densities. Figure 2 shows the electron density of $He(2s3s)^1S$ as a function of r_1 and r_2. In this graph the dependence of the wave function on θ_{12} has been neglected, but there are additional graphs that also illustrate the angular correlation of the electrons. Greene[33] finally has applied the method of hyperspherical coordinates to systems in which two electrons are around a closed-shell ionic core instead of a bare nucleus. By this autoionization states of alkaline earth atoms can be treated.

A quite different approach for the characterization of doubly excited states was given by Kellman and Herrick.[34-37] They realized the similarities between a linear triatomic molecule of the type A–B–A and a system of two electrons moving about their core in a strongly correlated way. Just as in the A–B–A molecule the movement of one atom A about the center atom B is not characterized by constants of motion, so also the characterization of the electron motion about the core by independent quantum numbers $(nl, n'l')$ is questionable. Herrick and Kellman suggested characterizing doubly excited states in analogy to the vibrational and rotational modes of an A–B–A molecule and arranged the known states of doubly excited He into new "supermultiplets." Yuh et al.[38] have investigated charge distributions in doubly excited He by using accurate wave functions and have

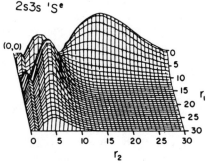

2s3s $^1S^e$

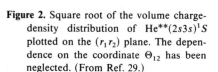

Figure 2. Square root of the volume charge-density distribution of $He^{**}(2s3s)^1S$ plotted on the $(r_1 r_2)$ plane. The dependence on the coordinate Θ_{12} has been neglected. (From Ref. 29.)

found that they indeed resemble to a large extent the stretching and binding modes of an A–B–A molecule. This strongly supports the Kellman–Herrick quantization.

Correlation between the excited electrons is important not only for the formation of quasistationary autoionizing states but also for their decay. Rehmus and Berry analyzed the mechanism of autoionization[39] and found that mainly two processes contribute to the decay: one is due to two-body collisions of the electrons and depends strongly on the correlation between the electrons; the other can be imagined as being due to a screening of the nucleus by one electron, allowing the other electron to escape. Both contributions can add constructively or destructively to yield the effective decay width. However, the second process becomes less important if none of the electrons has an appreciable probability of being close to the nucleus. Therefore, with increasing n, l values for the excited electrons the correlation-dependent contribution to the autoionization amplitude becomes the most important one. Consequently the proper consideration of electron correlation plays an important role in the calculation of autoionization lifetimes.[40–43]

3. Experimental Methods

Since the photoabsorption measurements of Madden and Codling[5,6] autoionization processes have been investigated by various methods. Excitation has been initiated by electrons,[7–13,44–48] by heavy particles,[14–22,49–56] or by beam-foil interaction.[57] Whereas the number of states that can be excited by photon impact is limited by selection rules, this limitation is less stringent for electron collisions, especially at low impact energies. For ion–atom or atom–atom collisions it is possible to provoke or suppress the excitation of certain types of autoionization states by careful selection of the collision partners.[52,53]

For an analysis of the induced ionization processes either the absorption or scattering of the projectiles has been observed or the ejected electrons or the resulting ions have been detected.[44] In some cases also secondary processes were used for the analysis, such as fluorescence from Rydberg states, which were populated indirectly via an autoionization state.[47,48] In photon-absorption measurements excellent resolution can be obtained, which makes it possible to determine line shapes, linewidths, and line positions with high accuracy. In measurements of particle-beam attenuation or scattering, on the other hand, the accuracy is often limited by the energy spread of the projectile beam. The state of electron-scattering experiments was recently reviewed by Read[58]: the best energy width that can now be obtained in electron beams from thermal emitters is ~ 8 meV[59]; more typical values, however, are 20–50 meV. Ion and atom beams have energy widths that are typically larger than ~500 meV. Nevertheless, good energy resolution can be obtained by analysis of the ejected electrons. Typical widths of several He autoionization states are between 40 and 140 meV,[40-43] and this can easily be resolved by electron spectroscopy.

In heavy-particle collisions it is often not the resolving power of the electron spectrometer that limits the accuracy with which structures in ejected electron spectra can be measured, but the Doppler effect: broadenings and shifts of the electron peaks severely influence the spectra and have to be taken into account in the analysis, or to be avoided by special means in the experimental setup. Since electron spectra due to heavy-particle collisions are treated in some detail in the following sections, we will briefly discuss the various kinematical effects.

3.1. Kinematical Effects

An electron, emitted with a velocity $\mathbf{v}_e$ with respect to the emitter, which in turn moves at a velocity $\mathbf{V}$ in the laboratory, will have a velocity $\mathbf{v} = \mathbf{v}_e + \mathbf{V}$ and a corresponding energy of

$$\varepsilon = \varepsilon_e \pm 2(\cos \gamma)\{E[\varepsilon_e - (\sin^2 \gamma)Em/M]m/M\}^{1/2}$$
$$+ (\cos 2\gamma)Em/M \tag{3}$$

where E is the kinetic energy of the emitter, γ is the angle between $\mathbf{v}_e$ and $\mathbf{V}$, and m and M are the masses of electron and emitter, respectively. The energy in the emitter frame can be determined by

$$\varepsilon_e = \varepsilon - 2(\cos \gamma)[E\varepsilon m/M]^{1/2} + Em/M \tag{4}$$

If electrons are emitted from atoms in fast beams their energy can be drastically changed because of this Doppler shift. At observation angles $\gamma \approx 90°$ this shift becomes small; however, the variation of ε with the observation angle is strongest, and together with the angular acceptance of an electron detector this leads to a kinematical broadening of electron emission lines, which is just at its maximum in this angular range. The broadening can be reduced by measuring at detection angles $\gamma \sim 0°$ or $\gamma \sim 180°$, as was done, for example, by Morgenstern et al.[60] Another way to reduce this Doppler broadening is a special design of the electron spectrometer, which was recently reported by Bachmann et al.[61]: the angle-dependent electron energy is taken into account in the focusing properties of the spectrometer and allows decreasing the peak widths in electron spectra from fast ion beams, without having to reduce the angular acceptance angle.

An additional broadening of electron peaks is caused by the fact that, after a collisional excitation process, the electron-emitting atoms do not move into one well-defined direction. If they are scattered (or recoiled) at an angle (θ, ϕ) with respect to the projectile beam direction, and the electrons are detected at (ϑ, φ), Eq. (3) can be written more precisely as

$$\varepsilon = \varepsilon_e + a + b \cos(\varphi - \phi) + c \cos^2(\phi - \varphi) \tag{5}$$

with

$$a = \pm 2(\cos \vartheta \cos \theta)\{E[\varepsilon_e - (\sin^2 \gamma)Em/M]m/M\}^{1/2}$$
$$+ E(2 \cos^2 \theta \cos^2 \vartheta - 1)m/M \tag{6}$$

$$b = 2(\sin \vartheta \cos \theta)\{E[\varepsilon_e - (\sin^2 \gamma)Em/M]m/M\}^{1/2}$$
$$+ 2E(\cos \vartheta \cos \theta)m/M \tag{7}$$

$$c = 2E(\sin^2 \vartheta \sin^2 \theta)m/M \tag{8}$$

Gordeev and Ogurtsov[69] were the first ones to point out that the term $b \cos(\varphi - \phi)$ in Eq. (5) is responsible for a considerable peak-broadening: if there is no selection of the azimuth angle ϕ of the atoms, all ϕ will occur with the same probability and the shape of the peaks of the energy scale, for the case $|v| \gg |V|$, is given by

$$I(\varepsilon) \propto I(\varphi - \phi)[b^2 - (\varepsilon - \varepsilon_e - a)^2]^{-1/2} \tag{9}$$

For constant $I(\varphi - \phi)$ this is a doubly peaked shape with its center at $(\varepsilon + a)$ and singularities at energy separations $\pm b$ from this center. Actually measured peaks consist of superpositions of such shapes with different b, depending on θ as given by (7).

Structures in spectra of electrons due to low-energy heavy-particle collisions are mainly broadened by this effect. Remarkably enough the "width" b increases when the collision energy decreases, because then the inelastic processes occur at larger scattering angles θ. With an approximately constant reduced scattering angle $E\theta$ for the inelastic processes[63] one obtains

$$b \propto \sin \theta \sqrt{E} \propto \text{const}/\sqrt{E} \tag{10}$$

This broadening can be avoided either by detecting the electrons at angles $\vartheta \approx 0°$ or $\vartheta \approx 180°$ or by selecting certain azimuth scattering angles ϕ of the atoms in the experiment, e.g., by a coincidence measurement.

Finally also the electron intensity $I(\varepsilon, \vartheta)$ as measured in the laboratory is different from the one in the emitter frame. Total peak intensities are related by

$$I(\gamma) = I_e(\gamma_e) \frac{d\Omega_e}{d\Omega} = \frac{\varepsilon}{\varepsilon_e}\left(1 - \frac{m}{M}\frac{E}{\varepsilon}\sin^2\gamma\right)^{-1/2} \approx \frac{\varepsilon}{\varepsilon_e} \qquad (11)$$

whereas the energy-resolved intensities are related by

$$I(\gamma, \varepsilon) = I_e(\gamma_e\varepsilon_e) \frac{d\Omega_e}{d\Omega}\frac{d\varepsilon_e}{d\varepsilon} = I_e(\gamma_e\varepsilon_e)(\varepsilon/\varepsilon_e)^{1/2} \qquad (12)$$

It is worth noticing that this relation does not explicitly contain any angle.

The electron emission amplitudes $C(\varepsilon, \vartheta, \varphi)$ valid in the laboratory frame are then connected to the $B(\varepsilon_e\vartheta_e\varphi_e)$ valid in the emitter frame by

$$C(\varepsilon, \vartheta, \varphi) = B(\varepsilon_e\vartheta_e\varphi_e)(\varepsilon/\varepsilon_e)^{1/4} \qquad (13)$$

The various kinematical effects can best be observed in energy spectra of electrons due to collision processes with well-defined kinematics. Figure 4, which will be discussed in the following section, shows an example.

More exhaustive descriptions of kinematical effects have been given by various groups in recent years.[64-66]

3.2. Coincidence Experiments

The most detailed information about collisionally excited autoionizing states can be obtained from coincidence measurements. First of all kinematical broadenings of structures in the spectra are largely reduced when the collision kinematics are well defined. More important, however, is the increase of information to be obtained from electron angular distributions which are not averaged over all orientations of the scattering plane. The rotational symmetry about the projectile beam direction in a noncoincidence experiment is replaced by the mirror symmetry with respect to the scattering plane in a coincidence experiment. This makes it possible to obtain information about coherences in the collisional population of different magnetic sublevels, as will be discussed in Section 6.

Up to now only a few experiments were performed in which autoionization electrons were analyzed in coincidence with scattered electrons[67-71] or scattered ions.[72-74] Figure 3 shows an experimental setup as used in our group. Electrons ejected at a variable angle ϑ with respect to the projectile beam direction are analyzed in coincidence with ions, scattered through (θ, ϕ). By fixing the scattering angle θ and the azimuthal orientation

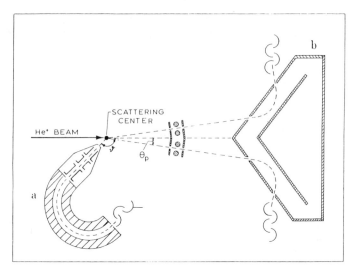

Figure 3. Experimental setup for the measurement of autoionization electrons in coincidence with scattered ions. a, electron spectrometer; b, ion spectrometer.

of the scattering plane ϕ, the kinematics is completely determined. Therefore kinematical peak broadenings are largely reduced in the corresponding spectra.

Figure 4 shows as an example a comparison of a noncoincident electron spectrum with coincident ones, due to He^+–He collisions. Electrons are due to excitation and decay of $He(2p^2)^1D$ in either the target atom or the charge exchanged projectile. Spectra taken at $\varphi = 0°$ and $\varphi = 180°$, i.e., with the electron detector in the scattering plane, show the kinematical shifts of target and projectile contributions as described by the term $b \cos(\varphi - \phi)$ in Eq. (5). Also the broadening of the electron peak due to projectile emission, which is caused by the finite acceptance angle of the electron detector, can be seen.

4. Line Shapes and Interference Effects in Autoionization Spectroscopy

Spectra of autoionization lines are often very different from normal optical spectra in that the line shapes are strongly influenced by interference effects. Symmetric line shapes on a low background signal are the exception. Two different types of processes can contribute to the ionization signal: (i) the direct ionization which leads to a continuous electron spectrum and which is especially important when projectiles with high velocities or photons are initiating the ionization process; (ii) excitation and decay of

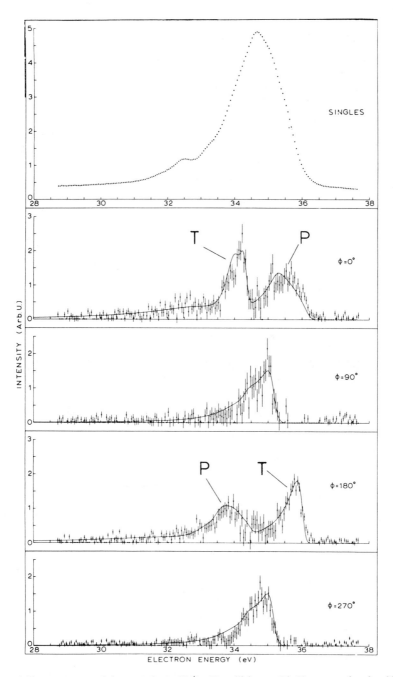

Figure 4. Energy spectra of electrons due to He$^+$ + He collisions at 2 keV, measured at $\vartheta = 90°$. On top, a noncoincident spectrum; below, spectra of electrons measured in coincidence with ions at azimuth angles $\phi = 0°$, $90°$, $180°$, and $270°$, respectively. T and P indicate that peaks are mainly due to emission from target or projectile.

autoionization states. If initial and final states in both cases are identical, the corresponding ionization amplitudes have to be added coherently and will interfere with each other. Also, ionization signals from different states may overlap each other and therefore have to be added coherently. The ionization signal can therefore be written as

$$I(\varepsilon, \vartheta, \varphi) = \sum_{\mu_i} w(\mu_i) \sum_{\mu_f} \left| C_d + \sum_{p} C^p(\varepsilon, \vartheta, \varphi) \right|^2 \qquad (14)$$

Here a summation over final states μ_f and an averaging over initial states μ_i with a weighting factor $w(\mu_i)$ has been performed. C_d is a direct and C^p are autoionization amplitudes, which depend on the energy ε (of impinging projectiles or ejected electrons) and detection angles (ϑ, φ) determined by the experiment. The energy-dependent relative phases of the complex amplitudes C' determine to a large extent the shape of the observed lines.

4.1. Direct Ionization and Autoionization

Figure 5 shows a typical energy spectrum of electrons characterized by the interference of direct and autoionization processes. In a more explicit

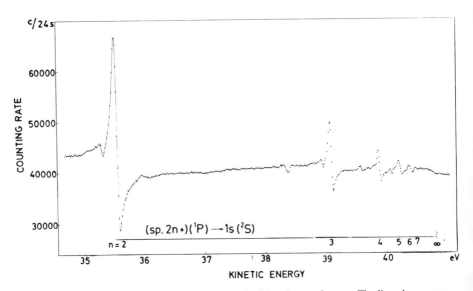

Figure 5. Autoionization spectrum from He, excited by electron impact. The line shapes are strongly influenced by interferences between direct ionization and autoionization. (From Ref. 10.)

way the line shapes can be described by

$$I = \left| |C_d| + |C^p| \frac{\exp(i\chi)}{\hat{\varepsilon} + i} \right|^2 \tag{15}$$

where $\hat{\varepsilon} = 2\tau(\varepsilon - \varepsilon_0)$ is a reduced energy, giving the separation between the measured electron energy ε and the resonance position ε_0 in multiples of the resonance-width $\Gamma = 1/\tau$, and χ is a relative phase. The phase of the autoionization amplitude changes by π when passing the resonance position. An often used parameterization has been given by Fano[75,76]

$$I = I_d + I_a \frac{(q + \hat{\varepsilon})^2}{1 + \hat{\varepsilon}^2} \tag{16}$$

where I_d and I_a are proportional to the cross sections of the direct and autoionization processes, respectively. Here the shape of the measured resonance is characterized by the parameter q.

In Fano's parameterization it is implied—differently from (14) and (15)—that autoionizing transitions from different states do not interfere with each other. The same is true for a parameterization given by Shore.[77] Except for this, however, the parameterizations are equivalent and the parameters are related by simple formulas. Autoionization spectra due to photoionization[5,6] as well as collisional ionization with fast electrons,[7-10] ions, and atoms,[19-22,55,56] have successfully been analyzed by means of Fano's parameterization. In collisions with low-velocity projectiles, however, other effects become important that influence the relative phases of the ionization amplitudes and thus the spectral shapes. Also interferences between different autoionization amplitudes have to be taken into account.

4.2. Postcollision Interaction (PCI) in Ion–Atom Collisions

If ionization processes are induced by ions at low velocities ($V \ll 1$ a.u.), direct ionization is negligible. In these cases the shapes of spectral lines, e.g., in ejected electron spectra, are mainly determined by the so-called "postcollision interaction" (PCI). As the name indicates, the ionization process is regarded as taking place in two steps: (i) the excitation of the autoionizing state, and (ii) the subsequent decay of this state, which is influenced by the interaction with the slowly receding collision partner. This interaction of the autoionizing atom with the nearby ion can influence energy and angular distribution of the ejected electrons as well as of the scattered projectiles.

This effect was first discussed by Barker and Berry[15] in connection with autoionization electron spectra from He$^+$–He collision. In a potential curve diagram as given in Figure 6a the autoionization process corresponds to a transition from the He** + He curve to the He$^+$–He$^+$ Coulomb curve. While transitions at infinite internuclear distances would yield an electron energy ε_0, decay processes at finite R lead to lower energies ε. With a constant decay time τ one obtains an electron distribution

$$I(\varepsilon, \vartheta, \varphi) = \frac{\sigma G(\vartheta \varphi)}{V\tau(\varepsilon_0 - \varepsilon)} \exp\left[-\varepsilon / V\tau\varepsilon_0(\varepsilon_0 - \varepsilon)\right] \tag{17}$$

with V the relative velocity of the collision partners, σ the population of the observed state, and $G(\vartheta\varphi)$ the angular distribution of the ejected electrons. The resulting spectral shape is shown in Figure 6b. It has a "width" of $1.07/V\tau$ and its maximum is shifted by $1/2V\tau$ with respect to ε_0. Since the width is larger than the natural linewidth by a factor of $1.07/V$, one can expect this effect to be important as long as the relative velocities of the collision partners are lower than one atomic unit. In the electron spectra of Barker and Berry[15] this effect was masked by kinematical effects and low resolution of the spectrometer, but the validity of the ideas was demonstrated later on.[51]

For collisions with highly charged ions this effect becomes even more pronounced. Strongly PCI-influenced line shapes with widths on the order of 1 eV were observed, for example, in energy spectra of autoionization electrons from Ne^{+**}, excited by Ne^{3+}–H$_2$ collisions.[78] In this case an

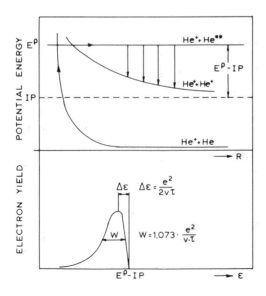

Figure 6. (a) Schematic potential energy diagram for the explanation of postcollision interaction. Autoionization processes at intermediate internuclear distances yield electrons at energies lower than the nominal energy. (b) Resulting line shape (schematically).

analysis by means of Eq. (17) allowed a determination of the Ne^{+**} lifetime.

Another peculiarity of the line shape described by Eq. (17) is its slope on the high energy side, which is steeper than that of the Lorentzian line shape corresponding to the natural lifetime τ of this state. Qualitatively this can be understood as being due to the fact that autoionization electrons with nearly the nominal energy ε_0 are due to transitions at large internuclear distances. For velocities lower than 1 a.u. the atom must have survived for longer than the natural lifetime τ to reach these distances and therefore a line sharpening effect occurs. In ion-atom collisions this effect has not yet been observed owing to other limitations of the effective resolution; however, in electron scattering experiments this effect has been measured and will be discussed in Section 4.4.

In an energy range where PCI broadened electron peaks due to autoionization from different states overlap each other, it cannot in principle be distinguished which state has been excited intermediately. Therefore the corresponding autoionization amplitudes C^p have to be superimposed coherently. Figure 7 shows an energy spectrum of electrons due to $Li^+ + He$ collisions[79] which between 30 and 36 eV has more peaks than there are autoionization states in the corresponding energy range. These peaks can be explained in terms of interferences between different autoionizing transitions.

Morgenstern *et al.*[60,80,81] have given a semiclassical description of the excitation and decay process, which is again based on the potential curve diagram of Figure 6. The transition amplitudes B^p in the emitter frame—which are connected to the C^p in the laboratory frame by Eq. (13)—are in

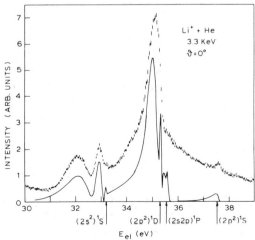

Figure 7. Autoionization electron spectrum from $Li^+ + He$ collisions. The structures are due to interferences of autoionizing transitions from different states. Solid line: fit calculation to reproduce the structures of the measured spectrum.

this description given by the general expression

$$B^p(\varepsilon, \vartheta, \varphi) = \int_0^\infty A^p(\vartheta, \varphi) \exp\left[-i \int_0^t [E^p - E^f(\varepsilon)] \, dt'\right] dt \quad (18)$$

Here $A^p(\vartheta, \varphi)$ represents the transition matrix element given in Eq. (13), which can be written as

$$A^p(\vartheta, \varphi) = \langle \mathbf{k}, \psi_{\text{ion}}| V_c |\psi^p\rangle = \|A^p\| f^p g^p(\vartheta, \varphi)(2L + 1)^{-1/2} \quad (19)$$

with f^p the population amplitude of the excited state, L its angular momentum, $g^p(\vartheta, \varphi)$ the angular electron distribution, and $\|A^p\|$ a reduced matrix element connected with the lifetime τ^p by

$$2\pi \|A^p\|^2 (2L + 1)^{-1} = (\tau^p)^{-1} \quad (20)$$

With a complex potential $E^p = \varepsilon_0^p - i/2\tau^p$ for the excited state (to allow for the decay) and a pure Coulombic potential $E^f(\varepsilon) = \varepsilon + 1/R$ for the final state, Eq. (18) can be evaluated by means of the stationary phase method, which yields

$$B^p(\varepsilon, \vartheta, \varphi) = f^p g^p(\vartheta, \varphi) \frac{\exp\left[-\varepsilon/2V\tau^p \varepsilon_0^p(\varepsilon_0^p - \varepsilon)\right]}{(\varepsilon_0^p - \varepsilon)(V\tau^p)^{1/2}} \exp i\varphi^p \quad (21)$$

where φ^p is the phase at the point of stationary phase, which is given by

$$\varphi^p = -[1 - (\varepsilon_0^p - \varepsilon)X + \ln(\varepsilon_0^p - \varepsilon)X]/V \quad (22)$$

with X the distance of closest approach of the collision partners, at which the excitation is supposed to take place. Since the phase of the transition amplitudes varies differently with the electron energy ε for the different states, the spectrum is characterized by interference structures in those regions where contributions from different states overlap. Figure 7 shows also a comparison of the measured spectrum with a calculated one in which the interference structures are fitted with the semiclassical formulas given above. The spectrum is dominated by interferences between autoionizing transitions from the doubly excited states $\text{He}(2p^2)^1 D$, $\text{He}(2s,2p)^1 P$, and $\text{He}(2s^2)^1 S$.

There is a strong analogy between these interference structures and the so-called quantum beats[82] that are observed in photon emission studies when atoms are coherently excited into different states by beam-foil, laser excitation, or other methods. Both phenomena are due to time-dependent interferences of transitions from different states. In quantum-beat studies the time scale is in most cases transformed into a scale of distances that the (fast) atoms have traveled since the excitation event. Typically quantum beat frequencies of $10^9 \sec^{-1}$ can be observed by that method. In case of PCI interferences the time scale is transformed into an energy scale of ejected electrons, and electron-beat frequencies of typically $\sim 10^{15} \sec^{-1}$ are observed by this method.

4.3. PCI in Electron–Atom Collisions

PCI effects can also be induced if an electron rather than an ion is the slowly receding collision partner. In this case the slow electron partly shields the charge of the remaining ion and therefore the ejected autoionization electron has a higher energy. However, whereas ions are "slow enough" ($V < 1$ a.u.) to induce such effects even at keV energies, electron energies have to be close to the excitation threshold of the autoionizing states. Electron collisions at threshold energies have been studied by several groups and an influence of PCI has been observed in the spectra of ejected or scattered electrons[12,45,83–90] as well as in the excitation functions of Rydberg states, which were excited by electron collisions.[45–48]

As can be expected from the classical model, one observed broadenings of the structures due to autoionization and shifts to higher energies. However, the classical model obviously could not explain all the observed features. Also an improved version of this model[91] did not give satisfactory agreement with the experiments and it was suspected that the observed disagreements reflected the inadequacy of a classical description for electrons at low energy. Nevertheless it turned out that within a semiclassical model as given by Morgenstern et al.[81] a quantitative explanation of the experimental results was possible. The model is based on a potential curve diagram as shown in Figure 8. The impinging electron with energy ε_i excites the state A^{**} with energy E^ρ and moves on slowly with an energy ε_s to larger separations R from A^{**}. If A^{**} decays at a distance $R^\rho(\varepsilon)$, the fast ejected autoionization electron will have an energy ε larger than the nominal

Figure 8. Schematic potential curve diagram for the explanation of postcollision interaction in electron–atom collision. Decay at a distance R^ρ between scattered electron and excited atom will yield an electron energy $\varepsilon > \varepsilon_0$. Subsequently the system follows the Coulomb curve, yielding a kinetic energy $-D$ for the scattered electron. Decay at smaller R^ρ will eventually lead to a trapping of the scattered electron.

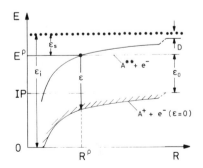

energy ε_0. The slow electron, on the other hand, finds itself in the Coulomb field of A^+ and is decelerated. Figure 8 shows a case in which the scattered electron remains unbound with an asymptotic energy $-D$. For decay at distances R below a certain value the scattered electron will find itself trapped in the field of A^+, forming a bound excited state A^* with a binding energy D. In this picture it is clear that the R-dependent transition amplitudes for autoionization vary smoothly between the cases $D > 0$ and $D < 0$. This implies that similar structures occur in the excitation functions of Rydberg states and in the spectra of scattered or ejected electrons, when an autoionization state is formed intermediately, which contributes to the measured signal. The quantitative treatment is analogous to that of ion–atom collisions except that now the deceleration of the slow electron in the final state has to be taken into account. This yields another expression for the phase φ^ρ at the point of stationary phase, which is now given by

$$\varphi^\rho = (2/D)^{1/2}\{\arctan\left[(\varepsilon^\rho/D)^{1/2}\right] - \pi/2\} \tag{23}$$

for $D > 0$, i.e., for the formation of bound states, and

$$\varphi^\rho = (-2D)^{-1/2}\ln\left[(\sqrt{\varepsilon_s^\rho} - \sqrt{-D})/(\sqrt{\varepsilon_s^\rho} + \sqrt{-D})\right] \tag{24}$$

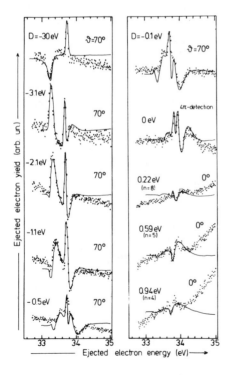

Ejected electron yield (arb. un.)

Ejected electron energy (eV) →

Figure 9. Comparison between measured electron spectra and calculated ones, using the semiclassical model.[81] Spectra with $D < 0$ correspond to ionization, leaving the scattered electron with a kinetic energy of $-D$. Spectra with $D > 0$ correspond to excitation of states with a binding energy D, which partly occurs via autoionizing states. (From Ref. 81.)

for $D < 0$, i.e., for the case of collisional ionization. (These formulas are simpler than those given in[81] but practically identical: here the excitation is supposed to take place at $R = 0$.) The phase φ^ρ varies smoothly when D changes from positive to negative values.

Figure 9 shows electron spectra measured by Hicks et al.[83] and by Smith et al.[92] for electron He collisions. They were obtained in the "constant-energy-loss" mode. The energy loss was adjusted such that D varied from -30 eV to $+0.94$ eV, corresponding to cases where the scattered electrons could escape with a kinetic energy of 30 eV or were trapped with a binding energy of 0.94 eV, respectively. Also shown are calculated curves obtained from a fit with the semiclassical formulas in which complex amplitudes for the population of the involved autoionization states $He(2s^2)^1S$ and $He(2s2p)^3P$ and for the direct ionization process were taken as fit parameters.[81] The interference structures and the PCI-induced high-energy tails of the autoionization peaks are well reproduced in this description. Also other types of measurements, e.g., excitation functions of Rydberg states, can be interpreted with the semiclassical formulas and in this way one can obtain quantitative information about the population of the various autoionizing states and the importance of the direct process.

An advantage of the semiclassical description for such an analysis is the fact that only simple analytical expressions are involved which can easily be evaluated numerically without the need of much computer time. On the other hand, the semiclassical model is not the only one that yields a proper description of the PCI effects. King et al.[45] developed the so-called "shake-down model" in analogy to the Auger effect. In this model it is assumed that the interaction between the ejected and the scattered electron causes a sudden transition of the scattered electron from its free state to the final bound or continuum state and a corresponding transition of the ejected electron to a state with higher energy. The transition amplitude is calculated as an overlap integral of the initial- and final-state wave functions. At first glance the semiclassical model with its time-dependent transition amplitudes and the shake-down model with its overlap integrals in R-space seem to be quite different approaches to the problem. However, it turns out that they are mathematically equivalent, as was pointed out by Niehaus.[93] This was also found later on by van de Water and Heideman,[94] who performed an exhaustive comparison of these two models. The relation between the "shake-down" model and the rigorous quantum mechanical treatment was discussed by Nienhuis and Heideman.[95] A comparison of measured spectra with calculated ones[96] using the shake-down model is shown in Figure 10. The excellent agreement between measured and calculated spectra is partly at the expense of a higher number of fit parameters which have been varied in this case: Read[96] has pointed out that one cannot assume a complete coherence between transitions from states with

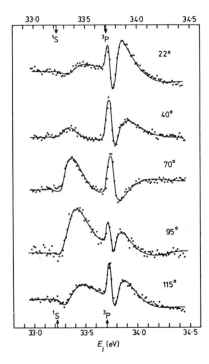

Figure 10. Autoionization electron spectra due to electron–He collisions at 60.2 eV, obtained at different detection angles. The solid curves show the best fits obtained using the shake-down model and assuming that autoionizing transitions from $(2s^2)^1S$ and $(2s2p)^3P$ do not interfere with each other but only with their respective parts of the direct ionization. (From Ref. 96.)

different total spin. In the calculations of Figure 10 it is therefore assumed that transitions from He ^1S and He ^3P do not interfere with each other, but only with different parts of the ionization continuum. Also the fit parameters were varied independently for each of the spectra at different detection angles.

The question of how far an interference between various autoionization amplitudes is possible has been analyzed in detail by Taulbjerg.[97] He showed that transitions from states with different spin multiplicity do not interfere, whereas transitions from coherently excited states with different orbital angular momentum are expected to interfere.

Other treatments of PCI have been given[98–100]; however, in these cases no comparisons with experimental results were performed.

4.4. PCI-Influenced Auger Electron Spectra

Also Auger processes can also be influenced by PCI effects: if they are induced by photon impact and the energy of the impinging photon is near threshold for the inner shell ionization, the subsequently emitted Auger electron will interact with the slowly receding photoelectron that was created in the initial ionization process. This can be observed in different ways: (i) The resulting Auger peak in the electron energy spectrum attains the typical asymmetric shape and is shifted to higher energies; correspondingly the peak of primary photoelectrons is shifted to lower energies. (ii) The double ionization cross section is not increasing stepwise when the photon energy is increased above the inner shell ionization energy, since there is a certain probability that the primary photoelectron is again captured after emission of the Auger electron. Such PCI effects have been observed experimentally in the last years.[101-105]

It is again possible to apply a semiclassical treatment and to develop analytical formulas for the amplitudes and the phases of the processes involved. This was done by Niehaus,[106] and measured shifts of Auger peaks[104] are in excellent agreement with this theory. Also the dependence of multiple ionization cross sections on the photon energy in the threshold region is well described. Figure 11 shows a comparison of a multiple ionization cross section as measured by van der Wiel et al.[102] and the one calculated with the semiclassical theory by Niehaus.

The situation is more complicated if Auger processes are induced by electron impact. In this case both the scattered electron e_s and the ionization electron e_i from the inner shell can interact with the Auger electron. Since the excess energy E_1 of the impinging electron can be shared in different ways between e_s and e_i, one of them can be very slow even at high initial

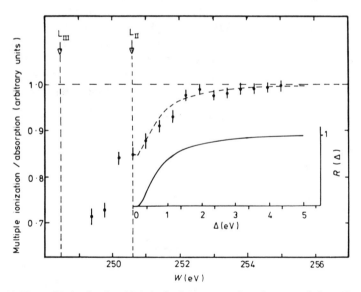

Figure 11. Normalized ratio of multiple ionization cross section of argon and absorption signal as a function of energy loss of fast electrons in the region of the L_2 threshold. The measured points are from Ref. 102. The insert shows the calculated ionization probability in the Auger decay, following an L_2-hole creation after $\tau = 100$ a.u. (From Ref. 106.)

excess energy E_1. Such an asymmetric sharing is in fact the most likely situation.[107] A rigorous treatment of this is difficult. For a description of the peakshifts $\Delta\varepsilon$ in the measured Auger spectra Huster and Mehlhorn[105] used the functional form

$$\Delta\varepsilon = cTE_1^\alpha \tag{25}$$

with E_1 the excess energy, T the width of the primary ionized state, and c and α as free fitting parameters. In this way good agreement with measured peak positions could be obtained.

Regarding peak shapes a first systematic investigation has been performed recently by Hedman et al.[108] As discussed in Section 4.2 the PCI induced line shape is steeper than the Lorentzian line shape on the side of the nominal energy ε_0. This line-sharpening effect was first observed by Wilden et al.[85] in Argon-autoionization spectra from electron collisions. Hedman et al.[108] now gave a quantitative analysis of PCI-influenced Ar LMM Auger line shapes. Figure 12 shows their experimental result: the Auger line has the typical PCI shape with a long tail on the high-energy side and a steep slope on the low-energy side, which reflects the long survival time of atoms yielding electrons at this energy. Also Eq. (17) and the corresponding equation in Niehaus' semiclassical description of PCI-influenced Auger spectra[106] yield a line-sharpening effect, and the Lorentzian line shape for the limiting case of high velocities is properly obtained from this semiclassical treatment.[81] However, the transition between these two limiting cases is not reproduced correctly. Hedman et al.[108] developed an improved version of the semiclassical model, which now describes properly the transition from the "sharp" PCI line shapes at low velocities

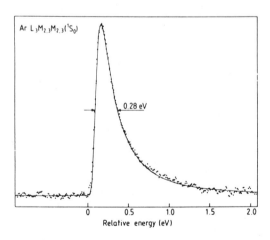

Figure 12. Comparison between measured and calculated PCI line shapes. The low-energy slope is steeper than the Lorentzian shape corresponding to the lifetime of the excited state. (From Ref. 108.)

to the Lorentzian shape at high velocities. Figure 12 shows an excellent agreement between this theoretical line shape and the experimental one.

4.5. PCI-induced Exchange of Angular Momentum

Up to now we have only considered an exchange of *energy* between the slowly receding charged collision partner and the fast ejected electron. However, there may also be an exchange of angular momentum. In the first place one would suspect that this can most conveniently be observed by its influence on angular electron distributions. However, such an observation has not been made yet. On the other hand, some indications have been reported that such an exchange of angular momentum does occur. Heideman and co-workers have investigated various possibilities[47,90,110] to detect this. One way is to analyze structures in the excitation functions of singly excited states with different l due to an indirect population via the capture of a slow scattered electron after excitation and decay of an autoionization state. With the assumption that the slow scattered electron initially has zero angular momentum, a population of Rydberg states with $l > 0$ is only possible via an exchange of angular momentum between captured and ejected electrons. However, the assumption of zero angular momentum for the scattered electron is questionable.

Another possibility is the analysis of ejected electron spectra due to electron impact excitation of autoionizing states via a negative-ion resonance. One example, investigated by van de Water and Heideman,[90] is the excitation of $He(2s^2)^1S$ via the negative resonance $He^-(2s2p^2)^2D$. Several processes can lead to the same final state:

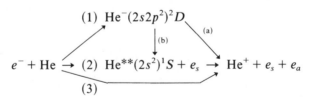

Energy spectra of autoionization electrons e_a can only be influenced by interferences between the various channels, if both electrons e_s and e_a have the same angular momentum. Reaction path 1b will result in scattered electrons with $l = 2$. Interferences with the other reaction paths—which have been observed experimentally—can only occur if those too result in scattered electrons with $l = 2$. With the assumptions that reaction path 1a is not important and that reactions 2 and 3 yield electrons with $l = 0$ only, an interference of channel 1 with channels 2 and 3 can therefore only occur by exchange of angular momentum. However, the assumptions that have to be made are questionable.

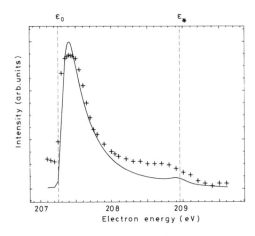

Figure 13. Comparison between measured and calculated line shapes for the Auger transition $ArL_2M_{23}M_{23}(^3P_{0,1,2})$. The structure at ε_* on the high-energy slope is due to exchange of angular momentum between scattered and ejected electron. (From Ref. 111.)

Recently some typical features in Auger spectra due to electron impact excitation could be interpreted as being due to exchange of angular momentum. Figure 13 shows an experimental electron spectrum. There occurs an extra peak on the high-energy slope of the Auger peak. This extra peak increases with decreasing excess energy of the primary electron and was interpreted by Niehaus and Zwakhals[111] in the following way: the normal PCI line shape is obtained from a potential curve diagram as given in Figure 8. If the electrons involved have angular momenta, the potential curves have to be modified in order to take the centrifugal potential into account. It turns out that for the case of angular momentum exchange the energy difference between the excited and the final state has an extremum at a separation of the collision partners, which corresponds to an ejected-electron energy of

$$\varepsilon_* = \varepsilon_0 + \{2[l_f(l_f + 1) - l_i(l_i + 1)]\}^{-1}$$

Because of this extremum in the difference potential the ejected electron intensity has a "rainbow" maximum at ε_* that can be described analytically within a refined semiclassical treatment,[111] which includes complex potentials and which is similar to the refined semiclassical treatment of Hedman *et al.* Figure 13 also includes a calculated spectrum, which clearly proves that the extra peak on the high-energy slope is due to an exchange of angular momentum.

5. Spectroscopic Data for Various Atoms

5.1. H^- and He

The two-electron systems H^- and He are the simplest atoms that can autoionize, and therefore they have been the subjects of many experimental and theoretical investigations. As was described in Section 2 of this chapter, they serve as test cases for a proper description of autoionizing states. A summary of the observed states of H^- and He is given, for example, by Williams in Part B, in Chapter 23, of this book.[112]

Since that chapter was completed new experiments have especially been performed on the H^- system: Bryant and co-workers[113,114] have investigated the photodetachment of H^- by shining laser light at different angles of incidence on a 600-MeV H^- beam. Owing to the relativistic velocity of the H^- atoms, effective photon energies between 1.8 and 11.5 eV could be obtained by variation of the angle between laser light and H^- beam. The photodetachment cross section shows sharp structures at photon energies where the formation of autodetaching ^1P states, e.g., $H^-(2s2p)^1P$, interferes with the direct photodetachment process. By this method much narrower resonances can be observed than by electron scattering on H, which also can be used to excite H^- resonances (at an electron energy that is lower than the corresponding photon energy by the binding energy of H^- of 0.754 eV).

The photodetachment signal, for example, shows a narrow Feshbach resonance at 10.93 eV—corresponding to an electron energy of 10.18 eV, i.e., just below the $n = 2$ level of H. This resonance, which is in agreement with theory,[115] is too narrow for a definite observation by electron scattering experiments. Also resonances in the vicinity of the $n = 3$ level of H, i.e., at photon energies of ~ 12.8 eV, show structures that yield more detailed information about doubly excited H^- states in this region than electron scattering experiments.

A specific advantage of the H^- photodetachment experiment is the possibility of investigating the behavior of autoionizing states in strong electric fields[113,116]: a moderate magnetic field of ~ 200 G in the laboratory corresponds to an electric field in the center-of-mass system as high as $\sim 10^5$ V/cm. This is a field strength similar to that caused by a singly charged ion at a distance of $\sim 200a_0$. Stark shifts and variations of lifetimes due to the presence of a close-by charged collision partner, e.g., in postcollision interaction, can therefore be expected to be similar to those observed in the H^- photoabsorption experiment.

5.2. Rare Gases Ne $\cdots$ Xe

Neon. There are two types of Ne autoionization states: those with configurations $Ne(1s^2 2s2p^6 nl)$, in which one $2s$ electron is excited into a higher level, and the doubly excited states with configurations $Ne(1s^2 2s^2 2p^4 nln'l')$. In photoabsorption spectra[117] and electron collisions[10] mainly the first type is excited. Since there is only a limited number of states belonging to one configuration, the identification is not too difficult. A comparison of various experimental data has been given by Spence.[118]

The identification of doubly excited states, on the other hand, is more complicated, since a large number of states corresponds to each configuration. Also it is not always clear which coupling scheme is the most suitable for the proper description of these states. Either the electrons are coupled in the "aufbau" scheme in order of their energy with levels, e.g., $Ne(2p^4)^3P(3s)^{2,4}P(3p)^3P$, or they are coupled in the external coupling scheme $Ne(2p^4)^3P[(3s3p)^{3,1}P]^3P$. Bandzaitis et al.[119] used the "aufbau" coupling scheme to calculate energies of autoionization states with configurations $Ne(2p^4)^3P(3s3p)$. Langlois and Sichel used a frozen-core-superposition-of-configurations method[120] to calculate the energies of a large number of odd-[121] and even-parity[122] levels of doubly excited Ne with an accuracy of 0.1 eV. These calculations can be used to identify the doubly excited states observed in photoabsorption,[117] electron collisions,[10,84,118,123,124] and heavy particle collisions.[53,125,126]

A very informative series of experiments has been performed by Olsen and Andersen,[125] who investigated energy spectra of electrons arising from collisions of different carefully selected ions or atoms with Ne. In this way

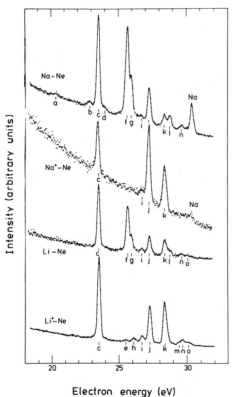

Figure 14. Electron spectra from 10-keV Na$^+$, Na$^+$ + Ne and 5-keV Li, Li$^+$ + Ne collisions, measured at 20° with respect to the projectile beam. Spin conservation implies that only singlet states can be excited in case of ion impact, whereas also triplet states may be excited for the neutral projectile. (From Ref. 125.)

Table 1. Comparison of Energies for Doubly Excited Ne Autoionization States as Determined Experimentally by Various Authors[a]

Peak label	Energy						Mean	Configuration $2p^4$(core)[nl, $n'l'$]
	Ref. 118	Ref. 124	Ref. 53	Ref. 125	Ref. 123	Ref. 117		
a	41.90		41.95		41.87		41.91	$(^3P)3s^2(^3P)$
b	44.45		44.50		44.48		44.48	$(^3P)[3s3p(^3P)](^3P)$ or (^3D)
c	45.00			45.02	44.97	44.98	44.98 (opt)	$(^3P)[3s3p(^3P)](^1P)$
			45.15		45.12		45.14	$(^1D)3s^2(^1D)$
			45.65				45.65	$(^3P)[3s3p(^1P)](^3P)$ or (^3D)
d	45.90	45.95					45.93	
			47.10				47.10	$(^3P)[3s4s(^3S)](^1P)$
		47.30	47.25				47.28	$(^1D)[3s3p(^3P)](^3F)$
			47.55				47.55	$(^1D)[3s3p(^3P)](^3P)$ or (^3D)
			47.70				47.70	$(^3P)[3s3d(^3D)](^1D,F,P)$
e	48.22						48.22	
f	48.35	48.40 (±0.1)	48.30				48.35	$(^3P)[3p^2(^3P)](^1D)$
g	48.65	48.60 (±0.1)					48.63	$(^3P)[3p^2(^3P)](^1S)$
h	48.90	48.90	48.90			48.907	48.91 (opt)	$(^1D)[3s3p(^1P)](^1P)$
i	49.11	49.15					49.13	$(^1S)[3s^2](^1S)$
j	49.35	49.45					49.40	$(^3P)[3p^2(^1D)]$
				49.88			49.88	
k	50.00		50.00				50.00	$(^1D)[3s3d(^1D)]$
l	50.50 (±0.1)	50.25		50.21			50.23	$(^3P)[3p^2(^1S)]$
		50.50 (±0.1)	50.45				50.48	$(^1D)[3s3d(^3D)]$
		50.95	51.05			51.00	51.00 (opt)	$(^3P)3d3p(^1P)$
m	51.30 (±0.1)	51.30 (±0.1)	51.25	51.24		51.309	51.31 (opt)	$(^3P)3d3p(^1P)$
		51.60 (±0.1)	51.70	51.61			51.64	$(^1D)[3p^2(^3P)]$
				51.81			51.81	
				52.04			52.04	
n	52.60 (±0.1)	52.60 (±0.1)		52.63		52.62	52.62 (opt)	$(^1S)[3s3p(^1P)](^1P)$

[a] The peak labels are identical with those of Figure 14. State identification is done by comparison with calculations of Langlois and Sichel.[121,122] (From Ref. 118.)

they could provoke or suppress the excitation of certain types of autoionizing states and therefore facilitate the identification of various states. Figure 14 shows some of their spectra. The peak positions were compared with the energies calculated by Langlois and Sichel.[121,122] Spence performed a detailed investigation of Ne autoionizing states by measuring near-threshold scattered-electron spectra. A comparison of his results with those of other experiments is shown in Table 1. The identification of the various lines is again that of Langlois and Sichel.[121,122]

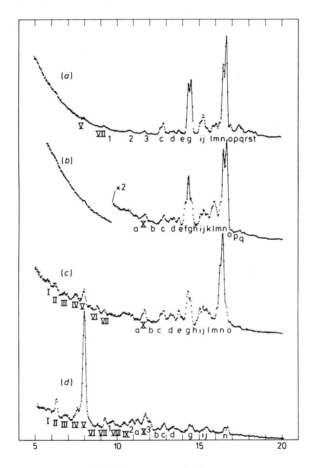

Electron energy (eV)

Figure 15. Electron spectra due to decay of Ar autoionizing states, excited by different collisions. (a) 5-keV He + Ar: only singlet Ar** states are excited, most of them doubly excited states. (b) 15-keV Li + Ar: also triplet states are excited. (c) 15-keV Li$^+$ + Ar: charge exchange leads to excitation of Ar^{+**}. (d) 5-keV He$^+$ + Ar: autoionization from Ar^{+**} dominates. (From Ref. 128.)

Autoionization states of Ne^+ ions were investigated by Morgenstern *et al.*[78] in collisions of multiply charged Ne^{q+} ions on various targets. Also here a large number of states was observed, which were difficult to identify, especially since in this case there are no detailed calculations as for neutral Ne states.

Argon. For Argon the situation is even more complicated than for Ne: (i) states with configurations $Ar(\cdots 3s3p^6nl)$ have energies in the same range as those with configurations $Ar(3s^23p^4nl,n'l')$; (ii) electrons in the $3d$ orbital have energies similar to those in the $4s$ or $4p$ orbital; (iii) the *LS* coupling scheme, which works reasonably well for the Ne autoionizing states, begins to break down in the case of Ar.[127] Also there are up to now no detailed calculations of autoionizing levels, and therefore a dependable interpretation of the complicated spectra is difficult. Nevertheless a large number of states could be identified in measurements of photonabsorption,[127] electron collisions,[10,89] and heavy particle collisions.[128] Jorgensen *et al.*[128] have again used the method of investigating collisions between Ar or Ar^+ and carefully selected collision partners. In this way they could assign various autoionization states in Ar and Ar^+. Figure 15 shows an energy spectrum of electrons due to 5-keV He + Ar collisions. Electrons

Table 2. Identification of Doubly Excited Ar Autoionization States as Given by Jørgensen *et al.*[a]

Peak label	Excitation energy (eV)	Identification
a	26.95	$(^3P)4s^2\,^3P$
b	28.11	$(^3P)4s4p$
c	28.66	$(^1D)4s^2\,^1D$
d	29.29	$(^3P)4s4p$
e	29.77	$(^3P)4s4p$
f	29.96	$(^1D)4s4p\,^3L$
g	30.39	$(^1D)4s3d\,^1L$
h	30.69	$(^3P)3d4p$
i	31.09	$(^3P)4p^2$ or $(^1D)4s4p$
j	31.34	$(^1D)4s4p$ or $(^3P)4p^2$?
k	31.74	$(^3P)3d^2$ or $3d4p\,^3L$?
l	31.88	$(^3P)3d^2$ or $3d4p$?
m	32.12	$(^3P)3d^2$?
n	32.54	$(^1D)3d4p$
o	32.76	$(^1D)3d4p$ or $(^1D)4p^2$?
p	33.02	$(^1D)3d4p$ or $(^1D)4p^2$?
q	33.47	$(^1D)3d^2$?
r	33.84	$(^1D)3d^2$?
s	34.01	$(^1D)3d^2$?
t	34.27	?

[a] Reference 128.

from doubly excited states with the configuration $Ar^{**}(3p^4nl,n'l')$ dominate the spectrum. Table 2 shows the identification of various lines as given by Jorgensen *et al.*

Krypton and Xenon. Also krypton and xenon autoionizing states were investigated by photoabsorption,[129,130] electron collisions,[10] and ion or atom collisions.[50,51] Whereas the identification of many singly excited states was possible in these experiments, the assignment of most of the doubly excited states has to be regarded as tentative. The complexity of the spectra is, for example, demonstrated in the electron collision experiments by Baxter *et al.*[132]

5.3. Alkali Atoms

Also the autoionizing states of alkali atoms were studied by the various methods in the last years. Ross and co-workers performed a systematic study of all alkali atoms, in which alkali vapor was bombarded with electrons at various energies and the ejected electrons were analyzed.[133-138] But also other methods such as light absorption from ground-state[139] or excited state[140] atoms, heavy particle collisions,[141] and beam-foil excitation[142,143] were used to excite the autoionizing states.

Lithium has attracted the most attention since this three-electron system is still simple enough to allow detailed theoretical calculations. Energies of Li autoionization states have been calculated by Wakid *et al.*[144] and—for a large number of doubly and also triply excited states—by Bruch *et al.*[143] The energies of several autoionizing states are precisely known from uv absorption spectroscopy.[139] The largest amount of experimental data, however, was obtained by collisional excitation of Li by electron impact[133,134] and ion impact.[145,146] The most detailed data are from Rødbro *et al.*,[146] who analyzed the projectile-emitted electrons in fast ion–atom collisions. In these experiments also several triply excited states of the type $Li(2l, n'l', n''l'')$ were observed for the first time. The energies of such states were later calculated by Chung,[147] in excellent agreement with the experimental values.

Special attention has been given to the lifetimes of Li autoionization states. There are two types of states that are metastable against autoionization: (i) those that cannot decay because of spin conservation, e.g., the state $Li(1s2s2p)^4P$, which was studied in detail by Feldman and Novik[148]; and (ii) those that cannot decay because of parity conservation, e.g., the state $(1s2p^2)^2P$ and generally all the doubly excited states with even parity and odd total orbital angular momentum (or vice versa). Whereas typical lifetimes of the "normal" Li autoionization states are between 10^{-14} and 10^{-13} sec,[149,150] the $Li(1s2s2p)^4P$ lives for $\tau = 5.1\ \mu sec$[148] and the $Li(1s2p^2)^2P$ has a lifetime in the order of 5.10^{-11} sec.[151-153] The main decay channels for these states are optical transitions to lower excited states. Such transitions from doubly excited even-parity ^2P states are promising candidates for lasers in the 20-nm region, and therefore they have been intensively studied in recent years. For example, Willison et al.[154] observed the $Li(1s2p^2)^2P \to Li(1s^22p)^2P$ transition in a microwave-heated Li plasma at a rate of 10^{16} photons per second.

Going to the higher Z alkali atoms the number of autoionization states is drastically increased since configurations of the type $Na(2p^5nln'l')$ can give rise to many states due to the various possible couplings of the excited electrons to the core and to each other. In *sodium* a large number of autoionization states were observed by light absorption from ground-state[155,156] and excited-state atoms[157,158] as well as by observation of ejected electron spectra due to electron[135,159] or heavy particle collisions.[141,160] Theoretical calculations of autoionizing Na states were performed by McGuire[161] and—for even-parity states—by Sugar et al.,[158] and this allows a positive identification for a large number of the observed states.

Also the other alkali atoms K, Rb, and Cs have been investigated by the different methods,[136–138,162,163] but in many cases only tentative assignments of the autoionizing states are possible.

5.4. Alkaline Earth Atoms

Autoionization states of alkaline earth atoms can either be obtained by excitation of a core electron or by excitation of both outer electrons. The configuration of the latter ones is similar to the He** or H⁻** configuration, now having a filled shell in place of a bare nucleus as a core. Greene[33] has extended the method of hyperspherical coordinates in order to describe such alkaline-earth autoionization states. Experimentally these states are attractive because they have relatively low excitation energies and some of them can be excited by stepwise laser excitation. This allows detailed studies of positions, configuration interactions, linewidths, and couplings to the ionization continuum. For example, Gallagher and co-workers have studied doubly excited states of Ba by stepwise laser excitation.[164–166] A point that has attracted special attention is the selective autoionization into excited ion states, which, e.g., was observed in Ba.[164,165] Such a selective autoionization can serve to obtain population inversion of an excited ionic state, which subsequently can be used to obtain laser action from these states. This is a promising concept for a new type of lasers. Amplified spontaneous emission in Ba^+ after selective autoionization of laser-excited $Ba(6p_{3/2}12p)$ into $Ba^+(6p)P_{1/2}$ has for the first time been observed by Bokor et al.[167]

A systematic analysis of autoionization electron spectra due to collisions of electrons on alkaline earth atoms has again been performed by Ross and co-workers[168–171] (see also references therein for other investigations).

Regarding core-excited states most of them are still unidentified. Only for core-excited Be, which was studied in detail by Rødbro et al.,[146] could a tentative assignment be given for many states.

6. Correlated and Uncorrelated Angular Distributions of Autoionization Electrons

Detailed information about collisionally excited autoionizing states can be obtained not only from the energy but also from the angular distribution of the ejected electrons. As opposed to photon emission the angular momentum of the emitted electrons is not limited to $l = 1$. Therefore not only the excitation parameters of P states can be determined, but also those of states with higher angular momentum. An advantage of detecting the photons, on the other hand is the possibility of observing their polarization properties, which contain information in addition to that from only the intensity: e.g., about the sign of the angular momentum in the excited atoms in addition to its absolute value.

In order to evaluate angular distribution measurements one must first choose an appropriate set of parameters by which the excited atoms are described. Secondly, the relation between the angular distributions and these parameters has to be established.

6.1. Excitation Amplitudes and Density Matrix of Excited Atoms

In an ideal experiment the excited atoms are prepared in pure states characterized by a wave function

$$\psi = \sum_M f_M |JM\rangle \qquad (26)$$

with complex excitation amplitudes f_M, which are related to the differential cross sections σ_M for population of the magnetic sublevels M by

$$\sigma_M = |f_M|^2 \qquad (27)$$

In a realistic experiment often the initial state of the collision system is not completely determined or the final state of the projectile which caused the excitation is not observed. Since the amplitudes f_M may depend on these quantities, the excited atoms in these cases cannot be characterized by a pure state but must be described by a density matrix $\rho = |\psi\rangle\langle\psi|$, the elements of which are averaged products of excitation amplitudes

$$\rho_{MM'} = \overline{f_M f_{M'}^*} \qquad (28)$$

In connection with the observation of angular distributions it is convenient to expand the density matrix in terms of irreducible spherical tensors by writing

$$\rho = \sum_{kq} \rho_{kq}(JJ) T_{kq}^{\dagger}(JJ) \tag{29}$$

where the T_{kq} are defined by[172]

$$T_{kq}(JJ') = \sum_{MM'} |JM\rangle(-)^{J-M} \begin{pmatrix} J & k & J' \\ -M & q & M' \end{pmatrix} (2k+1)^{1/2} \langle J'M'| \tag{30}$$

with the quantity in the large parentheses denoting a $3j$-symbol.

For the tensor components ρ_{kq} of the density matrix this yields

$$\rho_{kq}(JJ') = \mathrm{Tr}[\rho T_{kq}(JJ')]$$

$$= \sum_{MM'} \rho_{MM'}(JJ')(2k+1)^{1/2}(-1)^{-J-k-M'} \begin{pmatrix} J & k & J' \\ M & q & -M' \end{pmatrix} \tag{31}$$

Each of the components ρ_{kq} gives rise to a contribution to the angular electron distribution that is proportional to the spherical harmonic $Y_{kq}(\vartheta, \varphi)$. The rank k of these tensor components is limited by the triangle relation

$$k \le 2J \tag{32}$$

as can be seen from the $3j$-symbol in (30). Depending on the symmetry of the collision system not all of the excitation amplitudes f_M are independent, but are related to each other. The same is true for the density matrix elements $\rho_{MM'}$ and the tensor components ρ_{kq}. A basic discussion of invariance principles and their consequences has been given, for example, by Rodberg and Thaler[173] and a discussion of the symmetry properties of the density matrix has been given, for example, by Blum and Kleinpoppen[174] and by Blum.[175] An exploitation of these symmetries makes it possible to reduce the number of free parameters by which the atom can be described.

The highest symmetry of the collisionally excited atoms is obtained when the azimuth scattering angle of the scattered projectiles is not observed or if the projectiles are not observed at all. Then the excited atom has a rotational symmetry about the projectile beam direction. The scattering amplitudes f_M depend on the azimuth scattering angle ϕ as $\exp(-iM\phi)$ and therefore the averaged products $\overline{f_M f_{M'}^*}$ for $M \neq M'$ vanish, since all ϕ have the same probability. This means that the density matrix is diagonal in a coordinate frame, in which the symmetry axis is taken as quantization axis.

In those cases in which a subensemble of excited atoms is selected by observing them in coincidence with projectiles that were scattered through a certain azimuth angle ϕ, one no longer has rotational symmetry. However, one can still exploit the invariance of the atomic interaction with respect to the scattering plane.

For an actual exploitation of this symmetry one can consider two different coordinate frames: (a) the x-z plane is taken to be the scattering plane. A special case is the often adopted choice of the so-called collision frame, in which the projectile beam direction is taken as the z axis. (b) The z axis is chosen perpendicular to the scattering plane. This choice has several advantages, which are discussed by Herman and Hertel.[176] A transfer of angular momentum from the relative motion of the collision partners to the electronic motion can, for example, be realized much more directly in such a coordinate system. The symmetry relations between the parameters describing the excited atoms are different for these two cases.

If the x-z plane is chosen to be the scattering plane the invariance of the density matrix upon reflection yields

$$PR_y(\pi)\rho R_y^\dagger(\pi)P = \rho \tag{33}$$

with P the inversion operator and $R_y(\pi)$ representing a rotation about the y axis by an angle π. For the matrix elements this yields

$$\begin{aligned}\rho_{MM'} &= \langle JM|PR_y(\pi)\rho R_y^\dagger(\pi)P|JM'\rangle \\ &= (-1)^P(-1)^{J-M}\langle J,-M|\rho|J,-M'\rangle(-1)^P(-1)^{-J+M'} \\ &= (-1)^{M'-M}\rho_{-M-M'}\end{aligned} \tag{34}$$

Similarly one can obtain relations between the scattering amplitudes f_M, if there is a pure initial and a pure excited state to allow a description in terms of amplitudes. The symmetry relations between the f_M are dependent on the parity change $\Delta\pi$ from the initial to the excited state. If the atoms initially are in a $J = 0$ state and if again the z axis is chosen to be in the scattering plane one obtains

$$f_M = (-1)^{\Delta\pi+J-M}f_{-M} \tag{35}$$

In most cases of one-electron excitations $\Delta\pi$ and J are both odd or both even. Then (35) reduces to

$$f_M = (-1)^M f_{-M} \tag{36}$$

which is an often applied relation.

Table 3. Symmetry Properties of Excitation Amplitudes, Density Matrix
Elements, and Spherical Tensor Components of the Density Matrix for
Differently Chosen z Axis

	z in the scattering plane	z perpendicular to the scattering plane
Amplitudes	$f_M = (-)^{\Delta\pi + J - M} f_{-M}$ mostly $f_M = (-)^M f_{-M}$	$f_M = (-)^{\Delta\pi + M} f_M$
Density matrix elements	$\rho_{MM'} = (-)^{M'-M} \rho_{-M-M'}$	$\rho_{MM'} = (-)^{M'-M} \rho_{MM'}$
Statistical tensors	$\rho_{kq} = (-)^{k-q} \rho_{k-q}$	$\rho_{kq} = (-)^q \rho_{kq}$

In the case of two-electron excitations one has to be careful. For
excitation of $He(2p^2)^3P$ from the ground state, for example, $\Delta\pi + J$ is odd
and therefore $f_M = (-1)^{M+1} f_{-M}$. Finally, there are also relations between
the statistical tensors ρ_{kq}. One obtains

$$\rho_{k-q} = (-1)^{k-q} \rho_{kq} \tag{37}$$

Together with the Hermiticity relation $\rho_{kq}^* = (-1)^q \rho_{k-q}$ this means that the
ρ_{kq} are real for even k and purely imaginary for odd k. If the z axis is
chosen perpendicular to the scattering plane, relations analogous to (34)–
(37) can be obtained. All these relations are put together in Table 3. If z is
chosen perpendicular to the scattering plane one can see from the relations
in Table 3 that all density matrix elements with odd $M'-M$ and all statistical
tensors with odd q vanish.

6.2. Theoretical Shapes of Angular Electron Distributions

The angle-dependent electron intensity $I(\vartheta, \varphi)$ is in principle given by
the square of the matrix element (1). For the description of an actual
measurement one has to sum coherently over contributions from different
intermediate states (μ_a) and incoherently over those from different initial
and final states (μ_i) and (μ_f). This yields

$$I(\vartheta, \varphi) = \sum_{(\mu_i)} w(\mu_i) \sum_{(\mu_f)} \left| \sum_{(\mu_p)} \langle \mathbf{k}\psi_{ion} | V_c | \psi^p \rangle \right|^2 \tag{38}$$

where $w(\mu_i)$ are the probabilities that the states μ_i are initially represented
in the system, $\mathbf{k}$ represents the ejected electron, and ψ^p the autoionizing
state. For the general case it is convenient to use the density matrix formula-
tion,[177-179] which for the description of autoionization electron distributions was applied for the first time by Eichler and Fritsch.[177] With

$(\mu_p) = J, M$ and $(\mu_f) = J_f, M_f$ the electron intensity is then given by

$$I(\vartheta, \varphi) = \sum_{M_f} \langle kJ_fM_f | V_c\rho V_c^\dagger | kJ_fM_F \rangle \tag{39}$$

Expanding the ejected electron wave function $|k\rangle$ into partial waves and summing over the magnetic quantum numbers m_s of the ejected electron's spin—which is generally not observed—one can write the electron intensity in terms of multipole components of the density matrix[179]

$$I(\vartheta, \varphi) = \sum_{kq} \sum_{jj'll'} F_k(JJ'J_fjj'll')R(JJ'J_fjj'll') \cdot \rho_{kq}(JJ') Y^*_{kq}(\vartheta, \varphi) \tag{40}$$

In this way the angular distribution is determined by three types of quantities: ρ_{kq}, R, and F_k. The tensor components ρ_{kq} characterize the excited atom, each yielding a contribution to the electron intensity that is proportional to $Y^*_{kq}(\vartheta, \varphi)$. The quantities R are products of reduced matrix elements, describing the physics of the decay process

$$R(JJ'J_fjj'll') = \langle J_f(l\tfrac{1}{2})j \| V_c \| J \rangle \langle J_f(l'\tfrac{1}{2})j' \| V_c \| J' \rangle^* \tag{41}$$

with j, j' the angular momentum of the ejected electron. The coefficients F_k finally are geometry factors which depend on the angular momenta involved and characterize their couplings:

$$F_k(JJ'J_fjj'll') = (-1)^{J+J_f-1/2}$$

$$\times [(2J+1)(2J'+1)(2l+1)(2l'+1)(2j+1)(2j'+1)/4\pi]^{1/2}$$

$$\times \begin{pmatrix} l & k & l' \\ 0 & 0 & 0 \end{pmatrix} \begin{Bmatrix} l & l' & k \\ j' & j & \tfrac{1}{2} \end{Bmatrix} \begin{Bmatrix} j & J & J_f \\ J' & j' & k \end{Bmatrix} \tag{42}$$

In an actual process spin-dependent interactions are often unimportant and therefore the involved quantities depend only on the orbital angular momenta L rather than on J. For a comparison with experimental data one therefore has to consider what the quantities R, ρ_{kq}, and F_k are in such cases. First of all the autoionization process itself can often be regarded as spin independent, at least in those cases in which the normal autoionization due to Coulomb interaction between the electrons is allowed. In that case the reduced matrix elements and thus the coefficients R can be expressed in terms of L-dependent matrix elements by[177]

$$\langle J_fj \| V_c \| J \rangle = [(2j+1)(2J_f+1)(2L+1)(2S+1)]^{1/2}$$

$$\times \begin{Bmatrix} L_f & l & L \\ S_f & \tfrac{1}{2} & S \\ J_f & j & J \end{Bmatrix} \langle L_fl \| V_c \| L \rangle \tag{43}$$

If the autoionization process only depends on L rather than on J the excited atom has to be characterized by the L-dependent tensor components $\rho_{kq}(L)$ in place of the J-dependent ones. This makes sense anyway, since the excitation processes in heavy particle collisions are mostly spin independent. Therefore the spin can be regarded as spectator during the excitation process, and at time $t = 0$, when the excitation takes place, the density matrix factorizes into an L-dependent part and a spin-dependent one which is diagonal. The $\rho_{kq}(L, T = 0)$ therefore characterize the excitation process and are the quantities that one would like to know in order to obtain information about excitation mechanisms.

If the decay of the excited state takes place, before the spin is coupled to the orbital angular momentum the electron distribution is determined by $\rho_{kq}(L, t = 0)$. If, on the other hand, the lifetime is sufficiently large, then the spin is coupled and the anisotropy, which initially was only present in the L-subspace, is partly transferred into the spin subspace, where it does not influence the angular electron distribution. The time-dependent tensor components $\rho_{kq}(L, t)$ are then given by (see, e.g., the treatment of Blum,[175] Chapter 4.7.2)

$$\rho_{kq}(L, t) = \rho_{kq}(L, t = 0)G_k(L, t) \qquad (44)$$

with perturbation coefficients

$$G_k(L, t) = \frac{1}{2S + 1} \sum_{JJ'} (2J + 1)(2J' + 1) \begin{Bmatrix} J & J' & k \\ L & L & S \end{Bmatrix}^2 \cos(E_{J'} - E_J)t \qquad (45)$$

The $G_k(L, t)$ are oscillating quantities and describe the well-known phenomenon of quantum beats, which is caused by the fact that the anisotropy is to a certain extent periodically oscillating between the L and the S subspace.

This periodicity is generally not observed in electron detection experiments and therefore the $G_k(L, t)$ have to be appropriately averaged, weighted according to the decay intensity at each time t after the excitation. This average can be written in terms of the ratio $\delta_{JJ'} = \Delta E_{JJ'}/\Gamma$ of the fine-structure splitting $\Delta E_{JJ'}$ between the involved states and the decay width Γ as[179]

$$\overline{G_k(L)} = (2S + 1)^{-1} \sum_{JJ'} (2J + 1)(2J' + 1) \begin{Bmatrix} J & J' & k \\ L & L & S \end{Bmatrix}^2 \Big/ (1 + \delta_{JJ'}^2) \qquad (46)$$

If the fine-structure splitting is much smaller than the linewidth, i.e., $\delta \ll 1$, one obtains $\overline{G_k(L)} = 1$. In this case the decay occurs very fast and spin–orbit interaction cannot alter the initially formed $\rho_{kq}(L, t = 0)$. If, on the other hand, the fine-structure splitting is large, i.e., $\delta \gg 1$, the oscillating terms in (45) are completely averaged and one obtains

$$\overline{G_k(L)} = (2S + 1)^{-1} \sum_J (2J + 1)^2 \left\{ \begin{matrix} J & J & k \\ L & L & S \end{matrix} \right\}^2 \tag{47}$$

Owing to the precession of L about J the system is then characterized by averaged tensor components $\overline{\rho_{kq}(L, t)} = \rho_{kq}(L, t = 0)\overline{G_k(L)}$.

Finally expression (40) can be considerably simplified in those cases, in which the fine structure of the initial and final state is not resolved. The summation over j, j' and J_f can then be performed, yielding for the electron intensity

$$I(\vartheta, \varphi) = \sum_{kq} \sum_{ll'} F_k(LL_f ll') R(LL_f ll') \overline{\rho_{kq}(L)} Y^*_{kq}(\vartheta, \varphi) \tag{48}$$

with

$$R(LL_f ll') = \langle L_f l \| V_c \| L \rangle \langle L_f l' \| V_c \| L \rangle^* \tag{49}$$

and

$$F_k(LL_f ll') = [(2l + 1)(2l' + 1)/4\pi]^{1/2}(-1)^{L+l_f}$$
$$\times \begin{pmatrix} l' & k & l \\ 0 & 0 & 0 \end{pmatrix} \left\{ \begin{matrix} l & l' & k \\ L & L & L_f \end{matrix} \right\} \tag{50}$$

A more detailed discussion of angular electron distributions, in which not only the fine structure but also the hyperfine structure of the excited atoms is taken into account, was given by Bruch and Klar.[180,181]

The formulas (40) or (48) for the description of *correlated* angular distributions can also be used for the uncorrelated ones by only taking the appropriate values for the tensor components $\rho_{kq}(L)$ with $q = 0$. As was discussed in Section 6.1 the density matrix is diagonal for the case of noncoincident measurements and the ρ_{kq} vanish for $q \neq 0$, if the projectile beam direction is taken as quantization axis. Then equations (40) and (48) can simply be written as

$$I(\vartheta) = \sum_k \sum_{ll'(jj')} F_k R \rho_{k0}[(2k + 1)/4\pi]^{1/2} P_k(\cos \vartheta) \tag{51}$$

where the R and F_k are given by (41) and (42) or by (49) and (50), respectively. From the 3j-symbol in (50) one can infer that k can only have even values, since l and l' are both odd or both even owing to conservation of parity and angular momentum. Therefore the distribution described by (51) is symmetric with respect to $\vartheta = 90°$ and can be expressed in the more familiar form[182]

$$I(\vartheta) = I\left(1 + \sum_k A_k P_k(\cos \vartheta)\right) \tag{52}$$

with $k = 2, 4, 6, \ldots$. A deviation from the symmetry with respect to $90°$, which is found in some experiments, therefore indicates that the ionization process cannot properly be described by autoionization of an excited atom with well-defined parity, as was assumed in the derivation of (52). An example for such a deviation will be discussed in Section 6.4.

6.3. Coincidence Measurements to Determine Angular Correlations between Ejected Electrons and Scattered Projectiles

Coincidence measurements make it possible to obtain the most detailed information about the excited atoms since in this case the experimental data are not averaged over all orientations of the scattering plane and all scattering angles θ. Formally this means that the excited atoms can be described by a density matrix with nonvanishing nondiagonal elements. These elements or—equivalently for the case of pure states—the relative phases between the complex population amplitudes f_M for magnetic sublevels can be determined by angular correlation measurements.

As mentioned in Section 3.2 only a few such measurements have so far been performed for autoionization processes. In the case of electron-atom collisions the analysis of the data is complicated by the fact that direct ionization contributes considerably to the ejected electron intensity. With ions as projectiles the direct process can be neglected as long as the projectile is slow enough, i.e., <0.5 a.u. Only in such cases are the formulas given in Section 6.2 valid, and therefore we will discuss only such experiments in the following.

Ne** Excitation

The case of Ne excitation into the autoionizing state $Ne^{**}(2p^43s^2)^1D$ is a good example to see which information can be obtained from the analysis of angular correlations. Figure 16 shows an energy spectrum of electrons due to $He^+ + Ne$ collisions at an energy of 2 keV. Peak C is due to the autoionization of $Ne^{**}(2p^43s^2)^1D$. The electron intensity in this peak was measured[74] at various electron detection angles in coincidence with He^+ projectiles which were scattered inelastically through scattering angles (θ, Φ). Figure 17 shows the resulting electron intensity distributions in the scattering plane and in two other planes. Also a three-dimensional view of the total electron distribution is shown, which was calculated from the

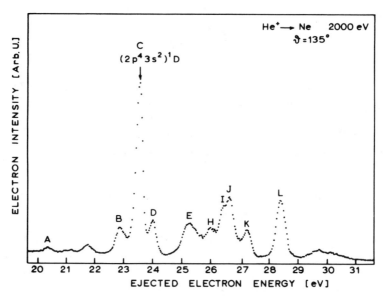

Figure 16. Energy spectrum of autoionization electrons from He$^+$ + Ne collisions. The angular distribution of electrons in peak C is investigated by coincidence measurements.

parameters obtained by fitting the theoretical distributions to the measured ones. A detailed analysis of the measurements has been given by Boskamp *et al.*[74] The coincident angular distributions can be described by equations (48)–(50), since the fine structure of the final $^2P_{1/2,3/2}$ state of Ne$^+$ was not resolved in the experiment. Equation (48) contains two types of unknown quantities, which one can determine by comparison with the experimental results: the tensor components ρ_{kq} of the density matrix and the reduced matrix elements $V_c(l)$ for the autoionization.

Regarding the matrix elements only two can contribute to the autoionization of the excited ^1D state into the final ^2P state of Ne$^+$, namely, $V_c(l = 3)$ and $V_c(l = 1)$, whereas $V_c(l = 2)$ vanishes owing to parity conservation. Therefore the ratio of the two matrix elements $V_c(l = 3)/V_c(l = 1)$ can be used as fitting parameter. From Figure 17 one is tempted to guess that this ratio is very small since the angular distributions look like *p*-distributions, implying a relatively small contribution with $l = 3$. However, it turns out[74] that the measurements are compatible with all values for this ratio, which lie on a circle in the complex plane, having its center on the real axis and intersecting this axis at 0 and at ~6. On the other hand, there are good reasons to believe that $V_c(l = 3)/V_c(l = 1) = 0$ is the actual value: autoionization is mainly caused by interaction of the two $3s$ electrons, and angular momentum conservation within this subsystem means that with $l = 1$ for the emitted electron also the other electron which jumps back to the $2p$

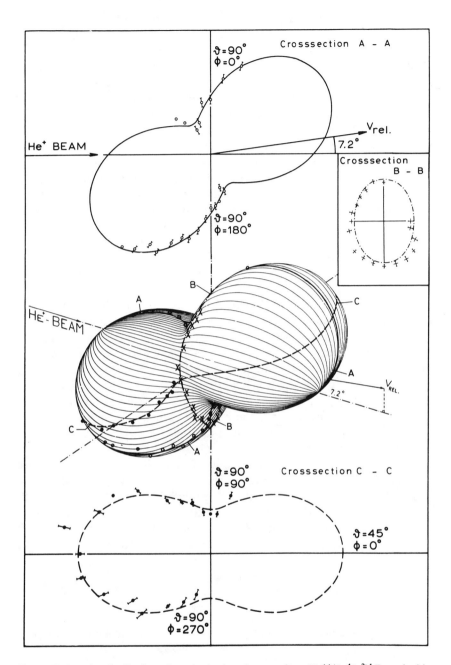

Figure 17. Angular distribution of autoionization electrons from Ne**$(2p^43s^2)^1D$, excited by He$^+$ impact. A–A shows the distribution in the scattering plane, B–B in the plane perpendicular to the He$^+$-beam direction, and C–C in a plane perpendicular to the scattering plane, tilted by 45° with respect to the He$^+$-beam direction. Solid curves represent the result of a fit calculation. Parameters obtained from the fit are used to calculate the complete distribution, which is shown in the three-dimensional view.

shell of Ne has $l = 1$. In this way it "fits" into a $2p$ hole, whereas with $l = 3$ for both electrons the process is only possible via correlated interactions between the $3s$ and the core electrons—a much less likely process.

Secondly, also the tensor components ρ_{kq} can be determined from the experiments. In those cases in which the excited atoms are in pure states this is equivalent to a determination of the complex population amplitudes f_M for magnetic sublevels, which are connected with the ρ_{kq} by (28) and (31). The ρ_{kq} are more convenient for the description of angular distributions. The amplitudes, on the other hand, are better suited for a physical understanding of excitation processes, since in slow heavy particle collisions the excitation occurs via certain potential curves which correlate to magnetic sublevels of the separated atoms. Since the investigated Ne** atoms are in a pure ^1D state a description in terms of amplitudes is justified. A fit of the measured distributions yields values for these amplitudes that are given in Table 4.

These amplitudes allow directly some conclusions about the physics of the excitation process. Regarding the numbers with respect to the coordinate frame in which the z-axis is chosen perpendicular to the scattering plane there are two points to note: (i) The fact that $|f_2|$ and $|f_{-2}|$ are unequal indicates that angular momentum was transferred from the motion of the heavy particles to the electron cloud. In this case one obtains $|L_\perp| = 2\{|f_2|^2 - |f_{-2}|^2\}/\sum_{i=0,\pm2} |f_i|^2 = 1.1\,\hbar$. The sign of $L_\perp$ cannot be determined from these measurements since the ρ_{kq} are symmetrically dependent on f_2 and f_{-2}. (ii) Since the phases of f_2 and f_{-2} are -0.36π and 0.34π, respectively, a rotation of the frame by $\gamma = 0.35\pi/2$ about the z-axis yields nearly vanishing phases for all three amplitudes. This indicates directly that there is a distinguished direction in the scattering plane which is tilted by $\gamma = 0.35\pi/2 \approx 31.5°$ with respect to the projectile beam direction. In Table 4b the excitation amplitudes are given in a frame with this direction as z-axis.

With respect to this axis it is mainly the $m = 0$ sublevel that is populated during the collision. Also one can see from Figure 17 that this z direction

Table 4. Population Amplitudes for the Magnetic Sublevels of Ne**$(2p^43s^2)^1D$, as Obtained from a Fit to the Data of Figure 17

(a) z Axis perpendicular to the scattering plane			(b) z Axis in the scattering plane, tilted by 31.5° with respect to the projectile beam direction		
	$|f_M|$	Phases		$|f_M|$	Phases
$M = -2$	0.79	0.34π	$M = 0$	0.93	0
$M = 0$	0.54	0	$M = 1$	0.26	0.5π
$M = 2$	0.28	-0.36π	$M = 2$	0.06	0.03π

is an axis of near rotational symmetry for the electron distribution. From these results one can infer the following excitation mechanism: at small internuclear distances a transition is induced from the initial Σ state of the quasimolecular system to a higher one by radial coupling near a potential curve crossing. Later on—at a much larger internuclear distance of $R_c \sim 2.4a_0$, which corresponds to a tilt angle of the internuclear axis of $31.5°$ with respect to the projectile beam direction—the rotation of the internuclear line is decoupled from the electronic motion. This can be ascribed to a mixing of the excited Σ-state with Π- and Δ-states which also correlate to the separated atom limit $He^+ + Ne^{**}(2p^43s^2)^1D$.

This example shows that angular correlation measurements make it possible to study several details of the excitation mechanism: (i) the type of coupling which leads to the excitation, (ii) the decoupling of electronic and nuclear motion at large distances, and (iii) the transfer of angular momentum. Even more information can be obtained in special cases like He^{**} excitation in $He^+ + He$ collisions.

*He** Excitation into States with Different L*

At first glance He^{**} looks like a particularly simple atom to study angular correlations. The angular momentum of the excited atom is completely transferred to the ejected electron, since the autoionization operator is a scalar which does not influence the angular momentum and since the remaining ion is in the isotropic $He^+(1s)^2S$ state. If the excited atom is in a pure state, characterized by angular momentum quantum numbers L, M, the electron distributions can therefore simply be written as

$$I(\vartheta, \varphi) \propto \left| \sum_M f_M Y_{LM}(\vartheta, \varphi) \right|^2 \qquad (53)$$

The actually measured distributions are complicated by two effects: (i) owing to the short He^{**} lifetimes the postcollision interaction can influence the electron distributions if He^{**} is excited by low-energy ion collisions; (ii) in those cases where He^+ is the projectile there occur charge exchange processes and also the fast projectile can be excited into autoionizing states. The simplicity of He autoionization, however, remains to a large extent if only one state is excited and if the electron signal is integrated over all electron energies.

In $He^+ + He$ collisions at collision energies of a few keV it is nearly exclusively one state that is excited, namely, $He^{**}(2p^2)^1D$. Coincident electron energy spectra from $He^+ + He$ collisions are shown in Figure 4. Such spectra were measured by Boskamp et al.[183] for different collision energies, scattering angles, and electron emission angles. These spectra were

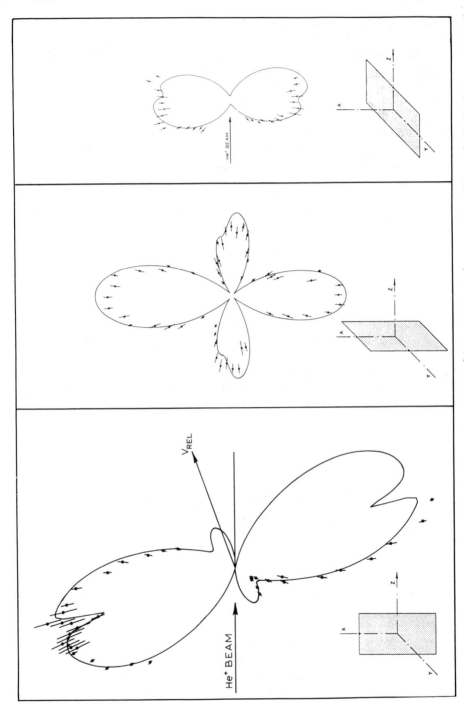

Figure 18. Angular distribution of autoionization electrons from He**$(2p^2)^1D$, excited by He⁺ impact at 2 keV ($\Theta = 10°$). Distributions are measured ... and is the figure. The x–z plane is the scattering plane. Solid curves represent the result of a fit calculation.

integrated over the electron energy in order to obtain the total electron intensity at a certain emission angle. The signal obtained in this way is plotted as a function of the electron emission angle in various polar plots in Figure 18. These plots can be regarded as cuts through the total angular electron distributions in different planes. It is not possible to use (53) in a straightforward way for an analysis of these distributions, since the electron signal contains contributions from target and projectile. The projectile-emitted electrons are influenced by the Doppler effect as described in Section 3.1, and therefore their angular distribution is slightly modified. Also the contributions from target and projectile can interfere with each other at electron emission angles, which are about perpendicular to the asymptotic relative velocity of the collision partners. This interference is responsible for the sharp structures which can be seen in the electron distribution within the scattering plane and will be discussed in more detail in Section 7. However, if these effects are properly taken into account the measured distributions can be used to obtain the population amplitudes f_M for the excited ^1D state from fit calculations. The solid lines in Figure 18 represent the best fits of such calculations and the resulting amplitudes are shown in Table 5. A complete angular electron distribution as would be ejected from one atom in a state described by the parameters of Table 5 is shown in a three-dimensional view in Figure 19. Again as in the case of Ne, these data can be used to obtain information about the excitation mechanism. With respect to the asymptotic internuclear axis it is now mainly the $M = 2$ sublevel that is excited during the collision. As opposed to Ne, in $He^+ + He$ collisions the excitation occurs by rotational coupling, i.e., at small internuclear distances there is a "sudden" rotation of the quasimolecular axis which cannot be followed by the cloud of the excited electrons. The $M = 0$ substate of the "united atom" $Be^+(1s2p^2)$, which was initially formed via the quasimolecular state $(1s\sigma_g)(2p\sigma_u^2)^2\Sigma_g$, is therefore projected on this

Table 5. Population Amplitudes for the Magnetic Sublevels of $He^{**}(2p^2)^1D$ as Obtained from a Fit to the Data of Figure 18

	$|f_M|$			Phases		
	$M = 2$	$M = 1$	$M = 0$	$M = 2$	$M = 1$	$M = 0$
(a) z Axis in the direction of the asymptotic molecular axis	0.62	0.19	0.39	1.19π	1.77π	0
(b) z Axis in the direction where the $m = 0$ amplitude has its maximum[a]	0.19	0.17	0.93	0.34π	1.50π	0

[a] The symmetry axis of the "clover leaf" in Figure 19.

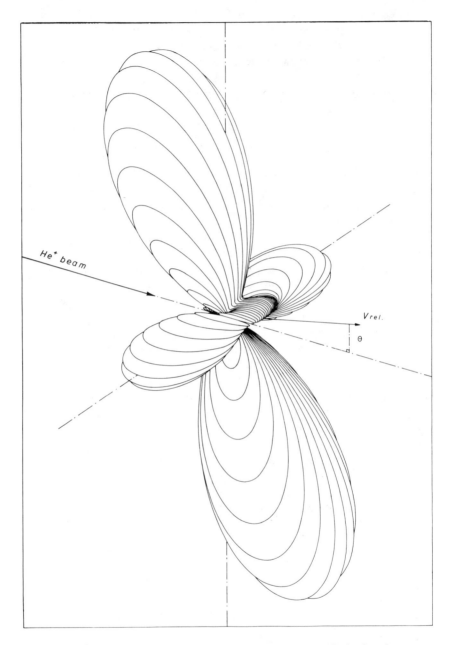

Figure 19. Three-dimensional view of the angular distribution of autoionization electrons as calculated with the population amplitudes for He** magnetic sublevels, obtained from fits of measured angular distributions.

rotated axis. The $M = 0$ part with respect to this new direction mainly correlates back to the ground state, whereas the $M = 2$ part correlates via the $(1s\sigma_g)(2p\sigma_u^2)^2\Delta_g$ molecular state to the $M = 2$ sublevel of the autoionizing He atom.

This is only a rough scheme of the excitation mechanism. A more detailed description is given by Boskamp *et al.*[(183)]

The situation is more complicated if several autoionizing states in He are excited by low-energy ion impact such that postcollision interaction leads to interferences between transitions from these states. As an example Figure 20 shows coincident electron energy spectra from Li$^+$ + He collisions.

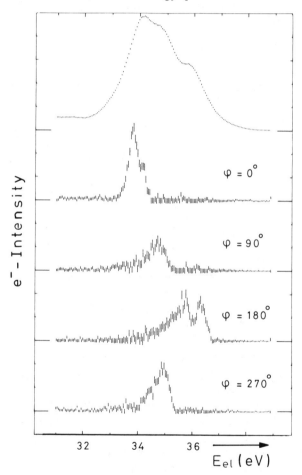

Figure 20. Energy spectra of He autoionization electrons from 2-keV Li$^+$ + He collisions, measured at $\vartheta = 90°$ with respect to the beam direction. On top, a noncoincident spectrum; below, spectra at different azimuthal angles φ. The coincident spectra exhibit structures that are due to interferences of transitions from the mainly excited He$(2p^2)^1D$ state with those from other atomic or quasimolecular states.

In this case the target–projectile symmetry of the collision system is removed and the peaks in the spectra are due to different He** states and their interferences. The analysis of such spectra makes it possible to determine relative phases of excitation amplitudes with different L or, equivalently, nondiagonal density matrix elements $\langle L'M'|\rho|LM\rangle$. To some extent this is analogous to the more familiar case of observing radiation from hydrogen in an electric field, where states with different L were excited coherently. Also in that case it is not possible to distinguish transitions from different L states, and the resulting implications have, for example, been discussed by Blum and Kleinpoppen.[174]

The mixing of transitions from different L states leads to a time dependence of the angular distributions. Owing to PCI the time scale is transformed into an electron-energy scale (see Section 4) and this leads to a dependence of the angular electron distribution on the electron energy. In this case the angle- and energy-dependent electron intensity is—for a resting emitter—given by the square over a coherent sum of transition amplitudes B^ρ as given by Eq. (21), where ρ now represents the angular momentum quantum numbers L, M and the angular distribution factors $g^\rho(\vartheta, \varphi)$ are simply given by the spherical harmonics $Y_{LM}(\vartheta, \varphi)$, which yields

$$I(\vartheta, \varphi) = \left| \sum_\rho B^\rho \right|^2$$

$$= \left| \sum_{LM} f_{LM} Y_{LM}(\vartheta, \varphi) \frac{\exp\left[-\varepsilon/2 V\tau^L \varepsilon_0^L (\varepsilon_0^L - \varepsilon)\right]}{(\varepsilon_0^L - \varepsilon)(V^L)^{1/2}} \exp i\varphi^L \right|^2 \quad (54)$$

where it is assumed that the energy and the lifetime of the excited state depend only on L and not on M. For emission from a moving atom the B^ρ have to be replaced by the C^ρ according to (13).

Kessel et al.[73] have used (54) for an analysis of coincident electron spectra arising from $He^+ + He$ collisions, thereby taking into account that besides the $(2p^2)^1D$ state also the $(2s2p)^1P$ state is excited to a small extent. As an interesting point it should be noted that a proper reproduction of P–D interference structures not only yields to absolute values of the relative phases between the various population amplitudes, but also the correct sign. A reversal of all phases would yield the same electron intensity for a certain fixed electron energy ε. However, since via the time dependence the ε dependence of the phases is determined by (22), such a change of phase signs would yield different intensities at other ε-values, i.e., different interference structures in the energy spectra. A well-defined sign of the phases corresponds to a well-defined sign of the angular momentum $L_\perp$ perpendicular to the scattering plane, which can be determined in this way. For

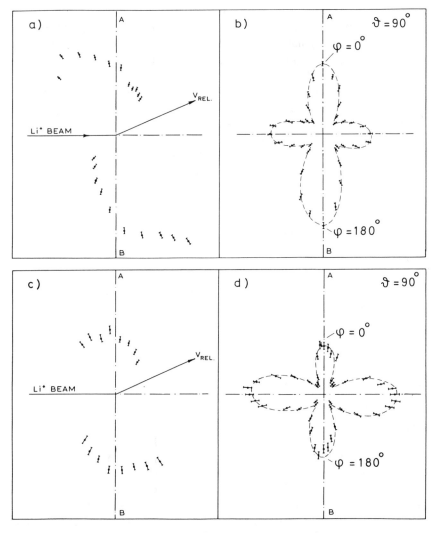

Figure 21. Angular distribution of autoionization electrons due to $Li^+ + He$ collisions at 2 keV (a and b) and 1.4 keV (c and d), measured at $\vartheta = 90°$ with respect to the beam direction. Distributions are shown in the scattering plane (a and c) and in the plane perpendicular to the beam direction (b and d). Owing to interference effects the distributions deviate considerably from those due to $He^+ + He$ collisions: the intensities at $\varphi = 0°$ and $\varphi = 180°$ are no longer equal, and the intensity perpendicular to the scattering plane ($\varphi = 90°$ and $\varphi = 270°$) is strongly increased.

He$^+$-He collisions at 2 keV Kessel *et al.* found that the angular momentum transferred to the electrons during the collision had the same sign as the angular momentum of the nuclear motion of the collision partners.

The coincident and noncoincident electron-energy spectra from Li$^+$ + He collisions as shown in Figures 20 and 7 demonstrate that it is in this case not just one state that is predominantly excited, and that interferences between the various transitions strongly influence the spectra. This collision system is therefore an interesting candidate for angular correlation measurements. Figure 21 shows angular electron distributions in which the data points again represent intensities integrated over the electron energy. These distributions are not yet analyzed in detail, but some interesting points can be stated: (i) The electron intensity at $\varphi = 180°$ is significantly higher than that at $\varphi = 0°$. This indicates that a coherent superposition of states with different parities [probably $(2s2p)$ and $(2p^2)$] is responsible for the autoionization. (ii) The electron intensities at $\varphi = 90°$ and $\varphi = 270°$, i.e., in directions perpendicular on the scattering plane, are significantly higher than in the case of excitation by He$^+$ projectiles, especially at a collision energy of 1.4 keV. Possibly an S-D interference is responsible for this.

6.4. Noncoincident Measurements of Angular Electron Distributions

If a coincident electron distribution $I_\Theta(\vartheta, \varphi)$ corresponding to a projectile scattering angle Θ is integrated over all azimuth angles φ and scattering angles Θ one obtains the noncoincident distribution

$$I(\vartheta) = \int_0^\pi \sin \Theta \, d\theta \int_0^{2\pi} I_\Theta(\vartheta, \varphi) \sin \vartheta \, d\varphi \qquad (55)$$

Generally the integration over scattering angles Θ does not strongly influence the distributions, since in most cases the overwhelming part of the excitation is due to scattering into a small range $\Delta\Theta$ of scattering angles. Integration over all azimuth angles φ, however, leads to a considerable loss of structures, as can, for example, be imagined from Figure 19. The remaining anisotropies reflect, of course, to some extent the details of the coincident distributions. Figure 22, for example, shows a noncoincident distribution of autoionization electrons from He$^{**}(2p^2)^1D$, excited by 10-keV He$^+$ impact. The same excitation mechanisms are active in He$^+$-He collisions at 2 and 10 keV and therefore the data of Figure 22 are comparable with those of Figure 18 and 19: the highest electron intensity is observed near $\vartheta \approx 90°$ and with respect to the beam direction as a z axis one obtains similar population probabilities for the magnetic sublevels. A similar comparison is possible between the coincident electron distributions from Ne$(2p^43s^2)^1D$ due to He$^+$ + Ne collisions in Figure 17 and noncoincident ones due to Li$^+$ + Ne collisions as

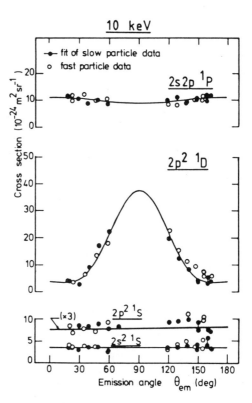

Figure 22. Noncoincident angular distributions of autoionization electrons from He⁺ + He collisions (from ref. 56). Detailed structures of coincident distributions (see, e.g., Figure 18) disappear owing to averaging over azimuthal angles.

shown in Figure 23. In this case the highest gain in electron intensity is observed at forward and backward angles. Also here similar population probabilities for magnetic sublevels are obtained in both cases. However, for noncoincident measurements one is limited to the choice of the beam direction as a z-axis, whereas Figures 17–21 clearly show that this is not optimal for the description of a single collision.

Deviations of the expected symmetry with respect to $\vartheta = 90°$ for the noncoincident distributions sometimes indicate directly that the electron emission is not due to autoionization of an isolated atom in a state with well-defined parity. At high collision energies contributions from direct ionization can cause an asymmetry, whereas at low collision energies the symmetry can be disturbed by quasimolecular effects. Figure 24 shows an example for the influence of direct ionization as measured by Bordenave-Montesquieu et al.[56] Electrons from He** due to 30-keV He⁺ + He collisions are asymmetrically ejected, whereas at 10 keV (Figure 22) they still have a symmetric distribution.

An example for deviations from symmetry due to quasimolecular effects was studied by Stolterfoht et al.[184] They analyzed shapes of He

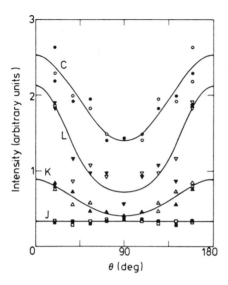

Figure 23. Noncoincident angular distributions of Ne** autoionization electrons due to Li$^+$ + Ne collisions (from Ref. 52). Curve C is for Ne$(2p^43s^2)^1D$ autoionization and can to some extent be compared with the distributions of Figure 17.

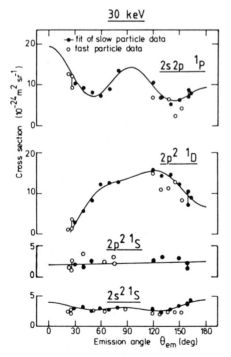

Figure 24. Noncoincident angular distributions of He** autoionization electrons. There is no longer a symmetry with respect to 90°. This indicates that interferences of direct ionization and autoionization processes become important. (From Ref. 56.)

autoionization lines in electron energy spectra from slow Li$^+$ + He collisions and found that these shapes were different for different observation angles.

An example for this is shown in Figure 25. The interpretation is based on the consideration that due to PCI the low-energy slope of the lines represents autoionization processes at small internuclear distances. Different low-energy tails for the lines observed at 30° and 150° therefore imply an asymmetric electron ejection for decay processes at small internuclear distances. This was ascribed to the fact that the eigenstates of the doubly excited He atom are at small separations influenced by the Stark effect caused by the Li$^+$ projectile. Because of this the He** eigenstates contain contributions with different parity and angular momentum and therefore the emitted electron wave will no longer contain partial waves with only odd or only even l, l' as discussed in Section 6.2, and this leads to an asymmetry in the angular distribution at small internuclear separations, i.e., in the low-energy tail of the autoionization lines.

Stolterfoht *et al.*[184] gave a qualitative estimate of this effect and showed that in slow Li$^+$ + He collisions this will definitely influence the electron spectra. The solid lines in Figure 25 show the calculated line shapes.

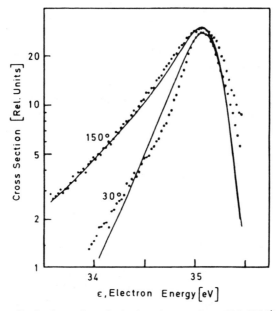

Figure 25. Energy distributions of autoionization electrons from 10-keV Li$^+$ + He collisions, measured at 30° and 150°. Different intensities of the low-energy line slope indicate that the angular distributions of electrons emitted at small internuclear distances are not symmetric with respect to 90°. (From Ref. 184.)

Unfortunately the electron spectra from $Li^+ + He$ collisions are also influenced by several other effects. Zwakhals *et al.*[185] have, for example, shown that a broad spectrum of "traveling velocities" has to be taken into account for a qualitative interpretation of these spectra. Therefore, measured line shapes are influenced by several effects which cannot clearly be separated from each other.

6.5. Electron Beats

In Section 6.2 it was discussed that the tensor components $\rho_{kq}(L)$ of the density matrix, which describe the subspace of orbital angular momenta, may vary periodically in time owing to the coupling of the initially excited orbital angular momentum with the spin. Consequently also the angular electron distributions may vary periodically in time, and observation at one specific angle will yield a signal with varying intensity. In order to observe these oscillations—often called electron beats in analogy to the quantum beats observed in photon emission—extremely short time intervals have to be resolved: since autoionization lifetimes are in the order of 10^{-12}–10^{-15} sec, oscillations within such time intervals have to be resolved. The interference structures in electron energy spectra caused by PCI (see, e.g., Figure 7 in Section 4.2) are in fact such electron beats on a microscopic time scale of $\sim 10^{-15}$ sec. Another method for observation of these beats would be the position-resolved detection of projectile-emitted autoionization electrons along the projectile beam downstreams of the excitation region. At beam energies of a few hundred keV and a position resolution of ~ 0.1 mm one can obtain a time resolution of $\sim 10^{-12}$ sec in this way. Electron beats have not yet been detected in this way; however, an analogous effect has been observed in field ionization: Leuchs and Walther[186] initiated the ionization of highly excited Na atoms by exposing them to an electric field at a certain time T after the excitation process. The resulting field-ionization signal showed periodic variations as a function of T which can be interpreted as electron beats and which allow to determine the fine-structure splitting of the J-levels connected with the initially excited L-state.

7. Electron Emission from Quasimolecules

Up to now we have mainly considered the autoionization of atoms which can be regarded as more or less isolated particles. Only for the case of a charged collision partner at low velocities was the influence of the charge on the final state potential curve taken into account in the description of postcollision interaction. A first step in describing the autoionization process as starting from quasimolecular levels is the treatment by Stolterfoht *et al.*,[184] in which Stark mixing of different states in the excited atom was taken into account in order to explain the disturbed angular distributions. In fact there are various cases in which quasimolecular effects have to be taken into account for a proper description of the autoionization process. Of course, this has to be done when the decay takes place at small internuclear distances, where the wave functions of the receding collision partners still overlap. Such cases will be discussed in Section 7.2. However, quasimolecular effects are even present in the case of decay at very large internuclear distances. An interesting example for this is again the $He^+ + He$ collision system.

7.1. Coherent Electron Emission from Two Separated Collision Partners

If an ion A^+ collides with its parent atom A in such a way that an autoionizing atom A^{**} is formed, there are two possible ways for a transition to the final state $A^+ + A^+ + e$, namely, (i) excitation of the target atom and (ii) charge exchange into an excited state of the projectile atom. For He^+-He collisions this can be written as

$$\underline{He^+} + He \begin{array}{c} \nearrow \underline{He^+} + He^{**} \searrow \\ \\ \searrow \underline{He^{**}} + He^+ \nearrow \end{array} \underline{He^+} + He^+ + e \qquad (56)$$

where the "fast" particle is indicated by underlining. Often the two paths can be distinguished from each other by means of the Doppler shift of the electron energies, which in general is different for the two paths. However, in those cases where electrons from target and projectile suffer the same Doppler shift, the two paths can in principle not be distinguished from each other, and the decay of the quasimolecule as a whole has to be considered. In that case the emission amplitudes C^T and C^P describing the electron emission from target and projectile, respectively, have to be superimposed coherently. The resulting interference should in principle be observable and thus demonstrate the quasimolecular character of the autoionization process, no matter at how large an internuclear distance the decay takes place.

Boskamp *et al.*[187] have analyzed such interferences experimentally and theoretically in great detail. Autoionization electrons due to $He^+ + He$ collisions were observed in coincidence with scattered projectiles at emission angles of ~90° with respect to the direction of the relative velocity of the collision partners after the collision. In this angular range the electron energy spectrum varies such that the peak due to projectile emission moves

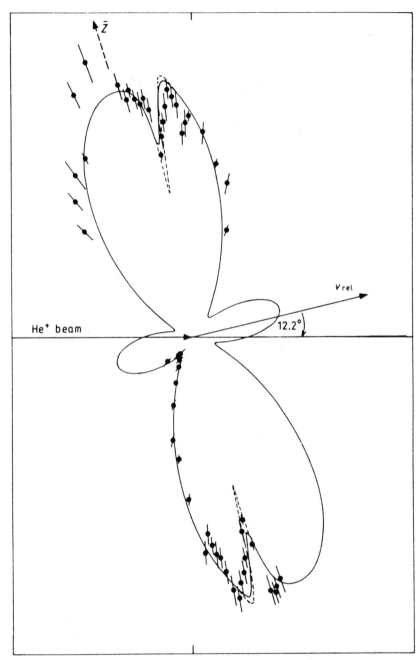

Figure 26. Angular distribution of autoionization electrons from $He^+ + He$ collisions in the scattering plane. The sharp structures at about right angles with respect to the asymptotic relative velocity are due to interferences between autoionizing transitions in target and projectile. The solid curve represents a fit calculation folded with the angular resolution of the electron spectrometer. The dotted line in the region of the structures is without that folding.

across the one due to target emission as a function of detection angle. Figure 26 shows a polar plot of the total electron intensity in the scattering plane which—as opposed to Figure 18—was measured with a better angular resolution. At angles of ~90° with respect to the asymptotic relative collision velocity characteristic dips can be seen, which are due to interference between target and projectile contributions. For a quantitative description of the observed interference, which seems to be destructive, one has to know moduli and phases of the amplitudes involved. Basic symmetry considerations as given by Niehaus[188] can be used for an understanding of the essential points: excitation of two electrons can only occur via one quasimolecular state, namely, $He_2^+(1s\sigma_g)(2p\sigma_u)^2 \, ^2\Sigma_g$. Therefore the cross section for charge exchange into doubly excited states does not oscillate as a function of scattering angle or collision energy as is, for example, the case for charge exchange into the ground state or for elastic scattering, where two potential curves are involved.[189] Target and projectile states are therefore populated with equal probabilities and with Eqs. (21) and (13) this makes it possible to determine the moduli of the transition amplitudes C^T and C^P, describing the emission from target and projectile, respectively, in the laboratory system. Their relative phases can be obtained by the following consideration, using a potential curve diagram as given in Figure 27. The $^2\Sigma_g$ state, responsible for the excitation, has even parity and, since parity is conserved, also the autoionizing quasimolecular state as well as the final state of the total system $He^+ + He^+ + e^-$ will have even parity. The

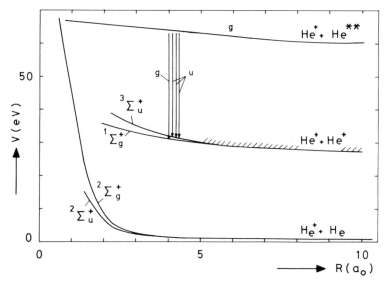

Figure 27. Potential curve diagram for excitation and decay of He** during $He^+ + He$ collisions. The emitted-electron wave can have odd or even parity depending on the parity of the final state $He^+ + He^+$.

subsystem $He^+ + He^+$ on the other hand, which is isoelectronic to H_2, can be in the two final states $^3\Sigma_u$ or $^1\Sigma_g$, which will be populated by autoioniz- ation according to their statistical weights 3/4 and 1/4 as indicated in Figure 27. In order to conserve the total even parity in each of these cases, the ejected electron wave function has to have odd or even parity, respectively, and therefore the electron intensity can be written as

$$I = \tfrac{3}{4}|C_u|^2 + \tfrac{1}{4}|C_g|^2 \tag{57}$$

with C_u and C_g the quasimolecular electron emission amplitudes with odd and even parity, respectively, which can be expressed by the atomic emission amplitudes by[187]

$$C_u(\varepsilon, \vartheta, \varphi) = \sum_L [C_L^T(\varepsilon, \vartheta, \varphi) - (-1)^{\Sigma l_i} C_L^P(\varepsilon, \vartheta, \varphi)]$$

$$\tag{58}$$

$$C_g(\varepsilon, \vartheta, \varphi) = \sum_L [C_L^T(\varepsilon, \vartheta, \varphi) + (-1)^{\Sigma l_i} C_L^P(\varepsilon, \vartheta, \varphi)]$$

By using (21), (22), and (13) to determine the atomic transition amplitudes $C^{T,P}$ with the population amplitudes f_M as fitting parameters one can in this way calculate the resulting electron energy spectra at certain angles (ϑ, φ). The solid lines in Figure 4 are in fact calculated in this way, and an integration over the electron energy yields the total calculated electron intensity at a certain angle (ϑ, φ). The solid line in Figure 26 represents such a fit calculation, and it can be seen that the interference structures are well reproduced. The interferences do not only occur in the scattering plane but in all directions perpendicular to the asymptotic relative velocity of the collision partners. Figure 28 gives a three-dimensional view of the resulting angular distribution. As opposed to Figure 19 here the interferences have been taken into account.

This demonstrates that the collision system has to be described as a quasimolecule even for very large internuclear distances. Although there is no longer any overlap between the electron wave functions of target and projectile, the coherence of the two emission amplitudes is conserved and is reflected in the interference structures of the angular distribution.

7.2. Quasimolecular Autoionization at Small Internuclear Distances

Electron emission from collision complexes at small internuclear dis- tances often yields complicated spectra for the following reasons: (i) The energy difference between excited and ionized state varies with the inter- nuclear distance and therefore no sharp lines can be expected. (ii) It is not clear how large a Doppler effect will be suffered by the electrons since it

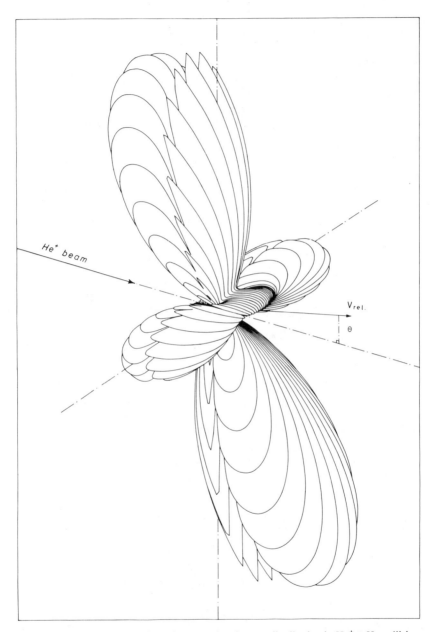

Figure 28. Three-dimensional view of the angular electron distribution in He$^+$ + He collisions, taking target–projectile interferences into account.

is not clear to what extent the electron emission can be regarded to take place from one or the other collision partner or from a molecule, moving with the center-of-mass velocity; (iii) finally, the transition probability for autoionization will in general depend on the internuclear distance, and together with (i) and (ii) this can severely influence the electron spectra.

A good example to study these problems is the collision system He + He at energies between 0.2 and 5 keV. Collisions at these energies lead to a quasimolecular complex which can only autoionize at small internuclear distances: promotion of two electrons gives rise to doubly excited states, which in the separated atom limit correlate to two single excited atoms. As opposed to one doubly excited and one ground-state atom this system is stable against autoionization at large R.

Figure 29 shows two electron energy spectra due to He + He collisions at 0.2 and 0.5 keV, which were reported by Gerber and Niehaus.[190] The peaks can be explained as being due to a Penning-type ionization of two He atoms, excited into the states indicated in the figure. This yields an electron energy of roughly the sum of the excitation energies minus the He

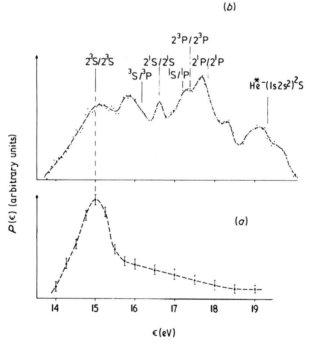

Figure 29. Electron energy spectra from He–He collisions at collision energies of (a) 200 eV and (b) 500 eV. Electrons are due to autoionization of quasimolecular states, correlating to two singly excited He atoms in the separated atom limit. The separated atom states are indicated in part (b).

ionization potential. For the ionization of the collision system in the state $He(2^3S) + He(2^3S)$ a quantitative analysis was possible since the difficulties mentioned above are only partly present: the involved potential curves are known—at least for the excited state $He(2^3S) + He(2^3S)$—and influences from the Doppler effect are nearly negligible at an observation angle of $\vartheta = 90°$, since broadenings due to potential-curve effects are obviously much larger. This made it possible to verify that a quasistationary description of the ionization process analogous to Penning ionization, which originally was developed for thermal collisions, is also valid at these much higher collision energies. This analysis yielded information about the range of internuclear distances in which autoionization takes place: the measurements could well be reproduced by assuming a functional dependence of the ionization rate $\Gamma(R)$ on the internuclear distance R of

$$\Gamma(R) = \Gamma(R^*) \exp\left[-(R - R^*)/R_n\right] \qquad (59)$$

with R^* the distance at the minimum of the excited state potential—in this case $R^* \sim 6a_0$—and R_n a free parameter. The best results were obtained with $R_n = 1.41a_0$.

The functional form (59) implies an increasing ionization rate for decreasing internuclear distance. This has proved to provide a good description of the reality in many cases.[191] Theoretical calculations of Kishinevsky et al.[192] give a somewhat modified picture. Decay rates were calculated for autoionizing quasimolecular states of the molecule $HeBe^{4+}$, consisting of two electrons and two Coulomb centers. The decay rates for the configurations $(2p)^2$ and $(2s)(2p)$, correlating in the separated-atom limit to $He(1s)^2 + Be^{4+}$ and $He^+(1s) + Be^{3+}(2s)$, respectively, were found to pass through a maximum at an internuclear distance of $\sim 0.6a_0$. The maximum ionization rates for these two processes, which can be characterized as transfer-ionization and Penning-ionization, respectively, were found to be $1.6 \cdot 10^{15}/\text{sec}$ and $1.1 \cdot 10^{15}/\text{sec}$, respectively. This is an order of magnitude higher than the value for the united atom. The electron spectra also provide information about the Doppler effect that has to be taken into account in this quasimolecular decay process. Electron energy spectra at various detection angles showed that each peak consists of two components, one of which exhibits a Doppler shift corresponding to the projectile velocity, whereas the other one is unshifted. The quasimolecular electronic wave function can in this case apparently be described by atomic eigenfunctions traveling with the velocities of the nuclei.

Another system that recently attracted much attention in connection with quasimolecular electron emission is the $(LiHe)^+$ collision complex. Electron energy spectra due to low energy $Li^+ + He$ collisions were measured by Yagishita et al.[193,194] for electron emission angles between $20°$ and $150°$

and by Zwakhals *et al.* for emission angles of 0° and 180°. Figure 30 shows spectra reported by Zwakhals *et al.* They consist of two parts: one due to atomic autoionization (AAI) of He, and one due to molecular autoionization (MAI) of the collision complex. Zwakhals *et al.* gave a detailed analysis of these spectra and showed that with a broad distribution of traveling wave factors ascribed to the electronic wave functions of the quasimolecule at each internuclear distance R one can account for the broad distribution of electron energies in the MAI part of the spectra. Also the variations of the

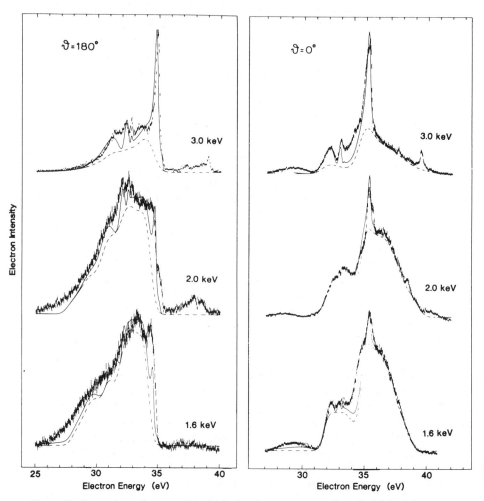

Figure 30. Comparison of measured and calculated energy spectra of autoionization electrons from Li$^+$ + He collisions. The calculated spectra contain two components: one due to molecular autoionization (dashed line) and one due to atomic autoionization, the sum of which is shown as a solid line.

spectra upon changing of the detection angle from 0° to 180° could quantitatively be explained by this interpretation.

Yagishita et al.[194] gave a somewhat different interpretation of the molecular-autoionization spectra: in the first place they assumed that only one emitter velocity is responsible for the Doppler shift, namely, the center-of-mass velocity of the system. The large width of the spectra was ascribed to potential curve effects as described by the Gerber–Niehaus theory.[190] In order to account for the variations of the spectra with the detection angle, the model was refined by Koike[195]: the emitter velocity V responsible for the Doppler shift of electron energies was assumed to vary depending on the internuclear distance R, where the decay takes place, being equal to the He velocity at large R and equal to the center-of-mass velocity at small R. In an elegant description the Doppler shift was introduced as an effective term in the interaction potentials involved. This made it possible to deduce analytical expressions for the Doppler-influenced MAI spectra, which also can account for interferences of different autoionizing transitions yielding the same electron energy.[196] For a quantitative description of the MAI spectra due to $Li^+ + He$ collisions, however, it had to be assumed that the difference potential between excited and ionized quasimolecule has a broad maximum at small internuclear distances, whereas *ab initio* calculations for this system[197] show a narrower maximum.

Two other groups of processes in which electron emission from quasimolecules takes place are Penning ionization and transfer ionization processes. However, they are different from those discussed here in that the electronic energy of the collision system in the initial state is high enough for an ionization process to occur. This leads to ionization even at large impact parameters and very low kinetic collision energies. For a more detailed discussion the reader is referred to recent review articles.[198–200]

Acknowledgment

It is a pleasure to acknowledge many fruitful discussions with several of my colleagues and co-workers in Utrecht, where this article was written in 1983. Especially, I would like to thank A. Niehaus and G. Nienhuis.

References

1. H. Beutler, Z. Phys. **93**, 177 (1935).
2. D. R. Hartree, The Calculation of Atomic Structures, Wiley, New York (1957).
3. F. H. Read, J. Phys. B **10**, 449 (1977).
4. C. A. Coulson, Valence, Oxford University Press, Oxford (1961).
5. R. P. Madden and K. Codling, Phys. Rev. Lett. **10**, 516 (1963).

6. R. P. Madden and K. Codling, *Astrophys. J.* **141**, 364 (1965).
7. S. M. Silverman and E. N. Lassettre, *J. Chem. Phys.* **40**, 1265 (1964).
8. J. A. Simpson, S. R. Mielczarek, and J. Cooper, *J. Opt. Soc. Am.* **54**, 269 (1964).
9. J. A. Simpson, G. E. Chamberlain and S. R. Mielczarek, *Phys. Rev.* **139**, A1039 (1965).
10. K. Siegbahn, C. Nordling, G. Johansson, J. Hedman, P. F. Hedén, K. Hamrin, U. Gelius, T. Bergmark, L. O. Werme, R. Manne, and Y. Baer, *ESCA Applied to Free Molecules*, North-Holland, Amsterdam (1969).
11. N. Oda, F. Nishimura, and S. Tahira, *Phys. Rev. Lett.* **24**, 42 (1970).
12. J. Comer and F. H. Read, *J. Electron Spectrosc.* **1**, 3 (1972).
13. P. J. Hicks and J. Comer, *J. Phys. B* **8**, 1866 (1975).
14. H. W. Berry, *Phys. Rev.* **121**, 1714 (1961).
15. R. B. Barker and H. W. Berry, *Phys. Rev.* **151**, 14 (1966).
16. M. E. Rudd, *Phys. Rev. Lett.* **13**, 503 (1964).
17. M. E. Rudd, *Phys. Rev. Lett.* **15**, 580 (1965).
18. F. D. Schowengerdt, S. R. Smart, and M. E. Rudd, *Phys. Rev. A* **7**, 560 (1973).
19. N. Stolterfoht, *Phys. Lett. A* **37**, 117 (1971).
20. N. Stolterfoht, D. Ridder, and P. Ziem, *Phys. Lett. A* **42**, 240 (1972).
21. A. Bordenave-Montesquieu and P. Benoit-Cattin, *Phys. Lett. A* **36**, 243 (1971).
22. A. Bordenave-Montesquieu, A. Gleizes, M. Rodiere, and P. Benoit-Cattin, *J. Phys. B* **6**, 1997 (1973).
23. J. W. Cooper, U. Fano, and F. Prats, *Phys. Rev. Lett.* **10**, 518 (1963).
24. J. Macek, *Phys. Rev.* **146**, 59 (1966).
25. J. Macek, *J. Phys. B* **1**, 831 (1968).
26. U. Fano, in *Atomic Physics*, Vol. I (edited by B. Bederson, V. W. Cohen, and F. M. J. Pichanik), p. 209, Plenum, New York (1969).
27. U. Fano and C. D. Lin, in *Atomic Physics*, Vol. IV (edited by G. zu Putlitz, E. W. Weber, and A. Winnacker), p. 47, Plenum, New York (1975).
28. U. Fano, in *Physics of Electronic and Atomic Collisions* (edited by J. Risley and R. Geballe), p. 27, University of Washington Press, Seattle (1976).
29. C. D. Lin, *Phys. Rev. A* **10**, 1986 (1974).
30. H. Klar and M. Klar, *J. Phys. B* **13**, 1057 (1980).
31. D. L. Miller and A. F. Starace, *J. Phys. B* **13**, L525 (1980).
32. C. D. Lin, *Phys. Rev. A* **25**, 76 (1982).
33. C. H. Greene, *Phys. Rev. A* **23**, 661 (1981).
34. M. E. Kellman and D. R. Herrick, *J. Phys. B* **11**, L755 (1978).
35. D. R. Herrick and M. E. Kellman, *Phys. Rev. A* **21**, 418 (1980).
36. D. R. Herrick, M. E. Kellman, and R. D. Poliak, *Phys. Rev. A* **22**, 1517 (1980).
37. M. E. Kellman and D. R. Herrick, *Phys. Rev. A* **22**, 1536 (1980).
38. H. J. Yuh, G. Ezra, P. Rehmus, and R. S. Berry, *Phys. Rev. Lett.* **47**, 497 (1981).
39. P. Rehmus and R. S. Berry, *Phys. Rev. A* **23**, 416 (1981).
40. A. K. Bhatia and A. Temkin, *Phys. Rev. A* **11**, 2018 (1975).
41. L. Lipsky and M. J. Conneely, *Phys. Rev. A* **14**, 2193 (1976).
42. Y. K. Ho, *Phys. Rev. A* **23**, 2137 (1981).
43. C. A. Nicolaides and E. Adamides, *Phys. Rev. A* **27**, 1691 (1983).
44. E. Bolduc, J. J. Quemener, and P. Marmet, *J. Chem. Phys.* **57**, 1957 (1972); E. Bolduc and P. Marmet, *Can. J. Phys.* **51**, 2108 (1973).
45. G. C. King, F. H. Read and R. C. Bradford, *J. Phys. B* **8**, 2210 (1975).
46. D. Roy, A. Delage, and J. D. Carette, *J. Phys. B* **9**, 1923 (1976).
47. T. van Ittersum, H. G. M. Heideman, G. Nienhuis, and J. Prins, *J. Phys. B* **9**, 1713 (1976).
48. A. Defrance, *J. Phys. B* **13**, 1229 (1980).
49. G. Gerber, R. Morgenstern, and A. Nienhaus, *Phys. Rev. Lett.* **23**, 511 (1969).

50. G. Gerber, R. Morgenstern, and A. Niehaus, *J. Phys. B* **5**, 1396 (1972).
51. G. Gerber, R. Morgenstern, and A. Niehaus, *J. Phys. B* **6**, 493 (1973).
52. J. O. Olsen and N. Andersen, *J. Phys. B* **10**, 101 (1977).
53. N. Andersen and J. O. Olsen, *J. Phys. B* **10**, L719 (1977).
54. J. O. Olsen, T. Andersen, M. Barat, Ch. Courbin-Gaussorgues, V. Sidis, J. Pommier, J. Agusti, N. Andersen, and A. Russek, *Phys. Rev. A* **19**, 1457 (1979).
55. A. Gleizes, A. Bordenave-Montesquieu, and P. Benoit-Cattin, *J. Phys. B* **14**, 4545 (1981).
56. A. Bordenave-Montesquieu, A. Gleizes, and P. Benoit-Cattin, *Phys. Rev. A* **25**, 245 (1982).
57. R. Bruch, G. Paul, J. Andrä, and L. Lipsky, *Phys. Rev. A* **12**, 1808 (1975).
58. F. H. Read, *Phys. Scr.* **27**, 103 (1983).
59. K. K. Jung, in *Electronic and Atomic Collisions* (edited by N. Oda and K. Takayanagi), p. 787, North-Holland, Amsterdam (1980).
60. R. Morgenstern, A. Niehaus, and U. Thielmann, *Phys. Rev. Lett.* **37**, 199 (1976).
61. A. Bachmann, A. Eberlein, and R. Bruch, *J. Phys. E* **15**, 207 (1982).
62. Yu. S. Gordeev and G. N. Ogurtsov, *Sov. Phys. JETP* **33**, 1105 (1971).
63. F. T. Smith, R. P. Marchi, W. Aberth, and D. C. Lorents, *Phys. Rev. A* **161**, 31 (1967).
64. M. E. Rudd and J. H. Macek, in *Case Studies in Atomic Physics* (edited by M. R. C. McDowell and E. W. McDaniel), p. 47, North-Holland, Amsterdam (1972).
65. P. Dahl, M. Rødbro, B. Fastrup, and M. E. Rudd, *J. Phys. B* **9**, 1567 (1976).
66. A. Gleizes, P. Benoit-Cattin, and A. Bordenave-Montesquieu, *J. Phys. B* **9**, 473 (1976).
67. E. Weigold, A. Ugbabe, and P. J. O. Teubner, *Phys. Rev. Lett.* **35**, 209 (1975).
68. M. A. Dillon and E. N. Lassettre, *J. Chem. Phys.* **67**, 680 (1977).
69. N. L. S. Martin, T. W. Ottley, and K. J. Ross, *J. Phys. B* **13**, 1867 (1980).
70. A. Pochart, R. J. Tweed, M. Doritch, and J. Peresse, *J. Phys. B* **15**, 2269 (1982).
71. E. C. Sewell and A. Crowe, *J. Phys. B* **15**, L357 (1982).
72. Q. C. Kessel, R. Morgenstern, B. Müller, and A. Niehaus, *Phys. Rev. Lett.* **40**, 645 (1978).
73. Q. C. Kessel, R. Morgenstern, B. Müller, and A. Niehaus, *Phys. Rev. A* **20**, 804 (1979).
74. E. Boskamp, O. Griebling, R. Morgenstern, and G. Nienhuis, *J. Phys. B* **15**, 3745 (1982).
75. U. Fano, *Phys. Rev.* **124**, 1866 (1961).
76. U. Fano and J. W. Cooper, *Phys. Rev.* **137**, A1364 (1965).
77. B. W. Shore, *Phys. Rev.* **171**, 43 (1968).
78. R. Morgenstern, A. Niehaus, and G. Zimmermann, *J. Phys. B* **13**, 4811 (1980).
79. C. J. Zwakhals, R. Morgenstern, and A. Niehaus, *Z. Phys. A* **307**, 41 (1982).
80. R. Morgenstern, A. Niehaus, and U. Thielmann, *J. Phys. B* **9**, L363 (1976).
81. R. Morgenstern, A. Niehaus, and U. Thielmann, *J. Phys. B* **10**, 1039 (1977).
82. H. J. Andrä, in *Progress in Atomic Spectroscopy*, Part B (edited by W. Hanle and H. Kleinpoppen), p. 829, Plenum, New York (1979).
83. P. J. Hicks, S. Cvejanović, J. Comer, F. H. Read, and J. M. Sharp, *Vacuum* **24**, 573 (1974).
84. J. M. Sharp, J. Comer, and P. J. Hicks, *J. Phys. B* **8**, 2512 (1975).
85. D. G. Wilden, J. Comer, and P. J. Hicks, *Nature* **273**, 651 (1978).
86. J. A. Baxter, J. Comer, P. J. Hicks, and J. W. McConkey, *J. Phys. B* **12**, 2031 (1979).
87. D. Spence, *Phys. Rev. A* **12**, 2353 (1975).
88. D. Spence, *Comments At. Mol. Phys.* **5**, 159 (1976).
89. J. Fryar and J. W. McConkey, *J. Phys. B* **9**, 619 (1976).
90. W. van de Water and H. G. M. Heideman, *J. Phys. B* **14**, 1065 (1981).
91. G. Nienhuis and H. G. M. Heideman, *J. Phys. B* **8**, 225 (1975).
92. A. J. Smith, P. J. Hicks, F. H. Read, S. Cvejanović, G. C. King, J. Comer, and J. M. Sharp, *J. Phys. B* **7**, L496 (1974).
93. A. Niehaus, in *Electronic and Atomic Collisions* (edited by G. Watel), p. 185, North-Holland, Amsterdam (1978).
94. W. van de Water and H. G. M. Heideman, *J. Phys. B* **13**, 4663 (1980).

95. G. Nienhuis and H. G. M. Heideman, *J. Phys. B* **9**, 2053 (1976).

96. F. H. Read, *J. Phys. B* **10**, L207 (1977).

97. K. Taulbjerg, *J. Phys. B* **13**, L337 (1980).

98. C. Bottcher and K. R. Schneider, *J. Phys. B* **9**, 911 (1976).

99. V. N. Ostrovskii, *Sov. Phys. JETP* **45**, 1092 (1978).

100. M. Ya. Amusia, M. Yu. Kuchiev, and S. A. Sheinerman, *Sov. Phys. JETP* **49**, 238 (1979).

101. V. Schmidt, N. Sandner, W. Mehlhorn, M. Y. Adam, and F. Wuilleumier, *Phys. Rev. Lett.* **38**, 63 (1977).

102. M. J. van der Wiel, G. R. Wight, and R. R. Tol, *J. Phys. B* **9**, L5 (1976).

103. M. K. Bahl, R. L. Watson, and K. J. Irgollic, *Phys. Rev. Lett.* **42**, 165 (1979).

104. H. Hanashiro, Y. Suzuki, T. Sasaki, A. Mikuni, T. Takayanagi, K. Wakiya, H. Suzuki, A. Danjo, T. Hino, and S. Ohtani, *J. Phys. B* **12**, L775 (1979).

105. R. Huster and W. Mehlhorn, *Z. Phys.* **307**, 67 (1982).

106. A. Niehaus, *J. Phys. B* **10**, 1845 (1977).

107. H. Ehrhardt, K. H. Hesselbacher, U. Jung, M. Schulz, T. Tekaat, and K. Willmann, *Z. Phys.* **244**, 254 (1971).

108. S. Hedman, K. Helenelund, L. Asplund, U. Gelius, and K. Siegbahn, *J. Phys. B* **15**, L799 (1982).

109. K. Helenelund, S. Hedman, L. Asplund, U. Gelius, and K. Siegbahn, *Phys. Scr.* **27**, 245 (1983).

110. H. G. M. Heideman, in *Coherence and Correlation in Atomic Collisions* (edited by H. Kleinpoppen and F. J. Williams), p. 493, Plenum, New York (1980).

111. A. Niehaus and C. J. Zwakhals, *J. Phys. B* **16**, L135 (1983).

112. J. F. Williams, in *Progress in Atomic Spectroscopy, Part B*, (edited by W. Hanle and H. Kleimpoppen), p. 1031, Plenum Press, New York (1979).

113. H. C. Bryant, B. D. Dieterle, J. Donahue, H. Sharifian, H. Tootoonchi, and D. M. Wolfe, *Phys. Rev. Lett.* **38**, 228 (1977),

114. H. C. Bryant, K. B. Buterfield, D. A. Clark, C. A. Frost, J. B. Donahue, P. A. M. Gram, M. E. Hamm, R. W. Hamm, and W. W. Smith, in *Atomic Physics 7* (edited by D. Kleppner and F. M. Pipkin), p. 29, Plenum Press, New York (1981).

115. J. T. Broad and W. P. Reinhardt, *Phys. Rev. A* **14**, 2159 (1976).

116. P. A. M. Gram, J. C. Pratt, M. A. Yates-Williams, H. C. Bryant, J. Donahue, H. Sharifian, and H. Tootoonchi, *Phys. Rev. Lett.* **40**, 107 (1978).

117. K. Codling, R. P. Madden, and D. L. Ederer, *Phys. Rev.* **155**, 26 (1967).

118. D. Spence. *J. Phys. B* **14**, 129 (1981).

119. A. Bandzaitis, A. Kupliauskiene, Z. Kupliauskis, V. Tutlys, *Liet. Fiz. Rinkinys* **19**, 187 (1979).

120. L. A. Parcell, J. Langlois, and J. M. Sichel, *J. Phys. B* **9**, 2385 (1976).

121. J. Langlois and J. M. Sichel, *J. Phys. B* **13**, 881 (1980).

122. J. Langlois and J. M. Sichel, *J. Phys. B* **13**, 3109 (1980).

123. D. G. Wilden, P. J. Hicks, and J. Comer, *J. Phys. B* **8**, 1477 (1977).

124. P. Veillette and P. Marchand, *Can. J. Phys.* **54**, 1208 (1976).

125. A. K. Edwards and M. E. Rudd, *Phys. Rev.* **170**, 140 (1968).

126. P. Bisgaard, J. O. Olsen, and N. Andersen, *J. Phys. B* **13**, 1403 (1980).

127. R. P. Madden, D. L. Ederer, and K. Codling, *Phys. Rev.* **177**, 136 (1969).

128. K. Jorgensen, N. Andersen, and J. O. Olsen, *J. Phys. B* **11**, 3951 (1978).

129. R. P. Madden and K. Codling, *J. Opt. Soc. Am.* **54**, 268 (1964).

130. K. Codling and R. P. Madden, *Phys. Rev. A* **4**, 2261 (1971).

131. D. Mathur and D. C. Frost, *J. Chem. Phys.* **75**, 5381 (1981).

132. J. A. Baxter, P. Mitchell, and J. Comer, *J. Phys. B* **15**, 1105 (1982).

133. D. Rassi, V. Pejčev, and K. J. Ross, *J. Phys. B* **10**, 3535 (1977).

134. V. Pejčev, K. J. Ross, and D. Rassi, *J. Phys. B* **10**, L597 (1977).
135. K. J. Ross, T. W. Ottley, V. Pejčev, and D. Rassi, *J. Phys. B* **9**, 3237 (1976).
136. G. Kawei, T. W. Ottley, V. Pejčev, and K. J. Ross, *J. Phys. B* **10**, 2923 (1977).
137. V. Pejčev, D. Rassi, K. J. Ross, and T. W. Ottley, *J. Phys. B* **10**, 1653 (1977).
138. V. Pejčev and K. J. Ross, *J. Phys. B* **10**, 2935 (1977).
139. D. L. Ederer, T. Lucatorto, and R. P. Madden, *Phys. Rev. Lett.* **25**, 1537 (1970).
140. T. J. McIlrath and T. B. Lucatorto, *Phys. Rev. Lett.* **38**, 1390 (1977).
141. D. L. Pegg, H. H. Haselton, R. S. Thoe, P. M. Griffin, M. D. Brown, and I. A. Sellin, *Phys. Rev. A* **12**, 1330 (1975).
142. H. G. Berry, *Phys. Scr.* **12**, 5 (1975).
143. R. Bruch, G. Paul, J. Andrä, and L. Lipsky, *Phys. Rev. A* **12**, 1808 (1975).
144. S. Wakid, A. K. Bhatia, and A. Temkin, *Phys. Rev. A* **21**, 496 (1980).
145. P. Ziem, R. Bruch, and N. Stolterfoht, *J. Phys. B* **8**, L480 (1975).
146. M. Rødbro, R. Bruch, and P. Bisgaard, *J. Phys. B* **12**, 2413 (1979).
147. K. T. Chung, *Phys. Rev. A* **25**, 1596 (1982).
148. P. Feldman and R. Novick, *Phys. Rev.* **160**, 143 (1967).
149. A. K. Bhatia, *Phys. Rev. A* **18**, 2523 (1978).
150. R. L. Simons, H. P. Kelly, and R. Bruch, *Phys. Rev. A* **19**, 682 (1979).
151. J. L. Fox and A. Dalgarno, *Phys. Rev. A* **16**, 283 (1977).
152. C. A. Nicolaides and D. R. Beck, *Phys. Rev. A* **17**, 2116 (1978).
153. C. F. Bunge, *Phys. Rev. A* **19**, 936 (1979).
154. J. R. Willison, R. W. Falcone, J. C. Wang, F. J. Young, and S. E. Harris, *Phys. Rev. Lett.* **44**, 1125 (1980).
155. J. P. Connerade, W. R. S. Garton, and M. W. D. Mansfield, *Astrophys. J.* **165**, 303 (1971).
156. H. W. Wolff, K. Radler, B. Sonntag, and R. Haensel, *Z. Phys.* **257**, 353 (1972).
157. T. B. Lucatorto and T. J. McIlrath, *Phys. Rev. Lett.* **37**, 428 (1976).
158. J. Sugar, T. B. Lucatorto, T. J. McIlrath, and A. W. Weiss, *Opt. Lett.* **4**, 109 (1979).
159. E. Breuckmann, B. Breuckmann, W. Mehlhorn, and W. Schmitz, *J. Phys. B* **10**, 3135 (1977).
160. P. Ziem, G. Wüstefeld, and N. Stolterfoht, Proc. 2nd Int. Conf. on Inner Shell Ionization Phenomena, Freiburg, Abstracts, p. 48 (1976).
161. E. J. McGuire, *Phys. Rev. A* **14**, 1402 (1976).
162. J. P. Connerade, *Astrophys. J.* **159**, 695 (1970).
163. J. P. Connerade, *Astrophys. J.* **159**, 685 (1970).
164. W. E. Cooke and T. F. Gallagher, *Phys. Rev. Lett.* **41**, 1648 (1978).
165. T. F. Gallagher, K. A. Safinya, and W. E. Cooke, *Phys. Rev. A* **21**, 148 (1980).
166. T. F. Gallagher, F. Gounand, R. Kachru, N. H. Tran, and P. Pillet, *Phys. Rev. A* **27**, 2485 (1983).
167. J. Bokor, R. R. Freeman, and W. E. Cooke, *Phys. Rev. Lett.* **48**, 1242 (1982).
168. V. Pejčev, D. Rassi, and K. J. Ross, *J. Phys. B* **13**, L305 (1980).
169. V. Pejčev, T. W. Ottley, D. Rassi, and K. J. Ross, *J. Phys. B* **11**, 531 (1978).
170. M. D. White, D. Rassi, and K. J. Ross, *J. Phys. B* **12**, 315 (1979).
171. D. Rassi and K. J. Ross, *J. Phys. B* **13**, 4683 (1980).
172. D. M. Brink and G. R. Satchler, *Angular Momentum*, second edition, Oxford University Press, Oxford (1968).
173. J. S. Rodberg and R. M. Thaler, *Introduction to Quantum Theory of Scattering*, Academic, New York (1967).
174. K. Blum and H. Kleinpoppen, *Phys. Rep.* **52**, 204 (1979).
175. K. Blum, *Density Matrix, Theory and Applications*, Plenum Press, New York (1981).
176. H. W. Hermann and I. V. Hertel, *Comments At. Mol. Phys.* **12**, 61 (1982).
177. J. Eichler and W. Fritsch, *J. Phys. B* **9**, 1477 (1976).
178. E. G. Berezhko, N. M. Kabachnik, and V. V. Sizov, *J. Phys. B* **11**, 1819 (1978).

179. W. Mehlhorn and K. Taulbjerg, *J. Phys. B* **13**, 445 (1980).
180. R. Bruch and H. Klar, *J. Phys. B* **13**, 1363 (1980).
181. R. Bruch and H. Klar, *J. Phys. B* **13**, 2885 (1980).
182. B. Cleff and W. Mehlhorn, *J. Phys. B* **7**, 593 (1974).
183. E. Boskamp, R. Morgenstern, P. v.d. Straten, and A. Niehaus, *J. Phys. B* **17**, 2823 (1984).
184. N. Stolterfoht, D. Brandt, and M. Prost, *Phys. Rev. Lett.* **43**, 1654 (1979).
185. C. J. Zwakhals, R. Morgenstern, and A. Niehaus, *Z. Phys. A* **307**, 41 (1982).
186. G. Leuchs and H. Walther, *Z. Phys. A* **293**, 93 (1979).
187. E. Boskamp, R. Morgenstern, and A. Niehaus, *J. Phys. B* **15**, 4577 (1982).
188. A. Niehaus, in Proceedings of the 5th European Conference on General Physics: "Trends in Physics," Istanbul 1981.
189. D. C. Lorents and W. Aberth, *Phys. Rev.* **139A**, 1017 (1965).
190. G. Gerber and A. Niehaus, *J. Phys. B* **9**, 123 (1976).
191. A. Niehaus, *Ber. Bunsenges. Phys. Chem.* **77**, 632 (1973).
192. L. M. Kishinevski, B. G. Krakov, and E. S. Parilis, *Phys. Lett.* **85A**, 141 (1981).
193. A. Yagishita, H. Oomoto, K. Wakiya, H. Suzuki, and F. Koike, *J. Phys. B* **11**, L111 (1978).
194. A. Yagishita, K. Wakiya, T. Takayanagi, H. Suzuki, S. Ohtani, and F. Koike, *Phys. Rev. A* **22**, 118 (1980).
195. F. Koike, *J. Phys. Soc. Jpn.* **51**, 618 (1982).
196. F. Koike, A. Yagishita, and M. Furune, *Phys. Rev. Lett.* **48**, 735 (1982).
197. K. Tanaka, private communication to F. Koike.
198. A. Niehaus, in *Physics of Electronic and Atomic Collisions* (edited by S. Datz), p. 237, North-Holland, Amsterdam (1982).
199. A. Niehaus, *Comments At. Mol. Phys.* **9**, 153 (1980).
200. A. Niehaus, in *The Excited State in Chemical Physics* (edited by J. Wm. McGowan), p. 399, Wiley, New York (1981).

9

Near Resonant Vacancy Exchange between Inner Shells of Colliding Heavy Particles

N. STOLTERFOHT

1. Introduction

During the last two decades, considerable interest has been devoted to studies of inner-shell excitation in slow ion–atom collisions. One of the fundamental processes that contributes to the production of inner-shell vacancies is charge exchange.[1-7] For inner shells close in energy, charge exchange is nearly resonant and, hence, available vacancies are transferred with high probability. When the projectile velocity is smaller than the velocity of the electron involved, the collision may be described using the molecular orbital (MO) model.[8-11] Within this model, inner-shell vacancy transfer proceeds via coupling of the MO's transiently formed during the collision. However, vacancy transfer takes place at relatively large internuclear distances in a fairly well localized coupling region where the molecular orbitals merge into atomic orbitals (AO). In this case the treatment of the collision within the framework of an atomic model is equivalent to that of a molecular model.

The outstanding feature of the near resonant charge exchange considered here is that the inner-shell orbitals under study are well separated from higher-lying orbitals. As a consequence, the theoretical analysis of the near-resonant charge exchange may be restricted to a few basis states.

N. STOLTERFOHT • Hahn-Meitner-Institut Berlin GmbH, Bereich Kern- und Strahlenphysik, D-1000 Berlin 39, Federal Republic of Germany.

415

Indeed, the collision systems with nearly matching inner shells constitute ideal cases to verify few-state models.

The application of few-state models has a long tradition in the field of outer-shell excitation, as described in detail in the work by Smirnov,[12] Rapp and Francis,[13] and Olson et al.[14] In the analysis, two-state models by Demkov,[15] Nikitin,[16,17] Landau,[18] and Zener[19] have been utilized with considerable success. Hence, it appeared challenging to apply these models in the field of inner-shell excitation.[20,21]

Different types of vacancy exchange processes are to be considered for inner shells. When an inner-shell vacancy is produced during the collision (e.g., in the higher-lying MO), the coupling region is passed once. For near zero impact parameters the single-passage process is denoted as vacancy sharing,[21] since the vacancy production takes place at distances small in comparison with the coupling radius. In the case, the charge exchange mechanism can be considered independently of the creation of the vacancy. In slow collisions, the transfer probability is primarily dependent on the energy difference of the states involved in the transition region. Hence, studies of single-passage processes yield information about molecular potential curves at relatively large internuclear distances.

When vacancies are present prior to the collision (e.g., by using highly stripped projectiles), the coupling region is passed twice. In this case, interference effects may occur so that the double passage process often exhibits oscillatory structures in the relevant excitation function.[15] The oscillations are governed by the molecular potentials at relatively small internuclear distances. Thus, the double-passage process reveals information about the molecular orbitals near the united atom limit.

During the last few years, considerable work has been performed in the field of near-resonant vacancy exchange between inner shells. Most emphasis has been given to processes involving K shells. Meyerhof[21] was first to investigate systematically K-vacancy sharing. Excellent agreement has been found between experimental data and theoretical results derived using analytical expressions from Demkov's model.[15] This showed that an analytical two-state model is well suited to describe K-vacancy sharing. In the past, several studies[22-28] have confirmed the Demkov–Meyerhof method. Similar inferences have been obtained from the work concerning double K-vacancy sharing.[29-32] More recent work has been devoted to cross sections for K-vacancy exchange[33-36] and to the detailed analysis of the double-passage process.[37-40] Related theoretical work has been carried out using both atomic basis states[41-43] and molecular basis states.[44-48] Also, further effort has been made to apply analytical model matrix elements.[11,27,49-53] Moreover, vacancy exchange between K and L shells has been studied experimentally[54-61] and theoretically.[62,63] In this case primarily single-passage processes have been investigated. In the work dealing with vacancy exchange between L shells both couplings between the π orbitals[64,65] and between the σ orbitals[66-68] have been considered.

The present article is devoted to experimental and theoretical studies concerning near-resonant charge exchange between inner shells. With regard to the significant amount of work performed in this field, no attempt is made to give a complete summary here. The present article is limited to processes involving K and L shells in relatively light systems, where relativistic effects are not dominant. Moreover, collision velocities smaller than or about equal to that of the inner-shell electron will be considered. For more information concerning charge-exchange studies the reader is referred to detailed articles.[3-7,69]

Here, emphasis is given to the application of few-state models in the description of the near-resonant vacancy exchange between inner shells. It is well known that the quantities relevant for inner-shell electrons may readily be scaled.[10] Therefore, the attempt is made to apply as much as possible analytic functional forms to describe the characteristic quantities of the collision system. In particular, analytic model matrix elements[50,63] derived from calculations with screened hydrogenic wave functions are applied. Hydrogenic wave functions are suitable for inner shells, since the electrons feel primarily the nuclear Coulomb field of the collision particles. Input for the analytic expressions is the standard information about atomic ionization potentials available in tabulated form. This procedure avoids a fresh numerical calculation for each new collision system.

Two aspects should be emphasized in connection with the use of the analytical matrix elements. First, transition probabilities may readily be evaluated by solving related coupled equations, once the matrix elements are determined. The effort to integrate the coupled equations is significantly smaller than that to evaluate numerically the matrix elements. Second, in a model calculation the mechanisms involved in the ion–atom collision process are more evident than in a numerical calculation. This evidence may contribute significantly to the qualitative understanding of the vacancy exchange process.

In the following Section 2 the basic theory of near-resonant vacancy exchange in few-state systems is presented. In Section 3, two-state processes are reviewed involving transitions between K shells and between L shells. Double-passage mechanisms are treated in addition to single-passage mechanisms. Section 4 is devoted to vacancy transfer in multistate systems. First, vacancy transfer between K and L shells is treated regarding various three-state systems. Then, coupling of four σ orbitals is considered producing vacancy transfer between adjacent L shells. Section 5 is devoted to final remarks.

Atomic units are used throughout this chapter.

2. Theoretical Methods

2.1. Vacancy Exchange Mechanisms

The theoretical analysis of near-resonant vacancy exchange between inner shells is performed using various approximations.[10] The nuclear motion is treated classically yielding the internuclear distance R as a function of time t. Furthermore, the independent particle model is used assuming one active electron in the varying two-center field of the collision partners. The remaining passive electrons have the effect of screening the nuclei involved in the collision. Hence, the electronic system is governed by the effective one-electron Hamiltonian $H = T + V_A + V_B$, where T is the kinetic energy operator of the electron and V_A and V_B refer to the potentials of the screened nuclei A and B, respectively. The Hamiltonian H depends parametrically on R and, thus, on t.

Accordingly, the molecular orbitals ϕ_n defined by the eigenvalue equation $H\phi_n = \varepsilon_n\phi_n$ depend on R. It is common use to denote the molecular orbitals also as adiabatic states. The atomic orbitals ϕ_n^0 are obtained from $H_0\phi_n^0 = \varepsilon_n^0\phi_n^0$, where H_0 stands for the Hamiltonians $H_A = T + V_A$ and $H_B = T + V_B$ of the atoms A and B, respectively. In this chapter, the atomic orbitals are also referred to as diabatic states, although it is realized that the term "diabatic" may be used for more general classes of states.[70] Important quantities in this chapter are the matrix elements $H_{nm} = \langle \phi_n^0|H|\phi_n^0 \rangle$ for potential coupling which govern the interaction of the atomic orbitals. In the following, the diagonal matrix elements H_{nn} are denoted as *diabatic potentials*,[41] since they represent the expectation value of the energy with respect to the diabatic states. Likewise, the nondiagonal matrix elements H_{nm} are also referred to as *interaction potentials*.

In the collision systems studied here, the molecular orbitals are formed by adiabatic interaction of a few atomic orbitals. As an example[11] the formation of molecular orbitals on the basis of two atomic states is shown schematically in Figure 1. Three ranges of the internuclear distance R are distinguished. They will be referred to as the overlap region, the distortion region, and the separated atom region occurring with increasing internuclear distance R. In the separated atom region ranging roughly down to about twice the outer shell radii of the collision partners, the orbitals are atomic in nature. The orbital energy curves are not affected, as the nuclear charges of the approaching particles are essentially screened. In the distortion region, the nuclei have dived into the outer shells and the screening is incomplete. The orbitals are still atomic in nature; however, their energies are distorted in the Coulomb field of the approaching nuclei.

At the distance R_m the inner atomic orbitals (AO) start to overlap and molecular orbitals are formed (for K shells $R_m \approx 5a_K$, where a_K is the

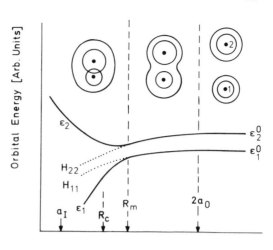

Figure 1. Schematic diagram of molecular orbital energies ε_1 and ε_2. The underlying diabatic potentials are H_{11} and H_{22}. The parameters a_I and R_c denote the inner shell radius and the coupling radius respectively. The quantities R_m and $2a_0$ separate typical regions of the internuclear distance. (From Ref. 11.)

mean K-shell radius). The interaction of the states causes a repulsion of the MO curves so that the difference in the energies ε_1 and ε_2 increases with decreasing R. The overlap region encloses also the coupling region where the charge transfer transitions take place during the collision. The coupling region which is centered around R_c ($\approx 3a_K$ for K shells) is located well outside the inner-shell radii involved.

In Figure 2 molecular orbital energies are shown for collision systems important in this chapter. In near symmetric collision systems the K shells of the collision partners are close in energy. The K-vacancy exchange takes place at distances where 1σ and 2σ orbitals merge into the $1s$ levels of the heavier and lighter collision partner, respectively. In asymmetric collision systems a near matching of K and L shells may occur. The $2s$ and $2p$ levels of the heavier partner and the $1s$ level of the lighter partner correlate with the 2σ, 3σ, and 4σ, respectively. Here, the charge transfer process is denoted as KL-vacancy exchange. In more symmetric systems where the $1s$ level is lower in energy than the $2p$ and $2s$ levels, the notation LK-vacancy exchange is used. Again, in near symmetric systems the L shells of the collision partner are close in energy. In this case the $2p$ and $2s$ levels of the particles correlate with four σ orbitals and two π orbitals. Here, L-vacancy exchange resulting from σ orbital coupling is to be considered as well as that produced by π orbital coupling.

The characteristic feature of the systems relevant for this article is that their diabatic potential curves are rather parallel. This feature is incorporated in the models by Demkov[15] and Nikitin[16,17] and it differs from the curve-crossing feature adopted in the Landau–Zener model.[18,19] Hence, the models by Demkov and Nikitin are favorable for the cases treated here.

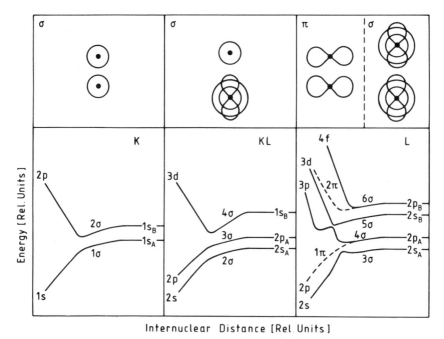

Internuclear Distance [Rel. Units]

Figure 2. Schematic molecular orbital diagrams for two-state, three-state, and four-state systems relevant for the coupling between K shells, K and L shells, and L shells, respectively. In the upper part of the figure, the shapes of the orbitals involved are indicated.

2.2. Basic Formalism

In this section the basic formalism for the time evolution of the vacancy exchange process is given. The analysis is made for atomic orbitals as well as for molecular orbitals, since the two treatments are nearly equivalent for the processes considered here. A basis transformation is introduced allowing for the change from atomic orbitals to molecular orbitals. This transformation is required, since the coupling matrix elements are deduced within the framework of an atomic treatment, whereas the integration of the coupled equations is performed within the framework of an adiabatic model.

The time evolution of the dynamic state Ψ of the relevant vacancy is determined by the time-dependent Schrödinger equation $id\Psi/dt = H\Psi$. It is convenient to expand Ψ in terms of appropriate basis states ψ_n:

$$\Psi = \sum_n c_n \psi_n \exp\left(-i \int^t \Omega_n \, dt'\right) \qquad (1)$$

where Ω_n is a suitable phase. It is noted that the basis states are not necessarily orthogonal. The Schrödinger equation results in the following system of coupled equations[6]:

$$i\dot{c}_n = \sum_m (\tilde{H}_{nm} - i\tilde{K}_{nm} - \Omega_m \delta_{nm}) c_m \exp\left(i \int^t \Delta\Omega_{nm} \, dt'\right) \qquad (2)$$

where $\tilde{H}_{nm} = \langle \tilde{\psi}_n | H | \psi_m \rangle$ and $\tilde{K}_{nm} = \langle \tilde{\psi}_n | \partial/\partial t | \psi_m \rangle$ are the matrix elements for potential and dynamic coupling, respectively. The arbitrary phase $\Delta\Omega_{nm} = \Omega_n - \Omega_m$ is fixed by setting $\Omega_n = \tilde{H}_{nn} - i\tilde{K}_{nn}$ and $\Omega_m = \tilde{H}_{mm} - i\tilde{K}_{mm}$ to remove the diagonal term ($m \neq n$) from the sum in Eq. (2).

In the derivation of the coupled equation (2) additional basis states $\tilde{\psi}_n$ are introduced being orthogonalized with respect to ψ_m, i.e., $\langle \tilde{\psi}_n | \psi_m \rangle = \delta_{nm}$. These states are obtained as $\tilde{\psi}_n \sum_i \tilde{S}_{ni} \psi_i$, where $\tilde{S}_{ni}$ is an element of the inverse overlap matrix: $\tilde{S}_{ni} = (S^{-1})_{ni}$ with $S_{ni} = \langle \psi_n | \psi_i \rangle$. For orthogonal states it follows that $\tilde{S}_{ni} = S_{ni} = \delta_{ni}$ and, thus, $\tilde{\psi}_n$ and ψ_n are identical.

The coupled equations are integrated using the initial conditions $c_n = \delta_{nm}$, which implies full occupation of the state m prior to the passage of the transition region. Let us denote the transition probability to state n for single and double passage of the transition region by W_n and P_n, respectively. These probabilities are obtained as a function of the impact parameter b from

$$W_n(b) = |c_n(0, \infty)|^2 \qquad (3a)$$

$$P_n(b) = |c_n(-\infty, \infty)|^2 \qquad (3b)$$

i.e., they are deduced for $t \to \infty$ after starting the integration of the coupled equations at $t = 0$ and $t = -\infty$, respectively. The total cross section for the excitation of state n is obtained by integrating over all impact parameters

$$\sigma_n = 2\pi \int P_n(b) b \, db \qquad (4)$$

Different aspects have to be considered when a choice is made with regard to the basis states ψ_n. First, care should be taken to truncate the basis, since the effort to solve the system of coupled equations increases significantly with increasing number of basis states. Second, since experimentally the occupation of a state ψ_n is observed after the particles have separated, care should be taken that ψ_m approaches asymptotically atomic states of the collision partners.

The fact that asymptotically the electron travels with one of the nuclei is generally taken into account by inclusion of translational factors given by Schneidermann and Russek[71]:

$$\chi = e^{-f(\mathbf{r}, \mathbf{R})\mathbf{v} \cdot \mathbf{r}} \qquad (5)$$

The proper choice of the function $f(\mathbf{r}, \mathbf{R})$ is still a matter of controversy[1,10,44] whose treatment is outside the scope of the present chapter. It should only be noted that $f = 1/2$ and $f = -1/2$ yield plane-wave translational factors[1] for atomic orbitals located at centers A and B, respectively. It should be added that the translational factor implies also the time-dependent phase $\exp(-iv^2t/8)$ where the origin is located at the midpoint between the collision partners. This phase cancels the following expressions of the coupled equations.

The asymptotic atomic behavior of the basis states is trivially achieved for the atomic orbitals, which form a convenient basis set because of their simplicity. Unfortunately, atomic orbitals that belong to different centers are not orthogonal at finite internuclear distances, since they originate from two different Hamiltonians H_A and H_B. Hence, the overlap of the basis states has to be accounted for in the coupled equations.

In the following, two methods will be used to treat the nonorthogonality of the atomic basis states. In the first method the original basis states are unchanged and an additional set of basis states is introduced as above. Then, the coupled equations are particularly simple, as they do not imply dynamic couplings. However, problems may arise with non-Hermitian symmetry of the coupling matrix. In the second method, the original basis states are changed such that they are mutually orthogonal. Then, the Hamiltonian matrix is Hermitian symmetric but additional dynamic couplings may occur in the corresponding equations.

When the atomic states ϕ_n^0 are used in conjunction with plane-wave translational factors,[1] the dynamic coupling matrix elements cancel in the coupled equation. For the occupation amplitudes a_n^0 of the atomic state n it follows that

$$i\dot{a}_n^0 = \sum_{m \neq n} \tilde{H}_{nm} a_n^0 \exp\left(i \int^t \Delta \tilde{H}_{nm}\, dt'\right) \tag{6}$$

where $\tilde{H}_{nm} = \sum_j \tilde{S}_{nj} H_{jm}$ with $H_{jm} = \langle \phi_j^0 | H | \phi_m^0 \rangle$ and $H_{jm} = \langle \phi_j^0 | \exp(i\mathbf{v}\mathbf{r}) H | \phi_m^0 \rangle$ for orbitals ϕ_j^0 and ϕ_m^0 located at the same center and at different centers, respectively. It is seen that the translational factors are important when the orbitals belong to different centers. The phase $\Delta\Omega_{nm}$ is set to be equal to $\Delta\tilde{H}_{nm} = \tilde{H}_{nn} - \tilde{H}_{mm}$.

Hermicity of the Hamiltonian H implies that the coupled equations conserve probability (unitarity) even with the approximation of using a finite basis set.[72] However, further approximations in solving the coupled equations may result in the loss of unitarity. Such loss may occur since the coupling matrix is not necessarily Hermitian symmetric.[72] Note that $\tilde{H}_{nm}$ is not equal to $\tilde{H}_{mn}^*$, in general. A well-known method[13,15,45,52] to ensure unitarity is to force the coupling matrix to be Hermitian symmetric by putting the corresponding matrix elements equal to some average value, e.g., to the arithmetic mean

$$\tilde{H}_{nm}^{av} = \tfrac{1}{2}(\tilde{H}_{nm} + \tilde{H}_{mn}^{*}) \tag{7}$$

It should be realized that certain approximations are adopted in conjunction with the averaging procedure.[72]

Clearly, for orthogonal basis states the coupling matrix is Hermitian symmetric. Hence, the procedure to achieve Hermitian symmetry in the coupling matrix becomes more transparent if one starts with orthogonal basis states. Let us denote orthogonal basis states by $\hat{\phi}_i^0$ and the corresponding Hamiltonian matrix by $\hat{H}_{nm}$. These states may be obtained by orthogonalizing the original states ϕ_i^0 (e.g., by using the well-known Schmidt method). In general, one may set

$$\hat{\phi}_i^0 = \sum_k q_{ik}\phi_k^0 \tag{8}$$

where q_{ik} are appropriate coefficients depending on the orthogonalization method used. The orthogonalization modifies the basis states so that they are not purely atomic any more. Contrary to the averaging procedure the orthogonalization method reveals the additional dynamical couplings so that they may properly be taken into account in the corresponding equations.

Further suitable choices for basis states are the molecular orbitals ϕ_n following from the Hamiltonian H. Hence, molecular basis states have the advantage that they are orthogonal. Also, they merge into atomic orbitals at large internuclear distances. Unfortunately, the inclusion of translational factors may impose nonorthogonality on the traveling molecular orbitals. However, when the same factor $f(\mathbf{r}, \mathbf{R})$ for each ϕ_n is used in Eq. (5), the MO basis retains orthogonality.[10] Here, such a basis will be adopted so that problems due to nonorthogonality are avoided.

When the molecular orbitals are used, the (nondiagonal) matrix elements H_{nm} cancel in the coupled equations so that the occupation amplitudes a_n of the state ϕ_n are obtained from

$$\dot{a}_n = -\sum_{m \neq n} K_{nm}a_m \exp\left(i \int^t \Delta\varepsilon_{nm} dt'\right) \tag{9}$$

with $K_{nm} = \langle\phi_n|\partial/\partial t|\phi_m\rangle$ and $\Delta\varepsilon_{nm} = \varepsilon_n - \varepsilon_m$, where the orbitals ϕ_n and ϕ_m are assumed to include the translational factors (5).

In the semiclassical approach where the internuclear distance R is obtained as a function of time it follows that $\partial/\partial t = \dot{\mathbf{R}}\nabla_R$. In particular, for transitions between states of equal symmetry treated exclusively in this article, the dynamic coupling matrix element reduces to

$$K_{nm} = v_R M_{nm} \tag{10}$$

where $M_{nm} = \langle \phi_n | \partial/\partial R | \phi_m \rangle$ is the radial coupling matrix element and v_R is the radial velocity $\partial R/\partial t$. For a straight line trajectory it follows that

$$V_R = v \left[1 + \left(\frac{b}{vt} \right)^2 \right]^{-1/2} = v \left[1 - \left(\frac{b}{R} \right)^2 \right]^{1/2} \tag{11}$$

For the collision systems studied in the chapter the molecular orbitals are generally well expressed as linear combinations of a limited number of atomic states, i.e.,

$$\phi_n = \sum_i u_{ni} \phi_i^0 \tag{12}$$

Equation (12) implies a basis transformation which yields the adiabatic energies:

$$\varepsilon_n = \sum_{ij} u_{ni}^* u_{nj} H_{ij} \tag{13}$$

and the dynamic coupling matrix elements

$$K_{nm} = \sum_{ij} (u_{ni}^* \dot{u}_{mj} S_{ij} + u_{ni}^* u_{mj} K_{ij}^0) \tag{14}$$

where $K_{ij}^0 = \langle \phi_i^0 | \partial/\partial t | \phi_j^0 \rangle$ and, again, $S_{ij} = \langle \phi_i^0 | \phi_j^0 \rangle$. The evaluation of the second term in Eq. (14) becomes easier if one makes use of the relations $K_{ii}^0 = 0$, $K_{ij}^0 = -K_{ji}^0$ for orbitals located at the same center, and $K_{ij}^0 = K_{ji}^0$ for orbitals located at different centers.

The basis transformation (12) is greatly simplified when orthogonalized basis states $\hat{\phi}_i^0$ are used. Let $\hat{u}_{ni}$ be the corresponding expansion coefficients. Then, in Eq. (14) S_{ij} and K_{ij}^0 are to be replaced by δ_{ij} and $\hat{K}_{ij}^0 = \langle \hat{\phi}_i^0 | \partial/\partial t | \hat{\phi}_j^0 \rangle$, respectively. It is noted that orthogonal basis states yield $\hat{K}_{ij}^0 = -\hat{K}_{ji}^0$ for any pair of orbitals. The important consequence of the orthogonalization is that the matrix $(\hat{u}_{ni})$ is unitary. Then, the coefficients $\hat{u}_{ni}$ may conveniently be obtained diagonalizing the potential coupling matrix $(\hat{H}_{ij})$. This method to obtain the coefficients $\hat{u}_{ni}$ will be used in conjunction with the model matrix elements given in the next section.

It should be emphasized that in cases where the molecular orbitals may be expressed in terms of a few atomic orbitals, the adiabatic and diabatic treatments of the collision are equivalent. Differences in the results from the two treatments originate primarily from differences in the applications of the translational factors.[43] For instance, the translational factors may be included in the atomic states. Alternatively, stationary atomic states may be used and molecular translational factors may be applied after the molecular states are formed.

2.3. Model Matrix Elements

For two states, Demkov[15] and Nikitin[17] introduced analytic model matrix elements which became widely used because of their convenient applicability. The matrix elements do not imply translational factors, so that they are real. Analytical matrix elements for which Hermitian symmetry is adopted may be assumed to be based on orthogonal states. Nevertheless, the notation H_{nm} will be used for the model matrix elements. With $\Delta H = H_{11} - H_{22}$ and $H_{21} = H_{12}$ the results are[17]

$$\Delta H = \Delta\varepsilon - D e^{-\alpha R} \tag{15a}$$

$$H_{12} = -\frac{C}{2} e^{-\alpha R} \tag{15b}$$

where $\Delta\varepsilon$, α, C, and D are adjustable model parameters. The quantity $\Delta\varepsilon$ is equal to the diabatic energy difference for large R, and α is a measure for the inverse radius of the electronic orbitals involved.

Nikitin[17] replaced C and D by more physical model parameters R_c and θ in setting $C = \Delta\varepsilon \sin\theta \exp(\alpha R_c)$ and $D = \Delta\varepsilon \cos\theta \exp(\alpha R_c)$. The angle θ obtained from $\tan\theta = C/D$ governs the mixing of the diabatic states, when adiabatic states are formed at small R. Again, R_c specifies the coupling radius, which may be evaluated from

$$R_c = \frac{1}{\alpha} \ln \frac{(C^2 + D^2)^{1/2}}{\Delta\varepsilon} \tag{16}$$

The matrix elements used by Demkov[15] are obtained with $D \ll C$ or $\theta \approx \pi/2$. Then, the second term in Eq. (15a), which models the distortion of the diabatic potentials, is small, so that $\Delta H \approx \Delta\varepsilon = \text{const}$, which is the specific feature of Demkov's model. In this case, the coupling radius is determined from $C \exp(-\alpha R_c) = \Delta\varepsilon$, i.e., the transition takes place near the distance where ΔH is equal to $2H_{12}$. On the contrary, for $C \ll D$ or $\theta \approx 0$ the coupling radius is determined by $D \exp(-\alpha R_c) = \Delta\varepsilon$. If this equation has a solution, it follows that $\Delta H(R_c) = 0$, i.e., the diabatic potential curves cross. The transitions take place at the crossing radius as adopted in the Landau–Zener model. This shows that the Demkov model and the Landau–Zener model refer to limiting cases for the two-state coupling mechanism.[27]

Parameters appropriate for Demkov's model have been discussed by Olson *et al.*[14] In Table 1 two sets of parameters are given. They are governed by the "velocity" parameters $\alpha_n = |2\varepsilon_n^0|^{1/2}$ following from the related orbital energies ε_n^0, whose values are available in tabulated form.[73,74] The set I of parameters has been used by Meyerhof[21] and Briggs.[49] Modified parameters more suitable for Demkov's model (set II in Table 1) will be discussed in Section 3.2.

It is noted that the logarithm in Eq. (16) is a function slowly varying with collision systems. Consequently, the coupling radius is essentially scaled by the inverse of $\alpha \leqslant \alpha_{12} = (\alpha_1 + \alpha_2)/2$. Thus, for rapid estimates of the coupling radius one may use the relation[14]

$$R_c = \frac{R_c^*}{\alpha_{12}} \tag{17}$$

where the normalized coupling radius R_c^* is expected to be rather independent of the collision system. Inspection of molecular orbital diagrams given in the following sections shows that $R_c^* = 3$ is a suitable mean value for $1s$ orbitals. Since α_{12}^{-1} is about equal to the inner-shell radii involved, Eq. (17) confirms readily that the coupling region is located well outside the inner-shell radii as suggested in Figure 1.

In the application of the matrix elements (15) to inner-shell electrons, problems arise with the exponential R dependence of the diabatic potential.[50] Inner-shell electrons are influenced by strong nuclear fields which involve a $1/R$ dependence following from the Coulomb law. Accordingly, several authors[50–53,75] have calculated the matrix elements H_{nm} analytically using (nonrelativistic) hydrogenic $1s$ wave functions. Similarly, the matrix elements H_{nm} may be calculated analytically using hydrogenic $2s$ and $2p$ wave functions.

Table 1. Parameters for Demkov's Matrix Elements Expressed in Eq. (15) with $D = 0$.[a]

Set	$\Delta\varepsilon$	α	C
I	$\Delta\varepsilon_{12}^0$	α_{12}	$\alpha_1^2 + \alpha_2^2$
II	$z\Delta\varepsilon_{12}^0$	$c\alpha_{12}$	$2k\bar{\alpha}_{12}^2$

[a] Set I refers to parameters $\Delta\varepsilon$ and α proposed by Meyerhof[21] and C by Briggs.[49] The quantities $\alpha_{12} = (\alpha_1 + \alpha_2)/2$, $\bar{\alpha}_{12} = (\alpha_1\alpha_2)^{1/2}$, and $\Delta\varepsilon_{12}^0 = \varepsilon_1^0 - \varepsilon_2^0$ are obtained from $\alpha_n = |2\varepsilon_n^0|$ with the orbital energy ε_n^0. Set II refers to modified parameters, where z, c, and k are given in Table 2.

Different types of matrix elements appear in the calculations, as the orbitals may be located at different centers. Let N and M be labels specifying the centers A and B. Then, one deals with the following categories: The diagonal matrix elements H_{nn}^{NN}, the nondiagonal matrix element H_{nm}^{NN} for one center, and the nondiagonal matrix element H_{nm}^{NM} for two centers ($H_{nm} = H_{mn}$, $n \neq m$, $N \neq M$).

The expressions deduced within the hydrogenic approximation do not exhibit exact exponential R dependences. The exponential functions in the matrix elements have polynomials as prefactors which are slowly increasing with R. Thus, to a good approximation the polynomials may be replaced by the constants when the exponents of the corresponding exponential functions are (slightly) decreased. Then, the matrix elements are particularly simple[51,63]:

$$H_{nn}^{NN} = \varepsilon_n^0 - \frac{1 - e^{-d\alpha_n R}}{R} Q_M \tag{18a}$$

$$H_{nm}^{NN} = \frac{3}{2} \frac{1 - P(d'\alpha_{nm}R)\, e^{-d'\alpha_{nm}R}}{\bar{\alpha}_{nm}R^2} Q_M \tag{18b}$$

$$H_{nm}^{NM} = k\bar{\alpha}_{nm}^2\, e^{-c\alpha_{nm}R} \tag{18c}$$

where $P(x) = 1 + x + x^2/2 + x^3/6 + x^4/24$. In the following the matrix elements (18) will be referred to as being deduced within the screened hydrogenic model (SHM).

The SHM matrix elements are primarily determined by the mean values $\alpha_{nm} = \frac{1}{2}(\alpha_n + \alpha_m)$ and $\bar{\alpha}_{nm} = (\alpha_n\alpha_m)^{1/2}$ following from the "velocity" parameters $\alpha_n = |2\varepsilon_n^0|^{1/2}$. Similarly, the quantity α_M is defined referring to center M only. It is equal to α_{1s} or to the mean value α_{2s2p} depending on whether K or L shell orbitals, respectively, are involved at center M.

The quantity α_M is a measure for the screened charge Z_M of particle M, i.e., $Z_M = n\alpha_M$, where n is the related principal quantum number. From Z_M the variable charge Q_M is deduced using exponential screening. For neutral systems $Q_M = Z_M \exp(-\alpha_0 R)$ is chosen, where the screening constant α_0 refers to the electrons outside the inner shell involved.[11] The screening constant is derived from $\alpha_0 = s\alpha_{NM}^t$ with $\alpha_{NM} = (\alpha_N + \alpha_M)/2$ and s and t being given in Table 2.

The parameters c, d, d', and k are constants that result from neglecting the polynomials mentioned above. Their values given in Table 2 have been obtained[51] by reproducing molecular orbital energies from other sources as shown in the following sections. The constants c, d, d', and k are dimensionless and their values are found to be independent of the collision system within a wide range of atomic numbers.[51] Hence, the matrix

Table 2. Parameter for the Model Matrix Element (18).[a]

Shells	Orbitals	c	d	d'	k	s (a.u.)	t	z
K-K	1σ-2σ	0.88	0.75	—	1.3	0.42	2/3	0.88
L-L	1π-2π	0.88	0.75	—	1.5	0.42	2/3	0.80
K-L	2σ-3σ-4σ	0.86	0.5[b]	1.5	2.1	0.65	1/2	—
L-L	3σ-4σ-5σ-6σ	0.86	0.5	1.5	3.6	0.42	2/3	—

[a] The parameter z is a measure of the binding effect; see text.
[b] This value is valid for $2s$ and $2p$ orbitals. For $1s$ orbitals $d = 0.75$ as in the first line.

elements (18) are essentially scaled by the parameters α_n so that they may be readily applied to different collision systems.

It is interesting to note that the R dependences of the diagonal matrix elements are equal for $1s$, $2s$, and $2p$ orbitals. The same is true for the nondiagonal matrix element for two centers. [Note also that the matrix element (18c) has an exponential R dependence as adopted in the models by Demkov[15] and Nikitin.[17]] The similarity of the matrix elements may be surprising, since the shapes of the orbitals involved in K and L shells are rather different (Figure 2). However, at relatively large distances where charge exchange takes place, the wave functions have nearly equal exponential R dependences. Differences between the shells are produced by the polynomials mentioned above. They are essentially replaced by the constant k within the present approximations of the matrix elements. Indeed, it is seen that the constant k increases significantly when $2s$ and $2p$ orbitals are used instead of $1s$ orbitals (Table 2).

In Eq. (18a) the second term has essentially the $1/R$ dependence mentioned above. Hereafter, this expression will be denoted as *binding term*, since it accounts for the increased binding of the electron in the Coulomb field of the approaching other nucleus.[76] It is noted that the binding effect produces a distortion of the atomic energies which is characteristic for inner-shell electrons (Figures 1 and 2). Generally, the binding effect results in a pulling together of the diabatic energy curves.[50]

In this case, the diabatic energy difference ΔH is reduced at the coupling radius R_c in comparison with the asymptotic value $\Delta \varepsilon_{12}^0$. To account for the reduction in the two-state systems the (dimensionless) parameter z is introduced by defining $\Delta H(R_c) = z \Delta \varepsilon_{12}^0$. Approximate values given in Table 2 show that for the 1π and 2π orbitals the factor z is larger than that for the 1σ and 2σ orbitals. This is understood from the fact that for a given α_M, the charge $Z_M = n\alpha_M$ increases by a factor of 2 when going from the $1s$ levels to the $2p$ levels owing to the change of the principal quantum number n. Consequently, the binding effect is enhanced. It is important to note that the increase of the binding effect is the principal difference between the σ and π states.

The specific feature for L-shell orbitals is the nondiagonal matrix element (18b). It accounts for the "Stark" interactions of the $2s$ and $2p$ orbitals located at one center when a charged particle approaches. It is noted that no attempt was made to simplify expression (18b). Nevertheless, the adjustable parameter d' was introduced to achieve some flexibility in its application to multielectron systems. (The one-electron calculation yields $d' = 2$). It is noted that the terms with the parameter d' vanish at large internuclear distances. There, asymptotically, expression (18b) approaches a value that is governed by the $2s$–$2p$ dipole matrix element obtained within the hydrogenic approximation.[77]

Finally, translational motion effects will be inspected. Pfeiffer and Garcia[52] derived analytic expressions for the averaged matrix element V_{12}^{av} using hydrogenic $1s$ wave functions with translational factors. (The diagonal matrix elements are not affected by translational factors.) The result is

$$V_{12}^{av} = B(v) \, e^{-(\alpha + iv/2)R} \tag{19}$$

where $B(v)$ is a complex function given by Eq. (10) of Ref. 52. Care should be taken with V_{12}^{av} as it does not account for the wave function overlap. Nevertheless, Eq. (19) exhibits the essential effects of the translational factors: Both the prefactor and the exponent of the exponential function are modified to complex quantities. It is shown further below that these modifications influence the resulting transition probabilities in a characteristic manner.

3. Two-State Systems

3.1. Two-State Formalism

The two-state system[78] is of particular importance, as the related equations may often be solved in closed form. Let ϕ_1^0 and ϕ_2^0 be the atomic states belonging to the centers A and B, respectively. From Eq. (6) it follows for the coupled equation that[2,79]

$$i\dot{a}_1^0 = \tilde{H}_{12} a_2^0 \exp\left(i \int^t \Delta\tilde{H} \, dt' \right) \tag{20a}$$

$$i\dot{a}_2^0 = \tilde{H}_{21} a_1^0 \exp\left(-i \int^t \Delta\tilde{H} \, dt' \right) \tag{20b}$$

where $\Delta\tilde{H} = \tilde{H}_{11} - \tilde{H}_{22}$. The matrix elements are obtained as ($n, m = 1, 2$; $n \neq m$)

$$\tilde{H}_{nn} = \varepsilon_n^0 + \frac{V_{nn} - S_{nm}V_{mn}}{1 - |S_{nm}|^2} \tag{21a}$$

$$\tilde{H}_{nm} = \frac{V_{nm} - S_{nm}V_{mm}}{1 - |S_{nm}|^2} \tag{21b}$$

where $V_{11} = \langle\phi_1^0|V_B|\phi_1^0\rangle$, $V_{22} = \langle\phi_2^0|V_A|\phi_2^0\rangle$, $V_{12} = \langle\phi_1^0|\exp(-i\mathbf{v}\mathbf{r})V_A|\phi_2^0\rangle$, $V_{21} = \langle\phi_2^0|\exp(i\mathbf{v}\mathbf{r})V_B|\phi_1^0\rangle$. Again, $S_{12} = S_{21}^* = \langle\phi_1^0|\exp(-i\mathbf{v}\mathbf{r})|\phi_2^0\rangle$ is the overlap matrix element. The phases in Eq. (20) are usually expressed $\int\Delta\tilde{H}\,dt' = \Delta\varepsilon_{12}^0 t + \delta$ with the distortion term δ and the asymptotic energy difference $\Delta\varepsilon_{12}^0 = \varepsilon_1^0 - \varepsilon_2^0$. The distortion term[80] $\delta = \int(\tilde{V}_{11} - \tilde{V}_{22})\,dt'$ implies the R-dependent terms $\tilde{V}_{nn}$ of the diabatic potentials $\tilde{H}_{nn}$ [$\tilde{V}_{nm}$ stands for the second expression in Eq. (21a)].

Although the matrix ($\tilde{H}_{nm}$) is not Hermitian symmetric, unitarity is retained when the coupled equations (20) are solved exactly. This has been done numerically by Lin and collaborators[41-43] using wave functions with translational factors and realistic interaction potentials. Similarly, Pfeiffer and Garcia[53] applied nonHermitian symmetric matrix elements under conditions preserving unitarity. They have been able to solve the coupled equations analytically on the basis of hydrogenic orbitals. However, in the calculation the overlap of the wave functions and the distortion in the diabatic potential have been neglected.

As pointed out above, Hermitian symmetry in the potential coupling matrix is achieved when orthogonalized basis states are used. For two states a convenient (symmetric) orthogonalization is given by[51]

$$\hat{\phi}_1^0 = p\phi_1^0 - q\phi_2^0 \tag{22a}$$

$$\hat{\phi}_2^0 = -q\phi_1^0 + p\phi_2^0 \tag{22b}$$

With the definitions[43]

$$p = \frac{\cos(\gamma/2)}{\cos\gamma} \quad \text{and} \quad q = \frac{\sin(\gamma/2)}{\cos\gamma}$$

it follows that $\sin\gamma = S_{12}$. Then, the matrix elements $\hat{H}_{nm} = \langle\hat{\phi}_n^0|H|\hat{\phi}_m^0\rangle$ are expressed as

$$\Delta\hat{H} = \frac{H_{11} - H_{22}}{[1 - |S_{12}|^2]^{1/2}} \tag{23a}$$

$$\hat{H}_{12} = \frac{H_{12}^{\text{av}} - S_{12}H^{\text{av}}}{1 - |S_{12}|^2} \tag{23b}$$

where $\Delta \hat{H} = \hat{H}_{11} - \hat{H}_{22}$, $H_{12}^{\text{av}} = (H_{12} + H_{21}^*)/2$, and $H^{\text{av}} = (H_{11} + H_{22})/2$. It is noted that $\hat{H}_{12}$ approaches $V_{12}^{\text{av}} = (V_{12} + V_{21}^*)/2$ when $S_{12} \to 0$. Comparison with Eq. (21) yields $\hat{H}_{12} = \tilde{H}_{12}^{\text{av}}$, so that for the nondiagonal matrix element the present orthogonalization is equivalent to the arithmetic averaging procedure expressed in Eq. (7).

Orthogonalized basis states have been used in the calculations by Fritsch et al.[43] and Stolterfoht.[51] Moreover, Hermitian symmetric matrix elements have been achieved by Pfeiffer and Garcia[52] using the arithmetic averaging procedure. Moreover, Hermitian symmetry is assumed for the analytic model matrix element.[15,17,51]

It is recalled that the matrix elements $\hat{H}_{nm}$ and $\tilde{H}_{nm}$ under consideration here account for the nonzero overlap of the wave functions. Actually, the effect of the wave function overlap is remarkable. This is shown in Figure 3, comparing $\hat{H}_{12}$ and V_{12}^{av} calculations with screened $1s$ wave functions for the system Ne + O. Figure 3 exhibits that $\hat{H}_{12}$ is substantially smaller than V_{12}^{av}. Moreover, it is found that $\hat{H}_{12}$ is closely approximated by an exponential curve representing SHM matrix element (18c). This shows that the matrix element $\hat{H}_{12}$ is well represented by an exponential function. Also, it

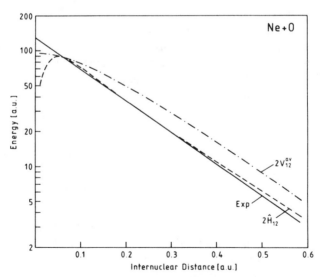

Figure 3. Potential coupling matrix elements calculated using screened hydrogenic wave function for the system Ne + O. The quantity $\hat{H}_{12}$ does and V_{12}^{av} does not account for the wave function overlap (see text). The curve labeled "Exp" follows from Eq. (18c).

is important to realize that model matrix elements may well account for the nonzero overlap of the wave functions.

For the adiabatic treatment of the collision, a basis transformation is performed in accordance with Eq. (12). Starting with orthogonal states, one obtains[78]

$$\phi_1 = u\hat{\phi}_1^0 + w\hat{\phi}_2^0 \tag{24a}$$

$$\phi_2 = -w\hat{\phi}_1^0 + u\hat{\phi}_2^0 \tag{24b}$$

With the conventional definitions $u = \cos \chi/2$ and $w = \sin \chi/2$ it follows from the diagonalization of $(\hat{H}_{nm})$ that $\tan \chi = 2\hat{H}_{12}/\Delta\hat{H}$.

Furthermore, from Eq. (13) it follows for the adiabatic energies $\varepsilon_{1,2} = \bar{\varepsilon} \pm \Delta\varepsilon_{12}/2$ that

$$\bar{\varepsilon} = \tfrac{1}{2}[\hat{H}_{11} + \hat{H}_{22}] \tag{25a}$$

$$\Delta\varepsilon_{12} = -[(\Delta\hat{H})^2 + |2\hat{H}_{12}|^2]^{1/2} \tag{25b}$$

The radial coupling matrix element $M_{12} = -0.5\partial\chi/\partial R$ yields

$$M_{12} = -\frac{1}{\Delta\varepsilon_{12}^2}\left[\Delta\hat{H}\frac{\partial\hat{H}_{12}}{\partial R} - \hat{H}_{12}\frac{\partial(\Delta\hat{H})}{\partial R}\right] \tag{26}$$

It is important to note that M_{12} is evaluated from the first term in Eq. (14) only. The second term cancels because the dynamic matrix elements $\hat{K}_{12}^0$ and $\hat{K}_{21}^0$ vanish. This is plausible since simultaneously $K_{12}^0 = -\hat{K}_{21}^0$ and $\hat{K}_{12}^0 = \hat{K}_{21}^0$ as $\hat{\Phi}_1^0$ and $\hat{\Phi}_2^0$ are orthogonal and they belong to different centers.

The coupled equations for the adiabatic states are obtained as

$$\dot{a}_1 = -v_R M_{12} a_2 \exp\left(i\int^t \Delta\varepsilon_{12}\, dt'\right) \tag{27a}$$

$$\dot{a}_2 = v_R M_{12} a_1 \exp\left(-i\int^t \Delta\varepsilon_{12}\, dt'\right) \tag{27b}$$

where v_R is the radial velocity given in Eq. (11). The coupled equations are integrated with the initial conditions $a_1 = 0$ and $a_2 = 1$ referring to one vacancy in the higher-lying orbital. Then, the probabilities W_n and P_n are obtained in accordance with Eq. (3).

If two vacancies are initially present in the higher-lying orbitals, either one or two vacancies are transferred during the collision. Let the corresponding probabilities be denoted by D_n and $D_n^{(2)}$, respectively. For nonrelaxed orbitals it follows from statistical rules[42] that

$$D_n = 2P_n(1 - P_n) \tag{28a}$$

$$D_n^{(2)} = P_n^2 \tag{28b}$$

Analogous relations[30] are used for the single passage probability W_n. Total cross sections for single and double vacancy transfer are obtained from Eq. (4) where the probability P_n should be replaced by D_n and $D_n^{(2)}$, respectively.

3.2. MO Energies and Radial Coupling Matrix Elements

To examine the model matrix elements, the adiabatic energies derived by means of Eq. (25) are compared with corresponding data obtained from Hartree–Fock calculations. Examples of molecular orbital diagrams relevant in this article are given in Figure 4 for C + Ne, Ne + Ar, and Ni + Ge. The calculations were performed using the HF code HONDO, which is part of the program package ALCHEMY for molecular structure calculations.[81] Conventional Gaussian basis sets were applied in the calculations. Basis functions at a third center (i.e., the charge centroid) were added to achieve accurate results at small internuclear distances.[51]

In Figure 5 the SHM matrix elements (18) are applied to study the formation in the 1σ and 2σ states in the system B + C. In Eq. (18) the parameters given in Table 2 have been taken making use of the fact that they are independent of the collision system. From the matrix elements ΔH and H_{12} the adiabatic energy difference $\Delta\varepsilon_{12}$ is derived by means of Eq. (25b). It is seen that the model results compare well with the HF calculation. Other examples for the applicability of the matrix elements (18) are given in Figure 6. They refer to the coupling of the $2p(m = 1)$ states in the systems S + Ar and Ni + Ge giving rise to the formation of the 1π and 2π molecular orbitals (Figure 2). As for the σ states it is seen that the model results compare well with the HF calculations.

Coming back to Figure 5, one may compare the models under consideration here. Different procedures are chosen to obtain values for the parameters required for the models. The parameters for Nikitin's model have been adjusted to fit the MO energies first proposed by Bøving.[27] It is seen that Nikitin's model is flexible enough to reproduce the MO results. For Demkov's model, parameters have been chosen as given in Table 1 as set I. These parameters have been selected since they have been extensively used in the past. It is noted, however, that the agreement between the HF calculations and Demkov's model may be improved when $\Delta\varepsilon$ and α are used as free fit parameters. Nevertheless, it was found[27] that Demkov's model is not flexible enough to reproduce fully the MO results.

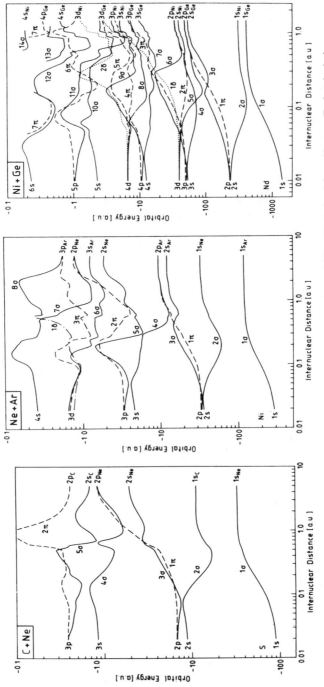

Figure 4. Molecular orbital energy diagrams for the systems C + Ne, Ne + Ar, and Ni + Ge derived from Hartree–Fock calculations.

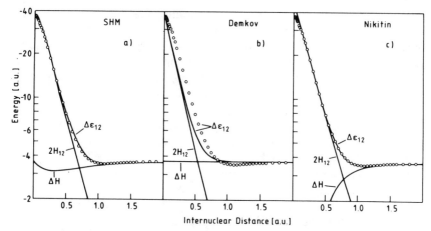

Figure 5. Molecular orbital energy difference $\Delta\varepsilon_{12}$ and $2H_{12}$ for the 1σ and 2σ MO's of the system $B + C$. The solid lines represent model matrix elements and the circles are from HF calculations. Results in (a) are from Eq. (18) and in (b) and (c) are from Eq. (16). (From Ref. 51.)

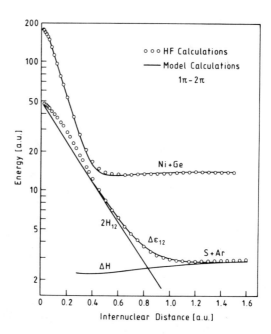

Figure 6. Molecular orbital energy difference $\Delta\varepsilon_{12}$ and matrix elements ΔH and $2H_{12}$ for the 1π MO's of the systems $S + Ar$ and $Ni + Ge$. Solid lines are from the SHM matrix elements (18) and circles are from HF calculations.

The curves in Figure 5 exhibit the binding effect, i.e., the decrease of the diabatic energy difference ΔH with decreasing R. It is seen that the models yield different results for the diabatic energy difference. Since Eq. (18) is based on realistic calculations for $1s$ orbitals, it is expected that ΔH is best described by the curves in Figure 5a. Then, it is seen from Figure 5c that Nikitin's model overestimates the binding effect.

Figure 7 shows the radial coupling matrix elements M_{12} derived from Eq. (26) by means of the different models. It is noted that M_{12} has a maximum near the coupling radius R_c, i.e., near the crossing of $2H_{12}$ and ΔH displayed in Figure 5. Nikitin's model yields values for M_{12} that coincide in their maximum with the SHM results.[51] The relatively high values for M_{12} from Nikitin's model may be understood from the overestimation of the binding effect. The matrix element from Demkov's model is shifted to small internuclear distances. Looking back at Figure 5b it is seen that this shift follows primarily from $2H_{12}$ being systematically too low.[50]

Hence, it appears desirable to readjust the parameters used for Demkov's model. Suitable parameters to approximate HF results are introduced in Table 1 as set II. The nondiagonal matrix element H_{12} is treated in accordance with Eq. (18c). Moreover, the parameter $\Delta\varepsilon$ is set equal to the diabatic energy difference at the coupling radius as justified explicitly in Section 3.4. This diabatic energy difference is obtained from the related asymptotic value $\Delta\varepsilon_{12}^0$ using the reduction factor z from Table 2. It may readily be verified that with the modified model parameters the radial

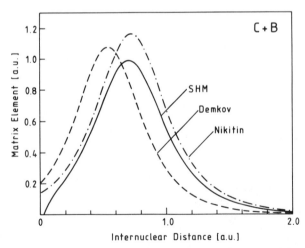

Figure 7. Radial coupling matrix elements derived from Eq. (26) using model matrix elements for the system B + C. The results labeled SHM, Demkov, and Nikitin refer to the matrix elements in Figures 5a, 5b, and 5c, respectively.

coupling matrix element from Demkov's model agrees well in position with the other model results.

Hence, the modifications of the model parameters influence primarily the coupling radius R_c. From Eq. (18) it is verified that R_c is remarkably larger (by ~40%) when the modified parameters (set II) are used instead of the original ones (set I). This shows that the latter parameters underestimate remarkably the coupling radius.

It should be added that the coupling radius R_c is about equal to the full width at half-maximum ΔR(FWHM) of the radial coupling matrix element (Figure 7). For Demkov's model it follows that ΔR(FWHM) $= \alpha^{-1} \ln (2 + \sqrt{3})^2 \approx 3\alpha_{12}^{-1}$, where α is replaced by $c\alpha_{12}$ (Table 1). Hence, the value $R_c^* = 3$ in Eq. (17) appears to be a reasonable choice for $1s$ orbitals.

3.3. Highly Ionized Collision Systems

Often, in the experiments multiply charged ions are used as projectiles.[34-39] Thus, it is important to study the effects of several electrons missing in the collision system. To achieve accurate energy curves, information is required about the dynamic behavior of the outer-shell electrons. Since the collision dynamics of the multielectron system are untractable at present, more or less restrictive model assumptions are needed to analyze the multiply ionized systems.[43,82,83] Progress about the charge flow in outer shells of imbalanced collision systems has been made recently by Eichler and Ho.[84]

Two extreme cases for the distribution of the outer shell electrons will be considered in the following. The cases refer to situations present in the incoming and outgoing part of the collision. In the incoming part of the collision where a highly ionized projectile is incident on a neutral target atom, the system is strongly imbalanced in charge. During the collision charge flow occurs from the target atom to the projectile. In addition electrons are removed from the collision system. However, this loss of electrons will be disregarded, since it does not alter the principles of the following considerations.

In the outgoing part the outer shell electrons are usually shared between the collision partners. Although complete charge balancing is not expected in energetic ion–atom collisions, it should be realized that the transitions of the inner-shell electrons take place at relatively small internuclear distances where the outer-shell electrons are not yet localized at one specific center. This situation is similar to that of complete charge balance in the outer shell.

Information about the charge balanced system may be obtained from Hartree–Fock calculations for ionized particles. The HF method yields MO energies for the minimum of the total energy. This implies that the cloud of the outer-shell electrons is relaxed, i.e., the outer-shell vacancies are (nearly) equally shared between the collision partners. In Figure 8 results of HF calculations[51] are shown for the systems $(Ne + F)^0$ and $(Ne + F)^{9+}$. It is seen that the MO energies ε_1 and ε_2 change considerably when the charge state of the system varies, whereas the energy difference $\Delta\varepsilon_{12}$ for $(NeF)^{9+}$ and that of the neutral system are practically equal. Since the probability for charge exchange depends primarily on the difference of the orbital energies [Eq. (27)], significant ionization effects are not expected for the charge balanced system.

With some modifications the model matrix elements (18) may also be applied to ionized systems. It appears reasonable to generalize the exponential screening function as follows:

$$Q_M = (Z_M - q_M)\, e^{-\alpha_0 R} + q_M \tag{29}$$

where q_M is the charge state of the particle M. An example for the application of Eq. (29) to the quasirelaxed system $F^{3+} + Ne^{4+}$ is given in Figure 9. The model calculations compare well with Hartree-Fock data for the system $(NeF)^{7+}$. Again, it is found that the energy difference $\Delta\varepsilon_{12}$ for $(NeF)^{7+}$ is nearly equal to that of the neutral system.

The fit of the HF data by means of Eq. (18) shows that α_0 is increased by about a factor of 2 (i.e., $s = 0.84$ a.u. in Table 1). This may be understood as the outer-shell electrons of the ionized systems being more tightly bound. For the orbital energies, ε_n^0 values of the ionized systems are taken. However,

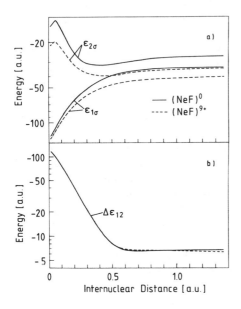

Figure 8. Hartree-Fock calculations (a) of 1σ and 2σ molecular orbitals for the neutral system NeF and the ionized system $(NeF)^{9+}$. In (b) the corresponding energy difference $\Delta\varepsilon_{12}$ is shown. (From Ref. 51.)

the fit values for the parameters α_n are found to be equal to those for the neutral system. The parameter α_n is a measure for the mean orbital velocity of the corresponding electron which appears to be rather insensitive to screening effects.

To achieve some insight about the charge imbalanced system, model calculations are performed for the system $F^{7+} + Ne$. In such a system, the nuclei of the collision partners are differently screened at large R. In the following, this effect will be referred to as *unilateral screening*. In fact, the unilateral screening produces a swapping of the atomic $1s$ levels of the collision partners, i.e., the $1s$ binding energy of F^{7+} is larger than that of Ne. Hence, intriguing questions arise with respect to the correlation between the levels of the separated and the united atoms.[82,83]

The results of the model calculations are given in Figure 9. The parameters in expression (18) are chosen to be the same as for the ionized relaxed system; only the parameters q_M were changed in the screening formula (29). It is noted that Eq. (29) implies the assumption that the outer-shell electrons are "frozen" at the collision partners. This assumption is reasonable for projectile velocities large in comparison with the velocity of the outer-shell electrons. The significant feature of the charge imbalanced system is the pseudocrossing seen at a distance well outside the K-shell coupling region. This crossing reverses the level swapping. Actually, near the coupling radius R_c, the charge imbalanced system is rather similar to the relaxed or neutral system. Hence, for the charge imbalanced system it is concluded also that no significant unilateral screening effects are present

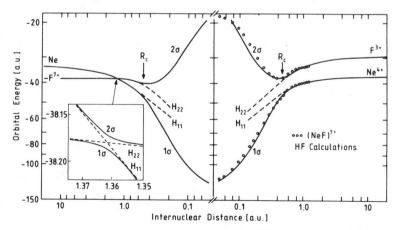

Figure 9. Molecular orbitals 1σ and 2σ derived from the SHM matrix elements (18) for the system $F^{7+} + Ne$ (incoming part) and $F^{3+} + Ne^{4+}$ (outgoing part). Circles represent Hartree–Fock calculations for the system $(NeF)^{7+}$. Underlying diabatic potentials are H_{11} and H_{22}. The coupling radius is R_c. The inset shows an enlarged figure of the pseudocrossing between the 1σ and 2σ orbitals in the incoming part.

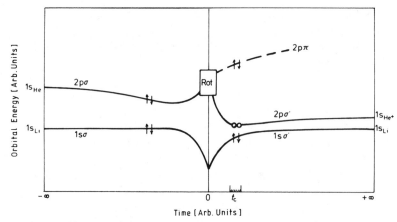

Figure 10. Molecular orbitals 1σ and 2σ based on Hartree–Fock calculations for the neutral Li + He system (incoming part) and the corresponding ionized system (outgoing part). The diagram illustrates qualitatively the relaxation effect subsequent to multiple 2σ excitation by rotational coupling. From Sidis *et al.*[47]

near the coupling radius. This finding provides some justification to apply the model matrix elements of the neutral system to ionized systems.

Finally, it should be pointed out that the situation is totally different for inner-shell screening effects which occur when an additional vacancy is produced in the inner orbital. At the coupling radius the additional vacancy is practically localized at one nucleus, resulting in a decrease of the corresponding orbital energy due to unilateral screening effects. This is shown in Figure 10 for the Li + He systems,[47] where 2σ vacancies are produced near the united atom limit. Thus, the 2σ orbital energy in the outgoing part of the collision is significantly smaller than that in the ingoing part. In this case, the transition probabilities are significantly affected as shown in the next section.

3.4. Vacancy Sharing

When the vacancy in the higher-lying molecular orbital is produced near the united atom limit, vacancy transfer may occur in the outgoing part of the collisions.[21] This single-passage process is reversed when electrons in the higher-lying MO are missing prior to the collision and vacancies are "detected" at the united atom limit.[64,65] The single-passage process taking place at near-zero impact parameters is denoted as vacancy sharing. Because of its simplicity, two-state vacancy sharing is specifically suited to verify model calculations. The methods to achieve approximate solutions for the two-state coupled equations have been outlined by Crothers.[85,86]

Analytical solutions of the coupled equations have been evaluated by Demkov[15] and Nikitin[17] for $R = vt$, which follows for zero impact parameter. The evaluation procedure is somewhat elaborate, yielding the following result for the sharing probabilities[17]

$$W_n(0) = \frac{e^{2\xi \cos^2(\theta/2)} - 1}{e^{2\xi} - 1} \tag{30}$$

with the "Massey parameter"

$$\xi = \pi \frac{\Delta\varepsilon}{\alpha v} \tag{31}$$

It is interesting to note that the sharing probability is independent of the coupling radius R_c. The particular solution by Demkov[15] is obtained for $\theta = \pi/2$:

$$W_n(0) = \frac{e^{-\xi}}{1 + e^{-\xi}} \tag{32}$$

It is instructive to verify the sharing probability in the low-velocity region where $\xi > 1$ so that $\exp(2\xi) \gg 1$. Then, Eq. (30) reduces to

$$W_n(0) = e^{-\xi(1 - \cos\theta)} \tag{33}$$

It is noted that for low velocities the sharing probability is not influenced by the slope of the diabatic potential curves. From Eq. (32) follows the same expression, but the factor $1 - \cos\theta$ is missing in the exponent. This factor determines the diabatic energy difference in the coupling region, since $\Delta H(R_c) = \Delta\varepsilon(1 - \cos\theta)$, as may readily be deduced from Eq. (15). Thus, in the low-velocity range the treatments by Nikitin and Demkov are identical provided that in Demkov's formula the diabatic energy difference $\Delta H(R_c)$ is inserted instead of the asymptotic value $\Delta\varepsilon_{12}^0$. Thus, the modified value for $\Delta\varepsilon$ is justified in Table 1.

The situation is different when the diabatic potential curves cross as assumed in the Landau–Zener model.[18,19] In this case the diabatic energy difference vanishes and it loses significance as a model parameter. Then, the expression (33) should be interpreted differently. As noted in Section 3.2, the Landau–Zener model implies $\theta \ll 1$ so that $1 - \cos\theta \approx [\sin^2\theta]/2 = 2[H_{12}(R_c)/\Delta\varepsilon]^2$. Here, it is useful to introduce the derivative of the diabatic energy difference $F = \partial\Delta H/\partial R$, which may be regarded as the force "holding" the system in the diabatic state. Likewise, the interaction H_{12} "pushes"

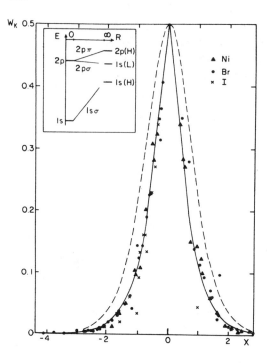

Figure 11. K-vacancy sharing probabilities W for various collision systems involving Ni, Br, and I projectiles as a function of the scaling parameters X from Eq. (35). The solid line follows from Eq. (32) with $\xi = 2X$. From Meyerhof.[22]

the system out of the diabatic state. It may readily be shown that

$$W_n(0) = \exp\left[-2\pi \frac{H_{12}^2(R_c)}{vF(R_c)} \cos \theta\right] \qquad (34)$$

For vanishing angle θ this expression is identified as the Landau–Zener formula.[18,19] Small values of θ imply that at $R < R_c$ the adiabatic states merge into the diabatic states, similarly as for $R > R_c$. A typical example for a Landau–Zener crossing is given in the inset of Figure 9.

Meyerhof[21] was the first to apply the formula (32) to K-vacancy sharing and to verify its scaling ability. The results are shown in Figure 11. The experimental sharing probabilities are deduced from $W = \sigma_H/(\sigma_H + \sigma_L)$, where σ_H and σ_L are the cross sections for K excitation of the heavier and lighter collision partner, respectively. The experimental data compare well with the universal curve following from Eq. (32) by means of the scaling parameter

$$X = \pi \frac{\Delta \varepsilon_{12}^0}{2\alpha_{12} v} \qquad (35)$$

Here the model parameters have been chosen to be $\Delta \varepsilon = \Delta \varepsilon_{12}^0$ and $\alpha = \alpha_{12}$

(set I in Table 1) so that the exponent in Eq. (32) yields $\xi = 2X$. With $\Delta\alpha = \alpha_1 - \alpha_2$ it follows from $\Delta\varepsilon_{12}^0 = \alpha_{12}\Delta\alpha$ that $\xi = \pi\Delta\alpha/v$. This shows that the probability $W_n(0)$ scales very conveniently. It depends on the properties of the collision system essentially through the quantity $\Delta\alpha$. It is noted that for K shells $\Delta\alpha$ is about equal to the difference ΔZ in the nuclear charges of the collision partners, i.e., $\Delta\alpha = \gamma\Delta Z$ with $\gamma \lesssim 1$. Hence, instead of $\pi\Delta\alpha/v$ one may use also $\pi\Delta Z/v$ as an appropriate scaling parameter for K shells.[22]

Several subsequent studies have confirmed Meyerhof's analysis. In particular, precision measurements using the method of Auger spectroscopy have shown that Eq. (32) is able to predict K-vacancy sharing probabilities with accuracies of typically $\pm10\%$. This is seen in Figure 12, where Auger data from different laboratories[23-25] are summarized. The experimental results are in excellent agreement with the Meyerhof–Demkov formula (32). Only the relatively light system B + C shows remarkable discrepancies from the theoretical curve.

Light systems may be influenced by multiple ionization effects from inner shells. When the 2σ MO becomes multiply ionized, unilateral changes of the binding energies are expected (Section 3.3). This has been verified

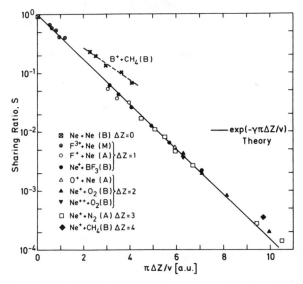

Figure 12. Sharing ratio $S = W_n/(1 - W_n)$ for K-vacancy as a function of $\pi\Delta Z/v$, where ΔZ is the difference of the nuclear charges of the collision partners. Labels A, B, and M refer to measurements by Fastrup *et al.*,[23] Schneider and Stolterfoht,[24] and Woods *et al.*,[25] respectively. The theoretical results are from Eq. (32) using $\xi = \pi\gamma\Delta Z/v$ with $\gamma = 0.87$. Because of $\gamma\Delta Z = \Delta\alpha$, the present exponent ξ is equal to that used by Meyerhof[21] (Figure 11).

for the Li + He system[26] shown in Figure 13. The experimental data are seen to be orders of magnitude higher than theoretical results obtained with orbital energies for the neutral system (curve 1). On the other hand, use of orbital energies modified by an additional 2σ vacancy yields results (curve 2) in good agreement with the experimental data. Probably, unilateral screening similar to that for Li + He is responsible also for enhanced vacancy sharing observed in the B + C system.[51]

Screening effects from the inner shell have been examined also in the work devoted to double K-vacancy sharing. For the system S + Ar McDonald *et al.*[32] have detected a 20% difference in the one-electron probability for transitions into the 2σ MO with one and two vacancies. However, for most heavier systems such effects have not been observed. The experimental results[29-31] for double K-vacancy sharing have been shown to be in accordance with statistical rules for nonrelaxed orbitals such as those given in Eq. (28). Hence, as expected, the screening effects from inner shells diminish for heavier collision partners.

The studies of Auger electrons have shown that K-vacancy sharing is practically independent of outer-shell ionization.[24] Although the K-binding energies are significantly influenced by the removal of the outer-shell electrons, the corresponding transition probabilities are found to be unaffected (Fig. 8). This may be explained by the fact that $W_n(0)$ depends primarily on the binding energy difference which is nearly unaltered in the charge balanced system. Outer-shell charge balance is expected in the outgoing part of the collision as discussed in Section 3.3.

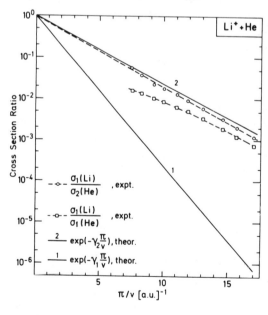

Figure 13. Cross section ratio for single $1s$ excitation of Li and double $1s$ excitation of He in Li$^+$ + He collisions. Theoretical curves 1 and 2 are derived from Eq. (32) with binding energies referring to single and double excitation of He, respectively. From Stolterfoht and Leithäuser.[26]

The relatively high accuracy of the sharing formula (32) is rather noteworthy with regard to its simplicity. Actually, the success of formula (32) has never been fully explained. It has been suggested that the neglect of different effects cancel each other in its deduction.[44] For instance, translational factors are not included in the analysis. It is possible that the model parameters are chosen such that they compensate partially for the missing translational factors.

Moreover, as pointed out in Section 3.2, the parameters $\Delta\varepsilon$ and α (set I in Table 1) generally used in the Meyerhof–Demkov formula are not quite correct. The more accurate set II of parameters yields with Eq. (31) the exponent

$$\xi = \frac{z}{c}2X \tag{36}$$

where X is the scaling parameter from Eq. (35). For K shells, the factors z and c (Table 2) are equal, so that $\xi = 2X$ in accordance with Meyerhof's analysis.[21] Hence, the inaccuracies of $\Delta\varepsilon$ and α cancel in the previous treatment of K-vacancy sharing.

It is emphasized that this cancellation does not occur for L-vacancy sharing. With set II in Table 1, one obtains $\xi = 1.82X_L$ from Eq. (36) where the index refers to the L shell. This exponent is in good agreement with the observation by Lennard et al.,[65] who studied extensively vacancy sharing between 1π and 2π orbitals. The results are shown in Figure 14 indicating that the experimental results are well fitted by Eq. (32) with the exponent $\xi = 1.79X_L$.

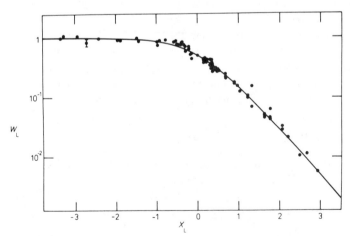

Figure 14. Probability W_L for L-vacancy sharing as a function of the scaling parameter X_L from Eq. (35). Solid line is from Eq. (32) using the exponent $\xi = 1.79X_L$. From Lennard et al.,[65] where also the key to the experimental data is given.

Lennard et al.[65] suggested additional mechanisms such as coupling to higher-lying MO's to explain the difference observed. The present analysis shows that the origin for the difference between K and L shells should be searched in the sharing mechanism itself. The difference observed in the exponent ξ is primarily produced by the reduction factor z (Tables 1 and 2), which, in turn is a measure for the binding effect. It was pointed out in Section 3.2 that the binding effect for the π states is more important than that for the σ states. Hence, it is concluded that the reduced exponent ξ for L-vacancy sharing is produced by the enhancement of the corresponding binding effect.

Finally, the influence of the translational factor on the exchange probabilities will be estimated. The effect of the translational factors on $W_n(0)$ for intermediate and small impact velocities may be verified by means of the expressions given by Pfeiffer and Garcia.[52] From Eq. (19) it is seen that for $R = vt$ one obtains the phase $v^2 t/2$ which adds $v^2/2$ to the diabatic energy difference $\Delta \tilde{H}$ in the coupled equations (15). Consequently, in the Massey formula (31) the parameter $\Delta \varepsilon$ is to be replaced by the total energy difference $\Delta \varepsilon + v^2/2$, which implies apart from the change of the orbital energy of the electron also the change of its translational energy. Thus, the inclusion of the translational factors decreases the exponential functions in Eq. (32) by the factor $\exp(-\pi v/2\alpha)$. Hence, the translational factors may approximately be taken into account by multiplying $W_n(0)$ by

$$f_{\text{TF}} = e^{-\pi v/2\alpha}(1 + e^{-\xi}) \tag{37}$$

It is interesting to note that this correction factor scales primarily with α. It creates a maximum in $W_n(0)$ for $\Delta \varepsilon = v^2/2$ in accordance with the Massey criterion.[53] It is added than when the correction factor (37) is applied to the sharing formula (32), the results are reduced such that the agreement between experiment and theory often deteriorates. This finding supports the suggestion made above that the model parameters in Eq. (32) are chosen such that they compensate partially for the missing translational factors.

3.5. Impact Parameter Dependence

Extension of Demkov's model to nonzero impact parameters has been performed by Briggs[49] using an approximate solution of the coupled equations.[85,86] It is found that the convenient formula (32) retains its validity for nonzero impact parameter b provided the exponent ξ is modified. The result is

$$W_n(b) = \frac{e^{-\xi y}}{1 + e^{-\xi y}} \tag{38}$$

where

$$y(b) = [(f^2 + 2g^2)^{1/2} - f]^{1/2}$$

with

$$f = g^2 \left[1 - \left(\frac{b}{R_c} \right)^2 \right] - \frac{1}{2} \quad \text{and} \quad g = \sqrt{2} \frac{\alpha R_c}{\pi}$$

The coupling radius R_c follows from Eq. (16) with $D = 0$. For $1s$ orbitals $g \approx 1$ as may be seen from Eq. (17). For $b = 0$ it follows that $y = 1$ yielding the Demkov–Meyerhof result (32) at zero impact parameter.[49]

To give a plausible interpretation for the function $y(b)$ it is pointed out that the probability for charge exchange is governed by the radial velocity relevant during the coupling of the states.[15,17] For a sufficiently narrow coupling region, the radial velocity may be obtained from Eq. (11) setting $R = R_c$. However, generally, the coupling region relevant to near resonant charge exchange under consideration here has a considerable width. In this case, one should search for some average radial velocity effective near the coupling region. The quantity y may be used to introduce the effective radial velocity

$$\bar{v}_R = v/y \tag{39}$$

Hence, the extension of the Meyerhof–Demkov formula (32) to the Briggs formula (38) may be regarded as being put forward through the replacement of the incident velocity v by the radial velocity $\bar{v}_R$.

Briggs[49] applied Eq. (38) to derive K-vacancy exchange probabilities for Ne + O in comparison with more elaborate calculations using the HF method.[45,46] The results were found to fall off too rapidly with increasing impact parameter in comparison with the HF calculations. Similar effects have been noted by Schuch et al.,[37] who measured K-vacancy transfer in the S + Ar system. The rapid decrease of the data follows from set I in Table 1 used as model parameters which has been shown to underestimate the coupling radius (Section 3.2). Hence, it is noted that the inaccuracies of these parameters do not cancel at finite impact parameters.[50] Figure 15 shows revised calculations from Eq. (38) using the modified set II of model parameters (Table 1). The results are seen to be in reasonable agreement with the HF calculations by Briggs.[49]

It is interesting to note that the HF results exhibit some oscillatory structures which may be due to weak interferences occurring during the single passage of the coupling region. Similar oscillations are seen in numerical solutions[50] of the coupled equations using the model matrix elements (18). It appears that these structures show up only if the coupled equations are solved exactly. When the approximate integration methods by Crothers[86] are applied as for Eq. (38), the oscillatory structures are smoothed out (Figure 15). It can be questioned whether the interference effects in the single-passage process may be observed experimentally. The structures seen in the calculations may be due to spurious effects of the idealized two-state systems.

To account for the translational factor effects at nonzero impact parameter it would be desirable to find a correction factor such as that given by Eq. (37). There, however, the influence of the translational factors is difficult to verify. The analysis by Pfeiffer and Garcia[52] suggests that expression (37) overestimates the effect of the translational factors at nonzero impact parameter. Hence, in this case the correction (37) may be used as an upper limit estimate. More realistic correction values are expected when the projectile velocity v is replaced by the radial velocity $\bar{v}_R$ similar to that given by Eq. (39). It should be realized that this replacement implies uncertainties in the treatment of the translational factors. However, these uncertainties lose importance when the correction due to the translational factors is small.

3.6. Double Passage Process

Following the analysis of Crothers,[86] the transfer probability for the double passage of the transition region is obtained as

$$P_n = 4W_n(1 - W_n)\sin^2\left(\frac{1}{2}\int_{-t_c}^{t_c} \Delta\varepsilon_{12}\,dt\right) \qquad (40)$$

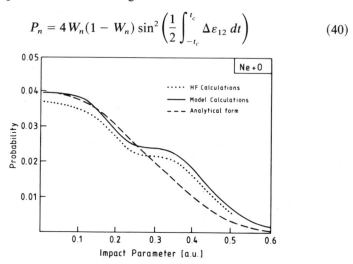

Figure 15. Single passage probability for 2σ–1σ vacancy transfer in the system Ne + O. Hartree–Fock calculations are from Briggs.[49] Model calculations are from Stolterfoht[50] using the SHM matrix elements (18). Analytical results are from Eq. (38).

where W_n is the corresponding single-passage probability and $\Delta\varepsilon_{12}$ follows from Eq. (25b). The time t_c corresponds to a distance near the coupling region. It is emphasized that the $\sin^2$ function produces oscillatory structures typical for the double passage probability. For rapid oscillations, the $\sin^2$ function may be replaced by its mean value $1/2$ yielding $P_n = 2W_n(1 - W_n)$.

The double-passage process has been studied by Schuch *et al.*[37,38] for $S^{15+} + Ar$ and by Hagmann *et al.*[39,40] for $F^{8+} + Ne$. The data clearly exhibit the oscillatory structures created by interference effects. In Figure 16 the experimental results for $F^{8+} + Ne$ are compared with calculations obtained utilizing the atomic two-state expansion method by Lin and Tunell.[42] In that work elaborate calculations have been made to obtain numerically matrix elements $\tilde{H}_{nm}$ including translational factors. Also shown are model results[51] deduced from the matrix elements (18) using parameters for the neutral system (Table 1).

For $b > 0$ 1 a.u. the two sets of theoretical results differ primarily in the frequency of the oscillations. The difference is probably produced by the influence of translational factors. As mentioned above, the single-passage probability W_n is influenced by translational motion effects through the phase in Eq. (19). In addition, the double passage probability P_n is affected by the prefactor $B(v)$. The influence of $B(v)$ on the transition probability is not obvious. Here, it should only be noticed that with increasing velocity v the absolute value $|B(v)|$ decreases so that $|H_{12}|$ is reduced. Then, it follows from Eq. (40) that the oscillation frequency of the double passage probability is reduced.

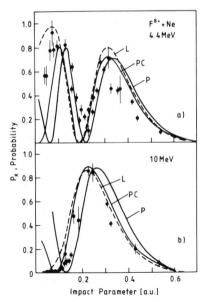

Figure 16. Double passage probability $P_K(b)$ for 2σ-1σ vacancy transfer in the system $F^{8+} +$ Ne. In (a) and (b) projectile energies are 4.4 and 10 MeV, respectively. Experimental data are from Hagmann *et al.*[39] The curve labeled L is from calculations[39] using the method by Lin and Tunell.[42] The curve labeled P is derived[46] using the SHM matrix elements (18). For the curve labeled PC the MO energies were modified to account roughly for effects of translational factors (see text). (From Ref. 51.)

Indeed, for $b > 0.1$ a.u. the theoretical results given by Hagmann *et al.*[39] are well reproduced by the model calculations using the SHM matrix elements (18) after the parameter k, and thus $|H_{12}|$ was reduced by 0.88 and 0.82 for the impact energies of 4.4 and 10 MeV, respectively (see the curve labeled PC in Figure 16). This shows that the oscillatory structure of $P(b)$ is sensitively dependent on the absolute value of H_{12}.

When the impact parameter decreases below about 0.1 a.u., the discrepancies between the theoretical curves increase significantly. These discrepancies cannot be accounted for by translational factors. Rather, they are produced by basic differences in the theoretical approaches. It was found that $|H_{12}|$ evaluated with atomic $1s$ orbitals decreases strongly at small internuclear distances (see the curve labeled $2\hat{H}_{12}$ in Figure 3). In the model calculations, however, the nondiagonal matrix element is chosen to fit the 1σ-2σ MO energy difference. This energy difference increases when R approaches zero (see the curve labeled Exp in Figure 3). Hence, while the model SHM matrix elements (18) are based on a two-state atomic expansion near the united atom limit, it exhibits characteristics of a two-state molecular expansion. On the other hand, the treatment by Lin and Tunnell[42] retains the features of a two-state atomic expansion near the united atom limit.

Figure 16 shows that for 10 MeV incident energy at small impact parameters the experimental data favor the atomic expansion model. However, for 4.4 MeV impact energy both sets of the theoretical data agree reasonably well with the experimental results. Hence, as expected the molecular expansion model gains importance at lower impact energies. This demonstrates that the interference structures yield information about the formation of molecular orbitals near the united atom limit.

Similar conclusions have been reached recently by Schuch *et al.*[38] studying the system S + Ar in a wide range of incident energies. In Figure 17 results are given for the relatively low projectile energy of 7.9 MeV. It is seen that the positions of the minima and maxima of the experimental data are well reproduced by the SHM calculations. However, the amplitude of the oscillations of the experimental data is smaller than that of the theoretical results. An explanation[38] of this finding may be the fact that the system is not symmetric with respect to the incoming and outgoing parts of the collision (Figure 9), so that the constructive and destructive interferences are diminished. Then, the conclusion that unilateral screening is not essential for K-vacancy exchange (Section 3.3) should be refined. Obviously, the overall transition probability averaged over the oscillations is not much affected (Figure 17). However, it may well be that the oscillations are influenced by small differences in the MO energy curves for the incoming and outgoing parts of the collision.

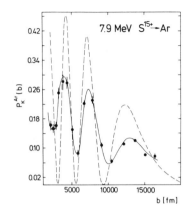

Figure 17. Double passage probability P_K^{Ar} for $2\sigma-1\sigma$ vacancy transfer in the system $S^{15+} + Ar$. Experimental results are from Schuch et al.[38] The solid line is drawn to guide the eye. The dashed curve represents calculations using the SHM matrix elements (18).

3.7. Total Cross Sections

Total cross sections for vacancy transfer are obtained after integration of the corresponding probabilities over the impact parameter. The transfer of K vacancies has been studied by Hall et al.[34] using various collision systems, e.g., $Si^{14+} + Ti$. Here, the projectile carries two K vacancies so that either one or two vacancies may be transferred during the collision. In Figure 18 experimental results are shown for the transfer of one vacancy. Also given are theoretical cross sections evaluated as $\sigma_K = 2\pi \int D_K b\, db$, where D_K is defined by Eq. (28a). (Here, the index refers to the K shell.) Theoretical results for D_K have been obtained by using the SHM matrix elements (18) and by using the numerical methods[42] which incorporate translational factors.

In the model results, translational factors have roughly been taken into account by application of the correction factor (37). For cross sections, the velocity v should be replaced by the effective radial velocity $\bar{v}_R$ averaged also over the impact parameter. With the reasonable choice of R_c for the mean impact parameter one obtains $\bar{v}_R = 0.7v$ from Eq. (39). However, it should be kept in mind that significant uncertainties may enter in this treatment of the translational factors. Thus, the good agreement between the model results and the experimental data might be accidental.

For convenient evaluation of the total cross section one may use the simple formula

$$\sigma_n = P\pi R_x^2 \tag{41}$$

where the assumption is made that the transition probability is represented by a step function curve of height P and extension R_x. Different choices

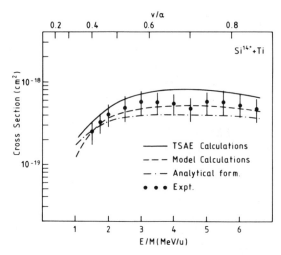

Figure 18. Total cross section for K-vacancy transfer in Si^{14} + Ti collisions. The experimental data are from Hall *et al.*[34] The curve labeled TSAE originates from calculations[34] using the method by Lin and Tunnell.[42] The model calculations are obtained using the SHM matrix elements (18) and the correction function (37). The analytic results are from Eq. (41) with corrections from Eq. (37).

are possible for the parameters P and R_x. For one electron missing in the 2σ orbital, it is reasonable to set P equal to $P_n = 2W_n(1 - W_n)$ in accordance with Eq. (40) where the $\sin^2$ function is assumed to cancel through the integration over the impact parameter. For two electrons missing in the 2σ orbital, P is set to be equal to D_n from Eq. (28a).

It appears reasonable to set R_x equal to the coupling radius R_c from Eq. (16). However, verification of the experimental results by Hall *et al.*[34] shows that a more suitable choice for R_x is the approximate coupling radius given by Eq. (17) with $R_c^* = 3$. This may be due to the fact that the major part of the transition does not necessarily take place at the exact coupling radius. As an example, results from Eq. (41) are shown in Figure 18 for the Si^{14+} + Ti system. The data are corrected using the factor f_{TF} from Eq. (37) as in the model calculations. It is seen that the analytical results are in reasonable agreement with the experimental data.

The applicability of the simple formula (41) should not be overestimated for K-vacancy exchange. It applies favorably for processes taking place in a well-localized transition region. This is not the case for the vacancy exchange between inner shells. However, for rapid estimates, formula (41) may serve as a useful tool to determine total cross sections for inner-shell vacancy exchange.[33,35]

4. Multistate Systems

4.1. General Considerations

Charge exchange processes involving shells other than the K shell generally imply more than two states. In nonrelativistic systems 3 and 4 relevant σ orbitals are to be considered for KL-vacancy exchange and L-vacancy exchange, respectively. When the SHM matrix elements (18) are applied to multistate systems, a procedure is chosen similar to that for $1s$ orbitals. First, the Hamiltonian matrix (H_{nm}) is diagonalized to obtain adiabatic energies, which, in turn, are compared with MO data from HF calculations. For more than two states, the diagonalization is performed using numerical methods. Second, the diagonalization procedure yields the coefficients $\hat{u}_{nm}$ for the expansion of the molecular orbitals in terms of atomic orbitals. These coefficients are differentiated to deduce the radial coupling matrix elements by means of Eq. (14). Here, also numerical methods are used.

In the calculations the second term of the right-hand side of Eq. (14) is neglected. This term vanishes for molecular orbitals located asymptotically at different centers as already noted for K shells. For molecular orbitals with the $2s$ and $2p$ orbitals at one center, the second term in Eq. (14) is not zero even at asymptotically large internuclear distances. Hence, it produces the spurious couplings which are typical for matrix elements calculated without translational factors.[10] Thus, the neglect of the second term in Eq. (14) has the advantage that spurious couplings are avoided.

Finally, the coupled equations (9) for adiabatic states are integrated to determine probabilities for vacancy exchange. In the following sections, results are discussed for systems with three and four states. Prior to this study, the possibility is examined of applying the two-state models in the description of KL- and LK-vacancy sharing.

4.2. KL- and LK-Vacancy Sharing

4.2.1. Application of Two-State Models

For the analysis of systems with more than two states, no analytical solutions exist to describe vacancy exchange. Thus, it appears useful to find out whether two-state models are applicable. Several authors[11,54,57-61] have made the attempt to analyze KL- and LK-vacancy exchange within the framework of two-state models. In particular, the multistate coupling has been described in terms of a series of successive two-state processes.[54,61]

In the application of two-state models to KL- and LK-vacancy sharing, controversial results are encountered. It is found that the Demkov–Meyerhof formula (32) applied with the usual model parameters[21] underestimates the experimental data for KL-vacancy sharing,[57] whereas it overestimates the data for the LK systems.[11] This observation may be understood from the influence of the binding effect, which is found to be very strong in systems with matching K and L levels. Owing to the asymmetry of the system, the K orbital is more affected than the L orbital. This results in a pulling together of the corresponding energy curves in KL systems as shown schematically in Figure 19. (A smaller pulling together is observed for K shells, too.) Also, it is seen in LK systems that the K orbital is more affected than the L orbitals. In this case, however, the binding effect results in a repulsion of the related energy curves. Consequently, the LK-sharing data are reduced, whereas the KL-sharing data are enhanced in comparison with the results of the usual Meyerhof–Demkov method,[21] which does not account for the binding effect.

It seems that Demkov's model is not well suited for the treatment of asymmetric collision systems. Thus, following the suggestion by Meyerhof et al.,[54] various studies of KL- and LK-vacancy sharing have been performed in terms of Nikitin's model,[17] which has a greater flexibility in describing the orbital energies. Often, the model parameters have been extracted[57-59] from the fit of the 3σ–4σ MO energy difference obtained from independent molecular orbital calculations.

As an example, the 3σ–4σ MO energy difference is shown in Figure 20 for the Ne + Kr system. The data clearly reveal the strong influence of the binding effect resulting in a pulling together of the MO energy curves. The MO data are compared with results from Nikitin's model using the fit

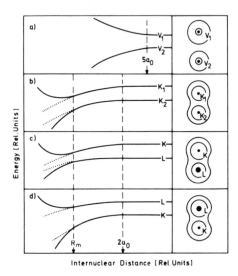

Figure 19. Schematic plot of inner-shell MO's involved in K-vacancy sharing (b), KL-vacancy sharing (c), and LK-vacancy sharing (d). For comparison, valence orbitals V_1 and V_2 are also shown (a). Notation as in Figure 1. (From Ref. 11.)

procedure and from the SHM matrix elements (18) using the parameters in Table 2.

In Figure 20 the respective data from the two models indicate characteristic discrepancies. Firstly, it is seen that the diabatic energy difference ΔH from Nikitin's model decreases too rapidly with decreasing internuclear distance R. This means that Nikitin's model overestimates strongly the binding effect as already noted for the K-vacancy sharing system $B + C$ (Figure 5c). Secondly, the slope of the curve representing $2H_{1s2p}$ from Nikitin's model is obviously too small. Reed et al.[59] pointed out that the values for the exponent α_p from the fit procedure are systematically smaller than those generally used in Meyerhof's analysis.[21] This finding may be explained by the influence of the 2s orbital, which is disregarded in the two-state analysis. In a detailed study of the adiabatic mixing of the related atomic orbitals, it has been shown by Fritsch and Wille[62] that the states 2s and 2p of Kr and 1s of Ne are all strongly involved in the formation of the molecular orbitals. In this case the fit procedure by means of a diabatic two-state model is questionable. In fact, without the 2s state the MO data at small R would follow the $2H_{1s2p}$ curve representing the SHM matrix element (18c) as shown further below.

The underestimations of ΔH in the coupling region and of the exponent α have opposite effects on the vacancy sharing probability. The decrease of α reduces the transition probability, whereas the decrease of ΔH enhances it; see Eq. (32). For instance, the binding effect is significant in the strongly asymmetric systems, so that the underestimation of ΔH dominates. Thus, for KL systems (e.g., Ne + Kr) relatively high vacancy sharing data follow from Nikitin's model. On the contrary, in LK systems (e.g., O + Ar) the binding effect is not essential. Here, the underestimation of α dominates

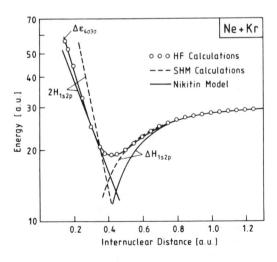

Figure 20. Calculations of the 4σ–3σ MO energy difference $\Delta\varepsilon_{4\sigma3\sigma}$ and matrix elements ΔH_{1s2p} and $2H_{1s2p}$ for the system Ne + Kr. The circles are from HF calculations. The solid and the dashed lines refer to results from Nikitin's model[17] and the SHM matrix elements (18), respectively.

and thus the relatively small vacancy sharing data follow from Nikitin's model. The systems Ne + Kr and O + Ar will be treated in some detail in the following sections.

The present discussion shows that a great deal of the problems involved in the application of the two-state models originates from inappropriate methods of determining the model parameters. It is felt that with some care the two-state models may be used for fair estimates of KL-vacancy sharing probabilities. In particular, it is pointed out that both the Meyerhof–Demkov method and the Nikitin method may be applied with nearly equal results. As shown in Section 3.4, the results from the two models are identical for low impact energies provided the diabatic energy difference $\Delta H(R_c)$ is inserted for $\Delta\varepsilon$ in Eq. (32). Also, a value should be chosen for the exponent α not too different from that originally proposed by Meyerhof.[21] Thus, it is suggested to apply the set II of parameters given in Table 1. The energy difference $\Delta H(R_c)$ may be estimated from related MO energy diagrams or by solving $2H_{12}(R_c) = \Delta H(R_c)$ with the SHM matrix elements (18).

Once the diabatic energy difference $\Delta H(R_c)$ is known, Nikitin's model may be utilized as well. When $\Delta\varepsilon$ is set to be equal to the asymptotic value $\Delta\varepsilon_{12}^0$, the Nikitin angle θ is obtained from

$$\cos\theta = 1 - z \tag{42}$$

where $z = \Delta H(R_c)/\Delta\varepsilon_{12}^0$ as in Table 1. It is noted that $z < 1$ for KL systems and $z > 1$ for LK systems (Figure 19). An analysis of various asymmetric systems by means of Nikitin's model has been made previously[11] using an expression similar to that given in Eq. (42). The results for vacancy sharing have been found to be in reasonable agreement with experimental data.

Certainly, the analytical models should be applied with caution to systems with more than two states. High accuracies are not expected from the two-state formulas. For a more adequate treatment of (nonrelativistic) systems with matching K and L shells, a three-state analysis is required to describe vacancy exchange process.

4.2.2. The System Ne + Kr

In the past, the KL system Ne + Kr has received particular attention.[57,62] In Figure 21 the MO energy diagram for the neutral system $(Ne + Kr)^0$ is given as obtained by means of HF calculations. Also, the highly ionized system $(Ne + Kr)^{35+}$ is shown for comparison. It is seen that the $2s$ and $2p$ levels of Kr are lower in energy than the $1s$ levels of Ne. The $1s$ level correlates with the 4σ molecular orbital, which exhibits a significant decrease in energy as the internuclear distance decreases. This

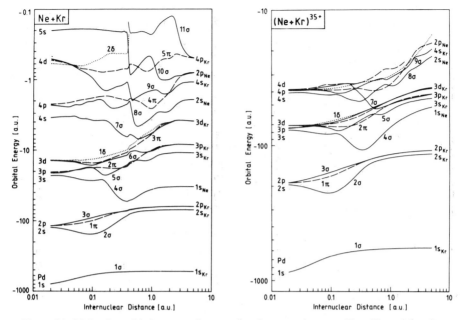

Figure 21. Molecular orbital energy diagrams for the neutral system Ne + Kr and for the ionized system $(Ne + Kr)^{35+}$ derived from HF calculations.

variation is due to the binding effect, which was noted to be particularly strong for the KL system Ne + Kr (Section 4.3.1).

Figure 22a shows the 2σ, 3σ, and 4σ MO's involved in KL-vacancy exchange. The HF data are compared with model calculations using the previously mentioned method to diagonalize the SHM matrix elements (18). It is seen that the HF data are well reproduced by the model results. It is interesting to note that the model calculations reveal quantitatively the effect of the strong binding in the KL systems. It is due to the effective nuclear charge Z_M of the center with the L-shell orbitals being about a factor of 2 larger than that with the K-shell orbital. (Recall the relation $Z_M = n\alpha_M$ with n being the principal quantum number.) Therefore, the K electron is more influenced with respect to binding effects than the L-shell electron (Fig. 19). Thus, the difference of the corresponding orbital energies is strongly affected.

Figure 22b shows the 4σ–2σ and 4σ–3σ MO energy differences together with the corresponding potential matrix elements H_{nm}. Generally, the MO energy differences do not coincide with the related nondiagonal matrix element at small internuclear distances. For instance, at small R the matrix element $2|H_{1s2p}|$ is significantly larger than the 4σ–3σ MO energy difference. Hence, it is unrealistic to adjust $2|H_{1s2p}|$ to the 3σ–4σ energy difference as is usually done in the two-state analysis.[57,59] In such an analysis, the slope of the $2|H_{1s2p}|$ curve is remarkably underestimated as mentioned in Section

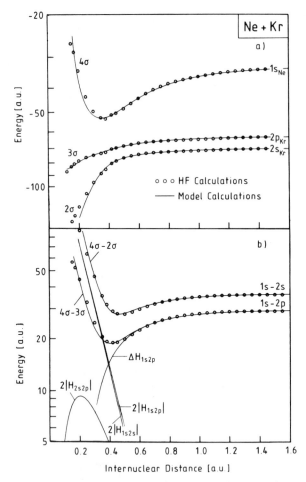

Figure 22. Energies for the molecular orbitals 2σ, 3σ, and 4σ for Ne + Kr (upper part). In the lower part are given the 4σ–2σ and 4σ–3σ MO energy differences in comparison with the matrix elements ΔH_{1s2p}, $2|H_{1s2s}|$, $2|H_{1s2p}|$, and $2|H_{2s2p}|$. The circles represent HF calculations and the solid line refers to model calculations using the SHM matrix elements (18).

4.3.1. Moreover, Figure 22b indicates that the "Stark" matrix element H_{2s2p} has no significant effect on the formation of the molecular orbitals. It should be noted that this matrix element gains importance in LK systems. Nevertheless, it is difficult to test its values sensitively by the MO fitting procedure.

To deduce radial coupling matrix elements, Eq. (14) is applied. It is expected that KL-vacancy sharing is primarily determined by the 3σ–4σ radial coupling matrix element shown in Figure 23. The model results compare well with the data from the more elaborate three-state calcula-

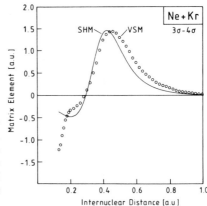

Figure 23. Radial coupling matrix element for the 3σ and 4σ MO's in the system Ne + Kr. The curve labeled SHM refers to calculations using the model matrix elements (18) and the circles labeled VSM are calculated by Fritsch and Wille[62] on the basis of the variable screening model.[87]

tions using the variable screening model (VSM).[62] Also, the 2σ–3σ radial coupling matrix element $M_{2\sigma3\sigma}$ is found to be in reasonable agreement with the VSM calculations,[62] which include translational factors. It should be recalled from Section 4.1 that $M_{2\sigma3\sigma}$ is particularly affected by the translational factors.

It has been pointed out by Fritsch and Wille[62] that the functional form of $M_{3\sigma4\sigma}$ from a three-state analysis differs considerably from that of the two-state model,[17] which does not account for the $2s$ state. The shape of $M_{3\sigma4\sigma}$ from the three-state analysis may be understood from the contribution of the different atomic orbitals involved. At internuclear distances $R > 0.3$ a.u. the interaction between the $1s$ and $2p$ orbitals dominates, whereas at $R < 0.3$ a.u. the $1s$–$2s$ interaction gains importance. Since the quantities H_{1s2s} and H_{1s2p} which account for the atomic interactions have different signs, the radial coupling matrix element passes through zero near 0.3 a.u. (Figure 23). Thus, the $2s$ orbital has primarily the effect of reducing the 3σ–4σ radial coupling.

In Figure 24 results for the sharing probabilities from the three-state calculations are shown. The impact parameter is chosen to be equal to 0.15 a.u. where the vacancy production in the 4σ MO is expected to occur.[62] The SHM results are seen to agree quite well with the VSM calculations.[62] Also shown are results from two-state calculations performed using Nikitin's model and SHM matrix elements (from Figure 19). It is interesting to note that the data from the three-state models are systematically lower than the results from two-state calculations. Obviously, the decrease of $M_{3\sigma4\sigma}$ resulting from the influence of the $2s$ orbital reduces the corresponding transition probability.

Finally, outer-shell ionization effects will be studied for asymmetric systems. In Section 3.3 it was shown that in symmetric systems the $1s$ energies are equally shifted with increasing degree of ionization so that the

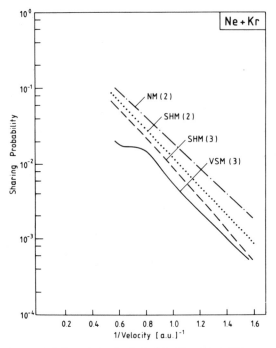

Figure 24. Excitation probability of the Kr L shell in Ne + Kr collisions versus the inverse projectile velocity. The curve labeled NM(2) follows from two-state calculations using Nikitin's model.[17] The curves labeled SHM(2) and SHM(3) represent model calculations using matrix elements (18) for two and three states, respectively. The curve labeled VSM(3) refers to three-state calculations by Fritsch and Wille[62] using the variable screening model.[87]

corresponding energy difference is not affected. This follows from the fact that in the relaxed system the outer-shell vacancies are (nearly) equally shared between the collision partners. Trivially, such sharing is not possible any more in strongly asymmetric systems if the degree of ionization becomes large. After the lighter collision partner is fully stripped, the additional vacancies have to go to the heavier one. Then, unilateral screening effects are expected.

For instance, in the system $(Ne + Kr)^{35+}$ the vacancy distribution ratio may become as large as $25 : 10$ as R increases. Hence, asymptotically the L electrons of Kr are expected to be more affected than the K electrons of Ne. Indeed, it is found that the corresponding energy difference for $(Ne + Kr)^{35+}$ is enhanced by a factor of 2. This may be seen from Figure 25, which compares the relevant energy differences $\Delta\varepsilon_{nm}$ for the neutral and ionized systems. However, it is also seen that the respective ε_{nm}'s, which differ significantly at large R, approach each other near the coupling region, i.e., near 0.4 a.u. This approach may be explained by the binding effect, which is strongly enhanced in highly ionized systems. Again, as for symmetric systems, it may be argued that at distances near the coupling radius an outer-shell vacancy is not yet localized at one specific center and, hence, unilateral screening effects are expected to be reduced.

In Figure 25 the HF calculations for $(Ne + Kr)^{35+}$ are compared with model results from Eq. (18). Similarly as for $1s$ orbitals (Section 3.3), the nondiagonal matrix elements are taken to be the same as for neutral systems. In particular, Q_M is not altered in Eq. (18b). Changes are made only in the diagonal matrix elements by adjusting the ε_n^0's to the atomic energies relevant in the ionized system. There, also, Q_M's from Eq. (29) are used, with α_0 being increased by a factor of 2. Using the particle charge states as free fit parameters, $q_{Ne} = 8.5$ and $q_{Kr} = 26.5$ are obtained. This shows that in $(Ne + Kr)^{35+}$ the distribution of vacancies is indeed strongly asymmetric.

In Figure 26 the SHM results for vacancy sharing are shown in comparison with the data for the neutral system (see also Figure 24). It is seen that the results for the ionized system are significantly lower than those for the neutral system. Thus, despite the increased binding effect, the diabatic energy difference of the ionized system remains smaller than that for the neutral system. Hence, it is concluded that in asymmetric systems, unilateral screening effects play a certain role so that the corresponding transition probabilities are reduced.

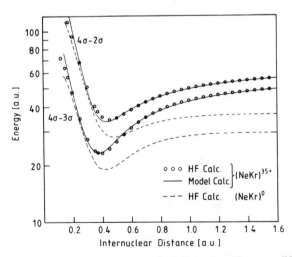

Figure 25. Hartree–Fock calculations for the 4σ–2σ and 4σ–3σ MO energy differences of the neutral system $(Ne + Kr)^0$ and the ionized system $(Ne + Kr)^{35+}$. The solid line refers to model calculations using the SHM matrix elements (18).

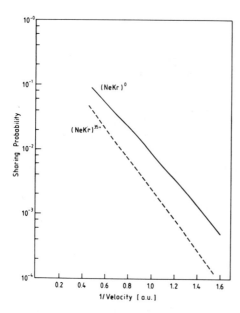

Figure 26. Model calculations of the Kr L excitation probability for the neutral collision system $(Ne + Kr)^0$ and the ionized system $(Ne + Kr)^{35+}$ using the SHM matrix elements (18).

4.2.3. The Systems B + Ar, C + Ar, N + Ar, and O + Ar

Systematic experimental studies of KL and LK vacancy sharing have been made by Reed et al.[59] applying the method of Auger spectroscopy. In the experiments B, C, N, and O have been used as projectiles incident on Ar. These projectiles have been chosen to study the transition from a KL system (B + Ar) via an intermediate case (C + Ar), to LK systems (N + Ar). The MO energy diagrams for these systems are given in Figure 27 obtained from HF calculations. It is seen that in the B + Ar system the $1s$ levels of B are higher in energy than the $2s$ and $2p$ levels from Ar. In C + Ar the $1s$ level is located just in between the $2s$ and $2p$ levels. In N + Ar the $1s$ level falls below the $2s$ and $2p$ levels.

In Figure 28 the HF results are given for the MO's involved in KL-vacancy exchange. The characteristic influence of the binding effect (Figure 19) may be inspected by means of the curves referring asymptotically to the $1s$-$2p$ energy difference. The 4σ-3σ data for the KL system B + Ar exhibit a minimum owing to the pulling together of the diabatic potential curves. On the contrary, as R decreases the repulsion of the diabatic potential curves produces a monotonic increase of the 4σ-3σ data for C + Ar and of the 4σ-2σ data for O + Ar. Figure 28 shows also results from model calculations using the SHM matrix elements (18). Excellent agreement is observed between the model calculations and the HF results. It should be recalled that a unique set of dimensionless parameters is utilized in the model calculations (Table 2).

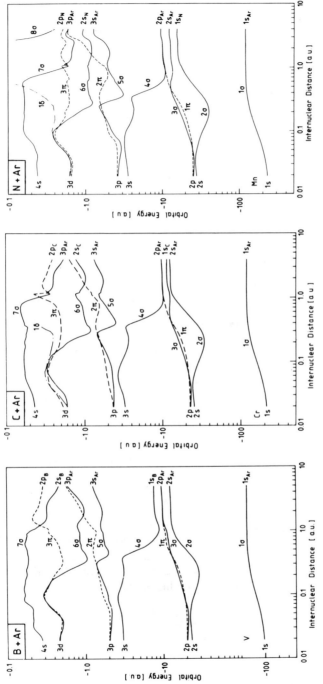

Figure 27. Molecular orbital energy diagrams for the systems B + Ar, C + Ar, and N + Ar derived from HF calculations.

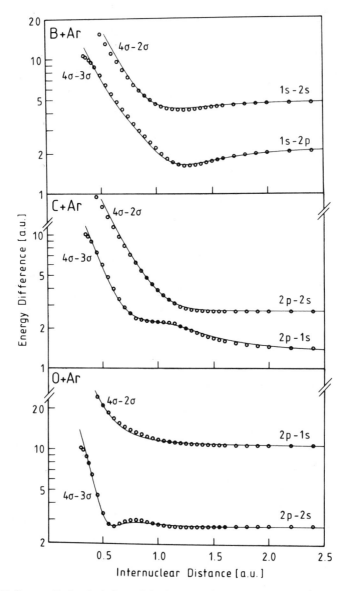

Figure 28. Hartree–Fock calculations of the 4σ–2σ and 4σ–3σ MO energy differences for the systems B + Ar, C + Ar, and O + Ar (circles). The solid lines represent model calculations using the SHM matrix elements (18).

To derive vacancy sharing probabilities for B, C, N, and O impact on Ar, three-state coupled equations are solved. As initial condition, one vacancy is assumed to be located in the highest-lying orbital, i.e., the 4σ MO. In Figure 29 the Ar L excitation probability is displayed for the system B + Ar. For the other systems the K-excitation probability of the lighter atom is given. In case of the LK systems N + Ar and O + Ar the three-state model calculations are found to be orders of magnitude higher than the previous two-state model calculations using Nikitin's model.[59] In Figure 29 an example is given for O + Ar. This finding may primarily be explained by the underestimation of parameter α in the two-state analysis as discussed in Section 4.2.1. However, even when α is set to more realistic values (set II in Table 2), the results from Nikitin's two-state model remain smaller

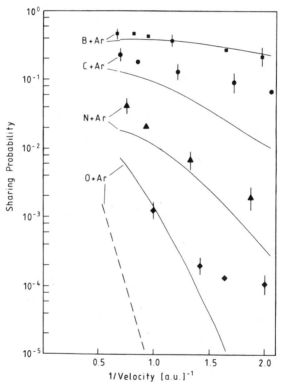

Figure 29. Vacancy sharing probabilities for the K and L shells in the systems B + Ar, C + Ar, N + Ar, and O + Ar as a function of the inverse projectile velocity. The data refer to the excitation of the lower-lying levels, i.e., the Ar L-shell orbitals for B + Ar and the K-shell orbitals of the lighter particle in the other systems. The dots refer to experimental results by Reed *et al.*[59] and the solid lines follow from three-state model calculations on the basis of the SHM matrix elements (18). The dashed line represents two-state calculations[59] for O + Ar by means of Nikitin's model.[17]

than the data from the three-state model. It appears that in LK systems the $2s$ state plays a significant role. It has the effect of a "bridge" which favors the transitions to the $1s$ state of the lighter collision partner.[56]

In Figure 29, significant discrepancies between the SHM calculations and experiment are still observed. These discrepancies are partially understood. The transition probability in the LK systems is relatively small in the range of small incident velocities, so that, there, mechanisms other than vacancy sharing become important. Unilateral screening effects due to double vacancy production in the 4σ MO and/or two-electron transitions filling[88,89] the double hole states are expected to be important.

The theoretical results of the intermediate system C + Ar are sensitively dependent on the MO energies used in the analysis. For instance, previous three-state calculations similar to the present one but using another set of MO energies[11] yield good agreement with experiment.[60,63] It appears that for the system C + Ar further theoretical work is required.

The model calculations of LK-vacancy sharing yield probabilities seperately for the excitation of $2s$ and $2p$ states. These data may be used to gain information about diabatic correlation rules. For LK systems two different rules have been proposed with respect to the diabatic state $3d\sigma$ which corresponds to the 4σ MO. Barat and Lichten[9] proposed the correlation of the $3d\sigma$ state with the $2p$ level, whereas Eichler et al.[90] suggested the correlation with the $2s$ level.

In Figure 30 the excitation probabilities of the $2s$ and $2p$ states are compared for the LK systems N + Ar and O + Ar. It should be pointed out that in the calculations one vacancy is assumed to be present in the 4σ MO prior to the collision. Thus, as expected, at the low velocity limit the vacancy goes to the $2p$ state. However, with increasing velocity v the excitation of

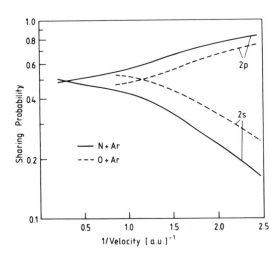

Figure 30. Model calculations of the excitation probabilities for the Ar $2s$ and $2p$ levels in the systems N + Ar and O + Ar as a function of the inverse projectile velocity. The calculations are based on the SHM matrix elements (18).

the $2s$ state gains importance and it becomes larger than that of the $2p$ state at $v = 0.8$ a.u. and $v = 3$ a.u. for O and N impact, respectively. It is noted that the average velocity of the K and L shell electrons involved in the systems N + Ar and O + Ar is about 5 a.u. Hence, to retain the present MO concept, one should consider incident energies smaller than, say, 1 a.u. Then, Figure 30 indicates that for N + Ar the Barat–Lichten rule[9] is favored, whereas for O + Ar the rule by Eichler et al.[90] becomes important. Hence, it is concluded that for the present collision system none of the correlation rules is clearly supported.

4.3. L-Vacancy Sharing

In near symmetric collision systems where the L shells of the particles are close in energy, L-vacancy exchange may be studied under near resonant conditions. The states involved in L-vacancy exchange may be observed in Figure 31 displaying the MO diagrams of the systems Al + Ar, Si + Ar, and S + Ar obtained from HF calculations. The $2p(m = 1)$ state of the heavier particle and the $2p(m = 1)$ of the lighter particle correlate with the 1π and 2π orbitals, respectively. Since the vacancy exchange between the π states has already been discussed in Section 3.4, attention is focused on the σ orbitals. It is seen that the 3σ, 4σ, 5σ, and 6σ MO's correlate with the $2s$, $2p(m = 0)$ levels of the lighter particle and the $2s$ and $2p(m = 0)$ of the heavier particle, respectively.

The sizes of the atomic orbitals referring to the σ states differ significantly from those for the π states as indicated schematically in Figure 2. Therefore, the overlaps of these atomic states are expected to occur at different distances. Indeed, the interaction radius of the $2p(m = 0)$ states is remarkably larger than that of the $2p(m = 1)$ states. In Figures 6 and 32 this may be examined by means of the repulsion of the molecular orbitals, which is an indication of overlap of the underlying atomic states. It is seen that the 1π and 2π MO's repel each other at $R \lesssim 0.8$ a.u., whereas the σ MO's involved repel each other at $R \lesssim 1.2$ a.u.

Applications of the SHM matrix elements (18) to L-vacancy exchange systems are given in Figure 32 for Si + Ar and S + Ar. Again, the unique set of dimensionless model parameters from Table 2 is used in the model calculations. Furthermore, transition probabilities between the σ orbitals are derived by solving four-state coupled equations (9). The results of the Ar L-shell excitation probability are shown in Figure 33 for the system Si + Ar and S + Ar. The data are compared with corresponding results for transitions between π orbitals obtained by using the two-state SHM (Section 4.3). It is recalled that both sets of data are based on the same initial condition, i.e., one vacancy in the highest-lying molecular orbital, which is the 6σ MO and 2π MO for transitions between σ and π orbitals, respec-

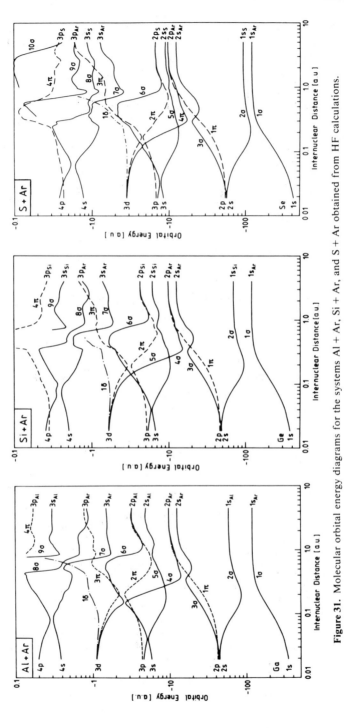

Figure 31. Molecular orbital energy diagrams for the systems Al + Ar, Si + Ar, and S + Ar obtained from HF calculations.

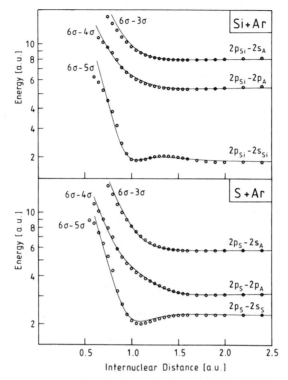

Figure 32. Hartree–Fock calculations for the 6σ–3σ, 6σ–4σ, and 6σ–5σ MO energy differences for the systems Si + Ar and S + Ar (circles). The solid lines follow from model calculations using the SHM matrix elements (18).

tively. Figure 33 shows that in general the transition probability for σ orbitals is significantly higher than that for π orbitals. It appears unlikely that this difference is due to the difference in the interaction radii mentioned above. Rather, it is expected that the presence of the $2s$ state of the lighter particle enhances the transition probability for the σ orbitals. As noted for LK-sharing systems, the $2s$ state may act as a "bridge" which favors the transition from the 6σ MO to the 4σ and 3σ MO.

The comparison in Figure 33 is unrealistic, since in a collision the initial conditions for the σ and π orbitals are not the same. In fact, the data shown for transitions via the σ and π orbitals are to be weighted with the probabilities to create vacancies in the 6σ and 2π MO's, respectively. However, then the concept of vacancy-sharing probabilities loses significance, since it is based on the initial condition of having one vacancy in the highest-lying MO. In this case it is favorable to use the concept of cross sections. Generally, the sharing of the vacancy is independent of its creation, and, thus, it is independent of the impact parameter within the

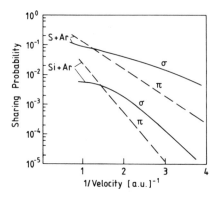

Figure 33. Model calculations of vacancy-sharing probabilities for the systems Si + Ar and S + Ar as a function of the inverse projectile velocity. The calculations are based on the SHM matrix elements (18). The solid and dashed lines refer to Ar L excitation via the coupling of σ and π orbitals, respectively.

range where the vacancy is created. Then, the cross sections for excitation of the heavier particle via σ and π states are obtained by weighting the related probabilities (Figure 33) with the cross section for production of vacancies in the 6σ and 2π orbitals, respectively.

The results of this weighting procedure[68] are given in Figure 34, where the contributions from the σ and π states are labeled RAD(σ) and RAD(π), respectively. In the analysis experimental data are taken for the cross section

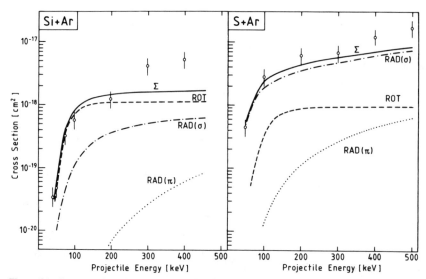

Figure 34. Cross sections for Ar L excitation in the collision systems Si + Ar and S + Ar as a function of the projectile energy. The experimental data are from Schneider *et al.*[67] The curve labeled ROT refers to 1δ-2π-4σ rotational coupling calculations by Wille.[91] The curves labeled RAD(σ) and RAD(π) follow from the probabilities for sharing transitions via σ and π orbitals (Figure 33) weighted with cross sections for vacancy production in the 6σ orbital[67] and the 2π orbital,[91] respectively. The curve labeled $\sum$ represents the sum of the theoretical data. (From Ref. 68.)

of vacancy production in the 6σ state, which was set to be equal to the cross section for the excitation of the lighter particle. Vacancies in the 2π orbital are produced by transitions from the 1δ MO via a rotational coupling mechanism. In this case theoretical data are taken as derived by Wille.[91] From Figure 34 it is concluded that the contribution of the RAD(π) process is negligible in the cases studied here.

Figure 34 shows also experimental results obtained by Schneider et al.[67] using the method of Auger spectroscopy. For the system S + Ar it is seen that the RAD(σ) process provides the major contribution to the Ar L excitation. However, for Si + Ar the RAD(σ) process becomes less important. In this case an additional process different from vacancy sharing has to be considered. This process, analyzed by Wille,[91] is attributed to the 1δ-2π-4σ rotational coupling mechanism (Figure 31). As expected the respective cross sections from this process, labeled ROT in Figure 34, are practically the same for the systems S + Ar and Si + Ar. On the contrary, with increasing asymmetry of the system the RAD(σ) process loses importance, since the energy gap between the L shells of the collision partners increases (Figure 31). Hence, it is concluded that pure L-vacancy sharing may be studied only in rather symmetric collision systems.

5. Conclusions

Near-resonant vacancy exchange between inner shells is a fundamental process in ion–atom collisions that has extensively been studied in the past. It is attractive for a theoretical study, since its analysis may be based on a few atomic or molecular states. Thus, relatively simple models may be applied in the calculations. The important advantage of the application of few-state models is that the mechanisms under study become readily transparent. Hence, the few-state analysis reveals detailed information about the properties of the collision system.

The application of few-state models has a good tradition in the field of outer-shell excitation. In particular, analytic matrix elements have been used with considerable success. For inner shells, model matrix elements are even more appropriate because of their scaling properties. An analytic matrix element that has been shown to be applicable to one collision system may usually be scaled to another system. Thus, no individual numerical calculation is required for each collision system.

The parameters of the model matrix elements used here are adjusted to fit Hartree–Fock energies. Inner-shell electrons are well described by means of the independent particle model within which the Hartree–Fock method yields accurate one-electron (MO) energies. At low incident velocities, where the MO concept is applicable, the vacancy transfer probabilities are sensitively dependent on the one-electron energies involved.

Hence, accurate MO energies are needed. Fortunately, for the SHM matrix elements it is not required to perform the adjustment procedure for each collision system separately. Once the dimensionless model parameters are fixed, the matrix elements may readily be applied to other systems. Thus, for various systems, MO energies were calculated in excellent agreement with the Hartree–Fock results.

For two states, the coupled equation describing the dynamics of the collision system may be solved in closed form. Apart from quantitative results, the two-state formulas yield considerable contributions to the qualitative understanding of the mechanisms involved in vacancy exchange. In particular, the formulas show that each quantity measured in conjunction with vacancy exchange provides a sensitive test of specific atomic properties. The related processes are vacancy sharing, single passage transfer at nonzero impact parameter, and double passage transitions. Studying these processes in this order, one may gain progressively new information about the properties of the collision system.

The study of vacancy sharing provides a sensitive test of the gap between the diabatic energy curves and the slope of the interaction potential curve. This slope is determined by the exponent α, which is found to be rather insensitive to various effects, e.g., multiple ionization. Thus, vacancy sharing data are primarily used to obtain information about the diabatic potentials. Here, the binding effect which governs the distortion of the diabatic potentials plays an important role. For example, relatively small differences between the binding effects on σ and π orbitals are visible in the sharing probabilities. With respect to this sensitivity to the diabatic potentials it is important to note that the vacancy sharing probability is independent of multiple ionization in outer shells. However, the vacancy sharing probability is found to be affected by multiple inner-shell ionization.

The vacancy sharing probability is independent of the coupling radius. Information about the location and size of the coupling region may be obtained from studying the impact parameter dependence of the single-passage process. Here, the quantity for which new information is achieved is the absolute magnitude of the interaction potential within the coupling region.

The same quantity but at smaller internuclear distances is tested, when the double-passage process for vacancy transfer is examined. The transition probability exhibits oscillatory structures due to interferences between ingoing and outgoing parts of the collision. The related two-state formula shows that the oscillatory structure is superimposed on a smooth function already known from the single-passage process. Thus, the new information is implied in the oscillatory structure. The locations of the minima and maxima are determined by the difference of the related MO energies. Hence, the oscillatory structure allows for studying the formation of MO's at small internuclear distances. The oscillation is expected to be very sensitive on the variation of the MO curves. Recent data show indications that the interference structures are influenced by differences between the MO curves for the incoming and outgoing parts of the collision.[38]

Most of the comments made here about the two-state systems retain their validity for systems with more than two states. The particular feature of the system with matching K and L shell is dominance of the binding effect. Therefore, in a model calculation care must be taken that the distortion of the diabatic potentials is well accounted for.

In KL and LK systems three atomic states are involved in the formation of the molecular orbitals. Thus, the method of obtaining the parameters for the two-state models from fitting molecular orbital energies appears to be questionable. However, this does not mean that two-state models are completely useless in the analysis of three-state systems. For instance, with some caution the two-state model may be used to describe vacancy sharing in the KL system. There, the $2s$ energy level is located outside the gap between the $1s$ and $2p$ energy levels, which are most important for the KL-sharing process. In this case, the $2s$ state plays more the role of a passive spectator. However, in LK systems the $2s$ energy level is located within the $1s$–$2p$ energy gap. The $2s$ level may act as a "bridge" which favors the transitions between the $1s$ and $2p$ levels.

The same is true in L-sharing systems, where always the $2s$ level of the lighter partner is located between the $2p$ energy levels. Indeed, for Si + Ar and S + Ar an order of magnitude difference is observed between the results of the two-state calculation and the corresponding four-state calculation. Here, definitely an analysis with more than two states is required.

Acknowledgment

I am much indebted to Professor J. Eichler, Dr. W. Fritsch, and Dr. U. Wille for many stimulating discussions. I thank Dr. A. Itoh, Dr. D. Schneider, and Dr. Th. Zouros for critically reading the manuscript and for helpful suggestions. Valuable comments and communication of results prior to publication by Dr. R. Schuch is greatfully acknowledged.

References

1. D. R. Bates and R. McCarroll, *Proc. R. Soc. London Ser. A* **245**, 175 (1958).
2. M. R. C. McDowell and J. P. Coleman, *Introduction to the Theory of Atomic Collisions*, North Holland, Amsterdam (1970).

3. D. Basu, S. C. Mukherjee, and D. P. Sural, *Phys. Rep.* **42C**, 145 (1978).
4. Belkić, R. Gayet, and A. Salin, *Phys. Rep.* **56C**, 279 (1979).
5. R. K. Janev and L. P. Presnyakov, *Phys. Rep.* **70**, 1 (1981).
6. P. T. Greenland, *Phys. Rep.* **81**, 131 (1982).
7. C. D. Lin and P. Richards, *Adv. At. Mol. Phys.* **17**, 275 (1981).
8. U. Fano and W. Lichten, *Phys. Rev. Lett.* **14**, 627 (1965).
9. M. Barat and W. Lichten, *Phys. Rev. A* **6**, 211 (1972).
10. J. S. Briggs, *Rep. Prog. Phys.* **39**, 217 (1976).
11. N. Stolterfoht, *The Physics of Ionized Gases* (edited by R. K. Janev), Institute of Physics, Beograd (1979), p. 93.
12. B. M. Smirnov, *Sov. Phys. Dokl.* **12**, 242 (1967).
13. D. Rapp and W. E. Francis, *J. Chem. Phys.* **37**, 2631 (1962).
14. R. E. Olson, F. T. Smith, and E. Bauer, *Appl. Opt.* **10**, 1848 (1971).
15. Yu. N. Demkov, *Zh. Eksp. Teor. Fiz.* **45**, 195 (1963) [*Sov. Phys.-JETP* **18**, 138 (1964)].
16. E. E. Nikitin, *Opt. Spektrosk.* **13**, 341 (1962).
17. E. E. Nikitin, *Advances in Quantum Chemistry*, Vol. 5, Academic, New York (1970), p. 135.
18. C. D. Landau, *J. Phys. (USSR)* **2**, 46 (1932).
19. C. Zener, *Proc. R. Soc. London Ser. A* **137**, 696 (1932).
20. R. C. Der, R. J. Fortner, T. M. Kavanagh, and J. M. Khan, *Phys. Rev. Lett.* **24**, 1272 (1970).
21. W. E. Meyerhof, *Phys. Rev. Lett.* **31**, 1341 (1973).
22. N. Stolterfoht, P. Ziem, and D. Ridder, *J. Phys. B* **7**, L409 (1974).
23. B. Fastrup, B. Aagaard, E. Bøving, D. Schneider, P. Ziem, and N. Stolterfoht, *IXth International Conference on the Physics of Electronic and Atomic Collisions*, Abstracts of Papers (edited by J. S. Risley and R. Geballe), University of Washington Press, Seattle (1975), p. 1058.
24. D. Schneider and N. Stolterfoht, *Phys. Rev. A* **19**, 55 (1979).
25. C. W. Woods, R. L. Kauffman, N. Stolterfoht, and P. Richard, *Phys. Rev. A* **13**, 1358 (1976).
26. N. Stolterfoht and U. Leithäuser, *Phys. Rev. Lett.* **36**, 186 (1976).
27. E. G. Bøving, *J. Phys. B* **10**, L63 (1977).
28. W. E. Meyerhof, R. Anholt, and T. K. Saylor, *Phys. Rev. A* **16**, 169 (1977).
29. Ch. Stoller, W. Wölfli, G. Bonani, M. Stöckli, and M. Suter, *Phys. Rev. A* **15**, 990 (1977).
30. W. N. Lennard, I. V. Mitchell, and D. Phillips, *J. Phys. B* **11**, 1283 (1978).
31. J. S. Greenberg, P. Vincent, and W. Lichten, *Phys. Rev. A* **16**, 964 (1977).
32. J. R. McDonald, R. Schulé, R. Schuch, H. Schmidt-Böcking, and D. Liesen, *Phys. Rev. Lett.* **40**, 1330 (1978).
33. J. D. Garcia, R. J. Fortner, H. C. Werner, D. Schneider, N. Stolterfoht, and D. Ridder, *Phys. Rev. A* **22**, 1884 (1980).
34. J. Hall, P. Richard, T. Gray, C. D. Lin, K. Jones, B. Johnson, and D. Gregory, *Phys. Rev. A* **24**, 2416 (1981).
35. A. Chetioui, J. P. Rozet, J. P. Briand, and C. Stephen, *J. Phys. B* **14**, 1625 (1981).
36. K. Wohrer, A. Chetioui, J. P. Rozet, A. Jolly, and C. Stephen, *J. Phys. B* **17**, 1575 (1984).
37. R. Schuch, G. Nolte, H. Schmidt-Böcking, and W. Lichtenberg, *Phys. Rev. Lett.* **43**, 1104 (1979).
38. R. Schuch, H. Ingwersen, E. Justiniano, H. Schmidt-Böcking, M. Schulz, and F. Ziegler, *J. Phys. B* **17**, 2319 (1984).
39. S. Hagmann, C. L. Cocke, J. R. McDonald, P. Richard, H. Schmidt-Böcking, and R. Schuch, *Phys. Rev. A* **25**, 1918 (1982).
40. S. Hagmann, C. L. Cocke, P. Richard, A. Skutlartz, S. Kelbch, H. Schmidt-Böcking, and R. Schuch, *XIIIth International Conference on the Physics of Electronic and Atomic Collisions*, Book of Invited Papers (edited by J. Eichler, I. V. Hertel, and N. Stolterfoht), Berlin (1984), p. 385.

41. C. D. Lin, S. C. Soong, and L. N. Tunnell, *Phys. Rev. A* **17**, 1646 (1978).
42. C. D. Lin and L. N. Tunnell, *Phys. Rev. A* **22**, 76 (1980).
43. W. Fritsch, C. D. Lin, and L. N. Tunnell, *J. Phys. B* **14**, 2861 (1981).
44. K. Taulbjerg, S. Vaaben, and B. Fastrup, *Phys. Rev. A* **12**, 2325 (1975).
45. K. Taulbjerg and J. S. Briggs, *J. Phys. B* **8**, 1895 (1975).
46. J. S. Briggs and K. Taulbjerg, *J. Phys. B* **8**, 1909 (1975).
47. V. Sidis, N. Stolterfoht, and M. Barat, *J. Phys. B* **10**, 2815 (1977).
48. A. Macias, A. Riera, and A. Salin, *J. Phys. B* **12**, 447 (1979).
49. J. S. Briggs, Atomic Energy Research Establishment Technical Report No. 594 (1974).
50. N. Stolterfoht, *J. Phys. B* **13**, L559 (1980).
51. N. Stolterfoht, *J. Phys. B* **16**, 2385 (1983).
52. S. J. Pfeiffer and J. D. Garcia, *Phys. Rev. A* **23**, 2267 (1981).
53. S. J. Pfeiffer and J. D. Garcia, *J. Phys. B* **15**, 1275 (1982).
54. W. E. Meyerhof, R. Anholt, J. Eichler, and A. Salop, *Phys. Rev. A* **17**, 108 (1978).
55. E. M. Middelsworth, Jr., D. J. Donahue, L. C. McIntyre, Jr., and E. M. Bernstein, *Phys. Rev. A* **18**, 1765 (1978).
56. N. Stolterfoht, D. Schneider, and D. Brandt, *Xth International Conference on the Physics of Electronic and Atomic Collisions*, Abstracts of Papers (edited by G. Watel), Paris (1977), p. 902.
57. P. H. Woerlee, R. J. Fortner, S. Doorn, Th. P. Hoogkamer, and F. W. Saris, *J. Phys. B* **11**, L425 (1978).
58. R. J. Fortner, *IEEE Trans. Nucl. Sci.* **26**, 1016 (1979).
59. K. J. Reed, J. D. Garcia, R. J. Fortner, N. Stolterfoht, and D. Schneider, *Phys. Rev. A* **22**, 903 (1980).
60. R. Shanker, R. Hippler, R. Bilau, and H. O. Lutz, *Phys. Lett.* **99A**, 313 (1983).
61. A. Warczak, D. Liesen, D. Maor, P. H. Mokler, and W. A. Schönfeldt, *J. Phys. B* **16**, 1575 (1983).
62. W. Fritsch and U. Willie, *J. Phys. B* **12**, L645 (1979).
63. N. Stolterfoht, *European Conference on Atomic Physics*, Book of Abstracts (edited by J. Kowalski, G. zu Putlitz, and H. G. Weber), Heidelberg (1981), p. 880.
64. W. N. Lennard and I. V. Mitchell, *J. Phys. B* **9**, L317 (1976).
65. W. N. Lennard, I. V. Mitchell, J. S. Forster, and D. Phillips, *J. Phys. B* **10**, 2199 (1977).
66. W. E. Meyerhof, A. Rüetschi, Ch. Stoller, M. Stöckli, and W. Wölfli, *Phys. Rev. A* **20**, 154 (1979).
67. D. Schneider, G. Nolte, N. Stolterfoht, and U. Wille, *XIIth International Conference on the Physics of Electronic and Atomic Collisions*, Book of Abstracts (edited by S. Datz), Gatlinburg (1981), p. 828.
68. N. Stolterfoht, D. Schneider, G. Nolte, and U. Wille, *8th International Conference on Atomic Physics*, Program and Abstracts (edited by I. Lindgren, A. Rosen, and S. Swanberg), Göteborg (1982), p. B67.
69. T. Iwai, Y. Kaneko, M. Kimura, N. Kobayashi, S. Ohtani, K. Okuno, S. Takagi, H. Tawara, and S. Tsurubuchi, *Phys. Rev. A* **26**, 105 (1982).
70. J. B. Delos and W. R. Thorson, *J. Chem. Phys.* **70**, 1774 (1979).
71. S. B. Schneidermann and A. Russek, *Phys. Rev.* **181**, 311 (1969).
72. T. A. Green, *Proc. Phys. Soc.* **86**, 1017 (1965).
73. J. A. Bearden and A. F. Burr, *Rev. Mod. Phys.* **39**, 125 (1967).
74. F. Froese-Fischer, *At. Data* **4**, 301 (1872).
75. N. Coulson, *Proc. R. Soc. Edinburgh A* **61**, 210 (1941).
76. W. Brandt, R. Laubert, and I. A. Sellin, *Phys. Lett.* **21**, 518 (1966).
77. Bethe and Salpeter, *Quantum Mechanics of One- and Two-Electron Atoms*, Springer, Berlin (1975).

78. M. S. Child, *Advances in Atomic and Molecular Physics*, Vol. 14 (edited by D. R. Bates and B. Bederson), Academic, New York (1978), p. 225.
79. D. R. Bates, *Proc. R. Soc. London Ser. A* **247**, 294 (1958).
80. H. K. Macomber and T. G. Webb, *Proc. Phys. Soc.* **92**, 839 (1967).
81. P. S. Bagus, *Selected Topics in Molecular Physics*, Springer, Berlin (1972).
82. J. Eichler, *Phys. Rev. Lett.* **40**, 1560 (1978).
83. J. Eichler, W. Fritsch, and U. Wille, *Phys. Rev. A* **20**, 1448 (1979).
84. J. Eichler and Tak San Ho, *Z. Phys.* **311**, 19 (1983).
85. D. S. Crothers, *Adv. Phys.* **20**, 405 (1971).
86. D. S. Crothers, *J. Phys. B* **6**, 1418 (1973).
87. J. Eichler and U. Wille, *Physica A* **11**, 1973 (1975).
88. M. Barat, private communication (1975).
89. V. V. Afrosimov, Y. U. S. Goordeev, A. N. Zinoviev, D. H. Rasulov, and A. P. Shergin, *IXth International Conference on the Physics of Electronic and Atomic Collisions*, Abstract of Papers (edited by J. S. Risley and R. Geballe), University of Washington Press, Seattle (1975), p. 1066.
90. J. Eichler, U. Wille, B. Fastrup, and K. Taulbjerg, *Phys. Rev. A* **14**, 707 (1976).
91. U. Wille, *Proceedings of the European Conference on Atomic Physics*, Contributed Papers (edited by J. Kowalski, G. zu Putlitz, and H. G. Weber), Heidelberg (1981), p. 878.

10

Polarization Correlation in the Two-Photon Decay of Atoms

A. J. DUNCAN

1. Introduction

Interest in the polarization correlation of photons goes back to the early measurements[1] of the linear polarization correlation of the two photons produced in the annihilation of *para*-positronium which were carried out as a result of a suggestion by Wheeler[2] that these photons, when detected, have orthogonal polarizations. Yang[3] subsequently pointed out that such measurements are capable of giving information on the parity state of nuclear particles that decay into two photons. In addition, the polarization correlation observed in the two-photon decay of atoms is considered to be one of the few phenomena where semiclassical theories of radiation are inadequate[4,5] and it is necessary to invoke a full quantum theory of radiation. The effect has also been used to demonstrate the phenomenon of quantum interference.[6]

The polarization correlation in two-photon processes has thus proved a topic of considerable interest in its own right. However, without doubt, the main stimulus to the performance of polarization correlation measurements came first from the *Gedankenexperiment* of Bohm[7] and the paper of Bohm and Aharonov[8] in which the so-called paradox of Einstein, Podolsky, and Rosen[9] (EPR) was put in terms of the polarization of photons and subsequently from the work of Bell[10] and its interpretation in experimental terms by Clauser, Horne, Shimony, and Holt,[11] and Clauser and Horne.[12]

A. J. DUNCAN • Physics Department, University of Stirling, Stirling, FK9 4LA, Scotland.

The experiments using the annihilation of *para*-positronium have been reviewed elsewhere[13,14] and this article concentrates on those in which the two photons are emitted from an atomic source. In order to understand fully the significance of these experiments, however, it is necessary to give some consideration to the underlying theory. The following discussion is not intended to be exhaustive and the reader is referred to the review articles by Clauser and Shimony,[13] Pipkin[14] and to the work of Aspect[15,16] for further details.

2. Theoretical Considerations

2.1. Polarization Correlation

Consider first the ideal situation where pairs of photons, frequencies ν_1 and ν_2, emitted in the $-z$ direction and $+z$ direction, respectively, from an atomic source, are analyzed by two-channel polarizers π_1 and π_2 as shown in Figure 1. The detectors D_{ij} $(i, j = 1, 2)$ are assumed to be 100% efficient. The transmission axes of the polarizers are set in the directions $\hat{a}$ and $\hat{b}$, where $\hat{a}$ and $\hat{b}$ are unit vectors parallel to the x-y plane. The use of two-channel polarizers allows the polarization components of radiation both parallel to and perpendicular to the transmission axis of each polarizer to be monitored simultaneously. For simplicity of notation we shall, from now on, omit the unit vector sign from $\hat{a}$ and $\hat{b}$.

It is normal in this type of experiment, where we are looking for correlations between the polarizations of the two photons, to assign the value $+1$ to detection in D_{11} or D_{21} and -1 to detection in D_{12} or D_{22}. Detection to the left of the source can thus be represented by a variable A, say, which can take on the values ± 1, and detection to the right by a variable

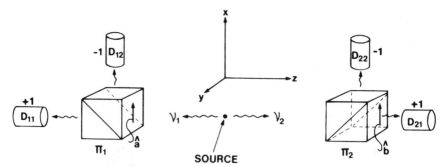

Figure 1. Diagram to illustrate the ideal measurement of polarization correlation. π_1 and π_2 are ideal two-channel polarizers set with their transmission axes in the directions $\hat{a}$ and $\hat{b}$, respectively. D_{ij} $(i, j = 1, 2)$ are 100% efficient detectors. The source emits pairs of photons, frequencies ν_1 and ν_2, in the $-z$ and $+z$ directions, respectively.

B, which can also take on the values ± 1. It follows that a measure of the extent of the correlation between A and B for given settings a and b of the polarizers is the correlation coefficient $E(a, b)$ defined as

$$E(a, b) = \frac{\overline{AB} - \bar{A}\bar{B}}{(\overline{A^2 B^2})^{1/2}} \tag{1}$$

where the bar denotes an average over an ensemble of emitted pairs. In the present case the expression reduces to

$$E(a, b) = \overline{AB} \tag{2}$$

since $\overline{A^2} = \overline{B^2} = 1$ and, since each variable is as likely to be positive as negative, we expect that $\bar{A} = \bar{B} = 0$.

If we denote by $P_{++}(a, b)$ the probability of a photon pair giving a result $+1$ to the left and $+1$ to the right with similar definitions for $P_{+-}(a, b)$, $P_{-+}(a, b)$, and $P_{--}(a, b)$ then

$$E(a, b) = P_{++}(a, b) + P_{--}(a, b) - P_{+-}(a, b) - P_{-+}(a, b) \tag{3}$$

Alternatively, if in a given time, N photon pairs are emitted resulting in $N_{++}(a, b)$ detection events in which $+1$ is registered to the left and $+1$ to the right, then, provided N is sufficiently large, $P_{++}(a, b) = N_{++}(a, b)/N$ with similar expressions for $P_{+-}(a, b)$, $P_{-+}(a, b)$, and $P_{--}(a, b)$ so that we can write

$$E(a, b) = \frac{N_{++}(a, b) + N_{--}(a, b) - N_{+-}(a, b) - N_{-+}(a, b)}{N} \tag{4}$$

In quantum mechanical terms, we can say that there is an observable, represented by the operator $A^*(a)$, with eigenvectors $|a^{\pm}\rangle$ and eigenvalues $A = \pm 1$, respectively, describing the results of measurement of photon ν_1 parallel and perpendicular to a and an observable, represented by the operator $B^*(b)$ with eigenvectors $|b^{\pm}\rangle$ and eigenvalues $B = \pm 1$, describing the result of measurement of photon ν_2 parallel and perpendicular to b. It is then easy to see that, in terms of the linear polarization basis vectors $|x\rangle$ and $|y\rangle$,

$$|a^+\rangle = \cos\theta_1 |x\rangle_1 + \sin\theta_1 |y\rangle_1, |a^-\rangle = -\sin\theta_1 |x\rangle_1 + \cos\theta_1 |y\rangle_1$$
$$\tag{5}$$
$$|b^+\rangle = \cos\theta_2 |x\rangle_2 + \sin\theta_2 |y\rangle_2, |b^-\rangle = -\sin\theta_2 |x\rangle_2 + \cos\theta_2 |y\rangle_2$$

where $|x\rangle_1$ and $|y\rangle_1$ denote polarization states on the left, $|x\rangle_2$ and $|y\rangle_2$ on the right, and θ_1 and θ_2 are the angles between the x axis and a and b, respectively. It follows that

$$A^*(a) = (+1)|a^+\rangle\langle a^+| + (-1)|a^-\rangle\langle a^-| \tag{6}$$

and

$$B^*(b) = (+1)|b^+\rangle\langle b^+| + (-1)|b^-\rangle\langle b^-| \tag{7}$$

since then

$$A^*(a)|a^+\rangle = (+1)|a^+\rangle \text{ etc.} \tag{8}$$

If the two-photon state vector is represented by $|\psi\rangle$ then we can calculate the expectation value for the product A^*B^* according to

$$\begin{aligned}
\langle\psi|A^*B^*|\psi\rangle &= |\langle b^+|\langle a^+|\psi\rangle|^2 + |\langle b^-|\langle a^-|\psi\rangle|^2 \\
&\quad - |\langle b^+|\langle a^-|\psi\rangle|^2 - |\langle b^-|\langle a^+|\psi\rangle|^2 \\
&= P_{++}(a, b) + P_{--}(a, b) - P_{+-}(a, b) - P_{-+}(a, b) \\
&= E(a, b)
\end{aligned} \tag{9}$$

The correlation coefficient $E(a, b)$ can thus be identified as the expectation value of the direct product operator A^*B^*.

2.2. The Two-Photon State Vector

In the situation illustrated in Figure 1 where the two photons are emitted in diametrically opposite directions, the polarization part of the associated state vector can be derived from simple consideration of conservation of angular momentum and parity. It should be noted that, in practice, for each photon pair, frequencies ν_1 and ν_2 emitted, respectively, in the $+z$ and $-z$ directions there is a complementary photon pair, frequencies ν_1 and ν_2, emitted, respectively, in the $-z$ and $+z$ directions. Because of this symmetry, in what follows it is only necessary to consider the polarization properties of the photon pair.

In order to conserve angular momentum when the two photons are emitted in opposite directions from a source whose constituents are isotropic before and after emission, it is necessary for the photons to have equal helicity. In terms of right-handed ($|R\rangle$) and left-handed ($|L\rangle$) helicity states we thus expect photon pairs to be represented by the vectors $|R\rangle_1|R\rangle_2$, $|L\rangle_1|L\rangle_2$ or a superposition of these, where $|R\rangle_1$ denotes a photon of right-handed helicity propagating to the left, $|R\rangle_2$ a photon of right-handed helicity propagating to the right, and so on. In addition, since, in the processes used experimentally, the initial and final states of the constituents of the source have definite parity, the two-photon state itself must also have definite parity. If P is the parity operator, $P|R\rangle_1 = |L\rangle_2$, $P|L\rangle_1 = |R\rangle_2$, etc. and hence, to ensure definite parity, we conclude that the two-photon state vector must be in one of the forms

$$|\psi\rangle_\pm = 2^{-1/2}[|R\rangle_1|R\rangle_2 \pm |L\rangle_1|L\rangle_2] \tag{10}$$

since then $P|\psi\rangle_+ = |\psi\rangle_+$ (even parity) and $P|\psi\rangle_- = -|\psi\rangle_-$ (odd parity). Most experiments have used a 0-1-0 cascade process in the source which gives rise to the $|\psi\rangle_+$ even parity form for the two-photon state vector. A few, however, have used a 1-1-0 cascade, in which case $|\psi\rangle_-$ is the appropriate form of the state vector.

Assuming the photons propagate in the $-z$ and $+z$ directions, it is also possible to describe the two-photon state vector in terms of the linear polarization states $|x\rangle$ and $|y\rangle$ using the relations

$$|R\rangle_1 = 2^{-1/2}(|x\rangle_1 - i|y\rangle_1), \qquad |R\rangle_2 = 2^{-1/2}(|x\rangle_2 + i|y\rangle_2)$$

$$|L\rangle_1 = 2^{-1/2}(|x\rangle_1 + i|y\rangle_1), \qquad |L\rangle_2 = 2^{-1/2}(|x\rangle_2 - i|y\rangle_2) \tag{11}$$

Substituting these relations into Eq. (10) gives

$$|\psi\rangle_+ = 2^{-1/2}(|x\rangle_1|x\rangle_2 + |y\rangle_1|y\rangle_2) \tag{12a}$$

and

$$|\psi\rangle_- = 2^{-1/2}(|x\rangle_1|y\rangle_2 - |y\rangle_1|x\rangle_2) \tag{12b}$$

The choice of orientation of the orthogonal axes x and y in the plane perpendicular to the z axis is, of course, completely arbitrary because of the cylindrical symmetry which is assumed to exist about the z axis.

Putting Eq. (12a) for $|\psi\rangle_+$ into Eq. (9) for $E(a, b)$ and using Eq. (5) for $|a^+\rangle$, $|b^+\rangle$, $|a^-\rangle$, and $|b^-\rangle$, we find

$$P_{++}(a, b) = P_{--}(a, b) = \tfrac{1}{2}\cos^2(\theta_1 - \theta_2)$$

$$P_{+-}(a, b) = P_{-+}(a, b) = \tfrac{1}{2}\sin^2(\theta_1 - \theta_2) \tag{13}$$

and hence

$$E_{QM}(a, b) = \cos 2(\theta_1 - \theta_2) = \cos 2(a, b) \tag{14}$$

where $E_{QM}(a, b)$ represents the quantum mechanical prediction for $E(a, b)$ and $(a, b) = \theta_1 - \theta_2$ denotes the relative angle between a and b. Similarly, using Eq. (12b) for $|\psi\rangle_-$, we find $E_{QM}(a, b) = -\cos 2(a, b)$. In both cases $E(a, b)$ ranges from -1 to $+1$, both extremes corresponding to complete correlation. This complete correlation, which results from the form of the state vector in Eq. (12) and persists even if the measurement events in the photodetectors are spatially separated in the relativistic sense, lies at the heart of the discussion concerning the completeness of quantum mechanics and the possible existence of hidden variables.

2.3. Bell's Inequalities for the Ideal Case

Examination of the two-photon state vector in Eq. (10) or (12) shows that it implies nonlocality and lack of realism. It implies nonlocality since a measurement causes a collapse of, say, $|\psi\rangle_+$, Eq. (12a), to either $|x\rangle_1|x\rangle_2$ or $|y\rangle_1|y\rangle_2$, each possibility occurring with probability one half. Thus, detection of photon 1 to the left with polarization in the x direction ensures that photon 2 to the right behaves as a photon polarized in the x direction also. But, as we have already seen, the choice of x direction is quite arbitrary, so the polarization state measured for photon 2 is, in fact, determined by the measurement we choose to make on photon 1 at a position that may be spatially separated, in the relativistic sense, from the position at which the measurement on photon 2 is carried out. Lack of realism also follows from this argument, since it then is impossible to think of the individual photons possessing properties, in this case polarization, which exist independently of any measurements which may be made on them.

This nonlocality and lack of realism inherent in quantum mechanics has inspired many attempts through the years to explain the results in terms of a theory that is both local and realistic. Without a specific local realistic theory it is, of course, not possible to predict a value for $E(a, b)$ to compare with the quantum mechanical value $E_{QM}(a, b)$ in Eq. (14). However, in 1964 J. S. Bell[10] showed for the first time that such theories place constraints on $E(a, b)$, or rather combinations of $E(a, b)$, for different values of a and b.

To understand Bell's approach consider the experimental situation represented in Figure 1. We assume that the initial state of the two photons can be described in terms of hidden variables λ with a probability density $\rho(\lambda)$. The variable λ may denote a single variable or a set of variables, which may be discrete or continuous. However, for simplicity we write as if λ is a single continuous variable. We also assume that

$$\int_\Lambda \rho(\lambda) \, d\lambda = 1 \tag{15}$$

where Λ is the space of the states λ. The result A to the left then depends on λ and on a, the result B to the right on λ and b, but if we wish our theory to be local, A cannot depend on b nor B on a. If λ determines uniquely the measurement outcome for each photon pair we can define

$$E(a, b) = \int_\Lambda A(\lambda, a) B(\lambda, b) \rho(\lambda)\, d\lambda \qquad (16)$$

in accord with our discussion regarding the correlation function, Eq. (2). More generally, for a given λ describing an emitted pair of photons, the quantities A and B may take on the values $+1$ or -1 with a probability depending on λ. In this case, λ does not determine uniquely the outcome of each measurement of A and B but only their average values $\bar{A}$ and $\bar{B}$ over an ensemble of emissions.[17] We can then define

$$E(a, b) = \int_\Lambda \bar{A}(\lambda, a) \bar{B}(\lambda, b) \rho(\lambda)\, d\lambda \qquad (17)$$

where now the averages $\bar{A}$ and $\bar{B}$ will be, because of locality, independent of b and a, respectively. It follows that instead of $A = \pm 1$ and $B = \pm 1$ we now require only that $|\bar{A}| \leq 1$ and $|\bar{B}| \leq 1$ and these latter conditions are sufficient to derive an interesting restriction on these local stochastic realistic theories.

Let a' and b' be alternative settings of the polarizers and consider the expression

$$E(a, b) - E(a, b') = \int_\Lambda [\bar{A}(\lambda, a) \bar{B}(\lambda, b) - \bar{A}(\lambda, a) \bar{B}(\lambda, b')] \rho(\lambda)\, d\lambda$$

This expression can be written in the form

$$E(a, b) - E(a, b') = \int_\Lambda \bar{A}(\lambda, a) \bar{B}(\lambda, b)[1 \pm \bar{A}(\lambda, a') \bar{B}(\lambda, b')] \rho(\lambda)\, d\lambda$$
$$- \int_\Lambda \bar{A}(\lambda, a) \bar{B}(\lambda, b')[1 \pm \bar{A}(\lambda, a') \bar{B}(\lambda, b)] \rho(\lambda)\, d\lambda$$

Using the fact that $|\bar{A}| \leq 1$, $|\bar{B}| \leq 1$ we obtain

$$|E(a, b) - E(a, b')| \leq \int_\Lambda [1 \pm \bar{A}(\lambda, a') \bar{B}(\lambda, b')] \rho(\lambda)\, d\lambda$$
$$+ \int_\Lambda [1 \pm \bar{A}(\lambda, a') \bar{B}(\lambda, b)] \rho(\lambda)\, d\lambda$$

or

$$|E(a, b) - E(a, b')| \leqslant \pm[E(a', b') + E(a', b)] + 2 \int_\Lambda \rho(\lambda) \, d\lambda$$

and hence

$$-2 \leqslant E(a, b) - E(a, b') + E(a', b) + E(a', b') \leqslant 2 \qquad (18)$$

If we write $S(a, b, a', b') = E(a, b) - E(a, b') + E(a', b) + E(a', b')$ the inequality (18) may be written

$$-2 \leqslant S(a, b, a', b') \leqslant 2 \qquad (19)$$

Essentially the same inequality was derived by Clauser, Horne, Shimony, and Holt[11] and it is sometimes referred to as the Bell, Clauser, Horne, Shimony, and Holt (BCHSH) inequality. Its importance lies in the fact that it represents a general restriction on the predictions of theories based on local realism.

For example, if we take the quantum mechanical form $E_{QM}(a, b)$ given by Eq. (14) and evaluate $S(a, b, a', b')$ for various orientations of the coplanar vectors a, b, a', and b', it is easy to show that $S(a, b, a', b')$ takes on extremum values for the situations, shown in Figure 2, where $(a, b) = (b, a') = (a', b') = 22.5°$, $(a, b') = 67.5°$ when $S = +2\sqrt{2}$ and where $(a, b) = (b, a') = (a', b') = 67.5°$, $(a, b') = 22.5°$ when $S = -2\sqrt{2}$. Both extreme values for $S(a, b, a', b')$ clearly violate the BCHSH inequality showing that

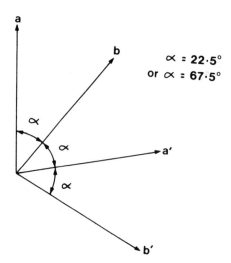

Figure 2. Orientations of the polarizers leading to extremum values of the function $S(a, b, a', b')$.

local realistic theories must necessarily disagree with quantum mechanics over at least some of the range of relative polarizer orientations. This situation may not at first seem surprising given the general success of quantum mechanics, but, in fact, before the development of the BCHSH inequality no necesssary conflict between local realism and quantum mechanics had been demonstrated. The BCHSH inequality showed that conflict did indeed exist and brought the question concerning the possibility of a local realistic description of the physical world into the experimental domain. It also pointed out an area where there might conceivably have been a breakdown in conventional quantum mechanics.

2.4. The BCHSH Inequality in Experimental Situations

In a real experiment, such as shown in Figure 3, the radiation emitted by the source is collected and collimated by a pair of lenses with a finite aperture before being analyzed by the polarizers. The various experiments that have been carried out differ mainly in their choice of source and type of polarizer used. Because of the angular correlation of the photon pairs emitted in two-photon decay processes from an atom, the finite solid angle of detection, and the low detection efficiency of the photodetectors, in practice, only a very small proportion of the photon pairs emitted by the source is actually detected. In this situation, in order to have any hope of violating the BCHSH inequality it is necessary to take N in Eq. (4) for $E(a, b)$ to be the total number of photon pairs that would have been detected in the absence of the polarizers, rather than the total number actually emitted by the source. In evaluating $E(a, b)$ we therefore take

$$N = N_{++}(a, b) + N_{+-}(a, b) + N_{-+}(a, b) + N_{--}(a, b) \qquad (20)$$

Adoption of the above procedure, however, implicitly assumes that, for each setting of the polarizers, the ensemble of detected pairs is a true representative sample of the ensemble of pairs emitted by the source.

The above form (19) of the BCHSH inequality applies to the situation where both orthogonal polarization states are detected on each side of the source. However, most experiments that have been carried out only detect the signals that are transmitted directly through the polarizers. One can

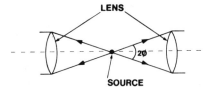

Figure 3. Diagram to illustrate collection and collimation of radiation by lenses in a real experiment. The lenses each subtend a half-angle ϕ at the source.

imagine that in Figure 1 only the detectors D_{11} and D_{21} are in place. In these circumstances, a new form for the BCHSH inequality is required since the quantities $N_{+-}(a, b)$, $N_{-+}(a, b)$, and $N_{--}(a, b)$ cannot now be directly measured. These quantities, however, can be deduced from measurements with either one or both polarizers removed. If we denote by ∞ the absence of a polarizer then we expect that

$$N_{++}(a, \infty) = N_{++}(a, b) + N_{+-}(a, b)$$

$$N_{++}(\infty, b) = N_{++}(a, b) + N_{-+}(a, b) \tag{21}$$

and

$$N_{++}(\infty, \infty) = N_{++}(a, b) + N_{+-}(a, b) + N_{-+}(a, b) + N_{--}(a, b) \tag{22}$$

From Eqs. (21) and (22) it follows that the inequality (19) can be rewritten in the form

$$-1 \leqslant S' \leqslant 0 \tag{23}$$

with

$$S' = [N(a, b) - N(a, b') + N(a', b) + N(a', b') - N(a', \infty)$$
$$- N(\infty, b)]/N(\infty, \infty) \tag{24}$$

omitting, for clarity, the subscripts on the N_{++}.

Finally, if we assume that the results of measurement depend only on the relative angle between the axes of the polarizers and write the inequality (23) for the two sets of angles which give extreme values for S', i.e., $(a, b) = (b, a') = (a', b') = 22.5°$, $(a, b') = 67.5°$ and $(a, b) = (b, a') = (a', b') = 67.5°$, $(a, b') = 22.5°$, then it is easy to derive the inequality

$$\eta \equiv \left| \frac{N(22.5°) - N(67.5°)}{N(\infty, \infty)} \right| \leqslant 0.25 \tag{25}$$

originally put forward by Freedman.[18] This form of inequality has proved to be one of the most useful experimentally since it only requires three measurements to be made.

2.5. Quantum Mechanical Predictions

In the situation discussed up to now the conflict between the predictions of local realism and quantum mechanics is quite clear. However, in a real experiment, several factors act to reduce the strength of the quantum mechanically predicted correlations, in certain circumstances to the extent that a violation of the BCHSH inequality can no longer be expected to occur. The finite solid angle of detection and the efficiencies of the polarizers are the two factors that reduce the expected correlation in this way. If ϕ is the half-angle subtended at the source by the lenses as shown in Figure 3 and ε_{M1}, and ε_{m1} are, respectively, the transmission efficiencies for light polarized parallel to and perpendicular to the axis a of polarizer π_1 and ε_{M2} and ε_{m2} the corresponding quantities for polarizer π_2 with its axis orientated in the direction b, then it can be shown[11] that

$$\frac{N(a, b)}{N(\infty, \infty)} = \tfrac{1}{4}\{(\varepsilon_{M1} + \varepsilon_{m1})(\varepsilon_{M2} + \varepsilon_{m2})$$

$$\pm F(\phi)(\varepsilon_{M1} - \varepsilon_{m1})(\varepsilon_{M2} - \varepsilon_{m2}) \cos 2(a, b)\} \qquad (26)$$

$$\frac{N(a, \infty)}{N(\infty, \infty)} = \tfrac{1}{2}(\varepsilon_{M1} + \varepsilon_{m1}), \qquad \frac{N(\infty, b)}{N(\infty, \infty)} = \tfrac{1}{2}(\varepsilon_{M2} + \varepsilon_{m2}) \qquad (27)$$

and, hence, assuming symmetrical two-channel polarizers, that

$$E_{\text{QM}}(a, b) = \pm F(\phi) \frac{(\varepsilon_{M1} - \varepsilon_{m1})(\varepsilon_{M2} - \varepsilon_{m2})}{(\varepsilon_{M1} + \varepsilon_{m1})(\varepsilon_{M2} + \varepsilon_{m2})} \cos 2(a, b) \qquad (28)$$

The $\pm$ sign applies, respectively, to the situation where the photon pairs result from a 0-1-0 or 1-1-0 type of cascade. The quantity $F(\phi)$, which has a different mathematical form in the two cases and is equal to unity when $\phi = 0$, takes into account the depolarizing effect of the noncollinear emission of the photon pairs.

In order to carry out a successful test of the BCHSH inequality, $F(\phi)$ must be greater than some minimum value which depends on the transmission efficiencies of the polarizers. Let us assume for simplicity that the transmission efficiency ε_M is the same for both polarizers; then, since $F(\phi)$ is a monotonically decreasing function of ϕ, there is an upper limit, which depends on ε_M, on the detector half-angle necessary for a test of the BCHSH inequality, as shown in Figure 4. Clearly, the use of a 0-1-0 cascade places a less stringent requirement on the apparatus parameters than does a 1-1-0 cascade.

It should be noted that it is not necessary to know the results (26), (27), and (28) in order to test the BCHSH inequality. They must be used, of course, if it is required to compare the experimental results with the quantum mechanical predictions.

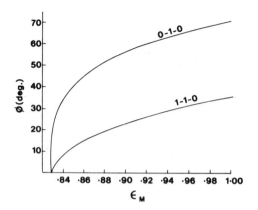

Figure 4. Upper limits on detector half-angle ϕ as a function of polarizer efficiency ε_M. To test the BCHSH inequality, the experiment must be performed with apparatus parameters chosen in the region below the appropriate curve—the upper curve for a 0-1-0 cascade, the lower for a 1-1-0 cascade. (From Clauser et al.[11])

2.6. The No-Enhancement Hypothesis

In deriving the BCHSH inequality in the forms (23) and (25) which are applicable in real experimental situations, it was necessary to assume that, for each setting of the polarizers, the ensemble of detected pairs is a true representative sample of the ensemble of pairs emitted by the source. This assumption is known as the "no-enhancement" hypothesis and was first stated by Clauser, Horne, Shimony, and Holt[11] in the following form: If a pair of photons emerges from the polarizers, the probability of their joint detection is independent of the orientations a and b of the polarizers' axes. Subsequently, Clauser and Horne[12] stated the hypothesis in the more general form: For every atomic emission, the probability of a count with a polarizer in place is less than or equal to the probability with the polarizer removed.

Although both forms of the hypothesis can be regarded as physically plausible, neither has been subjected to an experimental test.

2.7. The Schrödinger–Furry Hypothesis

The long-range correlation implied by the form of the state vector shown in Eq. (12) has led some physicists to question the range of validity of quantum mechanics. For example, it was suggested by Schrödinger[19] and Furry[20] that over a large enough distance the state vector $|\psi\rangle_+$, say, might collapse spontaneously to a mixture of states of the form $|x\rangle_1|x\rangle_2$, where x is any direction perpendicular to the z-axis. Such a collapse if it happened would cause a change in the polarization correlation and might possibly be expected to occur over distances greater than the coherence length of the photons. However, as we shall see later, there is no experimental evidence that such an effect takes place over the distances so far investigated. Indeed, Selleri and Tarozzi[21] have argued that, in quantum mechanical terms, such a collapse of the state vector would be inconsistent with conservation of angular momentum.

3. Experimental Work

So far the results of 12 experimental measurements of the polarization correlation in the two-photon decay of atoms have been published. All but two of these have been carried out in order to test the BCHSH inequality in one form or another.

3.1. Experiment of Kocher and Commins (1967)

The experiment of Kocher and Commins[22] is the original experiment of the type being considered but predates most of the theoretical work on the BCHSH inequality, in particular the important paper by Clauser, Horne, Shimony, and Holt,[11] which showed how the work of Bell could be applied in a real experiment. However, their arrangement of the apparatus is typical of almost all subsequent experiments, which differ, mainly, only in the nature of the source, its method of excitation, and the type of polarizers used to analyze the emitted photons.

In this experiment, as shown in Figures 5 and 6, the $3d4p\ ^1P_1$ state of calcium in a beam was excited by radiation of wavelength 227.5 nm from a hydrogen arc lamp. About 10% of the atoms that do not return directly to the ground state go to the $4p^2\ ^1S_0$ state, which is the initial state of the $4p^2\ ^1S_0$-$4s4p\ ^1P_1$-$4s^2\ ^1S_0$ cascade emitting photons of wavelengths 551.3 and 422.7 nm. Since the naturally occurring calcium used in this experiment contained 99.855% of the isotope with zero nuclear spin, there was no significant reduction to be expected in the polarization correlation due to the presence of hyperfine structure. It should also be noted that the calcium beam, whose density was about $3 \times 10^{10}\ cm^{-3}$, was orientated at an acute angle to the observation axis, thus ensuring that Doppler broadening of the resonance line reduced any effects due to resonance trapping of the 422.7-nm

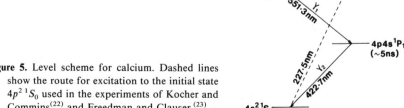

Figure 5. Level scheme for calcium. Dashed lines show the route for excitation to the initial state $4p^2\ ^1S_0$ used in the experiments of Kocher and Commins[22] and Freedman and Clauser.[23]

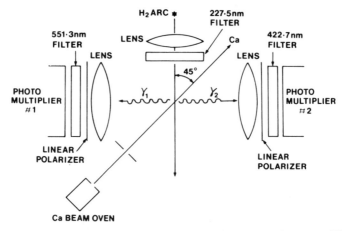

Figure 6. Schematic diagram of the apparatus of Kocher and Commins.[22]

photon. On each side of the source as shown in Figure 6 the photons were collected and collimated by a lens, then passed through a linear polarizer of the sheet polaroid type and a filter before being detected by a photomultiplier with good timing characteristics and low noise. The resulting photomultiplier pulses were fed to a coincidence circuit which allowed the events of interest to be distinguished from the comparatively large background singles rate in each photomultiplier.

Unfortunately, Kocher and Commins only made measurements for relative angles of 0° and 90° between the transmission axes of the polarizers and, in addition, the transmission characteristics of their polarizers did not satisfy the criterion illustrated in Figure 4 for a satisfactory test of the BCHSH inequality. However, Clauser[4] was able to use their result to demonstrate the inadequacy of semiclassical radiation theory in this situation and to show that the Schrödinger–Furry hypothesis was not tenable.

The paper of Kocher and Commins is also noteworthy in that it suggested an experiment in which the effect of a magnetic field on the decay could be studied. This suggestion was later taken up in the experiment by Aspect, Dalibard, and Roger,[6] which will be described later.

3.2. Experiment of Freedman and Clauser (1972)

Freedman and Clauser[23] used essentially the same arrangement as Kocher and Commins, exciting the calcium beam by 227.5-nm radiation from a deuterium arc. However, in order to satisfy the requirement for large efficient linear polarizers, they used pile-of-plates polarizers, each of which was about 1 m in length and consisted of ten 0.3-mm-thick glass sheets

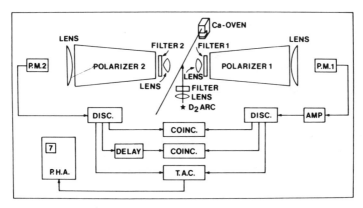

Figure 7. Schematic diagram of the apparatus of Freedman and Clauser[23] showing a typical arrangement for the electronics in this kind of experiment.

inclined nearly at Brewster's angle. The sheets were mounted on hinged frames so that they could be folded out of the optical path. The transmission efficiencies were measured to be $\varepsilon_{M1} = 0.97 \pm 0.01$, $\varepsilon_{m1} = 0.038 \pm 0.004$, $\varepsilon_{M2} = 0.96 \pm 0.01$, $\varepsilon_{m2} = 0.037 \pm 0.004$ and the half angle subtended by the lenses at the source was 30° giving $F(\phi) = 0.99$. Coincidence measurements were made for 100-sec periods, the periods during which all the plates were removed alternating with periods in which the plates were inserted. A schematic diagram of the apparatus is shown in Figure 7 which shows, in particular, the typical electronic counting circuit used in these experiments.

The results obtained as the relative orientation of the transmission axes of the polarizers was varied from 0° to 90°, were found to be in agreement with the quantum mechanical prediction as expressed in Eq. (26). Also, the results at $(a, b) = 22.5°$ and $(a, b) = 67.5°$ combined with those with both sets of polarizer plates removed gave $\eta = 0.300 \pm 0.008$, in clear violation of the Freedman form of the BCHSH inequality expressed in Eq. (25), and in agreement with the quantum mechanical prediction $\eta_{QM} = 0.301 \pm 0.007$ obtained from Eq. (26).

3.3. Experiment of Holt and Pipkin (1973)

Holt and Pipkin[24] observed the 567.6-nm and 404.7-nm photons emitted in the $9^1P_1 - 7^3S_1 - 6^3P_0$ cascade of the zero nuclear spin isotope ^{198}Hg of mercury. The relevant transitions are shown in Figure 8, from which it can be seen that the final cascade level is not the ground state of the atom. Thus in this experiment no precautions had to be taken to avoid the effects of resonance trapping.

To produce the required radiation, mercury vapor was excited to the 9^1P_1 state by a 100-eV electron beam, both the beam and the vapor being

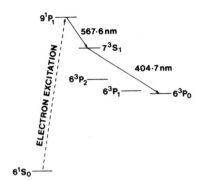

Figure 8. Level scheme for mercury showing the states used in the Holt and Pipkin experiment.[24]

contained in an encapsulated source made from Pyrex glass. Since the source was of the 1-1-0 variety, the requirements on polarizer efficiency and collection solid angle were more stringent than in the 0-1-0 case. In addition, any lack of isotropy among the excited atoms in the 9^1P_1 state could have a significant effect on the results. A third photomultiplier viewed the 435.8-nm photons from the 7^3S_1-6^3P_1 transition to monitor the lamp intensity and to produce a correction signal for the lamp stabilization circuitry.

In contrast to the two previous experiments, calcite-type polarizers were used with transmission efficiencies $\varepsilon_{M1} = 0.910 \pm 0.001$, $\varepsilon_{M2} = 0.880 \pm 0.001$, ε_{m1}, $\varepsilon_{m2} < 10^{-4}$. This type of polarizer has a much better extinction ratio than pile-of-plates polarizers, but the values of ε_{M1} and ε_{M2} are not particularly high. What is more, since a 1-1-0 cascade is being used, the factor $F(\phi)$ takes on the relatively low value 0.951 even with an acceptance half-angle of only 13°.

Experimentally, it was found that $\eta = 0.216 \pm 0.013$, a result that disagrees with the quantum mechanical prediction $\eta_{QM} = 0.266$ and clearly does not violate the BCHSH inequality. Although this discrepancy has never been completely explained, it is thought that the low value of η may have arisen as a result of stress-induced optical activity in the walls of the Pyrex glass envelope.[25] However, since this result is, in fact, the only one of its kind consistent with local realistic interpretation of these long-range two-photon polarization correlations, proponents of such theories have suggested that there may be some significance to be attached to the use of calcite polarizers or to the fact that the final state of the cascade is not the ground state of the atom.

3.4. Experiments of Clauser (1976)

In view of the unexpected result obtained by Holt and Pipkin,[24] Clauser[25] repeated their experiment using the same cascade but in the even isotope ^{202}Hg of mercury. Also, instead of calcite polarizers, he used pile-of-plates polarizers of the type used previously in the experiment described in Section 3.2, but with 15 rather than 10 plates to give transmission efficiencies $\varepsilon_{M1} = 0.965$, $\varepsilon_{m1} = 0.011$, $\varepsilon_{M2} = 0.972$, $\varepsilon_{m2} = 0.0084$. With a collection half-angle of 18.6°, it is expected from quantum mechanics that $\eta_{QM} = 0.2841$ while the experiment gave $\eta = 0.2885 \pm 0.0093$, violating the BCHSH inequality and in close agreement with the quantum mechanical result. In addition, a calculation of the variation of $N(a, b)/N(\infty, \infty)$ from Eq. (26) with angle (a, b), taking into account the measured polarizer efficiencies, the collection half-angle, depolarization due to the presence of ^{199}Hg and ^{201}Hg isotopes, and alignment of the 9^1P_1 state, showed close agreement between the experimental results and quantum mechanics.

In an extension to the above experiment Clauser[26] measured the circular polarization correlation by inserting quarter-wave plates between each linear polarizer and the source. The quarter-wave plates were constructed by applying pressure to bars of commercial grade quartz. Assuming ideal quarter-wave plates, quantum mechanics predicts that Eq. (26) still holds and thus Eq. (25) remains a valid form of the BCHSH inequality. From the experimental results Clauser found $\eta = 0.235 \pm 0.025$, while, taking into account the transmission efficiencies of the linear polarizers, the collection half-angle, and the lack of stability of the quarter-wave plates, he predicted $\eta_{QM} = 0.252$. Thus, although, within the limits of experimental error, these circular polarization results were in agreement with quantum mechanics, they failed to provide a conclusive test of the BCHSH inequality. However, Clauser did show that the results were not consistent with the Schrödinger–Furry hypothesis.

3.5. Experiment of Fry and Thompson (1976)

Fry and Thompson[27] used the 435.8-nm and 253.7-nm photons emitted in the $7^3S_1-6^3P_1-6^1S_0$ cascade in the zero nuclear spin isotope ^{200}Hg of mercury. The relevant transitions are shown in Figure 9. The 7^3S_1 state in a mercury beam was populated in a two-step process with electron bombardment excitation of the 6^3P_2 metastable state being followed downstream,

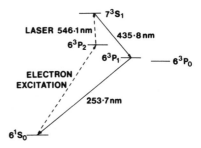

Figure 9. Level scheme for mercury showing the states used in the Fry and Thompson experiment.[27]

where all short-lived states had decayed, by absorption of resonant 546.1-nm radiation from a tunable dye laser the output of which was polarized with its electric field vector in the direction of the observation axis. The magnetic field in the interaction volume was reduced to less than 5 mG in all directions. Although mercury of natural isotopic abundance was used, the laser band-width was narrow enough (15 MHz) that the ^{200}Hg isotope could be selectively excited. In addition, since there was a one-to-one correspondence between all 435.8-nm and 257.3-nm photons, data accumulation rates were obtained that were high compared to those achieved in previous experiments, a typical run lasting about 80 min.

The polarizers used in this experiment were of some interest, being of the pile-of-plates variety, with each polarizer consisting of two sets of seven plates symmetrically arranged so as to cancel out transverse ray displacements. A typical time correlation spectrum for this type of experiment using an atomic cascade is shown in Figure 10.

Since the initial state of the cascade had $J = 1$ it was necessary to take into account possible effects resulting from unequal population of, and coherence between, the initial Zeeman sublevels, which Fry and Thompson did by measuring the polarization of the 435.8-nm fluorescence at appropriate angles. Allowing for these effects along with the transmission efficiencies of the polarizers $\varepsilon_{M1} = 0.98 \pm 0.01$, $\varepsilon_{M2} = 0.97 \pm 0.01$, $\varepsilon_{m1} = \varepsilon_{m2} = 0.02 \pm 0.005$, and half-angle $19.9° \pm 0.3°$ of the collection optics, it was predicted on the basis of quantum mechanics that $\eta_{QM} = 0.294 \pm 0.007$, whereas from the experiment the value $\eta = 0.296 \pm 0.014$ was found in agreement with the quantum mechanical result but clearly violating the BCHSH inequality. Fry and Thompson also made a least-squares fit of the

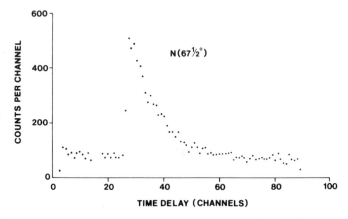

Figure 10. A typical time correlation spectrum for a setting $(a, b) = 67.5°$ between the transmission axes of the polarizers as obtained in the experiment of Fry and Thompson.[27] The total accumulation time is 80 min.

form $A + B \cos 2\phi + C \sin 2\phi$ to their data and obtained $A = 0.242 \pm 0.003$, $B = -0.212 \pm 0.004$, $C = -0.003 \pm 0.004$.

3.6. Experiment of Aspect, Grangier, and Roger (1981)

Following Kocher and Commins (Section 3.1) and Clauser and Freedman (Section 3.2), Aspect, Grangier, and Roger[28] used the 551.3-nm and 422.7-nm photons from the $4p^2\,{}^1S_0$–$4s4p^1P_1$–$4s^2\,{}^1S_0$ cascade in calcium. However, in their case the calcium atoms were excited to the $4p^2\,{}^1S_0$ state by a two-photon absorption process using a krypton-ion laser beam of wavelength 406 nm and a dye laser beam tuned to 581 nm, both laser beams being at right angles to the calcium atomic beam emitted from a tantalum oven. The laser beams had parallel polarizations and were focused at the interaction region to provide a source about 60 μm in diameter by 1 mm long. The density was about 3×10^{10} cm^{-3}, which resulted in a typical cascade rate of 4×10^7 sec^{-1}. The narrow resonance of the excitation process (less than 50 MHz) allowed selective excitation of the even ^{40}Ca isotope of calcium, thus preventing the strength of polarization correlation from being reduced by the effects of hyperfine structure. Feedback loops were used to control the wavelength of the tunable dye laser and the krypton-ion laser power output.

After collection and collimation by a lens system which subtended a half-angle of about 32° at the source, the photons from the cascade were analyzed by polarizers and filters in much the same way as in previous experiments. The polarizers were of the pile-of-plates type, each consisting of 10 optically flat plates set nearly at Brewster's angle, with efficiencies measured to be $\varepsilon_{M1} = 0.971 \pm 0.005$, $\varepsilon_{m1} = 0.029 \pm 0.005$, $\varepsilon_{M2} = 0.968 \pm 0.005$, and $\varepsilon_{m2} = 0.028 \pm 0.005$.

This experiment was noteworthy with regard to the strength of the source, with coincidence rates of up to 100 sec^{-1} allowing 1% statistical accuracy to be obtained in only 100 sec counting time. The result of measurement of $N(a, b)/N(\infty, \infty)$ against the angle (a, b) between the transmission axes of the two polarizers is shown in Figure 11, which also shows the predicted quantum mechanical curve. The agreement between the theory and experiment is clearly excellent. Using the experimental results at 22.5° and 67.5° gave $\eta = 0.3072 \pm 0.0043$, in agreement with the quantum mechanical prediction $\eta_{QM} = 0.308 \pm 0.002$ calculated using Eq. (26), and violating the BCHSH inequality by more than 13 standard deviations.

Aspect, Grangier, and Roger also used this experiment to test the BCHSH inequality in the form of Eq. (23), $-1 \leq S \leq 0$, which does not assume the rotational invariance required for the Freedman form of the inequality. They found $S' = 0.126 \pm 0.014$ violating inequality (23) by nine

$N(a,b)/N(\infty,\infty)$

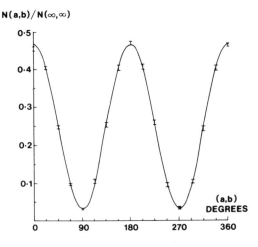

Figure 11. Normalized coincidence rate as a function of the relative orientation of the transmission axes of the polarizers in the experiment of Aspect, Grangier, and Roger.[28] The solid curve represents the quantum mechanical prediction.

standard deviations and in good agreement with the quantum mechanical prediction $S'_{QM} = 0.118 \pm 0.005$.

Finally, it was observed that moving each polarizer up to 6.5 m from the source, i.e., to four coherence lengths of the wave packet associated with the lifetime of the intermediate state of the cascade (5 nsec), produced no change in the results, thus providing further strong evidence against the Schrödinger–Furry hypothesis.

3.7. Experiment of Aspect, Grangier, and Roger (1982)

As discussed in Section 2.4 the experiments described up till now, in which only the N_{++} signals are detected, depart from the ideal arrangement discussed in Section 2.3. To remedy this situation, in 1982 Aspect, Grangier, and Roger[29] performed an experiment using the same source as described in Section 3.6 but with two-channel polarizers instead of the previous one-channel pile-of-plates type. Their apparatus is essentially that of Figure 1 with the addition of collecting and collimating lenses. Each polarizer, in the form of a polarizing cube constructed using the properties of dielectric thin films and antireflection coated, was rotatable about the observation axis. This arrangement allowed the quantity $E(a, b)$ defined in Eqs. (4) and (20) to be measured directly in a single run using a fourfold coincidence technique for each of the four relative orientations of the polarizers $(a, b) = (b, a') = (a', b') = 22.5°$, $(a, b') = 67.5°$. In this way the BCHSH inequality could be tested in the form of Eq. (19), $-2 \leq S \leq 2$ derived in Section 2.3.

From the experimental results the value $S = 2.697 \pm 0.015$ was found, whereas the quantum mechanical prediction obtained using Eq. (28) for $E(a, b)$ with $\varepsilon_{M1} = 0.950 \pm 0.005$, $\varepsilon_{M2} = 0.930 \pm 0.005$, $\varepsilon_{m1} = \varepsilon_{m2} = 0.007 \pm 0.005$ gave $S_{QM} = 2.70 \pm 0.05$, in good agreement with the experi-

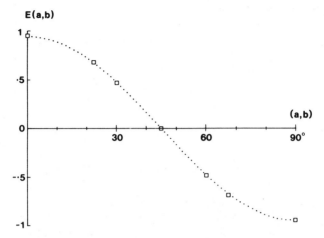

Figure 12. Correlation function $E(a, b)$ as a function of the relative angle (a, b) for the experiment of Aspect, Grangier, and Roger.[29] The indicated errors are ± 2 standard deviations. The dashed curve is the quantum mechanical prediction.

mental result. In addition, as shown in Figure 12, a measurement of $E(a, b)$ as a function of the angle (a, b) between the transmission axes of the polarizers gave close agreement with the quantum mechanical prediction based on Eq. (28).

As already pointed out in Section 2.6, for the experiment to provide a satisfactory test of the BCHSH inequality, it is necessary, of course, to assume that the ensemble of pairs actually detected is a true representative sample of the ensemble of pairs which are emitted. Although it is not possible to prove that the detected sample is unbiased in this way, nevertheless added confidence is given to the result of the experiment by the observation that the quantity $N_{++} + N_{--} + N_{+-} + N_{-+}$ was indeed constant as the polarizers were rotated about the observation axis.

It should be noted here that an experiment similar to the one described in this section is being carried out by a group at the University of Catania who have published[30] descriptions of their apparatus.

3.8. Experiment of Aspect, Dalibard, Grangier, and Roger (1984)

In an interesting extension to the previous experiment Aspect, Dalibard, Grangier, and Roger[6] observed the atomic quantum beats produced when a magnetic field is applied to the source in a direction at right angles to the observation axis thus removing the degeneracy of the intermediate $4s4p\,^1P_1$ state and allowing three possible decay channels as shown in Figure 13.

Their results, shown in Figure 14, where $e_\perp$ and $e_\parallel$ are unit vectors perpendicular and parallel, respectively, to the magnetic field, can be easily

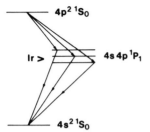

Figure 13. Interfering decay channels in a calcium atomic cascade in the presence of an applied magnetic field.[6]

interpreted in terms of interfering quantum paths, or perhaps more graphically as a process in which the emission of the first photon of the cascade prepares the atom in a coherent superposition of the Zeeman sublevels $|r; m = 0, \pm 1\rangle$ of the intermediate state, which then evolve in the magnetic field. If the first photon is detected with polarization in the direction $e_\perp$ (σ polarization), the subsequent atomic state is $1/\sqrt{2}(|r; m = +1\rangle + |r; m = -1\rangle)$, which describes a dipole aligned along $e_\perp$. This dipole then precesses around the magnetic field at the Larmor frequency so that the light emitted in the second stage of the cascade is polarized also in the $e_\perp$ direction and is modulated at the Larmor frequency. On the other hand, if the first photon is detected with polarization in the $e_\parallel$ direction (π polarization) the intermediate atomic state is $|r; m = 0\rangle$ describing a dipole aligned along $e_\parallel$ which does not precess in the magnetic field and results in the emission of light

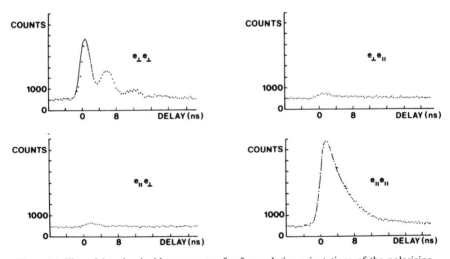

Figure 14. Time delayed coincidence spectra for four relative orientations of the polarizing cubes parallel to and perpendicular to the magnetic field which is applied in a direction at right angles to the observation axis.[6] The solid lines are theoretical fits corresponding to a Landé factor of the intermediate state $g_J = 1$.

in the second stage of the cascade, which is polarized along $e_{\parallel}$ but not modulated.

3.9. Experiment of Aspect, Dalibard, and Roger (1982)

In all the experiments described so far the orientations of the transmission axes of the polarizers have been fixed at various angles during the measurements. Thus, it could be argued that, in some way, the polarizers and the process of emission of photon pairs could reach some mutual rapport by exchange of signals with speed less than or equal to the speed of light. Such a possibility could be ruled out if the settings of the polarizers were changed in a time that was short compared to the time of flight of photons from the source to each polarizer. A possible scheme to achieve this ideal was suggested by Aspect[31] in 1976 and the experiment was realized in 1982 by Aspect, Dalibard, and Roger.[32]

In their experiment, which used the same source as described in Section 3.6, an optical switch rapidly redirected the light incident from the source to one of two polarizing cubes on each side of the source as shown in Figure 15. In contrast to the previous experiment described in Section 3.7, however, only the transmitting channels of the polarizing cubes were used. The switching of the light was effected by what is essentially a Bragg reflection from an ultrasonic standing wave in water. The light was completely transmitted without deflection when the amplitude of the standing wave was

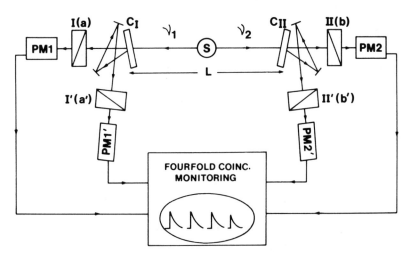

Figure 15. The experiment of Aspect, Dalibard, and Roger[32] with optical switches. Each switching device (C_I, C_{II}) is followed by two polarizers in two different orientations. The arrangement is equivalent to one in which a single polarizer on each side is switched quickly between two orientations. $L = 12$ m.

zero and was almost fully deflected through 10 mrad when the amplitude was a maximum. Switching between the two channels occurred about once every 10 nsec and since this time, as well as the lifetime of the intermediate state of the cascade (5 nsec), was small compared to L/c (40 nsec), where L was the separation between the switches (12 m) and c the speed of light, a detection event on one side and the corresponding change of orientation on the other side were separated by a spacelike interval. The results were registered in a fourfold coincidence monitoring system as shown in Figure 15, but because of the necessity to reduce the beam divergence in the optical system to achieve good switching, the coincidence rates were reduced to only a few per second, with an accidental background of about one per second.

If the two switches work at random, it is possible to write the BCHSH inequality in the slightly modified form of Eq. (23):

$$-1 \leq S'' \leq 0 \tag{29}$$

where

$$S'' = \frac{N(a, b)}{N(\infty, \infty)} - \frac{N(a, b')}{N(\infty, \infty')} + \frac{N(a', b)}{N(\infty', \infty)} + \frac{N(a', b')}{N(\infty', \infty')}$$

$$- \frac{N(a', \infty)}{N(\infty', \infty)} - \frac{N(\infty, b)}{N(\infty, \infty)}$$

Although the switching was, in practice, periodic rather than random, the switches on the two sides were driven by different generators at different frequencies and it was assumed that they functioned in an uncorrelated way. Experimentally, it was found that $S'' = 0.101 \pm 0.020$, in violation of the BCHSH inequalities, while taking into account the solid angle of detection and the efficiencies of the polarizers gave the quantum mechanical prediction $S''_{QM} = 0.112$. A measurement of the normalized coincidence rate as a function of the relative orientation of the polarizers also showed good agreement with quantum mechanics.

Finally, it should be noted that some criticisms[33-35] have been made of the experiments using a high-density calcium source on the grounds that there may have been significant effects due to resonance trapping. However, these criticisms have been countered by Aspect and Grangier.[36]

3.10. Experiments of Perrie, Duncan, Beyer, and Kleinpoppen (1985)

Perrie, Duncan, Beyer, and Kleinpoppen[37] measured for the first time the polarization correlation of the two photons emitted simultaneously by

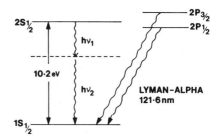

Figure 16. Level diagram for atomic deuterium, neglecting hyperfine structure (not to scale). The two photons, frequencies ν_1 and ν_2, can have any energy provided $h\nu_1 + h\nu_2 = 10.2$ eV (h is Planck's constant).

metastable atomic deuterium in a true second-order decay process and used the results to test the BCHSH inequality. Single-photon decay from the $2S_{1/2}$ state of deuterium is forbidden and, as illustrated in Figure 16, the main channel for the spontaneous deexcitation of this state is by the simultaneous emission of two photons, which can have any wavelength consistent with conservation of energy for the pair, the most probable occurrence being the emission of two photons each of wavelength 243 nm. Since the decay proceeds through virtual intermediate states, the effects of hyperfine structure can be neglected, and hence the angular and polarization correlations are predicted to be identical to those resulting from a 0–1–0 cascade in an atom with zero nuclear spin.

In the experiment, illustrated in Figure 17, a 1-keV metastable atomic deuterium beam of density about 10^4 cm^{-3} was produced by charge exchange, in cesium vapor, of deuterons extracted from a radio-frequency

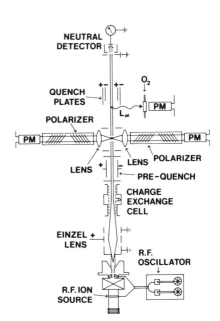

Figure 17. Schematic diagram of the apparatus of Perrie, Duncan, Beyer, and Kleinpoppen.[37]

ion source. Electric field prequench plates upstream from the observation region allowed the $2S_{1/2}$ component of the beam to be switched on and off by Stark mixing the $2S_{1/2}$ and $2P_{1/2}$ states, and, at the end of the apparatus, the beam was fully quenched so that the resulting Lyman-α signal could be used to normalize the two-photon coincidence signal.

As in previous experiments, the two-photon radiation was collected and collimated by a pair of lenses, each lens subtending a half-angle of 23° at the source. The polarizers were of the pile-of-plates type with 12 plates set nearly at Brewster's angle. In order to have a high transmission in the neighborhood of 243 nm, both the lenses and the plates of the polarizers were made from high-quality fused silica with a short wavelength cutoff at 160 nm. However, in practice, because of the absorption in oxygen, the short-wavelength cutoff occurred at 185 nm, which, in turn, implied a long-wavelength cutoff at 355 nm and hence an observation window between 185 and 355 nm. The transmission efficiencies of the polarizers were measured to be $\varepsilon_M = 0.908 \pm 0.013$ and $\varepsilon_m = 0.0299 \pm 0.0020$.

On each side of the source the pulses from the photomultipliers were fed to a standard coincidence circuit with the time correlation spectra obtained with the metastable atoms present and quenched being stored in separate segments of a multichannel analyzer memory and then subtracted at the end of a run. A typical spectrum obtained in this way is shown in Figure 18, from which it can be seen that the coincidence peak is symmetrical as expected for a simultaneous emission process in contrast to the situation illustrated in Figure 10 for a cascade process.

The results of measurement are shown in Figure 19, and clearly agree with the quantum mechanical prediction calculated in the usual way taking into account the half-angle subtended by the lenses at the source and the efficiencies of the polarizers. In addition, using the results at 22.5° and 67.5° gave $\eta = 0.268 \pm 0.010$, in violation of the BCHSH inequality but in agreement with the quantum mechanical result $\eta_{QM} = 0.272 \pm 0.008$.

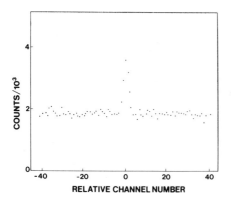

Figure 18. A typical time correlation spectrum for the experiment of Perrie, Duncan, Beyer, and Kleinpoppen[37] after subtraction of the spectrum obtained with the metastable component of the beam quenched. Polarizer plates removed. Time delay per channel 0.8 nsec. Total collection time 21.5 h. Singles rate with metastables present (quenched) about 1.15×10^4 sec^{-1} $(0.85 \times 10^4$ sec$^{-1})$. True two-photon coincidence rate 490 h^{-1}.

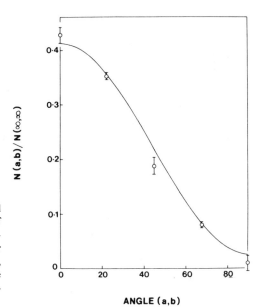

Figure 19. Coincidence signal $N(a, b)/N(\infty, \infty)$ as a function of the angle (a, b) between the transmission axes of the polarizers for the experiment of Perrie, Duncan, Beyer, and Kleinpoppen.[37] The solid curve represents the quantum mechanical prediction.

In an extension to the above experiment[38] the circular polarization correlation was measured by placing achromatic quarter-wave plates in each detection arm between linear polarizer and source. The results obtained were consistent with conservation of angular momentum for the photon pair along the observation axis but did not violate the BCHSH inequality. It seems likely, however, that this failure to violate the BCHSH inequality was due to the fact that the retardation of the plates can vary by ±10% over the wavelength range 185–355 nm even for a parallel light beam, whereas in the experiment the light rays from the source, after collimation, could be at an angle of up to ±2° to the observation axis.

4. Discussion

The overwhelming number of experiments that have been carried out to test the BCHSH inequality have produced results violating the BCHSH inequality in one form or another. The results also provide strong evidence against the Schrödinger–Furry hypothesis, and, in addition, in what is essentially a separate issue, they agree, with one exception, with the quantum mechanical prediction. The fact that the results violate the BCHSH inequality is important in that it forces us to accept the nonlocality and lack of realism inherent in quantum mechanics and to abandon any hope of finding a local realistic alternative to quantum mechanics, which was always a possibility prior to the discovery of the BCHSH inequality and the performance of the experiments described herein.

However, the low efficiency of the photomultipliers used in all these experiments leaves open the possibility that the results could be explained in terms of local realistic theories if the no-enhancement hypothesis is not made. Consequently, no experiment to date can be considered to have provided grounds for a completely unequivocal rejection of such theories, and, indeed, if such an assumption is not made, it has been shown by several authors[39-43] that it is possible to explain all existing experimental results in local realistic terms. Thus it seems likely that, in future, experimental work will be carried out either in situations where it is not necessary to make the no-enhancement hypothesis, for example in experiments of the type suggested by Lo and Shimony[44] or Selleri,[45] or in situations where the no-enhancement hypothesis can be subjected to direct experimental test, for example in experiments of the type suggested by Garuccio and Selleri.[42]

References

1. C. S. Wu and I. Shaknov, *Phys. Rev.* **77**, 136 (1950).
2. J. A. Wheeler, *Ann. N.Y. Acad. Sci.* **48**, 219 (1946).
3. C. N. Yang, *Phys. Rev.* **77**, 242 (1949).
4. J. F. Clauser, *Phys. Rev. A* **6**, 49 (1972).
5. L. Mandel, in *Progress in Optics* (edited by E. Wolf), Vol. 13, pp. 27-68, North-Holland, Amsterdam (1976).
6. A. Aspect, J. Dalibard, P. Grangier, and G. Roger, *Opt. Commun.* **49**, 429 (1984).
7. D. Bohm, *Quantum Theory*, Prentice-Hall, Englewood Cliffs, New Jersey (1951).
8. D. Bohm and Y. Aharonov, *Phys. Rev.* **108**, 1070 (1957).
9. A. Einstein, B. Podolsky, and N. Rosen, *Phys. Rev.* **47**, 777 (1935).
10. J. S. Bell, *Physics* **1**, 195 (1964).
11. J. F. Clauser, M. A. Horne, A. Shimony, and R. A. Holt, *Phys. Rev. Lett.* **23**, 880 (1969).
12. J. F. Clauser and M. A. Horne, *Phys. Rev. D* **10**, 526 (1974).
13. J. F. Clauser and A. Shimony, *Rep. Prog. Phys.* **41**, 1881 (1978).
14. F. M. Pipkin, in *Advances in Atomic and Molecular Physics* (edited by D. R. Bates and B. Bederson), Vol. 14, pp. 281-340, Academic, New York (1978).
15. A. Aspect, in *The Wave-Particle Dualism* (edited by S. Diner, D. Fargue, G. Lochak, and F. Selleri), pp. 377-390, D. Reidel, Dordrecht (1984).
16. A. Aspect, "Trois tests expérimentaux des inégalités de Bell par mesure de correlation de polarisation de photons", Thèse, Université de Paris-Sud, Centre d'Orsay, Paris, France (1983).
17. J. S. Bell, in *Proceedings of the International School of Physics "Enrico Fermi", Course 1L* (edited by B. d'Espagnat), pp. 171-181, Academic, New York (1971).
18. S. J. Freedman, Ph.D. thesis, University of California, Berkeley (1972).
19. E. Schrödinger, *Proc. Cambridge Phil. Soc.* **31**, 555 (1935).
20. W. H. Furry, *Phys. Rev.* **49**, 393 (1936).
21. F. Selleri and G. Tarozzi, *Riv. Nuovo Cimento* **4**, 1 (1980).

22. C. A. Kocher and E. D. Commins, *Phys. Rev. Lett.* **18**, 575 (1967).

23. S. J. Freedman and J. F. Clauser, *Phys. Rev. Lett.* **28**, 938 (1972).

24. R. A. Holt and F. M. Pipkin, Harvard University preprint (1974); see also Ref. 14.

25. J. F. Clauser, *Phys. Rev. Lett.* **36**, 1223 (1976).

26. J. F. Clauser, *Nuovo Cimento B* **33**, 740 (1976).

27. E. S. Fry and R. C. Thompson, *Phys. Rev. Lett.* **37**, 465 (1976).

28. A. Aspect, P. Grangier, and G. Roger, *Phys. Rev. Lett.* **47**, 460 (1981).

29. A. Aspect, P. Grangier, and G. Roger, *Phys. Rev. Lett.* **49**, 91 (1982).

30. F. Falciglia, L. Fornari, A. Garuccio, G. Iaci, and L. Pappalardo, in *The Wave-Particle Dualism* (edited by S. Diner, G. Lochak, and F. Selleri), pp. 397–412, D. Reidel, Dordrecht (1984).

31. A. Aspect, *Phys. Rev. D* **14**, 1944 (1976).

32. A. Aspect, J. Dalibard, and G. Roger, *Phys. Rev. Lett.* **49**, 1804 (1982).

33. T. W. Marshall, E. Santos, and F. Selleri, *Lett. Nuovo Cimento* **38**, 417 (1983).

34. F. Selleri, *Lett. Nuovo Cimento* **39**, 252 (1984).

35. S. Pascazio, *Nuovo Cimento D* **5**, 23 (1985).

36. A. Aspect and P. Grangier, *Lett. Nuovo Cimento* **43**, 345 (1985).

37. W. Perrie, A. J. Duncan, H. J. Beyer, and H. Kleinpoppen, *Phys. Rev. Lett.* **54**, 1790 (1985).

38. A. J. Duncan, W. Perrie, H. J. Beyer, and H. Kleinpoppen, in *Book of Abstracts, Second European Conference on Atomic and Molecular Physics* (edited by A. E. de Vries and M. J. van der Wiel), p. 116, Free University, Amsterdam, The Netherlands, April 15–19 (1985).

39. T. W. Marshall, E. Santos, and F. Selleri, *Phys. Lett* **98A**, 5 (1983).

40. T. W. Marshall, *Phys. Lett.* **99A**, 163 (1983).

41. T. W. Marshall, *Phys. Lett.* **100A**, 225 (1984).

42. A. Garuccio and F. Selleri, *Phys. Lett.* **103A**, 99 (1984).

43. T. W. Marshall and E. Santos, *Phys. Lett.* **107A**, 164 (1985).

44. T. K. Lo and A. Shimony, *Phys. Rev. A* **23**, 3003 (1981).

45. F. Selleri, *Phys. Lett.* **108A**, 197 (1985).

Index

507